ice

Institution of Civil Engineers

publishing

Highways

The location, design, construction and
maintenance of road pavements

Fifth edition

Edited by

Coleman O'Flaherty, AM

BE (NUI), MS, PhD (Iowa State), Hon LLD (Tas), CEng, FICE, FIEI, FIEAust, FIHT
Professor Emeritus, University of Tasmania, Australia

with

David Hughes

BSc (Hons) (QUB), PhD (Nott), CEng, MICE, MIEI, FCIHT, FHEA
Senior Lecturer in Geotechnical and Highway Engineering, Queen's University
Belfast, UK

Published by ICE Publishing, One Great George Street, Westminster, London SW1P 3AA

Full details of ICE Publishing sales representatives and distributors can be found at: www.icevirtuallibrary.com/info/printbooksales

First edition published in 1967 by Edward Arnold Ltd
Second edition published in 1974
Third edition published in 1986
Fourth edition published in 2002 by Butterworth Heinemann
This edition published by ICE Publishing in 2016

Other titles by ICE Publishing:
Transportation Engineering.
John Wright. ISBN 978-0-7277-5973-3
Practical Road Safety Auditing, Third Edition.
Belcher *et al.* ISBN 978-0-7277-6016-6
The Shell Bitumen Handbook, Sixth Edition.
Robert N. Hunter. ISBN 978-0-7277-5837-8
Principles of Pavement Engineering, Second Edition.
N. Thom. ISBN 978-0-7277-5853-8

www.icevirtuallibrary.com
A catalogue record for this book is available from the British Library.

ISBN 978-0-7277-5993-1

© Thomas Telford Limited 2016

ICE Publishing is a division of Thomas Telford Ltd, a wholly owned subsidiary of the Institution of Civil Engineers (ICE).

Commissioning Editor: Amber Thomas
Production Editor: Rebecca Taylor
Market Development Executive: Elizabeth Hobson

Typeset by Academic + Technical, Bristol
Index created by Indexing Specialists (UK) Ltd
Printed and bound in Great Britain by TJ International Ltd, Padstow

Contents

About the editors

Coleman O'Flaherty AM, BE (NUI), MS, PhD (Iowa State), Hon LLD (Tas), CEng, FICE, FIEI, FIEAust, FIHT

Professor Emeritus Coleman O'Flaherty graduated from the National University of Ireland (Galway) in 1954 and worked in Ireland, Canada and the USA before joining the Department of Civil Engineering, University of Leeds, UK, in 1962. He was Foundation Professor of Transport Engineering and Foundation Director of the Institute for Transport Studies at Leeds University before being invited to Australia as, initially, Commonwealth Visiting Professor, University of Melbourne, and, then, as First Assistant Commissioner (Chief Engineer) of the National Capital Development Commission, Canberra. Since retiring as Deputy Vice Chancellor at the University of Tasmania, he has been made a Professor Emeritus of the university, and awarded the honorary degree of Doctor of Laws. In 1999, he was appointed a Member of the Order of Australia for services to education and the community.

David Hughes BSc (Hons) (QUB), PhD (Nott), CEng, MICE, MIEI, FCIHT, FHEA
Senior Lecturer in Geotechnical and Highway Engineering, Queen's University Belfast, UK

Dr David Hughes graduated from Queen's University Belfast with a BSc (Hons) in civil engineering in 1982, and completed his doctorate with the Pavement Research Group at the University of Nottingham in 1986. He then worked with Roughton International Consulting Civil Engineers as a pavement and materials specialist on many road construction projects across the world, ranging from Nepal to Suriname. In 1990 he took up an academic position at Queen's University Belfast, where he is now Director of Education and Senior Lecturer in Geotechnical and Highway Engineering. His research is focused on pavement design and assessment, and on the performance of geotechnical infrastructure, in particular predicting and assessing the impact of climate change on infrastructure slopes.

About the contributors

Gordon Airey BSc, MSc, PhD (Nott), CEng, MICE, FIAT, MCIHT
Director, Nottingham Transportation Engineering Centre, UK

Professor Gordon Airey graduated from the University of Cape Town with a first-class honours degree in civil engineering in 1989, and worked for the Council for Scientific and Industrial Research in South Africa before taking a research associate position at the University of Nottingham, UK, in 1994. He obtained his doctorate from the University of Nottingham in 1997 before being appointed to the academic staff in the Department of Civil Engineering in 1998. Professor Airey is currently Director of the Nottingham Transportation Engineering Centre (NTEC) and a former director on the board of the International Society for Asphalt Pavements (ISAP).

Andrew Boyle MA (Cantab), CEng, MICE, FCIHT
Senior Partner, Andrew Boyle Associates Ltd, UK

Andrew Boyle graduated from the University of Cambridge in 1967, and worked on motorway and trunk road design and construction, before joining the UK Department for Transport in 1974. There, he worked on the management and maintenance of trunk roads in the West Midlands, before later managing the planning, design and construction of major schemes in the Midlands and London. After managing computer software development, he became Head of the Engineering Policy Division at the Highways Agency, where he had responsibility for the development of policy on value for money, quality, specifications, innovative traffic control systems and enforcement, and road safety, as well as European and international affairs. Since 1997 he has been an independent consultant, and has worked in the UK, Bahrain, Israel, Poland and Palestine, advising on a variety of highway matters. He is currently Chair of the UK National Committee of the World Road Association.

Michael Brennan BE (NUI), MSc (Leeds), DPhil (Ulster), Dip Comp Eng (Dublin), CEng, FIEI, MIAT
Consultant (formerly Senior Lecturer, National University of Ireland, Galway, Ireland)

Dr Michael Brennan graduated with first class honours from the National University of Ireland in 1970, and subsequently obtained a master's degree in transportation engineering at the University of Leeds. He worked with Ireland's national transport company (CIE) and as a research assistant at the Institute for Transport Studies at Leeds University, before starting a teaching career at University College, Galway, where he was appointed Statutory Lecturer in 1980. He spent sabbatical years at Purdue University in 1979–1980, and at Laboratoire Central des Ponts et Chaussées in 1987–1988. In 2000, the University of Ulster conferred the degree of DPhil upon him for his published research. Dr Brennan has collaborated with the asphalt industry in Ireland on research work into asphalt mix design and performance for four decades.

Seósamh Costello BE (NUI), MSc, PhD (Birm), CEng, MIEI, MIPENZ
Senior Lecturer, University of Auckland, New Zealand

Dr Seósamh Costello graduated from University College, Galway in 1993 before going on to obtain a master's degree in international highway engineering from the University of Birmingham in 1996. He worked as a research associate and then a research fellow at the University of Birmingham while simultaneously reading for his doctorate, which was conferred in 2001. He joined the lecturing staff at the University of Auckland in 2002, later serving as Associate Dean (Postgraduate) in the Faculty of Engineering. He has published widely in the field of transportation asset management, and has also consulted widely, having worked for a number of consultancies in the UK and Ireland.

Andrew Dawson BA (Lancaster), MSc (London), DIC, CEng, MICE, FGS
Associate Professor, Nottingham Transportation Centre, UK

Andrew Dawson commenced his career as a technician with major civil engineering contractors before attending the University of Lancaster and Imperial College London. He then spent 5 years with civil engineering consultants as a geotechnical engineer working on the design of earth dams, mine pit slope stability assessments, highway schemes, water resources and hydraulic studies. Since 1983 he has been Lecturer, Senior Lecturer and Associate Professor in Civil Engineering at the University of Nottingham, where he lectures in geotechnics and pavement engineering. His research interests have concentrated on pavement foundations and low-volume and unsealed road pavements, extending to the study of geosynthetics, repeated loading effects, pavement sub-drainage, secondary aggregates in un-, cement- and bituminous-bound pavements, trench reinstatement, effects of climate change, and pavements as solar energy collectors.

Gareth Hearn BSc, PhD (London), CEng, MICE, CGeol, FGS, CEnv
Director, Hearn Geoserve Ltd, UK

Dr Gareth Hearn is an independent consultant geomorphologist who was previously with Scott Wilson/URS as Technical Director for International Geotechnics and Geohazards. He graduated from Kingston Polytechnic, UK, with a first-class honours degree in geography, geology and pedology in 1981. In 1987 he completed his doctorate at the London School of Economics, following research undertaken in the Himalayas on the design and construction of mountain roads. Since then, he has been involved in environmental, geotechnical and geohazard assessment for infrastructure development projects in over 30 countries, focusing especially on the selection and design of alignments for road, rail and pipeline schemes through difficult terrain.

Kim Jenkins BSc, MSc (Natal), PhD (Stellenbosch), FSAT
SANRAL Chair in Pavement Engineering, Stellenbosch University, South Africa

Professor Kim Jenkins practised as a consultant in civil engineering for 12 years after qualifying from the University of Natal, South Africa, in 1983. His career in the private sector included investigations and design projects in geotechnical engineering, materials and pavement engineering. He then immersed himself in full-time research into pavement rehabilitation and recycling technologies, with a focus on performance of materials. The research was conducted at Stellenbosch University, South Africa, and Delft University of Technology, the Netherlands, culminating in a doctoral degree in 2000. He collaborates globally in pavement recycling and stabilisation technology. Professor Jenkins is currently the incumbent of the SANRAL Chair in Pavement Engineering at Stellenbosch University.

Alan Kavanagh BE (NUIG), MEngSc (NUIG), CEng, MIEI, MIAT
Technical Manager, Atlantic Bitumen Company Ltd, Ireland

Alan Kavanagh graduated with first-class honours in 1997 from the National University of Ireland, Galway, where he was subsequently also awarded the degree of

Master of Engineering Science. Since graduation, he has worked for the Colas Ireland Group of companies. In his current capacity as Technical Manager, he is responsible for the quality assurance and control of surface dressing, micro-surfacing and in situ recycling operations, the supply of bitumen for the production of hot-mix asphalts, and the supply of chemical emulsifiers, additives and adhesion agents for worldwide use. He also manages the Colas R&D Laboratory in Oranmore, Co. Galway. In the period 2001–2003, Alan Kavanagh was seconded to the parent company, Colas Inc., in Maryland, USA, to work as an estimator/project manager on highway maintenance.

Hussain Khalid BSc (Baghdad), MSc, PhD (Birmingham), CEng, MICE, MIAT
Senior Lecturer, School of Engineering, Liverpool University, UK

After completing his PhD at the University of Birmingham in 1985, Dr Hussain Khalid worked for 2 years at the University of Newcastle upon Tyne on research into modular pavements before moving to the University of Liverpool, where he is currently Senior Lecturer in Civil Engineering in the Centre for Engineering Sustainability. In 2008–2009 he worked on an industrial placement with A-one+ Managing Agent Contractor, where he was involved in the maintenance and management of the motorway network in north-west England. In 2010–2011, Hussain Khalid was Visiting Professor at Texas A&M University, where he investigated surface free energy and asphalt fatigue. Dr Khalid's research interests are in the characterisation and modelling of asphaltic materials behaviour, focusing on rheology, fatigue and fracture.

John Knapton BSc, PhD (Newcastle), DSc (Kwame Nkrumah), CEng, FICE, FIStructE, FCIHT
Engineering Consultant, UK

Professor John Knapton graduated in civil engineering from Newcastle University in 1970, and obtained his doctorate there in 1973. He has divided his career between academia and consulting, specialising in pavement engineering. He is the author of all four editions of the British Ports Association manual – *The Structural Design of Heavy Duty Pavements for Ports and Other Industries* – and was Chairman of the PIANC Working Group 165 that prepared guidelines on container terminal pavements. He established his consulting business in 1981, mainly as an expert witness in pavement disputes, and he held the Chair of Structural Engineering at Newcastle University from 1991 to 2001. Professor Knapton is currently Chair of the Small Element Pavement Technologists Council, which perpetuates a series of international conferences on concrete block paving. He is the author of three books on concrete floors, and sits on many committees in the British Standards Institution.

Derek McMullen BSc, MSc (QUB), CEng, MICE, MCIHT
Pavement Engineer, Atkins Ltd, UK

Derek McMullen graduated from Queen's University Belfast in 1976, completing a master's degree by research (on the applications of pavement analysis) in 1978. Early experience was gained with UK consulting engineers in southern England, working on the design and construction supervision of major roads. Following admission to the Institution of Civil Engineers in 1984, he was engaged on the design of pavement rehabilitation schemes and the implementation of pavement management systems in the Middle East, Africa and the UK. Since 2000, he has been based in the English Midlands, involved with highway asset management and pavement design in the UK, Eastern Europe and the Middle East.

Michael Maher MA, BAI, PhD (Dub), PEng, CEng, MIEI, Eur Ing
Principal, Golder Associates Ltd, Canada

After completing a doctorate in geotechnical engineering and pavement design at Trinity College Dublin, Michael Maher moved to Ontario, Canada, in 1978 to join Golder Associates. He is currently Golder's regional leader for large public infrastructure. In his 36 year career he has worked from Golder's offices in London and in St John's and Toronto, Canada, as well as in Qatar and Ireland. His practice areas span geotechnical and materials engineering, pavement design, construction aggregate performance, asset management and sustainability, highway drainage, and forensic engineering investigations. Michael Maher has undertaken assignments in North and South America, the Caribbean, Europe and the Middle East. His current emphasis is on the resolution of construction disputes, investigating premature failures, leading highway research studies, and acting as an expert witness.

Paul Nowak BSc (London), CEng, MICE, MIMMM, CGeol, FGS
Technical Director – Infrastructure Geotechnics, Atkins Ltd, UK

Paul Nowak is Technical Director at Atkins Ltd, with over 35 years' experience of the design and construction of infrastructure projects in the UK and overseas. He has been responsible for the investigation and design of new earthworks and existing earthworks assets. Latterly, he has been involved in design–build and public–private partnership schemes, predominantly as a designer to constructors and advisor to the lenders technical advisor.

Martin O'Connell BE (Hons), MEngSc (NUI), CEng, MIEI
Project Manager, Amey Airport Infrastructure, Amey Ltd, UK

Martin O'Connell graduated from the National University of Ireland (Dublin) in 2008 with an honours degree in civil engineering. He subsequently undertook a two year research master's degree on the performance of concrete incorporating ground granulated blastfurnace slag in aggressive wastewater environments, graduating in 2010. Martin has worked in the USA and the UK, latterly in the area of airport infrastructure construction and design with Amey plc. He has been involved in the development of Heathrow Airport's new Terminal 2 and has also undertaken consulting roles at a number of other British airports, including Glasgow, Stansted, Southampton and Cambridge. His interests include pavement design and the development of concrete and asphalt technology.

Dan Raynor BSc (Cardiff), MSc (Leeds) CEng, MICE, FGS
Associate Geotechnical Engineer, Ove Arup and Partners Ltd, UK

Dan Raynor graduated from Cardiff University in 1999 with an honours degree in geology and from the University of Leeds in 2001 with a master's degree in engineering geology. Following graduation, he worked as an engineering geologist at Hyder Consulting, where he focused on site investigations for various UK highways schemes before being seconded, in 2004, to Mowlem Civil Engineering and working as a site engineer during the construction of the Cwm Relief Road in South Wales. Since 2005, Dan Raynor has worked with the Arup Group. During this time he has provided geotechnical leadership for a wide variety of civil engineering, building and infrastructure projects, including ground investigations and geotechnical design for major highway schemes.

Martin Snaith OBE, MA, BAI, MSc (Dub), PhD (Nott), ScD (Dub), FREng, FICE, FCIHT
Engineering Consultant, UK

Professor Emeritus Martin Snaith graduated from Trinity College, Dublin, in 1968 with a first-class honours degree in engineering and a pass degree in arts. For most of his career he specialised in university teaching and research, particularly addressing the needs of roads in developing countries. Recently retired from the University of Birmingham, where he was Pro Vice Chancellor, he now advises a number of organisations in the UK and around the world on asset management issues. He is a director of Highway Management Services, which develops road management systems for use in countries as diverse as China and Cyprus. He is also Chair of the multinational consortium HDMGlobal, which is responsible for the development and dissemination of HDM-4, the de facto world standard economic evaluation model for roads, and still teaches aspiring and practising highway engineers in the UK and overseas.

Nick Thom MA (Cantab), PhD (Nott), MICE, MCIHT, MPWI
Lecturer, Nottingham Transportation Engineering Centre, UK

After joining Scott Wilson in 1978, Nick Thom worked both in the design office and on a highway construction site before carrying out his doctoral research into road foundations. He then joined Scott Wilson Pavement Engineering, continuing as a consultant until 2014. During this time he worked on the design and evaluation of highways, ports and airfields in many parts of the world. Since 1991 he has also been a member of the academic staff at the University of Nottingham, where he specialises in the analysis and design of pavements and rail track. In addition to teaching highway, airfield and railway engineering, his role at the university includes research on subjects as diverse as cold-mix asphalt, pavement management, fibre-reinforced concrete, railway track beds, and energy loss from rolling wheels.

Andrew Todd BSc, CEng, MICE, CMIWEM
Principal Engineer, Jacobs Ltd, UK

Andrew Todd graduated in 1977 and worked for 12 years in UK local government, mainly on the design and construction of major sewerage schemes: this included 8 years of construction supervision as a resident engineer. Following 6 years with consultants designing motorway drainage and sewerage schemes, he was appointed Research Principal for Pipelines and Drainage at the Transport Research Laboratory (TRL). At the TRL he undertook trials of innovative highway drainage systems, and wrote the guidance and specifications for their design and construction, as well as significantly contributing to the development of highway drainage asset condition determination and management. Andrew Todd is presently a consultant with Jacobs, working on the development of technical guidance for, and the design of, highway drainage systems.

David Woodward BSc, MPhil, DPhil (Ulster), MCIHT, MIAT, MIQ, MIEI
Reader in Infrastructure Engineering, University of Ulster, Ireland

Dr David Woodward graduated from the Ulster Polytechnic in 1982. He became involved with research into highway engineering materials at the University of Ulster, where he was appointed Lecturer in 1998 and Reader in Infrastructure Engineering in 2005. He is currently Head of the Highway Engineering Research Group, and in 2015 became Director of SABER (Studies Allied to Built Environment Research) at Ulster University.

Acknowledgements

I would like to express my very considerable thanks to all of my expert co-writers for agreeing to participate in this joint venture, and to my publishing 'supervisors' Gavin Jamieson, Amber Thomas and Rebecca Taylor, and their support team, who made this happen in a most courteous and amicable way. I would also like to express my appreciation to my previous publisher, Messrs Taylor & Francis, for agreeing to my wish to have this edition published by the Institution of Civil Engineers.

I would also like to thank Martin Snaith and David Hughes, without whose support and help this fifth edition would not have been initiated.

My colleagues and I are indebted to the many organisations and journals that allowed us to reproduce diagrams and tables from their publications. Most commonly, the reference is given in the title of each figure and table, but sometimes it is referred to in the text. The material that is quoted from government publications is Crown copyright and is reproduced by permission of The Stationery Office.

In relation to citations, the reader is urged to seek out the original material and, in particular, to consult the most up-to-date 'official' versions of recommended guides, practices, standards, etc., when actually involved in the location, design, construction and maintenance of roads.

Last, but far from least, I pay tribute to my wife Nuala, who helped me immeasurably in the development and preparation of the previous editions of this textbook. Her love, support and forbearance will never be forgotten by me.

Coleman O'Flaherty

Launceston, Tasmania
August 2015

Abbreviations and terminology

AADF	average annual daily flow
AADT	annual average daily traffic
AASHO	American Association of State Highway Officials
AASHTO	American Association of State Highway and Transportation Officials
AAV	aggregate abrasion value
AFNOR	Association Française de Normalisation
AGS	Association of Geotechnical Specialists
APL	actual point load
ASTER	Advanced Space-borne Thermal Emission and Reflection Radiometer
ASTM	American Society for Testing and Materials
AUTL	asphalt for ultra-thin layers
BBA	British Board of Agrément
BGS	British Geological Survey
BIM	building information modelling
BoQ	bill of quantities
BSI	British Standards Institution
CAD	computer-aided design
CBA	cost–benefit analysis
CBGM	cement bound granular material
CBR	California bearing ratio
CEC	cation exchange capacity
CEN	Comité Européen de Normalisation (European Committee for Standardization)
COBie	Construction Operations Building Information Exchange
CPIC	Construction Project Information Committee
CPT	cone penetration test
CRCB	continuously reinforced concrete base
CRCP	continuously reinforced concrete pavement
CT	circular texture
c_u	shear strength
D&B	design-and-build
DBFO	design, build, finance and operate
DCP	dynamic cone penetrometer
DEM	digital elevation model
DIN	Deutsches Institut für Normung, German Institute for Standardization
DfT	Department for Transport
DLP	Determinate-life pavement
DoP	declaration of performance
DUPV	dry unpolished value
EAL	equivalent axle load
ECI	early contractor involvement
EFTA	European Free Trade Association
EIAs	environmental impact assessments
EME	Enrobé à module élevé

EOTA	European Organisation for Technical Approvals
ESAL	equivalent standard axle load
ESPL	equivalent single point load
EVA	ethylene vinyl acetate
FAP	friction after polishing
FBA	furnace bottom ash
FEA	finite-element analysis
FPC	factory production control
FWD	falling-weight deflectometer
GGBS	ground granulated blast furnace slag
GIR	ground investigation report
GISs	geographical information systems
GNCA	granular base with no cementing action
G_{sa}	apparent specific gravity
G_{sb}	bulk specific gravity
G_{se}	effective specific gravity
HAGDMS	Highways Agency geotechnical database management system
HAPAS	Highway Authorities product approval scheme
HAPMS	Highways Agency pavement management system
HAWRAT	Highways Agency water risk assessment tool
HBM	hydraulically bound mixture
HCV	heavy commercial vehicle
HDM	heavy duty macadam
HFS	high-friction surfacing (BBTM, from its French name 'béton bitumineux très mince')
HMB	high-modulus base
HRA	hot rolled asphalts
HSE	Health and Safety Executive
I_c	consistency index
ICL	initial consumption of lime
ICS	initial consumption of stabiliser
I_d	density index
I_L	liquidity index
Ip or PI	plasticity index
IRI	international roughness index
ISO	International Organization for Standardization
ISOHDM	International Study of Highway Development and Management
ISSA	International Slurry Surfacing Association
ITT	initial type test
JRC	jointed reinforced concrete
LiDAR	light detecting and ranging
LLP	long-life pavement

LWD	lightweight deflectometer
LWT	loaded wheel test
MAAT	mean annual air temperature
MCV	moisture condition value test,
MDE	micro-Deval value
MEPDG	Mechanistic–Empirical Pavement Design Guide
MMS	minimum mass for sieving
MMTD	mean measured texture depth
MSW	municipal solid waste
MSWIBA	municipal solid waste incinerator bottom ash
MSWIFA	municipal solid waste incinerator fly ash
MTD	macro-texture depth
NPV	net present value
NSC	network structural condition
OMC	optimum moisture content
OS	Ordnance Survey
PA	porous asphalt
PE	polyethylene
PFA	Pulverised fuel ash
PI	penetration index for bitumen; plasticity index for soil
PIARC	Permanent International Association of Road Congresses
PMB	polymer-modified bitumen
PP	polypropylene
PPP	public–private partnership
PQC	pavement quality concrete
PSI	present serviceability index
PSMCs	performance-specified maintenance contracts
PSSR	preliminary sources study report
PSV	polished stone value
PTV	pendulum test value
PV	present value
PVC	polyvinylchloride
QP	quality protocol
QUADRO	QUeues And Delays at ROadworks
RAP	reclaimed asphalt pavement
RCA	recycled concrete aggregate
RCC	roller-compacted concrete
RD_{app}	apparent relative density
RD_{ssd}	saturated surface-dry relative density
RTFO	rolling thin-film oven test,
RTM	road test machine
SBS	styrene–butadiene–styrene
SCRIM	Sideway-force Coefficient Routine Investigation Machine
SEBS	styrene–ethylene–butadiene–styrene
SFC	sideways-force coefficient

SIS	styrene–isoprene–styrene
SMA	stone mastic asphalt
SN	structural number
SP	softening point
SPT	standard penetration test
SPZs	source protection zones
SRTM	Shuttle Radar Topographic Mission
SuDS	sustainable drainage system
SWEEP	Software for the Whole-life Economic Evaluation of Pavements
TRACS	TRAffic speed Condition Survey
TSD	traffic-speed deflectometer
TTBM	total thickness of bituminous material
UCS	unconfined compressive strength test
ULLP	Upgradeable to long-life pavement
URC	jointed unreinforced concrete
VFB	voids filled with bitumen
VMA	voids in the mineral aggregate
WIM	weigh-in-motion
w_L or LL	liquid limit
w_p or PL	plastic limit
WRAP	Waste and Resources Action Programme
WTAT	wet track abrasion test
WUPV	wet unpolished value

Highways
ISBN 978-0-7277-5993-1

ICE Publishing: All rights reserved
http://dx.doi.org/10.1680/h5e.59931.001

Introduction

Coleman O'Flaherty

Everybody travels, whether it be to work, shop, do business, play or visit people, and all commodities, foodstuffs, raw materials and manufactured goods must be carried from their places of origin to those of their adaption, sale or consumption. Nowadays, people and goods go by road to their destinations, or to docks for subsequent onward transport by water, to stations for movement by rail, and to airports for travel by air. However, whatever the intermediate travel mode used, when they arrive at their termini the great majority of people and goods must again travel by road vehicle to their final destinations.

At the end of 2014 there were 35.6 million vehicles (including 83.1% cars) licensed for use on the roads on the island of Great Britain (DfT, 2015a): this compares with 12.8 million vehicles (69.6% cars) when the first edition of this text (O'Flaherty, 1967) was published, and 4.2 million vehicles (50% cars) in 1951 during the recovery after World War II. Just as the computer is rapidly changing work and lifestyles today, the freedom of movement associated with the motor vehicle helped to bring sweeping social changes to the work and lives of most ordinary people: for example, in 2013 some 80% of English adults lived in a household with a car, and 68% of women and 81% of men had a full driving licence (DfT, 2014), and in 2014 women accounted for about 40% of the registered keepers of privately registered cars (DfT, 2015a).

Modern factories and stores no longer carry large inventories but instead rely on 'just-in-time' economical service by road. Trading towns depend for their livelihood on good road accessibility. For a nation to prosper in competition with other countries today, it must have a modern transport system so that people and goods can move, and be moved, quickly and economically, to and from its national boundaries: this now provides what is possibly the greatest incentive for developing a safe, efficient and environmentally acceptable national road system.

In 1951, it was estimated that there were 297 466 km of paved road (including 0 km of motorway) in Great Britain: it was not until 1958 that the first motorway (the 13 km Preston Bypass, now part of the M6) was completed. In 1967 there were 326 180 km of road, including 761 km of motorway, on the island, and by 2014 these figures had grown to 395 738 km and 3703 km, respectively (DfT, 2015b). While motorways amount to only 0.9% of the total road length, they currently carry 20.7% of the traffic (in vehicle-kilometres) travelled by private and commercial vehicles in Great Britain.

There were 474 000 heavy goods vehicles (HGVs) registered in the UK in 2014, and the average HGV gross vehicle weight was then 21.5 t (compared with 17.5 t in 1994). In 2014, 20% of HGVs had a gross vehicle weight >41 t, whereas the proportion in this category prior to 2001 (when the weight limit was raised from 41 to 44 t) was essentially zero.

Compared with 30 years ago, the percentage point (pp) change in the proportions of traffic accounted for by all vehicle types has decreased (i.e. HGVs decreased by 1.3 pp, cars and taxis by −0.5 pp, motor cycles by −2.3 pp, and buses and coaches by −0.3 pp) with the exception of light goods vehicles (LGVs), which increased by 4.5 pp. It is reported (DfT, 2013) that the growth in the proportion of LGV vehicle-km may be associated with changes in shopping habits toward more internet-based and home delivery retail over this time period.

The above statistics reflect one of the major continuing complexities associated with road pavement design – that of estimating the design traffic loads.

Current transport policy in the UK is primarily aimed at creating greater choice for personal travel by improving public transport services, encouraging cycling and walking, and significantly reducing movement by car. By comparison with countries in mainland Europe, the UK motorway network is overloaded and, at 60 km of motorway per million people (Greece has 67 km/million people, for example, and Germany has 148 km/million people), is less than half of the European average. This suggests that the UK, which has the world's fifth largest economy (by nominal gross domestic product), has an inadequate and inefficient road system that does not meet the commercial needs of industry. Major new road building in the UK has now almost halted. Instead (a) major new construction is focused on the rehabilitation and/or widening of (i.e. adding more traffic lanes to) existing major road sections, and (b) active traffic management techniques involving, for example, managed lanes, smart lanes, and managed/smart motorways, are being used to increase peak capacities and smooth traffic flows on exist-ing busy major roads.

It is now argued (e.g. by the prestigious Motorway Archive Trust) that this transport policy is failing to meet its targets and should be changed; that is, while public transport systems operate most effectively within hub (town centres) and spoke (radial route) patterns of movement, they do not have the flexibility to cope with the much greater dispersed patterns of movement that have resulted from the dispersal of homes and employment since the 1950s. Thus, in concept, it is postulated (Motorway Archive Trust, 2010) that the national road system developed in the UK through to 2060 should be divided into two connected networks: (a) a 'roads for movement' network, based on the current 50 310 km motorway and 'A' road system, for which the objective would be to provide reliable and predictable journey times by road to ports and airports, which is what industry needs and wants; and (b) a 'roads for access' network, comprising the rest of the roads, which would primarily serve communities, connect with the 'roads for movement' network, and be used by all forms of traffic (pedestrians, horses, cyclists, motor cycles, cars, buses, coaches, vans, lorries, etc.).

In the author's view, those things that are logical and sensible, and result in an improved national economy, tend to happen eventually.

Whatever the eventual outcome of the debate in the UK, there is little doubt but that much of the developing world is poised for significant economic expansion, and for associated growths in incomes. This has important implications for transport, trading, environmental and sustainability policies, not to mention the global oil market. While the technical literature is divided as to whether vehicle ownership rates in developing countries will reach the levels that are common in advanced economies, there is no doubt but that greater prosperity will ensure great increases in the numbers of road vehicles in those countries. Consequently, major new road programmes will need to be initiated to provide for the effective and efficient movement of the vehicles in these countries so that their industries and manufacturers can more successfully compete in the international marketplace. Current trends are such that these roads will be designed to have long lives and to maximise the reuse of existing road materials and the recycling of other appropriate materials in road pavements.

The scale of this potential is reflected in Table 1, which is based on selected data extracted from a major study (Dargay et al., 2007) of 45 countries that include 75% of the world's population. This study predicts that the world's total stock of vehicles will increase from some 812 million in 2002 to about 2.08 billion units in 2030 – by which time 56% of the world's vehicles will be owned by non-Organisation for Economic Co-operation and Development (OECD) countries, compared with 24% in 2002.

Table 1 Projections of population and vehicle ownership in selected OECD and non-OECD countries

Country	Vehicles/1000 population			Total number of vehicles			Population: millions		
	2002	2030	Annual growth: %	2002	2030	Annual growth: %	2002	2030	Annual growth: %
OECD									
UK	515	685	1.0	30.6	44	1.3	59	64	0.3
Germany	586	705	0.7	48.3	57.5	0.6	83	82	0.0
Ireland	472	812	2.0	1.9	3.9	2.7	4	5	0.7
South Korea	293	609	2.6	13.9	30.5	2.8	48	50	0.2
Non-OECD									
Brazil	121	377	4.1	20.8	83.7	5.1	171	222	0.9
China	16	269	10.6	20.5	390	11.1	1275	1451	0.4
India	17	110	7.0	17.4	156	8.1	1051	1417	1.1
Indonesia	29	166	6.5	6.2	46.1	7.4	216	278	0.9
South Africa	152	395	3.5	6.9	16.7	3.2	45	42	−0.3

Based on data from Dargay et al. (2007)

Note particularly that China's vehicle stocks are predicted to increase 19 times by 2030, and that its rate of vehicle ownership (269 vehicles/1000 people) will then only be at the level experienced by Western Europe in the mid-1970s and by South Korea in 2001.

The rapid increases in vehicle-ownership expansion imply rapid growth in both major and minor road construction in the countries in which they occur. They also emphasise why it is that the study of highway engineering should not be just about what is current practice in the UK but should be directed toward understanding fundamentals that are transferable across national boundaries. As with its predecessors, this is the aim of this fifth edition of this text on highways.

REFERENCES

Dargay J, Gately D and Sommer M (2007) Vehicle ownership and income growth, world-wide: 1960–2030. *Energy Journal* **28(4)**: 143–170.

DfT (Department for Transport) (2013) *Annual Road Traffic Estimates: Great Britain 2012*. DfT, London, UK.

DfT (2014) *National Travel Survey: England 2013*. DfT, London, UK.

DfT (2015a) *Vehicle Licensing Statistics: Quarter 4 (Oct–Dec) 2014*. DfT, London.

DfT (2015b) *Road Lengths in Great Britain: Statistical Release, May 2015*. DfT, London, UK.

Motorway Archive Trust (2010) *Evidence to the House of Commons Transport Select Committee's Inquiry into the Major Road Network: Eight Report of Session 2009–10*. Stationery Office, London, UK.

O'Flaherty CA (1967) *Highways*. Arnold, London, UK.

Highways
ISBN 978-0-7277-5993-1

ICE Publishing: All rights reserved
http://dx.doi.org/10.1680/h5e.59931.005

Chapter 1
The road development process: plans, specifications and contracts

Andrew Boyle   Senior Partner, Andrew Boyle Associates Ltd, UK

This chapter describes how road improvement schemes are developed, paying particular attention to the documentation required for construction. The procedures of Highways England (formerly the Highways Agency) are taken as the basis for this chapter but these are similar to those used for other roads in the UK as well as in other countries.

Before discussing documentation issues, it is useful to provide a brief overview of the road development process so as to put these in context. In this respect, it should be appreciated that while most highways are publicly owned and publicly operated, they can also be privately owned. Historically, most major road schemes were funded from central taxation, but some were built through public–private partnership arrangements where the initial finance was provided by financial institutions and was paid back over the life-time of the concession. A recent trend is to manage, improve, maintain and operate major highways by private companies and consortia under such licences or concessions.

1.1.   Overview

Figure 1.1 is a flow chart showing the basic classic steps in developing a road scheme. Note that the main phases in the development process of a scheme are planning, design and tender, followed by construction, maintenance and operation. The design phase may be divided into two or three sub-phases, namely concept, preliminary and detailed design.

As part of its normal responsibilities, a highway authority usually identifies a capacity or safety-related problem on its road network, either through regular surveys or as a result of information received from the public or from public authorities such as the police. The authority will then carry out a feasibility study, either in house or by employing external consultants, to identify and quantify the nature and scale of the problem and to determine whether there is a justified improvement scheme that will resolve or mitigate the problem. This study may result in proposals that can range from a simple scheme involving the installation of additional road signs or markings to complex ones involving the provision of an expensive motorway costing millions of pounds or proposals to use alternative transport modes to reduce road usage. If the authority decides to develop a significant scheme to solve the problem, it will establish a project management group to oversee this.

Figure 1.1 Classic steps in the preparation of a development programme for a new road

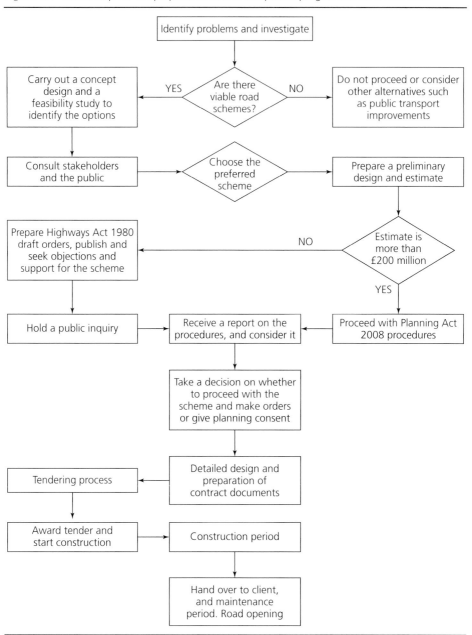

All authorities operate project management systems that, typically, define the roles and responsibilities of the various players both within the authority and in its design and construction chain. Usually, the project management group that is set up will comprise a project sponsor, a project manager and various supporting staff, as well as groups and/

or committees that oversee and give authority to these people, and they will operate according to a basic flow chart of the processes and work phases required. Between phases, the system will provide for 'gate' reviews so that at each stage the progress and completion of the work required is monitored and approved before progressing to the next stage.

Such policies and procedures are vital to achieving a successful outcome for any development, not just road schemes. For example, quality management is now universally applied to both design and construction work. Thus, the project management group will develop a quality management plan, and require the designer and (eventually) the successful contractor of the proposed scheme to each develop their own complementary plans. In the case of contractors, their quality management plan will, typically, include information on how they propose to carry out the works, who will be the site and management personnel responsible for the project, and the plant that they will use, together with criteria for checking and measurement so that it can be verified that their targets are being met. Similarly, it will be the project management group's responsibility to ensure that health and safety is taken seriously in both the design and construction phases of the project (e.g. in the UK the Construction, Design and Management Regulations (HMG, 2015) require certain safety activities to be undertaken by the project team).

The procedures for identifying major schemes in England are set down in the Project Control Manual (HA, 2008); these are similar to the procedures in the rest of the UK. During the concept design phase, appropriate options will be considered in terms of environmental impact assessments (EIAs), traffic forecasts, and economic costs and benefits. Details of the Highways England project management system are provided in its *Project Control Framework Handbook* (HA, 2013).

1.1.1 Planning procedures

All road improvement schemes have to comply with the statutory planning processes that apply in the country in which they are located. In some countries, this planning framework is provided by master plans that are agreed and promulgated either after consideration through consultation or by decree. In other countries, the basic decision as to whether to approve a major scheme may be taken by the government.

In the UK, the general approach with respect to major roads is for the government to announce its intention to proceed with a basket of road schemes, and then to proceed through public consultation, the publication of statutory orders, and the holding of public inquiries, to resolve any objections and decide on the merits and demerits of each of these schemes.

If it is considered that the proposed scheme should be supported the project management group may commission a consultant to prepare a preliminary design to enable a decision to be taken as to where the road should be located within the preferred route envelope. This will require topographical, environmental and geotechnical surveys to be undertaken that will provide the designer with sufficient detail regarding physical features (e.g. the contours of the land and the composition of the soils along the proposed alignment(s)) to enable the preparation of this design. The design output will need to be sufficiently robust to stand up to detailed scrutiny, and to enable options to be discarded that are likely to be too costly or too difficult to build. The impacts that the proposed horizontal and vertical alignments will have on side roads, accesses to private property, footpaths and so on that are affected by the proposed road alignment will also need to be considered at this stage. Additionally, an accurate assessment will also have to be made of the need for land purchases, including any draft compulsory purchase orders under the aegis of the Compulsory Purchase of Land Regulations (HMG, 1996).

If the cost of the proposed scheme is above the threshold (currently £200 million) for a nationally significant infrastructure project, its implementation will come under the provisions of the Planning Act 2008 (HMG, 2008), and an eminent planning inspectorate panel will need to be established to initiate a rigorous examination of a draft order for development consent. Highways England, as the developer of the proposal, will register as an interested party for these discussions, which will involve community consultations (including exhibitions). As part of this examination, updated cost estimates, EIAs, traffic forecasts, and economic benefits will be prepared, an outline business case produced, and a preferred route determined. After the public comments have been analysed, a report will be completed, and the planning inspectorate will offer advice and recommendations to the Secretary of State regarding its support (or otherwise) for the proposal.

In either of the above approaches, the preliminary design is used as the basis for obtaining planning permission. If the cost of the scheme is below £200 million, the publication of a notice of acceptance of a Planning Act 2008 draft development order is not required and, instead, draft orders must be published under Section 10 (for conventional trunk roads) or Section 16 (for special roads, i.e. motorways) of the Highways Act 1980 (HMG, 1980). In all cases, an EIA is required. In the case of special roads, if the proposal requires the alteration and possible closure of other affected roads, draft orders will need to be published for these also; known as draft side road orders for trunk roads, these are published under Section 14 of the Highways Act. In this case, interested parties are given a period of time in which to register their objections to, support for, or comment on, the orders after they have been published. They may also suggest alternatives, including those that were considered but rejected at an earlier stage. The road authority must consider these concerns, and seek to resolve them. If, however, some cannot be resolved, or the objectors have a valid argument against the proposal and the objections are statutory ones, a public inquiry must be held to consider these (and other non-statutory) objections. (Note: a statutory objection is one that involves somebody who is directly affected by the scheme.)

If a public inquiry is decided upon, an eminent person is appointed by the Lord Chancellor as the inspector to hear all the objectors, and the proposer's responses to them. This can be a long and complex process, particularly when the proposal relates to roads in urban areas. The inspector will hear all the concerns and submit a report with recommendations to the Secretary of State.

After the Secretary of State has considered the submitted report from the planning inspectorate (in the case of a Planning Act 2008 scheme) or from the inspector (in the case of an Highways Act 1980 scheme), the government department will issue a letter containing the Secretary of State's decision as to whether the draft development consent or the draft orders are to be confirmed as published, or rejected or amended. The Secretary of State (through the department) is required to set out the reasons for accepting, amending or rejecting the recommendations in the decision letter. A further opportunity is then available for objectors to challenge the decision before the orders are finally made; this challenge may subsequently be taken into the High Court, the Court of Appeal, the House of Lords, or the European Court of Human Rights.

As far as local authority roads in the UK are concerned, the powers to build these are slightly different in that the local authority has the power to grant itself planning permission relating to local (i.e. non-trunk) roads. However, if the proposed scheme is significant (e.g. if the roads involved are part of a major planning development such as a new town or if they relate to a power station or airport or seaport), the Secretary of State can, under the provisions of the Planning Act 2008, call for a similar process of public inquiry.

While the above very briefly summarises the planning process in the UK, similar scrutiny processes are generally employed in other countries in order to obtain planning consent for the construction of new major roads. It might be noted here, however, that some of these processes can be less confrontational, and more collaborative, than those used in the UK.

1.1.2 Detailed design
Once a road scheme has been granted planning permission, the detailed engineering design and the preparation of contract documents can begin.

While some indications of the topography and geological state of the route of the new road will have been obtained at the preliminary design stage, further detailed topographical, geological and geotechnical surveys will normally need to be carried out (see Chapter 3 – Site investigations – for more information) to confirm, and expand upon, the results obtained during the earlier surveys and to provide the greater level of detailed information required for the final design. The detailed design of bridges and other structures, drainage, earthworks and pavement is normally initiated at this stage, in compliance with appropriate design standards.

While health and safety issues are generally associated with construction activities, it is equally important that they be taken seriously in the design phase. In the UK, as else-where, there are now government regulations that ensure that proper attention is paid to health and safety issues by the design and construction project team(s).

It is appropriate to mention here that, upon completion of construction of the road scheme, the final design drawings used by the successful tenderer are corrected to produce as-built drawings that show exactly what was built and where it was built. These amended drawings are then provided to the client authority and kept for the record and for future maintenance usages.

1.2. Contract forms and types

Highways England has a procurement strategy (HA, 2009) that seeks to ensure that the best value for money is achieved in the shortest delivery time while sustainable targets are met. The mechanism chosen to deliver the construction of the project determines the degree of detail required in the design phase. The contract methods now generally used are termed construct only, public–private partnership (PPP) and early contractor involvement (ECI).

1.2.1 Construct-only contracts

Construct-only contracts reflect the historical approach to road procurement whereby separate contracts are awarded for the design and construction responsibilities. With this approach the highway design is either conducted in house by the road authority or is awarded to an engineering consultant who, typically, would be commissioned to assist in the statutory planning process, prepare detailed design drawings, specifications and tender documents, and then arrange for the calling of tenders for construction. This 'standard' way of road procurement is now increasingly being usurped by other methods.

The successful implementation of the construct-only contract relies on the designer to fully complete an accurate design and then prepare tender documents. This documentation typically comprises a suite of signed and approved drawings – which are often supplemented with computer-based string information – a priced bill of quantities (BoQ), and a job-specific specification and contractual conditions. The design would be fully completed prior to the issuance of the tender documents, and the final selection of the contractor would, typically, be based on the most competitive and conforming tender.

If there was any dispute between the contractor and the client authority after the award of the construction contract, the conditions of contract would be used to resolve contractual issues, the specification would deal with technical matters, and the BoQ would be relied upon for pricing concerns.

While the construct-only contract seems reasonable and logical, in reality contractual variations and associated disputes are common with this approach. Consequently, some disputes are not finally resolved, and the final cost of the project often cannot be finally determined, until well after the scheme has been constructed and the road is opened to traffic.

1.2.2 PPP contracts

In recent years there has been a great deal of interest in the UK regarding the use of other forms of contract in lieu of the standard construct-only contract. Examples of these are PPP contracts. There are a number of different types of PPP contractual arrangements in vogue today: these range from basic design-and-build (D&B) contracts through to the client simply determining the need for a new highway and leaving the whole of the planning, design and construction of the new facility to the contractor.

1.2.2.1 Design and build (D&B) contracts

As the name implies, the concept underlying a D&B contract is to bring together, in partnership at an early stage in the development of a major road scheme, the parties responsible for its design and the construction. With this type of arrangement, the design and construction responsibilities are awarded to a single contractor (or to a consortium of contractors or a joint venture), typically under a lump-sum contract. The successful contractor is chosen at an early stage in the development of a scheme; this is usually, but not exclusively, after the planning process has been completed. This contractual approach is seen as bringing the knowledge and skills of the construction industry into the design process at an early stage, thereby making the design more 'buildable', reducing the client's costs, and minimising potential disputes over the design and specification of the work.

To enable a contractor to bid for a D&B contract, it is normally sufficient for the client authority to provide preliminary design plans showing the preferred design solution but without the supporting detailed design of all road elements. Competing contractors then prepare tender proposals that demonstrate their intent and ability to both complete the design and undertake the construction of the scheme. The successful contractor takes responsibility for the preparation of the detailed design documents, and guarantees that they will comply with nominated design standards and guidelines, and that an as-built, free from error, complete set of design documents will be made available to the client upon completion of the project.

The D&B contract still has a client side. The contracting side includes not only the contractor (or group of contractors, for a very large scheme) but also an engineering designer, who is nominated by the client authority and joins the contracting team. However, with this form of contract the supervising engineer appointed by the client authority to oversee the implementation of the contract has a relatively reduced role.

1.2.2.2 Design, build, finance and operate (DBFO) contracts

Pressures on the public purse in recent years have led to an increasing consideration of the use of private instead of government finance for the procurement of major new roads. Many countries are now looking at involving non-government institutions to finance the provision of essential infrastructure previously funded from the public purse. These can be done in a number of ways, but the most common is in the form of a contractual arrangement whereby a company is given a contract to DBFO a length of road, or a portion of the road network in a geographical area, for a fixed time period (e.g. 30 years).

While the client road authority retains the basic legal responsibility for the road with this approach, the DBFO contractor agrees to design and pay for the construction of the scheme, and to maintain and operate it to agreed standards throughout the period of the concession. In return for its investment, the contractor may be given the right to impose tolls, under agreed conditions, on vehicles using the road during the concession period. Alternatively, if the direct payment of a toll is not acceptable (usually for political reasons), the client authority may pay for vehicle usage via 'shadow tolls' whereby the DBFO company is paid a variable rate, agreed at the onset of the contract, by the commissioning road authority: for example, as traffic using the road increases over time, an annual level of payment might be increased until a cut-off point is reached beyond which no more increases (except inflationary ones) are paid.

1.2.3   ECI contracts

This form of road scheme delivery utilises an agreement that is usually associated with a subsequent D&B contract. Generally, it involves responsibility for the scheme being awarded to a contractor very early in the design process, to undertake the design and construction of the scheme under a lump-sum contract.

The successful implementation of the ECI approach relies on well-scoped tender documents, performance criteria and realistic delivery time-frames. The characteristics of the ECI approach are that the selected contractor is involved in defining the scope of the scheme, resources are applied during the construction-planning phase to maximise benefits during construction, and the client exercises greater control over its implementation.

As with a normal D&B contract, it is usually sufficient for the client to provide potential contractors with preliminary design plans showing a preferred solution but without supporting detailed designs. The detailed designs are developed as a partnership between the nominated designer and the contractor, with input from the client. An agreed initial target cost is also agreed with the ECI contractor.

1.3.   Tender documents

The generic components of the tender documentation that would normally be prepared for a construct-only contract are the conditions of contract, specifications, methods of measurement, BoQs, and tender and contract drawings. With the more collaborative ECI and PPP contracts, the earlier the contractor is brought into the process, the more schematic the tender documents can be; that is, the later the contractor is involved, the more complete must be the documentation.

1.3.1   Conditions of contract

The conditions of contract set out the general requirements of the contract. Historically, the contract conditions that were commonly used for road schemes in the UK were those produced by the Institution of Civil Engineers (ICE, 1999); however, these are no longer supported by the ICE and have now been rebranded (ACE, 2011). More recently, the suites of conditions that form the New Engineering Contract (ICE, 2013), also known as NEC3, and promote a less confrontational and more collaborative approach, have become the norm. Internationally, the contract conditions developed by the Federation Internationale des Ingenieurs-Conseils (FIDIC, 1999) – which are not dissimilar to the old ICE conditions – are commonly used.

The conditions of contract seek to define the respective roles of the contractor, the supervising engineer, and the client, and to provide mediation methods (e.g. arbitration and adjudication) to resolve disputes. However, some participants in the contractual process consider that these conditions also tend to produce cost overruns and excess claims and, consequently, promote a confrontational stance between the participating groups. To overcome these problems, attempts are being made to encourage the concept of partnering, whereby contractual structures are put in place that encourage the development of a common intent by the participating parties to make the construction process more of a joint exercise that leads to a successful outcome. The need to finance road schemes that cannot be provided from the public purse has also led to other conditions and contract forms being tried out.

1.3.2   Specifications

Specifications deal with how road schemes are to be constructed. Their purpose is to ensure that clients get what they want when they contract to have projects built.

The specifications most commonly used for road construction in the UK (HA *et al.*, 1998a) comprise a large number of model clauses that are grouped together in various series, each of which deals with different aspects of construction (e.g. site clearance, fencing, safety barriers, drainage, earthworks, road pavements, signs, electrical works, steelwork, concrete structures and bridge works). The clauses provide detailed requirements for the manufacture and testing of components and materials used in the construction works, including details on how to erect sections of the works, how to excavate, transport and lay particular materials, and how to complete the works.

Generally, there are two main ways of specifying work in order to achieve the desired result. These involve using either 'method' (also known as 'recipe') specifications or 'end-product' (or 'performance') specifications.

With method/recipe specifications, reliance is placed on research and experience that shows that, if certain methods are correctly used to carry out particular works, the results can be expected to meet the requirements of the client. This approach is acceptable provided that the contractor complies with the exact requirements of the specification, and the works are carried out in the scheme exactly as they were carried out in the full-scale trials that resulted in the development of the specification. As the latter requirement is not always feasible, method specifications are normally permitted limited latitude to allow for deviations in their results. If the specification is faulty, the contractor can claim that the fault was not theirs if the work is deemed unsatisfactory.

With end-product/performance specifications the client must specify to the contractor the outcome desired from the completed construction (e.g. that the works, or certain components of it, meet specified requirements for durability and strength). Contractors generally like end-product specifications because they provide opportunities to devise and supply innovative, lower-cost solutions that will meet the performance requirements of the client.

Specifications may also refer to standards, design standards and advice notes.

A standard contains the agreed and correct composition for a product that is used throughout an industry. Some standards, termed codes of practice, also set out the required way to carry out construction processes. Increasingly, UK standards are derived from their equivalent European standards: in some cases, a British standard becomes a European standard, and vice versa.

A design standard is a document, most commonly produced by a government agency, that helps designers to prepare acceptable and safe designs. An advice note is a document that helps designers on how best to apply design standards (i.e. it contains information on how to interpret a design standard, as well as other information that may not be directly related to the design standard). Design standards are usually bound together in a design manual: for example, the *Design Manual for Roads and Bridges* is produced by Highways England (HA *et al.*, 1998b) for use in the UK. Similar design manuals are in use in many other countries.

1.3.3   Bills of quantities

A bill of quantities (BoQ) is that part of the tender documentation that lists the pricing information required for each of the items needed to form the constructed road. The function of a BoQ is to provide a structure whereby the tenderers can price these items so that, when they are summed, the tender price is obtained.

As part of the normal course of developing, say, a construct-only contract for a road scheme, the designer will produce a complete set of design drawings for inclusion in the contract documentation. Copies of this set become the contract drawings (subject to any changes identified and agreed during the tender period) and are used by the competing contractors in the preparation of their tender submissions.

Having completed the design drawings for the road scheme, the road designer uses these to estimate the quantities of all of the various types and classes of material required by the work. These are put together in the BoQ in the form of lists of items of work that, in the UK, are mostly compiled from a standard method of measurement publication (HA *et al.*, 1998c). These lists contain a quantity column (with measurements in, for example, linear metres, cubic metres or tonnes), a column for the tenderer's rate to be inserted against each item, and a column for the total price against each item.

Nowadays, BoQs are often sent to the tenderers as computer files so that, with the aid of various forms of proprietary software, they can use them to complete their tenders. While the files, with the completed tender information, can be returned to the client, a hard copy may also be required for legal reasons.

On some projects there may be items in the BoQ about which the designer is unclear. For example, it may be known that the soil in a given area has to be excavated but, as it is unsuitable for use as a fill material, it will have to be wasted: however, the exact volumes of waste soil cannot be determined until the construction is actually carried out. In such instances, it is usual for the designer to include a provisional item in the BoQ, and to seek from the successful tenderer information about the rate proposed to be charged for this work on the basis that an estimated volume of material will be moved.

The use of provisional items in BoQs should, however, be minimised because of the potential dangers that it poses for the client. For example, in the case of the waste soil noted above, tenderers may attach a high unit price, hoping that the actual quantity to be moved will be much larger than the estimated amount on which the rate is based: on the other hand, if the estimated volume of poor material is large and the amount actually moved is much lower, the successful contractor may subsequently claim a higher rate than was initially quoted on the grounds that misleading design information was provided so that, as the actual amount of poor soil removed was much less than expected, the construction programme was disrupted, hired equipment could not be used, and a heavy loss ensued.

It is appropriate to mention here the development of a recent UK information source that has had impacts on BoQs, namely building information modelling (BIM). The need to have good information about prices encouraged the construction industry to set up processes to obtain and use such information and, consequently, the Construction Project Information Committee (CPIC), which is formed by representatives of major industry institutions, has assumed responsibility for providing best-practice guidance on construction product information. This committee has produced a definition of what is meant by BIM: this ensured an agreed starting point, as different interpretations of the term were hampering its adoption.

In May 2011 the UK government's chief construction adviser called for BIM adoption on all construction projects of £5 million and over. In June 2011 the UK government published its BIM strategy, announcing its intention to require collaborative three-dimensional BIM (with all project and asset information, documentation and data being electronic) on its projects by the year 2016. Initially, compliance will require building data to be delivered in a vendor-neutral Construction Operations Building Information Exchange (COBie) format, thus overcoming the limited inter-operability of the BIM software suites currently available on the market. The UK government's BIM task group website clearly informs the supply chain about the BIM programme and its requirements.

1.4. The tendering process

Having completed the statutory and detailed design processes, the client next seeks interested contractors to competitively bid on the documented scheme so as to win the right to build it. The tendering process involves the contractors being invited to price the BoQ for the project, taking into account all the information and constraints that are included in the proposed contract.

In the European Community (EC), public sector clients have to comply with the public procurement directives of the EC: these require all schemes over a certain price threshold to be advertised in the *Official Journal of the European Union*. If a very large number of contractors express an interest in tendering for a major road project in the UK, Highways England may reduce the list to a manageable number, typically six, who are judged to have the capability to carry out the project. If short-listing such as this is not done, not only may the client authority have a very large number of tenders to evaluate but much time and money will be wasted by many tenderers in preparing submissions that, for various capability reasons, stand little chance of being accepted.

In the tendering process the contractors are provided with a complete set of tender documents, together with appropriate instructions to assist in their correct completion and submission. The task of compiling a tender typically begins with the documents being broken into sections, and invitations to quote for parts of the works are then sent out by each tendering contractor to various suppliers and subcontractors. Simultaneously, each contractor's planning engineers get together with their intended project manager, to start preparing a work programme to enable the scheme to be constructed in the most cost-effective way.

The work programme is essentially concerned with the smooth running of the construction work. It is developed very early in the preparation of the tender, so as to enable the allocation of enough people, equipment and other resources to individual parts of the scheme to ensure that each part can be completed within a scheduled time. The programme produced is vital to the task of estimating the cost of the works and the bringing together of the final tender proposal. Issues such as the 'buildability' of the scheme, the potential for claims and other considerations that might affect the final price submitted are normally considered in the course of developing this programme, so that a final price and timetable can be established for submission to the client's representative. Under the provisions of the conditions of contract, the successful tenderer is required to produce the work programme within a specified period of time after the award of the contract, so that the client is aware of the details of how the work is to be carried out and in what time-scale. This programme must be kept up to date during the work, so as to reflect the changes that result from coping with the uncertainties of a construction site.

During the tender preparation process it is not unusual for a contractor to notice anomalies in the documents, and to notify the client regarding these. If these include reasonable suggestions that are acceptable to the client authority, a tender amendment letter is sent out to all tenderers and, if the changes are important, the tender period may be extended to allow further time for their consideration.

Some tenderers may consider that they would be able to produce a cheaper tender if some aspects of the design or specification could be changed (e.g. the alternatives would better fit their construction processes or be more easily built). They are allowed to submit such alternatives, provided that a conforming tender is also submitted (a conforming tender is one that meets, in every respect, the unchanged requirements of the tender documents).

The completed tenders must be submitted to the client, by a certain time, to a location publicly nominated by the client authority. When received, the tenders are stamped to avoid subsequent controversy regarding whether or not the deadline for their receipt was met.

A tender board is normally established by the client authority to open the sealed tenders and decide on a ranking for the tenderers based on tender evaluation guidelines. The board also judges whether the tenders conform to the requirements, and whether to accept a contractor's alternative should one be proposed. The board's report is submitted to the client, who may then get the designers of the scheme to fully assess and prepare a consolidated report on the tenders, and recommend regarding the most appropriate one to accept. Historically, in the UK, the successful tenderer was the one who put in the lowest price. However, this is no longer the case, as quality assessment is increasingly included in the selection process.

The recommendation regarding the successful tenderer is made to whosoever the client has nominated, per its project management system, as responsible for taking the decision regarding the awarding of the contract.

Following the award of the contract to the successful tenderer, the construction documents, amended as necessary after appropriate negotiations, are issued as formal contract documents. These documents form the basis of the construction of the new road scheme.

The award of the contract does not mean that the client can forget about the project until construction is completed and the road is opened, as, almost inevitably, there will be alterations and variations during the on-site construction period. It has to be accepted that the variability of materials used in civil engineering works in general, as well as those encountered on any particular site, are such that the construction process may well necessitate changes that cannot be predicted in advance. Nonetheless, when developing their design, the designers of the scheme must seek to do all within their power to minimise the potential for changes and disputes using appropriate risk management techniques.

It is also essential that the form of contract chosen should be one that minimises the areas for dispute and maximises opportunities for the client and the contractor to work together in partnership.

1.5. Construction

In many ways, the keys to a successful contract are in the hands of the senior staff employed to carry out and supervise the works.

As discussed, until relatively recently the standard method of major road procurement in the UK typically involved the client authority employing an engineering consultant to design the scheme and assist in the statutory planning process, the development of contract documents, and the seeking of tenders. Once the decision is made regarding to whom the contract should be awarded, the standard contract would be based on the documentation described and on a formal exchange of letters between the client and the successful contractor. If a dispute subsequently arose, the conditions of contract were used to deal with contractual issues, the specification with technical matters, and the BoQ with pricing concerns.

With a standard construct-only contract, the contractor is responsible for the construction of the scheme to time, to specification and to an agreed cost. Thus, the contractors employ their own engineers to make sure that the scheme is built in accordance with the contractual requirements – and to maximise the contractor's profit in the process. To ensure that its contractual requirements are met, the client authority will normally employ a resident engineer to supervise the contract on its behalf (i.e. to check on what the contractor is doing, to agree variations, and to finally 'sign off' that the work has been carried out satisfactorily). The resident engineer is usually appointed by the designer of the scheme, but, occasionally, a specialist engineering firm experienced in construction supervision may be instructed to supervise the works. If the contractor's senior staff, and those employed to supervise the works on behalf of the client, are highly experienced, the project is more likely to be carried out economically, efficiently and safely so that, especially in complex projects, cost escalations and disputes are minimised.

The assessment of the risks involved in constructing a road scheme is a vital component of managing that project. Hence, client project managers as well as those employed by design consultants, contractors and concessionaires need to be well versed in risk assessment and its management and mitigation.

Contractual variations and claims, and disputes, are not uncommon in construction projects. A variation is a change to the agreed design that occurs after the contract has been awarded: for example, when the contractor finds something on site that is not included in the contracted tender price or, if it is included in the contract, cannot be built as intended for some reason. Alternatively, a variation may result from the supervising engineer or the client subsequently perceiving a design fault that has to be changed. A claim occurs when a variation is considered to affect the contractor's ability to construct the scheme as defined by the contract, and is costing extra money. Because of construction expediency, the claim costs of variations are not always agreed and/or properly documented at the onset of the change, and, consequently, much time and effort may be subsequently spent on arguments regarding their validity. Not uncommonly, either side in these disputes, or both, may then resort to the arbitration clauses in the conditions of contract in order to get a decision on a claim. In the more recent forms of contract (e.g. the NEC3 type), adjudication and mediation are used to speed up the process.

When all the claims are settled and the final measurement is made, the client and the contractor settle the final accounts, and most of the remaining money owed to the contractor is then paid. The only payment that normally remains outstanding after this settlement of the accounts is the retention money: this is a part of the tender sum that is kept by the client for, say, 12 months to pay for the rectification of possible defects in the completed work (and to ensure that they are carried out).

1.6. Maintenance

Once the construction phase of the contract is completed, the new road is handed over to the client authority for use by traffic. There may then be an opening ceremony, but these are less frequent now in countries where environmental issues have a high profile.

Following the hand-over of the road to the client there is a maintenance (i.e. rectification) period, usually no longer than 12 months, during which the contractor remains responsible for the road structure. Almost inevitably, there will be a list of minor problems identified at the final inspection before the hand-over, which are insufficiently serious to warrant their rectification before the road is opened. These will be put right at the contractor's expense, along with any other problems that come to light during the early life of the new road. The rectification work carried out by the contractor during this 'maintenance' period is separate from the normal road maintenance activities carried out by the road authority's maintenance workers or contractors.

1.7. Operation

When the form of contract provides for a concessionaire to operate the road after opening or as part of a long-term contract to manage, maintain, operate and improve a road (e.g. as with the M25 motorway in England), the contract will have to include requirements, targets and performance bonuses for the operation of the road. This task is usually the responsibility of the road authority but, increasingly, this is being contracted out, and the road authority remains as a procurement and monitoring organisation, passing the risks onto the contractor or concessionaire.

1.8. A note on supporting documentation

Underlying many of the standards used in the design and construction of roads are those developed by each country's standards institution or by regional or international standards bodies. In Europe, in order to facilitate the operation of the single market, the specifications for all construction products are being harmonised where possible. The European Committee for Standardization (Comité Européen de Normalisation, CEN) is the body that is mainly responsible for developing, through technical committees, European (EN) standards. The EC mandates the details of what levels of overall quality, and attestation of conformity, should be included within the harmonised portions of the EN standards. By attestation of conformity is meant the level of checking to be carried out to ensure that the product does what the specification states it can do: the attestation level may range from a simple manufacturer's declaration to a full third-party assessment. The harmonised EN standards are gradually replacing well-established national standards such as BS standards in the UK, German Institute for Standardization (Deutsches Institut für Normung, DIN) standards in Germany and Association Française de Normalisation (AFNOR) standards in France. Where developed and agreed, EN are mandatory for purchases dealing with public roads.

The worldwide standards body is the International Organization for Standardization (ISO). However, use of ISO standards is not mandatory unless national bodies adopt and use them in their own national series.

Another European body that is of importance in road construction is the European Organisation for Technical Approvals (EOTA), which gives technical approval to proprietary products. If, for example, a manufacturer develops a proprietary product such as a bridge-deck waterproofing system, then it would submit the product to EOTA for testing and certification that it met its general requirements, and that the manufacturer's claims can be substantiated. The current equivalent body in the UK is the British Board of Agrément (BBA).

REFERENCES

ACE (Association of Consultancy and Engineering) (2011) *Infrastructure Conditions of Contract*. ACE, London, UK.

FIDIC (Federation International des Ingenieurs-Conseils) (1999) *Conditions of Contract for Works of Civil Engineering Construction*. FIDIC, Lausanne, Switzerland.

HA (Highways Agency) (2008) *The Project Control Manual*. Stationery Office, London, UK. See www.highways.gov.uk (accessed 04/06/2015).

HA (2009) *Procurement Strategy 2009*. Stationery Office, London, UK. See www.highways.gov.uk (accessed 04/06/2015).

HA (2013) *The Project Control Framework Handbook*. Stationery Office, London, UK. See www.highways.gov.uk (accessed 04/06/2015).

HA, Transport Scotland, Welsh Government and Department for Regional Development Northern Ireland (1998a) *Manual of Contract Documents*, vol. 1. *Specification for Highway Works*. Stationery Office, London, UK.

HA, Transport Scotland, Welsh Government and Department for Regional Development Northern Ireland (1998b) *Design Manual for Roads and Bridges*. Stationery Office, London, UK.

HA, Transport Scotland, Welsh Government and Department for Regional Development Northern Ireland (1998c) In *Manual of Contract Documents*, vol. 3. *Method of Measurement*. Stationery Office, London, UK.

HMG (Her Majesty's Government) (1980) *The Highways Act 1980*. Stationery Office, London, UK.

HMG (1996) *Compulsory Purchase of Land Regulations, Cmnd 2145*. Stationery Office, London, UK.

HMG (2008) *The Planning Act 2008*. Stationery Office, London, UK.

HMG (2015) *Construction, Design and Management Regulations*. Stationery Office, London, UK.

ICE (Institution of Civil Engineers) (1999) *Conditions of Contract*, 7th edn. Thomas Telford, London, UK.

ICE (2013) *New Engineering Contract, NEC3*, 3rd edn, ICE, London, UK.

Highways
ISBN 978-0-7277-5993-1

http://dx.doi.org/10.1680/h5e.59931.021

Chapter 2
Route location

Gareth Hearn Director, Hearn Geoserve Ltd, UK

2.1.   Importance of route location

A wide range of factors control the location of the optimum route for a new road. The significance of these factors varies from project to project, region to region, and country to country, depending in each instance upon transport network and strategic planning, and socioeconomic, environmental, engineering and financial considerations. In all cases, the essential engineering requirements need to be satisfied; that is, a route needs to be selected and then designed to accommodate the anticipated traffic with an alignment that allows vehicles to travel safely at the design speed dictated by the category of road required. Route location is usually governed by environmental and social factors, public perception and stakeholder interests, the need to integrate the alignment with other transport infrastructure, and the desire to maximise socio-economic benefits and minimise route length and construction and operation costs.

Route location for a new road is a critical step in the development of any project, as it often determines its economic feasibility, and it affects the scope and detail of the studies that need to be carried out during subsequent investigation and design stages. The route selection process involves proceeding from the general to the specific, as route corridors, route options and eventually route alignments are identified, compared, selected and then designed. The main issues need to be identified early, so that they can be accommodated in decision-making as the corridor is selected and the route progressively refined. It is an iterative process which, if short-circuited or undertaken without due regard to critical factors, can result in the wrong decisions being made.

2.2.   Regional context

Before discussing the procedures for route selection, it is useful to comment briefly on some of the regional-specific factors that influence their application. Some of the greatest contrasts can be found between countries that already have well-established road networks and those whose networks are expanding rapidly to facilitate economic growth and regional development (HA *et al.*, 1982; Hearn and Hunt, 2013; TRL, 2005).

The average density of roads in the UK, for example, is currently about $172 \text{ km}/100 \text{ km}^2$, and the road network has reached a stage of maturity (Cook, 2011). The (now) Highways England (HA, 2013) has established route strategies in England that define operational and investment priorities, and the focus is primarily on the widening, dualling and operational management of the existing network. New road alignments are relatively short and usually aim to alleviate local congestion and improve road safety. Issues that govern the location of new roads include the supply of land and the safeguarding of important landscapes and sites of historical, ecological or scientific importance (Darrall, 2011; DCLG, 2012; HA *et al.*, 1982).

Public consultation plays a major role in the choice of alignment in the UK because of environmental and social issues associated with landscape and visual intrusion, open space and recreation, noise, air pollution and pedestrian access. The UK has a well-established system of planning controls and legal safeguards (see Chapter 1), and accessible historical data records (see Chapter 3), which can be called upon for guidance when locating a new road. The use of tunnels is being increasingly considered as a viable option for improving the flow of traffic while at the same time reducing, or avoiding altogether, the negative impacts associated with above-ground options, especially in environmentally sensitive areas.

By contrast, in the whole of sub-Sahara Africa, for example, the average road density in 2008 was only 6.9 km/100 km^2 (World Bank, 2010). The future expansion of trunk road networks and the connectivity of rural areas are seen as investment imperatives: thus, road construction is welcomed as a major contributor to economic development and livelihood improvement, and it is predicted (Coghlan, 2014) that the network of major roads will expand in size by a factor of between 6 and 10 by 2040. Low-intensity land uses predominate in African countries, and the generally low density of housing, commercial development and transport infrastructure means that, unlike in the UK, for example, there are greater opportunities to select routes that minimise economic and social disruption. While there are planning and environmental controls and safeguards in place in most African countries, the data sets with which to implement them effectively are often lacking, so that much primary data often need to be collected.

2.3. Generalised approach to route location: a four-stage process

The flow chart in Figure 2.1 shows the typical stages in route location, namely

1 project objectives
2 definition of the corridor and route options to be considered
3 route option comparison and selection of the preferred route within the corridor
4 preliminary design of the preferred route.

These four stages are indicative and iterative rather than definitive, and the details of each will vary according to the situation encountered. Ordinarily, the four stages should contain the following.

Stage 1: The proposed scheme should be supported by a robust case for change that fits within strategic transport policy objectives, and establishes an agreed set of objectives that define its desired outcomes. The latter objectives should include addressing appropriate issues of economic, environmental and social importance, as well as confirming

Figure 2.1 Principal stages and processes in route location

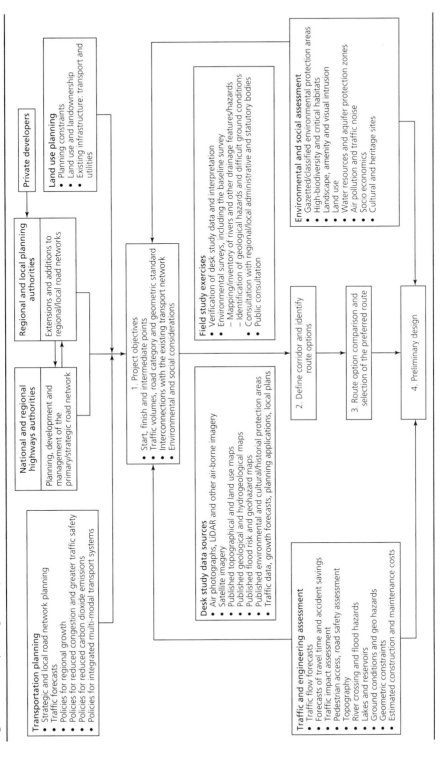

the key controls on the new alignment, including the start and finish points of the new road and any intermediate points to which it must be connected.

Stage 2: Selecting the route corridor involves defining a corridor that will include all feasible route options between the end termini; this may be very wide, or relatively narrow, depending on the circumstances. The corridor's limits may be defined by the need to avoid topographical barriers such as mountain ranges or major rivers and their flood plains. Corridor limits may also be constrained by urban areas and where planning controls prevent a change in land use: for example, environmental conservation areas, areas of historical and cultural importance, forest and 'green belt' areas, and/or areas set aside for residential and commercial development. This stage may also involve the initial identification of alternative route options within the corridor: these options may be entirely separate or they may be localised variants of an otherwise common route.

Stage 3: Figure 2.1 also broadly indicates the range of factors that need to be considered when comparing route options. There may be overriding considerations that dictate the preference of one route option over others: for example, environmental constraints, construction practicality, travel time savings, and/or construction and maintenance costs. Alternatively, the preferred route may not be obvious, and an integrated assessment may be required of all options (including the upgrading of an existing road, if appropriate) so that an aggregated and balanced comparison can be made.

The cost estimate for each option should combine the estimated construction cost with the costs of acquiring land and the measures that need to be put in place to compensate people who are affected, as well as provisions for mitigating difficult ground conditions and environmental and social impacts. People's time can be valued and, in some situations (depending upon the objectives underlying the decision to build the road), reduced journey times, accidents and vehicle-operating costs over the design life of the road may become critical factors when comparing route options.

To enable construction costs to be estimated to within reasonable limits, each route option is often developed to an outline or concept design using a suitable digital elevation model (DEM) and employing propriety highway design software. In such modelling, basic parameters such as design speed, carriageway width, maximum longitudinal gradient, minimum curve radius, and side-slope angles for cuttings and embankments are used to generate the alignments and to calculate earthwork quantities. Requirements for structures, such as overpasses, underpasses, bridges, tunnels, drainage and other such provisions, must also be taken into account.

Stage 4: This stage involves progressing the preferred route from a concept design to a preliminary design, defining the footprint of the road and requirements for land acquisition, and determining more closely its ultimate cost. Establishing the final centreline for the preferred route determined at stage 3 involves accommodating important control points (e.g. road junctions and bridge sites), the prescribed horizontal and vertical alignment parameters, and confirming that there is sufficient ground space available to construct the final road alignment (including for all intersections, structures and earthworks). In practice, additional topographical data obtained at this stage may show that local deviations are necessary in order to fit the centreline better to the topography and enable the alignment to be optimised. In hilly terrain, for example, the preliminary design may show that road safety can be improved by realigning the centreline slightly to avoid locating road sections where they are shielded from the sun (so that rainwater, snow and/or ice can more easily dissipate), or ensuring that horizontal curves are longer if they have to be superimposed on vertical curves, or minimising the use of horizontal and vertical curves as flat as possible at intersections with other roads. The exact positions of property boundaries, buildings and other artificial obstacles, as well as how property owners are affected by the proposed right-of-way, may also affect the final location of the centreline. In rural areas, minor deviations may allow the loss of farm-land to be minimised. The preliminary design is normally used as the basis for the prep-aration of the environmental impact assessment (EIA) for the scheme. As well as identifying the measures that need to be taken to minimise or offset anticipated negative impacts, the EIA is also a key contributor to the public consultation.

2.4. Selecting the preferred option

Stage 3 is a critical stage, as it requires consideration of a wide range of varied and sometimes conflicting factors, and it results in the selection and preliminary design of the preferred route. The remainder of this discussion therefore will focus on those aspects of route assessment and comparison that are commonly considered in stage 3. In accord-ance with Figure 2.1, these are grouped and summarised under the following headings:

- traffic
- engineering
- environmental and social issues
- comparative analyses.

2.4.1 Traffic assessment

Current traffic flow patterns (e.g. existing traffic flows and their composition, levels of congestion and accident statistics) need to be determined for each route option being evaluated. Predictions also need to be made as to how these might change over the proposed design life of the road, as a result of proposed land use changes and other infrastructure development initiatives, scheme-generated traffic, and economic growth in the area being serviced by the new road. While this information is needed to enable the necessary standards of design speed, road width, gradient and quality of pavement to be established, and the overall benefits of the scheme to be estimated, it will also have a bearing on the route option ultimately selected in terms of reduced journey times and traffic safety. In some cases, the traffic assessment can be based on desk study using traffic data routinely gathered by road authorities. However, in others, the existing data may be insufficient or out of date, and may need to be supplemented with route-option-specific information based on field surveys. Details of the implementation of traffic studies is beyond the scope of this chapter, and the reader is referred to the large body of information available in the literature (e.g. DfT, 2014a; HA *et al.*, 2006).

Traffic characteristics can be expected to vary significantly when different route options

- involve different road and other transport network connections
- have significantly different vertical and horizontal alignments that impact on the design speed, journey time and vehicle operating costs
- provide access connections to areas that may impact differently on traffic generation and road design (e.g. agricultural or mineral resource areas, built-up areas, and/or industrial and commercial facilities) and/or
- encounter significantly different traffic volumes and traffic flow patterns (e.g. options that involve building roads through existing built-up areas rather than bypassing them).

2.4.2   Engineering assessment

Appropriate details regarding the following will need to be determined from the desk study regarding each route option in order to provide information on the following:

- *topography and geological data*, including (a) terrain steepness and complexity, (b) lakes, rivers and major streams (including the locations of existing and potential crossings), (c) high water tables, flood plains and other sources of drainage hazard, (d) underlying rock that is close to the ground surface and (e) potential sources for the extraction of road-making materials
- *land usages*, including (a) information on existing infrastructure (e.g. roads, railways, canals, pipelines, above and below-ground utilities (such as stormwater pipes and sewers, water and gas lines, communication and power lines)) and other services, (b) delineation of built-up areas, and of industrial, commercial and agricultural areas, and (c) locations of mineral resource areas (including existing open-cast and underground mines), and areas identified for future resource extractions
- *hazardous ground conditions that should be avoided*, including areas of (a) geotechnical hazard (e.g. that are subject to landslides, solution or collapse, or involve poor subsoil conditions and/or expansive soils), (b) old mining areas, and (c) areas where contaminated ground and industrial legacies are likely to have health and safety and environmental mitigation cost implications.

Traditionally, the assessment of the terrain for route options was undertaken using paper mapping based on field surveys, but in recent decades increasing use has been made of digital topographical maps that can be manipulated using geographical information systems (GISs) and terrain modelling software in appropriate locales. Nonetheless, much information can be obtained from published topographical and geological maps, but the detail available varies considerably from country to country (see Chapter 3 for information regarding topographical and geological data that are available in the UK, especially through Ordnance Survey (OS) and British Geological Survey (BGS) sources).

There is no definitive scale at which route option studies should be undertaken: this depends upon the size of the scheme and its complexity. However, studies are often undertaken at scales of between 1 : 10 000 and 1 : 25 000, with more detailed investigations of critical areas carried out at larger scales. Many countries, however, do not have this range of mapping scales (i.e. 1 : 50 000 is not untypical of the largest scale available in some regions), so that recourse may have to be made to remote sensing techniques to derive terrain information that is otherwise unavailable.

When viewed through a stereoscope, stereo aerial photography can give a very good three-dimensional image in which the skilled interpreter can identify topographical features (e.g. cliffs and escarpments, steep slopes and flat areas), landslides, areas of erosion, river catchments, alluvial fans, river terraces and flood plains. Land use and details of villages, towns, field boundaries and existing transport infrastructure can also be identified, although some of these may have changed since the photography was taken. While stereo photography can greatly assist in landscape interpretation, it does not contain digital information on elevation, gradients or side slopes. Alternatively, new aerial photography can be combined with ground control to yield DEMs using photogrammetry, although this can be expensive, time-consuming and hard to organise in some countries. Consequently, airborne digital imagery data, primarily LiDAR (i.e. Light Detecting And Ranging), are increasingly used to generate DEMs for the planning and design of commercial development projects (including roads) worldwide. Constraints on the use of LiDAR include its cost and the logistical difficulty of acquiring data that are not available from archives.

Recent advances now enable satellite imagery to be used to generate digital elevation data to define, compare and select route options, especially where published maps provide only limited detail or are at too small a scale. As well as being used for topographical mapping, these data can also be used to develop drainage and catchment area mapping where GIS is able to automate the tasks. The DEMs can also be used as a base layer upon which other satellite imagery can be draped in order to enhance three-dimensional visualisation. Table 2.1 lists selective sources and accuracies of satellite-derived digital mapping. Satellite data can also be used to interpret ground conditions, including vegetation cover, drainage patterns and soil types using multi-spectral analysis.

Free-to-download satellite imagery and DEM data currently come in two main formats: SRTM (Shuttle Radar Topographic Mission) and ASTER (Advanced Space-borne Thermal Emission and Reflection Radiometer). Of these, ASTER is currently favoured for route location in many developing countries due to its greater resolution. The main issue relates to accuracy with respect to terrain type when comparing, for example, an alignment option on relatively flat terrain (for which calculated earthwork quantities using proprietary software should be reasonably accurate) with one in mountainous terrain (for which not only the earthwork quantities might not be particularly accurate but also the road alignment itself). SPOT5, which is of fairly high resolution and has good vertical accuracy, is available at reasonable cost, and is held by mapping agencies in many countries; it can provide ground modelling data that are likely to be of most value to route location in mountainous terrain.

The extent to which detailed geological mapping is available varies significantly. In many countries, 1 : 250 000 and even 1 : 1 million or 1 : 2 million mapping sometimes provides

Table 2.1 Overview of digital elevation data that can be derived from selected satellite sensors (accuracies were correct at the time of publication)

(a) Digital terrain modelling

Sensor	Resolution: m	Horizontal accuracy: m	Vertical accuracy: m	Availability/archive length
SRTM	3 arc seconds (90 m)	30	5–15 (terrain dependent)	Global coverage
ASTER	30	30	15–30	Global coverage
SPOT HRS DEM (SPOT5)	20–30	15	5–10 (terrain dependent)	Off-the-shelf product
WorldView-1 and 2	1	1–2 (with ground control, terrain dependent)	1–2 (with ground control, terrain dependent)	2008
GeoEye-1	1	1–2 (with ground control, terrain dependent)	1–2 (with ground control, terrain dependent)	2009

(b) Image interpretation

Sensor	Resolution: m		Scene size: km	Launch date
	Black and white	Colour		
WorldView-1	0.5	–	16 × 16	2007
WorldView-2	0.5	2	16 × 16	2009
GeoEye-1	0.5	1.6	15 × 15	2008
QuickBird	0.6	2.4	17 × 17	2002
IKONOS	1	4	11 × 11	1999
SPOT-5	2.5	10	60 × 60	2002
ASTER		15/30	60 × 60	2002
Landsat 7 ETM+	15	30	185 × 185	1999

Note: There are other platforms, such as SPOT-6 and WorldView-3, that provide higher resolution but their availability in archive is currently less, and their cost much higher, than their predecessors

Figure 2.2 Factors controlling route location on river flood plains (Hearn and Hunt, 2013)

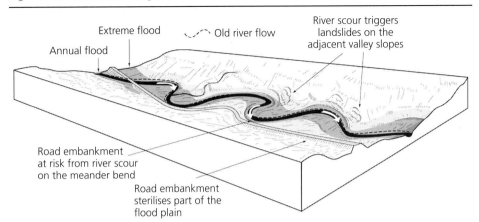

the only geological information available. Such small-scale maps often show only formation (i.e. age) level information rather than lithology, and are of limited engineering use. In many African countries, for example, there are particular conditions (e.g. faulted and fractured rocks and expansive, collapsing and dispersive soils) that can influence route location so that, if the published information on these conditions is inadequate, primary data will need to be derived through remote sensing and field investigation.

The numbers and sizes of river crossings and areas prone to flooding have a major impact on construction and operational costs, and, therefore, can affect route selection. While rainfall and runoff records are generally less relevant than flood-risk mapping for route selection, there are occasions when, for option comparison purposes, an assessment of the design flood is required to enable bridges, embankments and other drainage structures to be sized, and to help indicate a waterway's potential for scour. Aligning roads on flood plains not only exposes them to the risk of flooding and scour but may also increase flood risk through the 'sterilisation' of floodplain water storage (Figure 2.2). River scour is of most concern in mountain terrain or in environments where rivers frequently change course. Desk study sources for information on these issues include topographical maps, air photographs and other remote sensing data, and rainfall and river-gauge records that are usually held by statutory authorities. Unfortunately, in some countries the scales of published topographical maps are too small to show drainage networks with any accuracy, while river channels may have shifted since they were originally mapped. In these situations, the desk study will almost inevitably have to be supplemented by field observations in order to obtain the information required for route location.

Failure to take account of hazardous ground conditions during route selection can lead to serious consequences: for example, where slope movements along residual slip surfaces are reactivated during road construction (Weeks, 1969). Often these and other potential geohazards can be assessed from desk study sources, but primary data may also need to be collected using remote sensing and field investigations and, if the geotechnical ground conditions pose sufficient risk or have important cost implications, subsurface ground investigations will be required for route comparison.

2.4.3 Environmental and social impact assessment

Environmental and social impact assessment should contribute to all aspects of the development of a road scheme, from strategic planning through route location to road design, construction and operation. Route option assessments usually involve

- reviewing statutory and legal requirements relevant to potential environmental and social impacts
- identifying environmental issues and potential impacts on receptors (i.e. those elements of the landscape, hydrology, ecology and society that are potentially impacted)
- developing an aggregate assessment of the total environmental impact
- identifying mitigation measures and assessing the residual environmental impact after mitigation
- consulting with all stakeholders, including planning authorities and other regulators, service providers and social groups and communities likely to be affected by each option.

Most countries have a range of environmental and social data that are accessible for desk study purposes (e.g. maps that show the distribution of land use and the location of ecological, cultural and archaeological conservation areas are normally available from governmental agencies responsible for environmental protection). Nowadays, considerable amounts of digital data are also available online: for example, in the UK, these include maps showing zones of groundwater and surface water protection and groundwater vulnerability, nationally designated data sets that identify areas of natural beauty and important habitats and species sites, and areas of archaeological importance, scheduled monuments and listed buildings. Generally, however, it will be necessary to carry out field surveys in order to clarify and confirm the details of any environmental issues that might affect route options.

Particular social factors that might need to be addressed in some countries, depending on scheme location and circumstances, include the protection of indigenous peoples and disadvantaged groups and the societal role of women, especially in relation to land acquisition and compensation.

2.4.4 Comparative analyses

When all the required information has been gathered for the various route options, and all of the potential routes have been explored and issues clarified, a choice has to be made as to which should be the preferred route from the options being considered. Sometimes this selection process may result in a return to the base-case option (where one exists) of simply upgrading an existing highway; in other instances, it may result in the selection of a completely new alignment or one composed of part-existing/part-new alignments.

The methods used to assist in the selection process can vary considerably from country to country and from scheme to scheme, and are influenced by the need for, and the objectives underlying, the proposed road development. In the UK in the 1960s, much attention was paid to the use of cost–benefit analysis (CBA) to assist in choosing between options for public roads: this was at a time when maximising the economic return from scarce funds invested from the public purse was deemed to be a prime objective as governments attempted to meet the (then) overwhelming demands for new major highways. As time progressed and the limitations associated with the cost–benefit approach became more obvious, attention turned to the development of multi-criteria techniques that were more transparent, paid greater attention to the application of more inclusive approaches to environmental and social issues, and included public consultation and participation in the decision-making (e.g. see DfT (2014b, 2014c) and HA *et al.* (1982) for current UK guidance for locating major public roads).

2.4.4.1 Cost–benefit analysis (CBA)

CBA is a well-established technique for comparing the costs and benefits of a project. In brief, cost–benefit is a form of economic analysis that compares the total cost of each route option with the total benefits it would bring to current and anticipated road users over its design life, in order to derive a discounted-for-time economic measure of comparison. Benefits measured typically comprise savings in vehicle-operating costs and travel times, accident costs, and reduced maintenance requirements for existing roads (because lower numbers of vehicles may use them). Costs typically include those of construction, land acquisition and any required environmental mitigation as well as traffic delays during construction and the maintenance of the new road over its design life. A route option is deemed to be economically efficient if its benefits exceed its costs. The most economically desirable option is often identified by determining the net present value (NPV) of the benefits minus the costs associated with each option, dividing this by the NPV of the financial requirement associated with the option, and then ranking all options in the order of this overall indicator. The most efficient economic option is then deemed to be that for which the relative economic return is greatest. However, in some comparisons, the CBA results may well be similar, and not decisive, in terms of relative economic viability (e.g. where route options are close to each other).

Ideally, in CBA studies all the costs and the benefits associated with each route option are quantified in monetary terms to a specified price base over the design life of the proposed scheme. However, not all benefits and costs are easily defined for inclusion in economic analyses due to the difficulty in valuing them in monetary terms. Some examples of environmental and social indicators that are problematic in relation to their treatment in conventional CBA studies include the detrimental impacts on non-road users of noise, air pollution, visual intrusion and severance. Positive factors that are difficult to quantify relate to the strategic benefits of an expanded road network, the benefits of improved access to education and healthcare, and the wider implications for regional development. (An introductory overview of the use of CBA in evaluating road schemes is provided, for example, in Nash (2002).)

2.4.4.2 Multi-criteria analysis

Given the above, it is now generally accepted that road option decisions should not be based on economic appraisal alone, but should also consider impacts such as social and environmental ones. In contrast to CBA, multi-criteria analysis allows a range of factors to be included in the assessment of route options. Thus, as well as economic and engineering criteria, environmental, social and other non-quantifiable factors can be taken into account, including planning imperatives and public opinion.

The multi-criteria analysis approach generally involves

- establishing key objectives or indicators (e.g. for a public road these might relate to factors such as public acceptability, environmental impact, value for money, equity, accessibility, safety) and weighting them in terms of their relative importance to the success of the proposed scheme
- measuring the extent to which each route option performs against the key objectives
- ranking the route alternatives on the basis of their aggregated performance.

Advantages of the use of ranking procedures are their transparency, speed and simplicity, and potential for including social benefits and popular preference. The main criticisms against them relate to the fact that

- they involve qualitative assessment and thus provide an opportunity for the assignment of weightings that are based on personal preference (most usually on the judgement of professional staff planning the scheme)
- the degree to which particular objectives are achieved by each option is not assessed on the basis of measurement but, again, on the judgement of professional staff
- weightings are unlikely to be stable in the longer term, with the result that the highest-ranked option can become suboptimal.

The basis of a simple multi-criteria analysis approach is illustrated by a study, led by the author, to select the preferred route for a new haul road to link an existing mine with two new mine sites in the northern part of the Republic of Liberia in West Africa. The scheme site is relatively isolated and located in the sparsely populated, steep and complex topography of the Nimba Mountains that contain high-sensitivity floral and faunal habitats that had been previously mapped as part of a detailed environmental baseline survey for the wider mine project. As existing topographical mapping was limited, air-borne digital imagery (LiDAR) was commissioned to yield a DEM with 2 m contour intervals. Five route options (Figure 2.3), ranging in length from 8 to 13.5 km, were developed by manually plotting their approximate centrelines onto paper contour maps derived from the LiDAR DEM using ruling gradient (8–10%) and minimum radius curve (120–150 m) templates. These options were then digitised, and proprietary software was used to develop each route into an outline road design. Mass-haul earthworks quantities were calculated for each option, based on cross-sections at 20 m intervals.

Multi-criteria analysis was used to evaluate the five route options considered for this project, as summarised in Table 2.2. With this procedure, each indicator was given a weighting according to its perceived importance in achieving the overall objectives of

Figure 2.3 Example of map used in a study of alternative route options for the selection of a mining haul road in northern Liberia

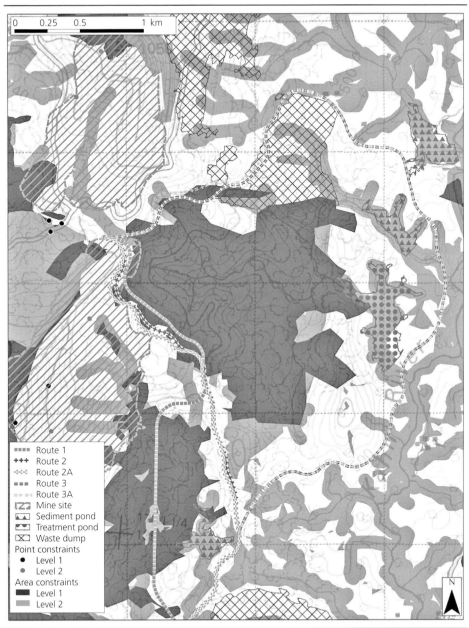

establishing a road with least environmental damage and least construction and operational cost. Thus, for example, the taking of land in the environmental constraint areas with highest biodiversity (level 1) was given a maximum weighting of 10. The construction cost and route length were given lesser weightings of 6 and 5, respectively,

Table 2.2 Ranks, weightings and scores used to develop the multi-criteria analysis for a mining haul road route selection in northern Liberia

Engineering factors	Rank	Weighting	Score	Environmental factors	Rank	Weighting	Score	Social impact factors	Rank	Weighting	Score	Total score
1 Cost	3.71	6	22.26	Level 1 areas	5	10	50	Areas of farmland removed	3.43	6	20.58	92.84
2	4.15	6	24.9		1.43	10	14.3		3.93	6	23.58	62.78
2A	3.74	6	22.44		0.86	10	8.6		3.93	6	23.58	54.62
3	5	6	30		1.33	10	13.3		5	6	30	73.3
3A	4.8	6	28.8		0.82	10	8.2		5	6	30	67
1 Length	3.48	5	17.4	Level 2 areas	2.76	5	13.8	No. of huts (temporary dwellings) removed	3	2	6	37.2
2	3.4	5	17		2.85	5	14.25		0	2	0	31.25
2A	3.43	5	17.15		2.74	5	13.7		0	2	0	30.85
3	5	5	25		4.92	5	24.6		0	2	0	49.6
3A	5	5	25		5	5	25		0	2	0	50
1 Compaction issues	3.47	2	6.94	Length above or through level 1	5	5	25	No. of cultural heritage sites affected	0	8	0	31.94
2	2.34	2	4.68		3.3	5	16.5		0	8	0	21.18
2A	2.63	2	5.26		2.09	5	10.45		0	8	0	15.71
3	5	2	10		1.95	5	9.75		0	8	0	19.75
3A	4.09	2	8.18		1.96	5	9.8		0	8	0	17.98
1 Mass haul	3.47	2	6.94	Excess of cut to be spoiled	3.79	5	18.95					25.89
2	2.34	2	4.68		4.94	5	24.7					29.38
2A	2.63	2	5.26		4.43	5	22.15					27.41
3	5	2	10		4.24	5	21.2					31.2
3A	4.09	2	8.18		5	5	25					33.18

Route	Geohazards (value)	Geohazards (weight)	Geohazards (score)	Need for borrow areas (value)	Need for borrow areas (weight)	Need for borrow areas (score)	Subtotal
1	5	5	25	0	4	0	25
2	5	5	25	0	4	0	25
2A	5	5	25	0	4	0	25
3	3	5	15	0	4	0	15
3A	3	5	15	0	4	0	15

Route	Mine facilities crossed (value)	Mine facilities crossed (weight)	Mine facilities crossed (score)	Number of stream crossings (value)	Number of stream crossings (weight)	Number of stream crossings (score)	Subtotal
1	0	5	0	4.32	3	12.96	12.96
2	0.07	5	0.35	2.46	3	7.38	7.73
2A	0.06	5	0.3	2.46	3	7.38	7.68
3	5	5	25	5	3	15	40
3A	3.57	5	17.85	5	3	15	32.85

Route	Total length of route (value)	Total length of route (weight)	Total length of route (score)	Subtotal
1	3.48	3	10.44	10.44
2	3.4	3	10.2	10.2
2A	3.43	3	10.29	10.29
3	5	3	15	15
3A	5	3	15	15

Route 1 final score	236
Route 2 final score	188
Route 2A final score	172
Route 3 final score	244
Route 3A final score	231

and so on for the other criteria. For each factor, the route option with the worst-case condition (the greatest level 1 land-take, greatest cost, longest length, etc.) was assigned a rank of 5. The other options were then assigned proportional ranks according to their value against the worst-case condition. For example, route option 1 was the least desirable in terms of level 1 environmental constraint, as it required a land-take of 12.20 ha: a rank of 5 was assigned to this route option for this factor. By contrast, route option 3A required a land-take of only 1.99 ha in areas of level 1 environmental constraint, and was assigned a rank of 0.82 (i.e. 1.99/12.20 × 5). By multiplying the rank and the weighting for each factor and then summing the total, an aggregate score was then obtained, and the route with the lowest score was deemed the preferred option. From Table 2.2 it can be seen that this was route option 2A. The multi-criteria analysis was then rerun using modified weightings to test the sensitivity of the outcome, and route 2A remained the preferred option.

2.5.   Some final comments

The most important point to take forward from this review is the fact that route selection is usually the most critical element of a new road development project and, hence, must contain sufficient investigation and analysis of the issues that ultimately determine the successful performance of the completed scheme. Thus, before initiating the route location study, the team leader must have a clear understanding of the rationale underlying the proposed road scheme and the objectives that need to be satisfied by it. Route options are then judged according to how well they satisfy these objectives.

Route selection is generally easiest in flat terrain, where existing infrastructure and building development are sparse, and where environmental and social issues are demonstrably of low significance. Under these relatively rare conditions, quantifiable traffic, engineering and cost issues are the predominant considerations. Route selection becomes progressively more complex where terrain difficulties, geohazards, existing infrastructure, land uses, and environmental and social issues are encountered, thus requiring an integrated and iterative multidisciplinary approach. The problems are probably at their most complex in and around urban areas, where community aspirations, interactions with existing transport infrastructure and public access, and environmental, social, economic and planning issues are dominant.

Because of these multivariable considerations, it should be recognised that there is no unique selection method that can be used to determine the preferred route, and that choosing between alternative options normally requires a combination of CBA and multi-criteria analysis, and the involvement of stakeholder consultations at critical stages in the selection process. The Liberian mine haul road illustration given in this chapter was based on a relatively simple multi-criteria analysis, although it still required the extensive (and expensive) derivation of environmental and social baseline data in order to be able to make informed comparisons. By contrast in countries with more mature road networks, such as the UK, much baseline data already exist, and the focus, in terms of locating a route for a public road, is more often on defining a scheme that optimises its integration with existing infrastructure, satisfies the often conflicting and challenging criteria of environmental and public acceptability, and demonstrates value for money in the use of taxpayer funds.

Route selection should be progressed in such a way that it allows critical issues to be identified and catered for as soon as possible during the investigation, and this is helped by identifying and involving key stakeholders and information sources early in the process. The use of desk study information, supplemented by field verification, should be maximised. In particular, remote sensing offers considerable data-collection opportunities that are often not entirely appreciated. These considerations apply equally to new roads planned for mature road networks as they do for rapidly expanding networks in less developed countries.

Finally, it should be noted that, while route location is critical to the success of a road scheme, it is important not to waste money by over-investing in the development of route options that are likely to be rejected later, for whatever reason. While a broad range of options should be considered in early phases of the study, these should be reviewed against a set of objective criteria as quickly as possible in order to obtain a shortlist of perhaps three options for more detailed investigation. This requires a holistic and informed approach from stage 1.

Acknowledgement

The author would like to thank the following for comments received on the draft of this chapter: T. Hunt, J. Mitchell, C. Bush and P. Tomlinson.

REFERENCES

Coghlan A (2014) Africa's road-building frenzy will transform continent. *New Scientist* **221(2951)**: 8–9.

Cook A (2011) *A Fresh Start for the Strategic Road Network*. Department for Transport, London, UK.

Darrall L (2011) Introduction to policy and practice. In *ICE Manual of Highway Design and Management* (Walsh ID (ed.)). Institution of Civil Engineers, London, UK, pp. 99–106.

DCLG (Department for Communities and Local Government) (2012) *National Planning Policy Framework*. DCLG, London, UK. See http://www.gov.uk/government/publications/national-planning-policy-framework–2 (accessed 02/06/2015).

DfT (Department for Transport) (2014a) *Tag Unit M1.2, Data Sources and Surveys*. DfT, London, UK. See http://www.gov.uk/transportanalysis-guidance-webtag (accessed 02/06/2015).

DfT (2014b) *Transport Analysis Guidance: An Overview of Transport Appraisal,* DfT, London, UK. See http://www.gov.uk/transport-analysis–guidance-webtag (accessed 02/06/2015).

DfT (2014c) *TAG Unit A3, Environmental Impact Appraisal*. DfT, London, UK. See http:// www.gov.uk/transport-analysis-guidance-webtag (accessed 02/06/2015).

HA (Highways Agency) (2013) *Route Strategies*. Stationery Office, London, UK. See http://www.highways.gov.uk/our-road-network/managing-our-roads/improving-our-network/route-strategies/ (accessed 02/06/2015).

HA, Transport Scotland, Welsh Government and Department for Regional Development Northern Ireland (1982) Section 1: Assessment of road schemes. Choice between options for trunk road schemes. In *Design Manual for Roads and Bridges*, vol. 5. *Assessment and Preparation of Road Schemes*. DfT, London, UK, TA 30/82.

HA, Transport Scotland, Welsh Government and Department for Regional Development Northern Ireland (2006) Section 2: pavement design and construction. Part 1: traffic assessment. In *Design Manual for Roads and Bridges*, vol. 7. *Pavement Design and Maintenance*. Stationery Office, London, UK, HD 24/06.

Hearn GJ and Hunt T (2013) *Route Selection Manual*. Ethiopian Roads Authority, Addis Ababa, Ethiopa.

Nash CA (2002) Economic and environmental appraisal of transport improvement projects. In *Transport Planning and Traffic Engineering* (O'Flaherty CA (ed.)), 4th edn. Arnold, London, UK, pp. 80–102.

TRL (Transport Research Laboratory) (2005) *Overseas Road Note No. 5: A Guide to Road Project Appraisal*. TRL, Crowthorne, UK.

Weeks AG (1969) The stability of natural slopes in SE England as affected by periglacial activity. *Quarterly Journal of Engineering Geology* **1**: 49–61.

World Bank (2010) *The Little Data Book on Africa. African Development Indicators 55201*. World Bank, Washington, DC, USA.

Highways
ISBN 978-0-7277-5993-1

ICE Publishing: All rights reserved
http://dx.doi.org/10.1680/h5e.59931.039

Chapter 3
Site investigations

Dan Raynor   Associate Geotechnical Engineer, Arup Ltd, UK

3.1.   Purpose of the site investigation

The purpose of the site investigation is to obtain information to inform the assessment of the options and costs of the proposed scheme, and, consistent with the objectives of the scheme, to allow safe and economical designs to be developed and constructed.

Consistent with definitions in the UK's code of practice for site investigations (BSI, 2010), the term site investigation denotes an investigation for the purposes of assessing the geotechnical suitability of a site for the construction of a highway scheme. This includes site walkover surveys and desk study assessments and ground investigations. The term ground investigation refers to the direct exploration of the ground (e.g. using intrusive or geophysical investigation techniques). Thus,

site investigation = site walkover + desk study + ground investigation

Key considerations of a site investigation include the definition of the ground and groundwater conditions, identification of ground hazards that could present a risk to the scheme, establishment of the engineering properties of the underlying soil and rock, and determination of suitable parameters for engineering design. It is critical that the site investigation be undertaken by a competent geotechnical specialist.

A comprehensive site investigation should provide all the necessary information to allow

■   an assessment of the overall geotechnical suitability of the road site and its environs for the proposed works, including the determination of the most suitable alignment option
■   safe and economical designs to be developed and implemented for both temporary and permanent works
■   an assessment of any changes that may arise in the ground, groundwater and environmental conditions as a result of the works
■   the determination of the optimum methods for safe construction
■   the fulfilment of the statutory and planning requirements associated with the scheme.

The first stage of the site investigation should be the desk study and site walkover survey, and the outcomes from these will inform the requirements of the ground investigation. Depending on the scale and complexity of the scheme, the ground investigation may be undertaken in a single phase or over a number of phases: the key consideration is that sufficient information is available at the right time to inform the different stages of the scheme as it is developed.

Once the need for a highway scheme or improvement strategy has been identified and approved, a number of alignment options may be considered, and a favoured option selected. The geology and stratigraphy, groundwater conditions, potential contamination and other ground hazards will have been considered during the route selection process, so that an assessment of the ground risks for each option can be made. For simple schemes where ground conditions are straightforward and well understood, the desk study and site walkover may be sufficient to enable the final route to be selected. For larger or more complex schemes, a phased ground investigation may be employed to inform the route selection. Where phased ground investigations are proposed, the earlier phases of the ground investigation will help refine the detailed scope of subsequent phases; that is, as the favoured alignment option is selected or the improvement strategy is better defined, scheme proposals are developed, and ground conditions and associated hazards, risks and uncertainties become clearer.

Overall, the scope of the site investigation and the reporting required for a given scheme will need to consider the anticipated nature and variability of the ground and groundwater conditions, the scale and complexity of the project, the environmental sensitivity of the site and adjacent areas, the amount and quality of the existing information that is available, and any particular planning requirements.

3.2. The value of adequate site investigation

Good site investigation reduces the number of uncertainties in the ground with consequent lower construction costs. Conversely, inadequate expenditure on site investigation results in greater uncertainty that can adversely impact on project risk, construction timescales and cost.

The importance of carrying out adequate site investigation has been highlighted by numerous studies. UK research (National Audit Office, 1989) highlighted 210 separate premature failures during highway works, at a total combined cost of some £260 million (1989 cost values): geotechnical failures as a result of a lack of understanding of the ground and groundwater conditions were identified as the major causes of these extra costs. The main geotechnical difficulties resulting in cost increases for road schemes were also highlighted by a later study (TRL, 1994), which identified the following as key issues:

- the classification of materials not being as anticipated
- the need to remove and replace additional unsuitable material below formations
- groundwater-related problems in cuttings.

A subsequent analysis (Clayton, 2001) of the 1994 data compared ground investigation costs against unplanned increases in construction costs, and concluded (Figure 3.1) that the less the investment in ground investigation works the greater the risk of increasing construction costs.

Figure 3.1 Comparison of site investigation expenditure and cost overruns for highways schemes (Clayton, 2001)

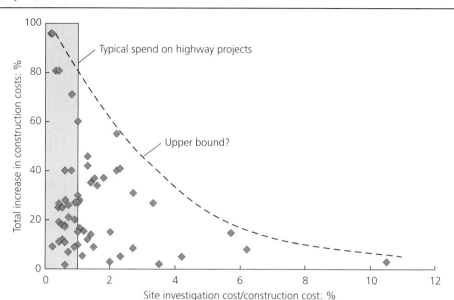

The costs associated with dealing with difficult ground conditions tend to be far greater if these are not identified until construction is underway, due to the cost implications of consequent delays to the programme if construction methods have to be changed or designs amended. Conversely, when risks are identified earlier, there is greater opportunity to consider them in the decision-making process (e.g. when considering alignment options). Figure 3.2 illustrates the general relationship between the opportunity to introduce change and the cost of change as the project proceeds.

Figure 3.2 Change opportunities and impact (Lazarus and Clifton, 2001)

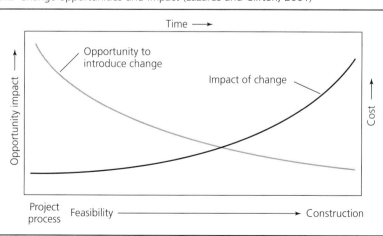

3.3. Key issues for site investigations

The site investigation should establish the properties of the ground for design and identify ground hazards that may impact on the design and construction methodology. Key considerations should include

- establishing the stratigraphy, understanding the geology of the site and establishing appropriate parameters for design
- understanding the groundwater conditions
- determining any contamination of soil and groundwater
- defining ground hazards and risks.

These issues are initially considered during the desk study and site walkover, and any further information required is then obtained during the ground investigation works.

3.3.1 Establishing the stratigraphy and understanding the geology of the site

Understanding the stratigraphy involves defining the depth, arrangement and nature of the underlying strata to allow a conceptual ground model to be developed for the site. The depth to the top of each stratum and its thickness and orientation, together with the variations in the properties for each stratum, laterally and with depth, should be determined to the investigation depths required by the design.

When developing the conceptual ground model, the importance of understanding the geological context of the site, including the depositional history and its impact on the geomorphology and engineering behaviour of the various strata, cannot be overstressed. Thus, the material properties (and variability) of each stratum, both laterally and with depth, should be described and classified: these properties include the strength and stiffness, consistency and particle grading, and chemical or electrochemical properties (e.g. pH and sulphate content) of the material.

The mass properties also need to be understood, as these can often be critical in controlling engineering behaviour and defining limit states for design. Examples include discontinuities within rock strata such as bedding planes, joints fractures and faults, all of which can form planes of weakness. The orientation, nature and extent of these discontinuities, and the nature of the materials that infill them, can be critical considerations in the stability assessment of rock cuttings. Similarly, if low-strength pre-existing shear surfaces are present within existing slopes formed of high-plasticity cohesive soils, they can become reactivated as a result of embankment loading or the excavation of cuttings.

Assessments of representative engineering parameters and coefficients for the various strata identified are needed for detailed engineering design, using a combination of in situ and laboratory test methods. For laboratory tests, it is important that samples of an appropriate quality and size are obtained for the particular tests being undertaken, to ensure that the parameters derived are representative of the in situ conditions.

Understanding the potential heterogeneity and anisotropy in soil and rock strata and the implications of these on engineering behaviour is critical. The nature of heterogeneity and the variations in properties of soil and rock laterally and with depth should be understood. Examples of anisotropy in soils include consolidation coefficients for soft clays and peats: these can differ significantly in their vertical and horizontal orientations, and this can have a significant impact on the consolidation behaviour of these materials under loading. Where the scheme proposals do not result in loading of soft materials and associated settlements, detailed information may not be required. However, if it is proposed that embankments be constructed over soft ground and, for whatever reason, it is not proposed that the soft materials be excavated and replaced, the scope of the investigation should be carefully considered.

For further background reading in relation to ground conditions, hazards and geological processes, a well-illustrated summary with examples of key issues that should be considered is available in the literature (Waltham, 2009).

3.3.2   Understanding the groundwater conditions

Understanding the hydrogeological conditions, including groundwater pressures and flows, is critical for the design of earthworks and retaining walls, structure foundations and highway drainage. Failure to understand and take due account of the groundwater conditions is a common reason for geotechnical failures on road schemes.

The estimation of potential groundwater flows in temporary and permanent works is necessary for the efficient sizing of permanent drainage, to enable the design of groundwater control measures during construction, and to support applications for discharge consents by local authorities and environmental regulators where these are required. The potential impacts of the proposed scheme on adjacent developments, and on surface water and groundwater, also need to be considered. These will be a particular concern if the scheme is within an area impacted by contamination, adjacent to sensitive water bodies or groundwater source protection zones, or if it has the potential to impact on the level of the water table or groundwater flow paths.

To allow these assessments to be made, it is vital that information about the groundwater conditions be gathered, including the location of the groundwater surface, the presence of perched or confined water bodies, seepages and springs, and information about the groundwater quality.

Groundwater flows may be controlled by material permeability in homogeneous sands and gravels, or concentrated along discontinuities, fissures or dissolution cavities in soluble strata. An understanding of the material and mass permeability of the strata is therefore required to allow an assessment of the potential groundwater flows and pressures. The presence of aquitards (i.e. low-permeability layers) can limit the movements of groundwater and result in perched groundwater above the level of the regional water table, or in confined groundwater bodies with elevated or artesian groundwater pressures at depth. Impacts of human activities such as mining, quarrying or dewatering can also significantly modify the hydrogeological conditions.

Long-term monitoring over a year or a number of years may be required to assess seasonal groundwater variations. For schemes in coastal or estuarine areas where there is the potential for tidal variations in the groundwater, continuous monitoring over a tidal cycle may be required. When installed in low-permeability cohesive materials, the time required for monitoring installations to reach equilibrium is an important ground investigation consideration.

3.3.3   Contamination of soil and groundwater

The assessment of contamination is a planning requirement for highway schemes in the UK. The two key items of legislation that currently define requirements are the Environmental Protection Act 1990 (HMG, 1990) and the Groundwater Directive (EC, 2006). Legislation and guidance regularly change, and for useful up-to-date guidance the Environment Agency's website (EA, 2015) represents a good starting point.

The Environmental Protection Act and its associated regulations require assessments to be made of land to determine whether significant harm will be, or might be, caused as a result of contamination of the soil and groundwater by the proposed scheme: potential impacts on controlled waters as a result of developments are also assessed. The Groundwater Directive aims to ensure that construction activities do not result in the pollution of controlled waters, including groundwater. Thus, the potential for existing conditions to be altered by the proposed road needs to be assessed to ensure that existing or new contamination pathways do not have detrimental impacts, within or adjacent to, the area of the scheme.

Where contaminated soil or groundwater is suspected, information pertaining to the physical and chemical properties of the soil and groundwater is needed to inform the conceptual model for the site and to allow the necessary characterisation and risk assessment to be undertaken. Where in situ materials are unsuitable for use within the scheme as a result of their physical and/or chemical properties, it may be necessary to dispose of them at landfill – in which case, chemical testing is required to classify and characterise this waste in accordance with regulatory requirements.

3.3.4   Ground hazards and risks

Ground conditions that might adversely impact on the stability or serviceability of the proposed scheme are considered to be potential hazards. An understanding of the ground conditions is critical to allow such hazards and associated risks to be understood so that they can be eliminated or, at least, addressed and controlled. Ground hazards commonly met on road schemes include

- areas of active or potential slope movements (including areas of existing slope instability or landslip), and those with a potential for failure (e.g. due to reactivation of existing relict shear surfaces in clay slopes, unfavourable rock discontinuities, cambering and valley bulging)
- natural cavities (including open fissures, solution cavities and swallow holes)
- soft compressible soils with low strength (including alluvial clays and silts and peat)

- collapsible soils, including loose or weakly cemented soils with the potential to collapse under loading or as a result of moisture content changes
- high or artesian groundwater conditions
- natural obstructions to piling (e.g. large boulders in glacial soils)
- ground gas emissions (e.g. methane, hydrogen sulphide, carbon dioxide and/or carbon monoxide can be an issue in areas of former mine workings, landfills or sites with organic soils, as can radon gas in certain geological conditions)
- expansive soils
- sensitive soils with the potential for strength loss (or even liquefaction) on disturbance.

The following human activities often result in modifications to the physical and chemical properties of the ground and groundwater, and these usually require special consideration in site investigations:

- mining-related features (e.g. shallow mine workings, and mine shafts and adits)
- areas of quarrying (including backfilled quarries)
- landfill sites
- loose or variable man-made ground
- buried obstructions (e.g. buried basements, foundations, tanks, tunnels, services and utilities)
- historical contaminative land uses
- unexploded ordnance.

3.4. Risk management and the sequencing of the ground investigation

The principle underlying effective ground risk management is that the risks and uncertainties associated with the ground conditions for the scheme are identified and either eliminated, controlled or reduced to acceptable levels of residual risk at various key stages of the scheme. The importance of undertaking a suitable scope of site investigation at the right stage to inform decision-making can be illustrated in the context of the key assessment stages adopted for assessing trunk road developments in the UK, as defined in the *Design Manual for Roads and Bridges* (DMRB) (HA *et al.*, 1993).

Three scheme assessment reporting stages are defined by Highways England (formerly the Highways Agency), so as to allow the impacts and benefits of UK schemes to be assessed, and to allow public and statutory bodies to comment on the proposals. At the stage 1 assessment, the need for a scheme or a broadly defined highway improvement strategy is explored: only a high-level desk study is usually undertaken at this stage, to define the broad ground conditions and identify the key hazards. The preferred route alignment or improvement strategy option is defined by the subsequent stage 2 assessment. Ground hazards identified at this stage can often be accounted for as part of the route selection process. A detailed desk study and, in some cases, early phases of the ground investigation may be used to inform this stage, depending on the complexity of the scheme.

Generally, the favoured route or improvement strategy option selected after the stage 2 assessment is developed into an outline scheme design during the stage 3 assessment. The advantages of the scheme in engineering, environmental, economic and traffic terms are also presented at this third stage, so that a decision can be made as to whether the scheme is justified. If the scheme is subject to a public inquiry, the stage 3 assessment will be prepared to inform this.

Preparation of an environmental statement is a key component of the stage 3 reporting that requires particular consideration during the site investigation. Information gathered during the desk study may be sufficient to inform the environmental statement: on the other hand, an early phase of ground investigation may also be required, so as to inform assessments of groundwater quality, contamination, required land-take, earthworks balance, embodied carbon dioxide and so on. Estimates of the cost and construction programme of the scheme are also required as part of the stage 3 assessment; however, the degree of certainty that can be achieved at this stage will be dependent on under-standing the ground conditions and any associated hazards, and on the engineering

Figure 3.3 The (now) Highways England's assessment (HA *et al.*, 1993) and geotechnical reporting (HA *et al.*, 2008) stages

Assessment stage (*DMRB* TD 37/93)	Scope of site investigation required	*DMRB* HD 22/08 reporting key stages
Stage 1 assessment report	High-level desk study required to provide overview of geology, geomorphology, ground hazards and engineering difficulties likely to arise from them. Usually a constraints map is prepared	**Key stage 1** Statement of Intent • Identify potential ground hazards and risks in the geotechnical risk register • Geotechnical classification of the scheme
Decision to proceed with the proposed new scheme (or improvements)		
Stage 2 assessment report	Detailed desk study and site walkover (preliminary sources study report, PSSR) and in some cases an early phase of ground investigation. Consideration of ground conditions and hazards in sufficient detail to inform decision on the preferred alignment option	**Key stage 2** Desk study report (PSSR) Ground investigation report (GIR)
Announce the preferred alignment (or improvements strategy)		• The PSSR. is useful to inform the stage 2 assessment, but this can alternatively be prepared at stage 3
Stage 3 assessment report	Consideration of ground conditions and hazards and associated risks for the preferred alignment option. Required information for the stage 3 assessment includes preliminary engineering design to allow the estimate of the scheme cost, earthworks balance, determination of necessary land take and information to inform the environmental statement	• The GIR is usually prepared at the stage 3 report stage • Update of the geotechnical risk register in both the PSSR and GIR
Publication of draft orders and environmental statement. Complete statutory process including a public inquiry		

solutions required to achieve the design needs of the scheme and allow for its safe and economical construction.

Highways England provides advice on the framework for geotechnical risk management, reporting and certification of trunk road schemes in England, Wales and Northern Ireland (HA *et al.*, 2008): its principles are also adopted in Scotland. This advice, which embodies a phased approach to site investigation and is linked with various key stages of reporting and certification, promotes the identification and communication of ground hazards and risk at various stages of the project. Its reporting stages are not explicitly defined to coincide with Highways England's scheme assessment process (HA *et al.*, 1993), as these vary between schemes depending on how the project is set up. Figure 3.3 illustrates in general terms how the geotechnical reporting stages can be coordinated to inform the scheme assessment reporting stages, and how this fits into the context of the later stages of detailed design and construction.

3.5.   The geotechnical category of a scheme

Eurocode 7 (BSI, 2007, 2013a) defines three geotechnical categories (Table 3.1), each of which is based on the complexity of the proposed geotechnical works and the geotechnical risk implications to health and safety. The geotechnical category influences the

Table 3.1 Geotechnical categories defined by Eurocode 7

Geotechnical category	Application	Reporting
1	Only small and relatively simple structures. Earthworks or geotechnical activities for which it is possible to ensure that the fundamental requirements will be satisfied on the basis of experience and qualitative geotechnical investigations	A combined desk study and ground investigation report is usually adequate. Ground investigation may not be required Usually negligible ground risk and routine geotechnical design and construction
2	Projects with conventional types of geotechnical structures, earthworks and activities, and with no exceptional geotechnical risks, unusual or difficult ground conditions or loading conditions	A separate desk study and ground investigation report is usually required. Ground investigation required Conditions and rules as defined in Eurocode 7
3	Projects with very large, unusual or complex geotechnical activities, earthworks and structures or those involving abnormal geotechnical risks or unusual or exceptionally difficult ground conditions	A separate desk study and ground investigation report is required. Ground investigation required Additional considerations over and above those in Eurocode 7 usually required

scope of the ground investigation and reporting that is normally required, and, hence, should be determined at the early stages of the project.

3.6. The desk study and site walkover

The objectives of the desk study and site walkover surveys are to

- review the geological, geotechnical, geomorphological, hydrological and geo-environmental aspects of the project site and the historical development of the area, including contamination risks, and prepare a ground model
- identify the stratigraphy, groundwater conditions and geotechnical risks, and the implications of these for all scheme options being considered
- provide a preliminary engineering assessment for the scheme and inform of any likely hazards to construction
- update the geotechnical hazard and risk register for the scheme
- develop objectives and methodology for the phased investigation of ground conditions.

The key to a successful desk study assessment is the ability to identify and locate for review all available existing relevant information. Sources that should be searched include

- published geological maps and memoirs
- groundwater and surface water information
- soil maps
- existing ground investigation data and construction information
- regulatory information and special site designations
- current and historical mapping and published records and papers
- historical and current aerial photography
- mining records, including records of mine entries and shallow workings
- natural cavities records
- unexploded ordnance information.

Packages of information including historical mapping, geological mapping and other ground risk-related information can be obtained from various commercial information providers. Many sources of mapping are available in geographic information system (GIS) format (e.g. as vector polygons), which can be input directly into GIS databases.

At appropriate times, site walkover surveys are undertaken to supplement the information from the desk study assessment. A preliminary engineering assessment of the preferred route alignment(s), including a geotechnical hazard and risk register associated with the route(s), should be a key output of the desk study assessment.

3.6.1 Published geological maps and memoirs

British Geological Survey (BGS) maps are available for the UK at scales of 1 : 10 000 (or, earlier, 1 : 10 650) and 1 : 50 000 (replacing the earlier 1 : 63 000 scale). Cross-sections showing the geological structure are often presented on the 1 : 50 000 and 1 : 63 360 maps, while details of the geology shown on these maps is given in accompanying explanatory memoirs, including further details of the stratigraphy and structure, descriptions of outcrops, and borehole logs and cross-sections. Older versions of these memoirs may have information not included in the more recent ones.

The BGS maps contain information relating to the geological structure, including the orientation and angle of dip of discontinuities and other features such as strata bedding, joint sets, faults and igneous intrusions. The 1 : 10 000 and 1 : 10 560 maps provide more accurate strata boundaries and additional information relating to the geological structure, borehole locations, quarry areas, mining-related features and so on. The depths and thicknesses of the strata encountered at outcrops are often annotated.

Solid geological maps are available that show the underlying rock types as they would appear if the cover of superficial deposits was removed. Drift maps show the distribution of the superficial deposits, and areas of landslip, unstable ground and made-ground may also be indicated in associated memoirs. Combined solid and drift maps are available for some areas. As well as providing information for the alignment of a scheme, the solid and drift geology maps can be used to locate potential sources of borrow material that can be used in embankments and road pavements.

The BGS British Regional Geology Guides describe the geology of larger regions of the UK, and provide general information that can be useful in understanding the regional context of the geology within the site area.

3.6.2   Groundwater and surface water information

In addition to the geological maps, the BGS also holds hydrogeological maps for the UK at scales ranging from 1 : 625 000 to 1 : 20 000 that provide information about aquifer properties, groundwater levels and pressures, and groundwater abstractions and quality. Surface water features are also usually indicated, including the positions of springs and seepages.

Aquifer designation maps are available that locate principal and secondary aquifers and non-productive strata at scales of 1 : 50 000 for both bedrock and superficial deposits (e.g. sands and gravels). These designations reflect the importance of the aquifers in terms of both drinking water supply and supporting surface-water base flows and wetland ecosystems.

Records of surface water and groundwater quality and abstraction points may be obtained from the environmental regulator and/or local authority.

3.6.3   Soil survey maps

Soil survey maps and their associated memoirs are published at various scales across the UK. The soils data for England and Wales are maintained by Cranfield University's Land Information System portal (Cranfield University, 2015), and similar information is available from the Soil Survey of Scotland (Scottish Government, 2015) and the Agri-Food and Biosciences Institute in Northern Ireland (AFBI, 2015). These data were prepared for agricultural and forestry purposes, and information is generally limited to profile depths of about 1.2 m, albeit some soils (e.g. peat) may be described to greater depths (sometimes >10 m).

The soil survey maps describe the parent material from which the soil was formed, as well as its texture, subsoil characteristics and drainage characteristics. The accompanying reports provide more detailed information about the physical and chemical properties of each soil. When compared with geological maps, soil survey maps can provide additional useful information, including the depth to rock, particle-size distribution, plasticity and soil moisture contents, occurrence of unconsolidated sands or compressible cohesive materials, permeability and drainage. The potential aggressiveness of a soil (due to low-pH or high-sulphate conditions) to buried concrete and steel may also be highlighted.

3.6.4 Existing investigation and construction information

Existing available data may include previous geotechnical desk study reports, factual ground investigation records from previous investigations, and associated interpretative reports.

The BGS maintains extensive unpublished records of strata encountered in boreholes, shafts and so on, and these data can be inspected, by arrangement, at BGS libraries. Field notebooks and other unpublished field reports are also available to be viewed.

The BGS also maintains a data set of exploratory holes, including boreholes and trial pits, that are available to view (free of charge) on its website via an interactive map tool (BGS, 2015). Other data that can be sought on the interactive online map include geophysical logs, site investigation reports, road reports, and mine and quarry plans and sections. The BGS also maintains a national geotechnical properties database that contains geotechnical information input into a digital Association of Geotechnical Specialists (AGS) format (AGS, 2015): this includes borehole and trial pit logs, together with results of laboratory and in situ test data extracted from site investigation records.

Highways England maintains a geotechnical database management system (HA GDMS) on its website (HA, 2015) that contains geotechnical information and inspection data for existing trunk roads and motorways in England. Information management systems are also maintained for Wales and Scotland, and other governmental agencies may also hold similar digital systems.

Maintenance records of existing geotechnical assets for roads (and railways) in the scheme area can help to identify areas where previous problems have occurred (e.g. frost heave, water seepage, embankment settlement and/or landslides). These can help focus assessments and assist with route selection.

Ground investigations previously carried out for other engineering projects in the vicinity can also provide useful information. Local knowledge gained in the execution of existing road, railway or building works in the area should also be sought, as should data available within local authority offices regarding previous road and other infra-structure developments in the area. Construction drawings for adjacent schemes and developments may also be available that contain useful ground-related data and information about existing foundations.

3.6.5 Regulatory information and special site designations

Details of particular site designations (e.g. sites of special scientific interest, areas of conservation and other protected sites) should be obtained as part of the desk study. Available records relating to landfill sites, pollution incidents and areas of contaminated land may be obtained from the local authority and from the environmental regulator.

3.6.6 Current and historical mapping and published records

Historical mapping is another source of information that may help to identify potential ground hazards and influence route selection. Former infrastructure and structures that might present a constraint to construction, or give clues about potential contamination, may no longer be indicated on current mapping. Changes to the alignment of rivers and canals, quarries that may have been subsequently backfilled, and mining-related features can often be identified from historical maps.

Ordnance Survey maps dating from the late 19th century to the current day are widely available for the UK at scales of 1 : 1250, 1 : 2500 and 1 : 10 000. Earlier mapping is also available in the form of the first-edition Ordnance Survey dating from the early 19th century. Tithe maps are available from as early as 1836 for England and Wales: these earlier maps, which can be most useful sources of information that are often overlooked, can be viewed at local authority archives, in the National Archives at Kew (for England) and, for the other UK regions, in the National Library of Wales at Aberystwyth, in the Public Record Office of Northern Ireland (PRONI) in Belfast, and in the National Archives of Scotland in Edinburgh.

Published papers may be available for previous developments in the vicinity that can provide useful summaries of ground conditions, ground hazards, and the design and construction solutions used to solve issues.

3.6.7 Historical and current aerial photography

Aerial photography can be viewed in national records offices in the UK and in similar records agencies in other countries. Air photographs can provide information in relation to the history of a site that is not shown on historical maps, including data relating to construction activity and industrial operations. Adjacent stereo-pairs of aerial photographs can be viewed using a stereoscope, which allows historical aerial photography to be viewed in three dimensions, typically with a significant vertical exaggeration: this allows landforms to be readily identified, and helps in understanding the surface topography and any changes to this over time. Air photographs can also be used to identify surface depressions that might be attributed to sinkhole features, breaks of slope, landslips and other geomorphological or man-made features.

The English Heritage Archives in Swindon hold the largest collection of historical aerial photography for England, with vertical and oblique images from 1920 to the present day: this includes RAF imagery from the 1940s, and Ordnance Survey photography used to assist map-making from the mid-1950s. Aerial photographs for Scotland can be viewed at the Royal Commission on the Ancient and Historical Monuments of Scotland (RCAHMS) in Edinburgh: also held there are the UK Aerial Reconnaissance Archives, which contain military intelligence photographs from around the world, ranging from World War 2 Allied and Luftwaffe reconnaissance photographs to post-war (up to the 1990s) aerial imagery. Records for Wales are held at the Central Register of Aerial Photography in Cardiff and, for Northern Ireland, in PRONI in Belfast.

The earlier wartime aerial photography is generally in monochrome, whereas later photography is a combination of monochrome and colour. Infra red, false-colour infra red, and multispectral photography may also be available: these can be used to identify, for example, changes in vegetation cover that provide clues about the underlying geology and groundwater conditions, positions of mine shafts and adits, underground cavities, unstable ground and other geological features (HA *et al.*, 2005a, 2005b).

3.6.8 Mining records

The Coal Authority holds records of abandoned coal workings and mine entries for the UK, and can provide information about workings and mine adits associated with coal seams and associated ironstone horizons. Details of previous treatment of old, shallow mine workings and entries may also be available. The Coal Authority's website provides details of available information (Coal Authority, 2015).

Mine abandonment plans are also available: these can be viewed at the Coal Authority records office in Mansfield or downloaded as PDFs. The available mining records are, however, incomplete in that the requirement to deposit abandonment plans dates from 1872, and earlier abandoned workings are often not recorded. Thus, the potential presence of unrecorded mine entries and shallow workings should always be considered wherever there were workable deposits.

The South Wales Mining Desk Study (Arup, 1986) is a good source of information on areas of coal mining risk and of subsidence features in the South Wales Coalfield. The solution-mining of salt (e.g. in the Cheshire Basin) can also result in significant subsidence risk: information on this can be obtained from the Cheshire Brine Subsidence Compensation Board and/or the Coal Authority. Other mineral workings that may need to be considered include ironstone and other metalliferous deposits, gravel pits, evaporate mining and chalk and limestone quarries.

3.6.9 Natural cavities

The dissolution of calcareous rocks such as limestone and chalk can result in significant subsurface voids. Solution cavities tend to be concentrated along natural discontinuities (e.g. joints and bedding planes, and significant fissures), so that gullies and even cave networks can result.

The published geological maps and memoirs identify areas that are potentially affected by natural cavities. Other sources of information include national databases relating to natural cavities.

3.6.10 Unexploded ordnance

Unexploded ordnance may be a risk where a site has a history of previous military use, or where it may have been a military target during wartime. If the ground is soft (e.g. an alluvial soil), there is a greater risk of buried ordnance being present, and if subsequent development of target areas has been limited, ordnance is less likely to have been detected and removed. Details of the risk assessment process that should be considered at such sites are available in the literature (Stone *et al.*, 2009).

3.6.11 Walkover surveys

A thorough visual inspection of any proposed route should be made in the early stages of a site investigation. As well as confirming the details of ground hazards identified in the desk study, the site walkover may highlight issues not identified from the published information: for example, more information about existing and historical site uses can be obtained, including the height and conditions of existing structures, the presence of basements and drainage infrastructure. Visual or olfactory evidence of contamination, or indirect evidence associated with areas of made-ground, surface or subsurface tanks and so on, can also be determined.

Visual inspections of old pits and quarries, existing road and railway cuttings, and rock outcrops provide extra data about soil and rock profiles, while measurements of the angles of dip and strike of discontinuities within rock outcrops provide information that is useful for stability assessments. While former shallow mining areas or sink-holes may have been detected from air photographs, their locations and condition can be confirmed and refined by visual inspection.

Areas of potential slope movements, including landslip and soil creep, are readily identified during site walkovers: these may be manifested as broken and terraced ground on hill slopes, small steps in steep slopes, inclined tree trunks and fence posts and so on. Similarly, damage to brickwork, buildings and roads, and disruption of drainage may provide evidence of subsidence or slope movements.

Geological faults can be identified from topographical features such as a step in the line of an escarpment or linear depressions where heavily fractured rock adjacent to a fault has been more rapidly eroded than the surrounding, more intact, rock. Areas of flat ground in upland areas with soft soils associated with former lakes or areas of upland peat, and the boundaries of soft alluvium in low-lying areas, can be established during the walkover. Marshland vegetation is indicative of poorly drained ground or high groundwater conditions, while the presence of bracken and gorse usually indicates well-drained conditions. Changes in the appearance or type of vegetation may provide additional clues about changes in the underlying conditions, or the locations of backfilled pits and quarries.

The site walkover also allows assessments to be made of any constraints or restrictions on the ground investigation, including safe locations for and access to exploratory-hole sites, any height or width restrictions on investigation equipment, constraints associated with current or former land use, potential obstructions, surface conditions and reinstatement requirements.

For further detailed advice regarding procedures to use when undertaking walkover surveys, see Annex C of BS 5930 (BSI, 2010).

3.6.12   Preliminary engineering assessment

The preparation of a preliminary engineering assessment is a key requirement of the desk study (HA *et al.*, 2008). Typically, this would involve the initial definition of the existing and proposed earthworks and structures along the preferred road alignment(s), including the required geometry of the proposed soil and rock cuttings and embankments, the forms and dimensions of any structure foundations, and options for any retaining structures and strengthened earthworks.

Estimates of the anticipated subgrade conditions of the highway should also be presented, including likely frost susceptibility and strength measurements, to inform the early estimation of pavement capping needs. Other key requirements include the feasibility of reusing materials excavated within the scheme boundaries, and construction considerations such as de-watering requirements and ease of rock excavation.

Where more than one route is being considered, information should be presented for each route option to allow comparisons and help inform the decision-making regarding the preferred route. Key features should be indicated in relation to the various route alignment options on a series of geotechnical constraints and location plans, and (if sufficient information is available to make them meaningful) longitudinal geological sections should be prepared.

A geotechnical hazard and risk register is a key output at this stage. This register, which is central to the risk management of the project, is embodied within the geotechnical certification process used in the UK (HA *et al.*, 2008): it defines the identified hazards and associated risks and presents details of how these can be eliminated, mitigated or controlled. Risks, which are usually quantified on the basis of the likelihood of occurrence of a given hazard and its severity if it should occur, can be assessed in terms of a variety of metrics, including cost, impact on the construction programme and safety. Examples illustrating the importance of the geotechnical hazard and risk register are available in the literature (e.g. Clayton, 2001).

The risk register is a live document that is updated throughout the lifetime of the scheme as additional information becomes available. Thus, any ground investigations that are carried out to reduce the uncertainty associated with identified risks, and improve the development of scheme designs, should be linked with the risk register.

3.7.   Planning the ground investigation

Any proposal to carry out a ground investigation should be based on the conclusions of the desk study and site walkover assessments. It should clearly define its objectives, and include a schedule of activities linked to a series of plans at a suitable scale. In document HD 22/08 of the *Design Manual for Roads and Bridges* (HA *et al.*, 2008), this is defined as 'Annex A' to the preliminary sources study report.

Guidance on the planning and execution of ground investigations for trunk roads and motorways in the UK is available in the literature (BSI, 2010; HA *et al.*, 2005a, 2005b; Perry and West, 1996). When planning these investigations, the techniques selected should be those that enable the desired quality of information to be obtained, taking account of the ground conditions and constraints already identified. Important considerations include the nature of the terrain and availability of adequate access for the proposed equipment. Constraints associated with current land uses must also be taken into account: for example, existing developments, the potential for buried structures, services and other obstructions, height restrictions imposed by overhead utilities and the presence and nature of livestock and crops.

Permissions must be obtained from landowners, and compensation for access agreed in advance of any ground investigation. Maintaining good public relations is important, and a public relations officer can assist in such negotiations.

Ground investigations should be targeted at eliminating, or at least reducing to acceptable levels, identified hazards and uncertainties. The information obtained should allow the most efficient and cost-effective designs and construction methods to be developed. Additional information will usually have to be sought to allow assessments of potential land contamination, and of the potential impacts of the scheme proposals on the environment.

In the UK, Eurocode 7 Part 2 (BSI, 2007) and associated National Annexes (BSI, 2009, 2013a) are used to define the requirements of, and give guidance in relation to, the planning and reporting of geotechnical investigations.

3.7.1   Scope of the ground investigation

The frequencies of investigation should be sufficient to allow safe and economical geotechnical designs to be developed for the road pavement, earthworks and highway structures. Sufficient exploratory holes should be sunk to enable the required level of detail regarding the stratigraphy, geological structure and groundwater conditions to be established, and representative design characteristic parameters to be derived. While Eurocode 7 suggests typical exploratory-hole spacings of 20–200 m as a guide for road schemes, the actual spacing used will be dependent on the ground variability (i.e. where significant ground variability is anticipated, closely spaced holes should be considered).

Exploratory holes should be arranged to include adequate offsets from the centreline: these offset distances will depend on the overall width of the structure, embankment or cutting and on the potential extent of their influence. Consideration should be given to the load distribution from the embankments and foundations and the extent of the ground area that could impact on, or be impacted by, the scheme. When positioning standpipes for monitoring groundwater, it may be desirable to install some of these beyond the immediate footprint of the scheme, to allow for ongoing monitoring during construction. The need for information for temporary works should also be considered (e.g. for temporary cuttings or for crane bases at structures).

Figure 3.4 Recommendations for minimum depths of ground investigation (h, height of the embankment or cutting; z_a, exploration depth). (© BSI (http://www.bsigroup.com), reproduced with permission from BS EN 1997-2:2007 (BSI, 2007))

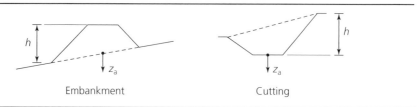

Embankment   Cutting

Ground investigations should be extended sufficiently below the proposed formation or structure locations, to provide the necessary information for the full depth of ground that might influence the design. Annex B of Eurocode 7 Part 2 (BSI, 2007) provides guidance regarding exploration depths below foundations, embankments and cuttings. If weak or voided strata are suspected to occur at depths that could affect the scheme, the investigations should be extended to the full depth of the potential influence, to allow assessment of settlement and stability. The depths of investigation required below the anticipated founding level need only be minimal in other instances: for example, if a foundation is to be carried out on a stratum known to have competent and laterally consistent properties.

For shallow foundations (e.g. for strip, pad or raft types), the depth of investigation below the foundation base should normally be at least 1.5 times the width of the loaded area below the base of the foundation. Where piled foundations are envisaged, the investigation depths should extend deeper than the anticipated maximum pile toe-depth (e.g. typically 5 m below the anticipated base of the piles), but this may be reduced if the piles are founded on strong rock.

For embankments, investigation depths of at least 6 m, or between $0.8h$ and $1.2h$ below the embankment base (where h is the embankment height), are recommended as a minimum, while at the locations of cuttings and at-grade sections, the ground investigation should extend to depths of at least 2 m below the proposed finished road sub-formation, or to $0.4h$ (where h is the cutting height) (Figure 3.4). If the vertical alignment of the proposed road, or the construction option, is still uncertain in a particular area, a greater depth of investigation may allow consideration of other options. Alternatively, the deferral of investigations in that area until a later phase may be appropriate.

In all cases, the planned depths of investigation and of any instrumentation, testing and sampling should be open to review, depending upon the actual conditions encountered in the field: for example, estimates of the thicknesses of soft ground may need to be reviewed during the investigation and borehole depths modified to suit the actual conditions encountered. The road designer should have a supervisory role to oversee the need for any modifications.

3.7.2 Phasing of the investigation

The main factors influencing the phasing of an investigation are the nature of the route being examined, the anticipated design of the proposed scheme, and the locations of its structures. Phased investigations may be considered for a number of reasons.

- *To optimise the initial route selection.* For some schemes, an initial ground investigation phase may have to be undertaken prior to the selection of the favoured alignment option. This may result from a need to have a better understanding of specific hazards for particular options, a better understanding of design requirements, and/or to reduce the cost or programme uncertainty. In this case, the preliminary engineering assessment (see Section 3.6.12) will have drawn attention to the key hazards and areas of uncertainty for the options being considered.
- *To ensure that the investigation is suitably targeted.* Where ground conditions are poorly understood or scheme proposals complex, there may be insufficient information to fully develop a scope of investigation that provides all the necessary information efficiently. In this case, an initial phase of investigation will help to focus subsequent phases.
- *Additional information may be required as the scheme proceeds.* For example, if additional information regarding locations and structural forms is needed for the detailed design of bridges or other highway structures.
- *To verify certain design assumptions.* Assumptions made concerning ground conditions may need to be confirmed during the construction phase. Installation of additional monitoring procedures may also be required prior to or during construction.

3.7.3 Specification for the investigation

In the UK, as in many other countries, ground investigations are normally carried out by specialist contractors, and this means that, for contractual reasons, the investigation requirements need to be formally defined (e.g. see HA *et al.*, 2005a). The key technical documentation required for this purpose typically includes

- a suitable specification to define general requirements
- a series of schedules to define particular requirements of the investigation
- a corresponding method of measurement to define how the investigation works are to be measured for payment purposes
- a bill of quantities to define the costs associated with each element of the investigation.

The purpose of the ground investigation specification is to provide a quality framework for the ground investigation, and define how it is to be carried out.

It is important to appreciate that there are specific requirements and responsibilities placed on all parties to a contract to contribute to the management of health and safety risks on site. Thus, when preparing a specification, the health and safety implications of the proposed investigation works must be identified, considered and eliminated where possible: where elimination is not possible (e.g. for risks relating to ground contamination or unstable ground), they must be communicated appropriately as part of the specification documentation.

Details of documentation requirements for ground investigations, and useful guidance regarding best practice principles that are applicable to all highways schemes, are readily available in the literature (BDA, 2008; HA *et al.*, 1997, 2005a; ICE, 1993; SISG, 2011). More information regarding health and safety issues are available on the Health and Safety Executive website (HSE, 2015).

3.8.   Undertaking the ground investigation

The techniques and methodologies used in ground investigations can essentially be divided into those that are intrusive (i.e. they involve disturbing the ground) and those that are non-intrusive (e.g. geophysical techniques). Investigation methods are constantly being improved and new ones introduced, and it is beyond the scope of this chapter to consider all available techniques. Thus, the emphasis in this section is placed on some of the more common methods, and the key issues that should be considered when deciding which method(s) to use.

Fundamentally, the technique(s) adopted in any given investigation should

- be capable of achieving the required depth of investigation
- allow accurate descriptions of the material and mass properties to be obtained as required
- provide soil samples of the required number, quality and size to enable laboratory testing to be carried out as required
- allow the required in situ field tests to be carried out
- enable parameters required for design to be established.

The most cost-effective techniques that will provide the required information within the required timescales should be used. Eurocode 7 Part 2 (BSI, 2007) and BS 5930 (BSI, 2010) provide guidance regarding ground investigation techniques available to meet particular requirements, including details of equipment, procedures and the appropriate evaluation of investigation data.

Appropriate technical supervision of the investigation ensures that the information needed for design is obtained, as the details of the investigation will normally need to be modified as it progresses. For example, as noted previously, the depth and/or frequency of exploratory holes may need to be increased where ground conditions experienced in the field differ from those anticipated, or the details of monitoring installations may need to be adjusted on the basis of the actual conditions encountered. The ability to have prompt communication between the road designer and the investigation contractor, and the keeping of detailed exploratory-hole log descriptions during the investigation, are key to ensuring that these decisions are made in a timely fashion.

It is vital that consistent and unambiguous descriptions of soil and rock samples are recorded as the ground investigation is carried out. Standard logging systems are available for soils and rocks (e.g. in the UK see BSI, 2003, 2006, 2010, 2013b, 2013c).

3.8.1   Intrusive investigation methods

The most commonly used intrusive ground investigation methods include trial pit excavations, cable percussion boreholes, rotary boreholes and window sample holes: these techniques allow material descriptions and the recovery of soil or rock samples of varying size and quality for subsequent laboratory testing. In situ testing techniques are also available that allow the properties of the ground to be recorded directly in the field, without the need to recover samples for laboratory tests: commonly used in situ tests include the standard penetration test (SPT), the dynamic probing test, the shear vane test, the cone penetration test (CPT), the pressuremeter test and the plate load test.

3.8.1.1 Exposures and trial excavations

Exposures and excavations offer the best opportunities for the accurate recording of surface or shallow strata, allowing good descriptive records to be made of both the material and mass properties. They allow useful details of discontinuities such as fractures, bedding and faults, of soil fabric such as sand or peat lenses, and of groundwater information to be obtained by direct observation. Sketches can be made of trial pit faces and relevant exposures, and photographs can be taken to supplement the descriptions.

Trial pits and trenches are simple and cheap, and they can be dug either by hand or using conventional excavation plant. Trial pits are typically 0.6–1.5 m wide and 3–4 m long, and are usually extended to depths of between 2 and 5 m, although deeper pits are sometimes undertaken using larger plant. They have the advantage that they allow the soil and weathered rock profiles to be inspected directly; they also allow a much better appreciation of lateral variations over short distances and of the orientations of mass fabric, including discontinuities and shear surfaces. Trial excavations enable buried obstructions (e.g. foundations and tanks) to be inspected and accurately surveyed, and visual and olfactory evidence of ground contamination is easily obtained. Groundwater flows into trial pits provide information on groundwater levels and permeability.

Disturbed samples representative of each stratum can be readily taken for contamination and basic geotechnical classification testing, while tube samples of reasonable quality can be recovered from the base of the pit using the excavator arm. Where safe access and suitable support of the pit sides is provided, undisturbed 'block samples' can be recovered, and in situ tests (e.g. hand shear vane tests) undertaken in the pit sides.

3.8.1.2 Cable percussion boreholes

Cable percussion boring (Figure 3.5) is a commonly used technique for investigating soil strata to depths of up to about 40 m.

With this method, boreholes are advanced in the ground by repeatedly raising and dropping a wire rope attached to a boring tool. Various boring tools can be used to advance the borehole through different ground conditions: for example, a chisel (to break boulders and other obstructions), a clay cutter (for use in cohesive soils) or a bailer-and-shell (for use in sands and gravels). Steel casings of, generally, 150–250 mm diameter are used to support the borehole sides as it is advanced. Continuous samples can be recovered, with disturbed samples taken directly from the boring tools, and undisturbed samples can also be recovered from cohesive strata using a variety of samplers, including U100 samplers and thin-walled sample tubes. A variety of in situ tests can be undertaken within the holes (e.g. the SPT, the shear vane test, the pressuremeter test

Figure 3.5 Cable percussion rig

and permeability tests), and soil samples can be taken for description and testing, as the boreholes are advanced. Boreholes allow the installation of groundwater monitoring and sampling standpipes, as well as other instrumentation (e.g. extensometers and inclinometers that are used to measure ground movements).

3.8.1.3 Rotary boreholes

While cable percussion techniques can obtain limited information in weak and weathered rocks, investigations in rock generally require the use of rotary boreholes. Rotary drilling usually involves either of two processes, namely open-hole drilling or rotary-core drilling.

Open-hole drilling is mainly used for the quick penetration of strata that are of limited interest to the investigation, as the material structure is almost totally destroyed in the drilling process. Typically, it might be used with weak rocks or soils to install test instruments, or as a probing technique within rocks to test for voids such as mine workings or solution cavities.

A variety of open-hole drilling rigs are available, including wheeled and tracked rigs. Small rigs are useful where access or headroom is restricted, and rigs for use on steep slopes are also available. Multi-purpose rigs are available that can be fitted with a variety of different boring, drilling, sampling and testing gear.

With conventional rotary-core drilling, a ring bit studded with industrial diamonds (for hard rocks) or tungsten carbide inserts (for soft rocks) is screwed onto the bottom of the outer barrel of two concentric cylindrical barrels. The outer barrel is fastened to, and rotated by, extension rod sections of hollow steel tubing. As the bit cuts into the rock, the isolated rock core feeds into the inner non-rotating barrel, and this is raised at regular intervals to the surface so that the core can be extracted. Drilling fluid is added to the core bit through the annulus between the two barrels, to lubricate and cool the drill head and aid with the removal of spoil from the borehole. A spring-steel 'core catcher' is located just above the core bit to prevent the rock from dropping out of the core barrel.

'Triple-tube' core barrels are also available that incorporate a removable sample tube or liner within the inner non-rotating barrel: this arrangement helps to maintain the integrity of the core and to maximise the quality of the sample recovered. Using triple-barrel systems, it is possible to recover core samples of weak fractured rocks and even soil. Various core diameters are available, and, in general, the larger the diameter of the core the better the quality of sample that can be recovered. Thus, for soils and weak or heavily fractured rocks, large-diameter cores should be used to maximise the potential for the recovery of good quality samples.

The sonic drilling technique is a relatively recent innovation that uses high-frequency resonant energy to advance the drill string. This allows much faster drilling rates to be achieved, and is capable of recovering continuous samples of both soil and rock.

3.8.1.4 Window samplers

A window sampler is a percussion drill that is either mounted on a small tracked rig or hand-held where access is more difficult. Window samplers, which are capable of recovering a continuous profile of disturbed samples for logging and basic classification tests, can also be used to undertake standard in situ testing such as SPTs. Generally used for shallow investigations less than 8 m deep, window sampling can be particularly effective in cohesive and fine granular soils, but the samplers cannot readily penetrate dense, coarse gravel or soils with any significant cobble or boulder content. Sample recovery can also be difficult in loose soils such as a sands and silts, particularly if they are below the water table.

As with boreholes, groundwater and gas-monitoring standpipes and movement-monitoring installations, such as extensometers and inclinometers, can be installed in window sample holes.

3.8.1.5 Sampling for laboratory testing

The technique adopted and the type of soil or rock sampled influence the size and quality of the sample(s) obtained for laboratory testing.

Table 3.2 Quality classes of soil samples for laboratory testing and the sampling categories to be used

Soil properties	Quality class				
	1	2	3	4	5
Unchanged soil properties:					
Particle size	X	X	X	X	
Water content	X	X	X		
Density, density index, permeability	X	X			
Compressibility, shear strength	X				
Properties that can be determined:					
Sequence of layers	X	X	X	X	X
Boundaries of strata – broad	X	X	X	X	
Boundaries of strata – fine	X	X			
Atterberg limits, particle density, organic content	X	X	X	X	
Water content	X	X	X		
Density, density index, porosity, permeability	X	X			
Compressibility, shear strength	X				
Sampling category according to BS EN ISO 22475-1	←——————— A ———————→				
			←—— B ——→		
					C

© BSI (http://www.bsigroup.com), reproduced with permission from BS EN 1997-2:2007 (BSI, 2007)

The quantity and quality of the sample required is dependent on the information sought from it. If a detailed description of the fabric of a sensitive soil or a weathered rock mass is sought, or the compressibility or shear strength testing of soils is required, high-quality undisturbed samples are needed. On the other hand, lower-quality disturbed samples are adequate for classification tests involving particle grading. (Details of different soil test methods, and the quality of the samples required to undertake them, are discussed in Chapter 4.)

A summary of five quality classes for soil samples and the properties that can be accurately determined from them is shown in Table 3.2. In this table, the higher the class number and the sampling category, the lower the quality of information that can be reliably obtained from the sample. Thus, piston samplers are usually needed to recover class 1 (highest quality) samples. Thin-walled or thick-walled open-tube samplers (e.g. U100 samplers) are capable of recovering samples of quality class 2 or 3, while bulk or tub samples are normally adequate for class 4 testing. Class 5 samples are recovered from open-hole drilling or washings, and cannot be reliably used for any form of testing: typically, they are mostly used to assess the approximate sequencing of the strata.

High-quality samples can usually only be recovered in soft clays or silts. It is not generally possible to recover undisturbed samples where cobbles or boulders are present;

however, in such soils it may be possible to obtain parameters for design from available in situ test methods.

BS EN ISO 22475 (BSI, 2006) provides detailed advice on the technical principles of sampling, and on the appropriate techniques to use for the recovery of soil and rock samples.

3.8.1.6 Standard penetration test (SPT)

The SPT is undertaken at the base of a borehole or a window sample hole, usually at regular intervals as the depth of hole is advanced. With fine granular and cohesive soils, a split-spoon SPT sampler is driven into the ground by a hammer of 63.5 kg mass dropped from a standard height and, once the sampler has been 'seated', the number of blows (N) necessary to achieve its penetration to a depth of 300 mm is recorded. The split-spoon sampler also allows small disturbed samples to be obtained for identification purposes. An SPT cone is generally used instead of the split-spoon sampler to test dense or coarse granular soils or in weathered rock.

The SPT is an effective test for determining various engineering properties of soils and weathered rock, and well-established empirical correlations with strength and stiffness are available in the literature (Clayton, 1995).

3.8.1.7 Dynamic probing

In a manner similar to the SPT, dynamic probing involves driving a cone into the ground using a drop-weight, and recording the number of blows required to achieve a set amount of penetration (typically 100 mm). The results obtained are then correlated with material properties such as strength and density, using established empirical relationships. Dynamic probing allows a continuous profile of blow count with depth to be obtained from ground surface.

Varying weights of hammer (i.e. light, medium, heavy and super-heavy) are available to carry out this test. The super-heavy hammer provides equivalent work per blow to that of an SPT drop-hammer, and this allows direct correlation of super-heavy results with SPT N-values.

3.8.1.8 Shear vane tests

The shear vane test can be used to determine the in situ undrained shear strengths of cohesive soils. The test works by inserting a cruciform vane, and rotating it until the soil fails. The undrained shear strength and sensitivity are determined on the basis of the measured torque applied and the size of vane used. Tests can be undertaken within boreholes by adding extension rods to the shear vane, and this enables profiles of undrained shear strength to be obtained as greater depths are attained.

A hand-turned shear vane test provides a quick and simple way of measuring the in situ undrained shear strength of clays in the sides or bases of trial pits. In this case, the test involves pushing a 19 or 33 mm vane into the soil and rotating the shearing dial at a slow, steady rate until failure occurs.

3.8.1.9   Cone penetration test (CPT)

The CPT provides a rapid and cost-effective means of assessing the stratigraphy and engineering properties of the ground. The cone penetrometer comprises a standard-sized cone and a cylindrical friction sleeve. The penetration resistance on both the cone and sleeve are measured as they are pushed into the ground at a constant rate of penetration. Depending on the ratios of the recorded cone and sleeve resistances, a CPT determination of the soil profile can be made by comparing the results obtained with relationships on empirically derived charts.

As with the SPT, CPT results can be used to empirically determine the density, strength, and stiffness of soils and weak rocks. The test also enables consolidation parameters and the sensitivity of soft clays and silts to be estimated.

Piezocones are standard cones that are equipped with additional instrumentation to allow continuous measurement of the porewater pressure. The data obtained provide further information on ground conditions (e.g. porewater pressure spikes indicate cohesive soil layers). Piezocones can also be used to undertake dissipation tests: these measure the rate of dissipation of porewater pressure with time, providing information about the permeability and consolidation characteristics of clays and silts. Piezoballs and T bars are more recent innovations that are used instead of standard cones to allow more accurate assessments to be made of the strength and stiffness properties of very soft soils.

CPTs require specialist interpretation. Their application is generally most effective when they are used in conjunction with other investigation methods (e.g. boreholes and trial pits).

3.8.1.10 Pressuremeter tests

Pressuremeters, which are installed vertically in the ground, measure the in situ deformation of soil or soft rock caused by the expansion of a cylindrical flexible membrane under pressure. There is a variety of different types of pressuremeter test available for use with different ground conditions: these include full-displacement types that are pushed into the ground, and are suitable for use with cohesive and fine-grained granular soils; self-boring types that are installed using a rotary drilling rig, and are suitable for soft clays, sands and weak rock; and pre-bored types that are installed in stiff clays or weak rock, and for which the test pocket is drilled prior to installation. Pressuremeter test results are used to derive horizontal stiffness and strength properties as well as in situ lateral stress (which is difficult to obtain by other means).

Certain types of pressuremeter, including self-boring pressuremeters and dilatometers, can also be used to derive consolidation parameters that inform assessments of settlement in ground that may be difficult to sample for laboratory testing. (Note: pressuremeters measure horizontal characteristics, and hence they may not be appropriate for the assessment of vertical stiffness and consolidation characteristics in anisotropic soils.)

3.8.1.11 Plate-loading tests

Plate-loading tests allow the vertical stiffness and strength properties of soil and rock masses to be determined by recording the load, and the corresponding plate settlement, when a rigid circular steel plate of 300 or 600 mm diameter is loaded. While these tests are usually undertaken at the ground surface or at shallow depths at the base of excavations, they can also be carried out on the bases of large diameter boreholes.

3.8.1.12 Groundwater assessments

The depth of groundwater strikes in exploratory holes should always be recorded, together with the rates of water ingress and the magnitudes of groundwater rise. Investigations may require the targeting of groundwater pressures or quality within specific strata, and, in such cases, careful review of the borehole logs is required to ensure that the instrumentation is installed at the required depths.

Standpipes or standpipe piezometers (Figure 3.6) may be installed within boreholes and window sample holes to allow design groundwater levels or porewater pressures to be measured, and to allow samples of water to be obtained for testing. Standpipe installations can also be used to monitor ground gases (e.g. carbon dioxide, carbon monoxide, methane and hydrogen sulphide) where these may present a potential risk.

When it is known that groundwater monitoring will need to be continued during construction, installations should be placed at suitable offsets from the construction alignment to avoid being damaged or causing restrictions to construction operations.

In situ permeability tests can be undertaken either directly in boreholes during drilling or within standpipe installations. Variable-head tests are the most commonly used permeability tests: these include rising-head and falling-head tests, for which the time required for the natural water level to recover following either pumping or the addition of water is recorded. Pumping tests involving the pumping of water from a pumping well and the recording of the groundwater response in a number of nearby observation wells are described in BS 5930 (BSI, 2010, 2013a).

3.8.2   Non-intrusive investigation methods

Geophysical investigations are non-intrusive ground investigations that measure variations in the physical properties of subsurface materials in order to allow an interpretation of the below-ground conditions. Geophysical investigation techniques allow large sites to be investigated quickly at relatively low cost, and can be used effectively at sites where excavation or probing is restricted due to logistical or operational constraints.

Subsequent intrusive investigations are usually required to confirm the interpretation of the ground conditions obtained from the geophysical survey. However, a well-planned and targeted geophysical survey can usually be correlated using a relatively limited intrusive investigation. When investigating subsurface voids, mine shafts, contamination plumes and so on, the judicial use of geophysical techniques can help to minimise the frequency of the exploratory holes required, by allowing targeted investigation of geophysical anomalies that may relate to these potential hazards.

There are four general civil engineering applications to which geophysical surveys are routinely applied (BSI, 2010), namely

Figure 3.6 A typical standpipe detail

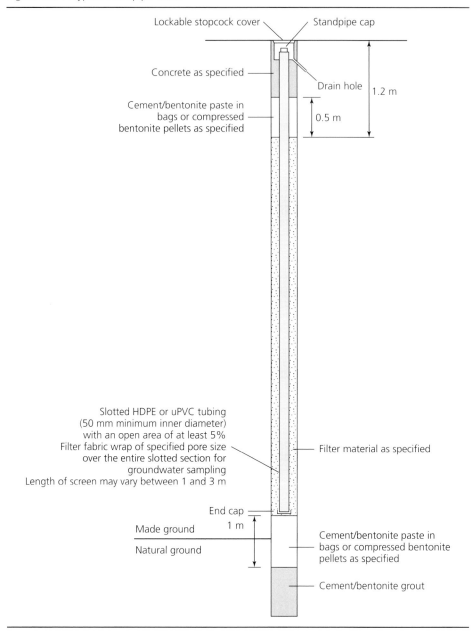

- *In geological investigations*: for example, mapping geological boundaries between layers; determining the thickness of superficial deposits and the depth to rockhead; establishing weathering profiles; and studying erosional and structural features such as the locations of buried channels, faults and zones of heavily fractured rock.

- *For resource assessments*: for example, locating aquifers and determining the groundwater elevation and quality; exploring for sand and gravel deposits and rocks for aggregate; and identifying clay deposits.
- *For hazard assessments*: for example, detecting voids and buried artefacts, natural cavities, old foundations, pipelines, mineshafts and adits, unexploded ordnance, contamination plumes, and the extent and depth of landslip deposits; and detecting leaks in barriers.
- *To assess engineering properties*: for example, determining properties such as dynamic elastic moduli, rock rippability and quality, and soil corrositivity.

The main geophysical survey techniques include

- *seismic refraction*, which involves measuring variations in the seismic velocity with depth
- *electrical resistivity*, which involves measuring variations in the electrical resistivity of the ground
- *ground-penetrating radar*, which requires detecting reflected radar pulses from discontinuities, obstructions or strata boundaries
- *electromagnetic conductivity*, which involves measuring variations in the electrical conductivity of the ground
- *magnetism*, which involves measuring variations in the magnetic properties of the ground
- *microgravity*, which requires measuring variations in the Earth's gravitational field.

The effective implementation of geophysical surveys and the interpretation of their data require the skills of experienced geophysical specialists. Predicting the materials and strata boundaries from the various geophysical measurements recorded requires a significant amount of data processing and specialist interpretation. Often, a number of techniques are best used together to allow the most accurate interpretations to be made, and specialist advice should always be sought when specifying these survey methods.

Some geophysical surveys can also be undertaken within boreholes: for example, wire-line geophysical logging involves the lowering of various instruments (i.e. sondes) on a cable down into cased or uncased boreholes to obtain continuous profiles of a range of properties. Assessments can be made of changes in the stratigraphy, density and porosity of the ground. Acoustic and optical televiewers are also available that can image the drill-hole wall and provide details of the stratigraphy and rock structure encountered.

3.8.2.1 Seismic refraction method

With the seismic refraction technique, shear waves (S waves) and primary waves (P waves) are generated in the ground by striking a steel plate with a hammer or falling weight, or by exploding a small explosive charge at the ground surface, and the times taken by the P or S waves to travel from the point of generation to geophones located at defined distances along a line on the ground surface are recorded. The P or S wave

Figure 3.7 Seismic values for natural materials (Stewart and Beaven, 1980)

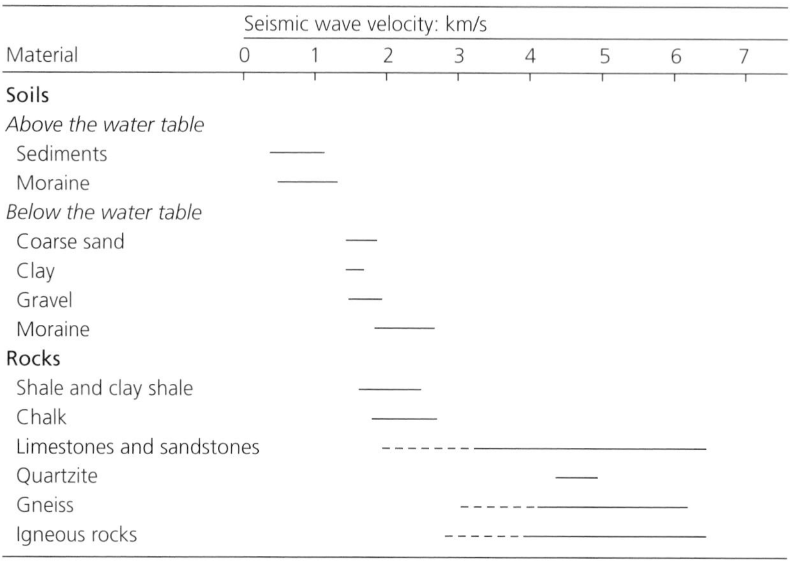

Material	Seismic wave velocity: km/s							
	0	1	2	3	4	5	6	7
Soils								
Above the water table								
Sediments	──							
Moraine	──							
Below the water table								
Coarse sand	─							
Clay	─							
Gravel	─							
Moraine	──							
Rocks								
Shale and clay shale	──							
Chalk	──							
Limestones and sandstones	─ ─ ─ ─ ─ ────────							
Quartzite	──							
Gneiss	─ ─ ─ ─ ────────							
Igneous rocks	─ ─ ─ ─ ────────							

velocities are then determined from measured distances and travel times, enabling the depth to the layer beneath to be estimated.

Seismic profiling can be an effective means of determining the depth to the bedrock. The method is most effectively used where there is a marked velocity contrast between the different strata, and where that velocity increases with depth. It is less effective where the seismic velocities of the strata are similar or where the deeper strata have a lower seismic velocity. The results obtained can also be used to assess the ease of excavation of the rock mass, by comparing them with established data in published tables. Typical ranges of seismic velocities for different materials are shown in Figure 3.7.

Down-hole seismic techniques are also available that allow the variations in seismic velocity with depth to be assessed by measuring the travel time at a point at the surface. Parallel seismic crosshole and crosshole tomography techniques involve the measurement of the travel times of seismic waves between two or more boreholes, allowing the seismic velocity of the soil and rock between the boreholes to be determined (McDowell *et al.*, 2002).

3.8.2.2 Electrical resistivity surveys

To carry out a resistivity survey, an electric current is passed into the ground via electrodes and both the current flowing through the ground and the reduction in voltage between the electrodes is measured, thereby allowing the resistivity of the ground to be determined.

With the exception of certain metallic minerals that are good conductors, the constituent minerals of the earth generally have a high resistivity. Resistivity surveys are effective for

Figure 3.8 Resistivity values for natural materials (ICE, 1994)

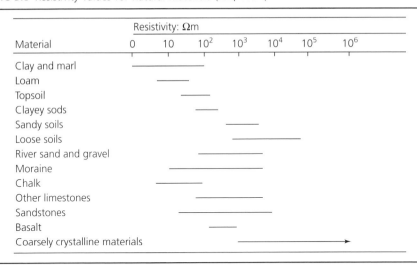

Material	Resistivity: Ωm						
	0	10	10^2	10^3	10^4	10^5	10^6
Clay and marl							
Loam							
Topsoil							
Clayey sods							
Sandy soils							
Loose soils							
River sand and gravel							
Moraine							
Chalk							
Other limestones							
Sandstones							
Basalt							
Coarsely crystalline materials							

picking up variations in moisture content or salinity. Changes in resistivity laterally and with depth can be determined and used to interpret the stratigraphy, determine the depth to groundwater, and locate particular ground hazards such as shallow mine workings, mine entries, solution features and hydrocarbon contamination plumes. Figure 3.8 presents typical resistivity values for a variety of different natural materials.

3.8.2.3   Ground-penetrating radar surveys
Ground-penetrating radar works by transmitting radio waves into the ground and measuring the transmission times through the ground of the generated electrical fields. Changes in the properties of the ground with depth result in the reflection of the radio wave pulses, with the strength of the reflected signal being relative to the size of the target and the nature of the discontinuity surface. The depths to the surfaces can be calibrated from the measured two-way travel times of the reflections, either by using appropriate velocity values of electromagnetic energy through the strata or by direct correlation with borehole logs.

3.8.2.4   Magnetic surveys
Magnetic surveys record distortions of the Earth's magnetic field. These surveys are most usefully employed in detecting underground metallic objects (e.g. buried mine shafts lined with ferrous supports, tunnels, buried services, etc.). However, if ferrous metals in the form of fences, cast iron drains or power lines are present at or near the ground surface, these can mask the signals from the targets. Iron-rich ground fills can also prevent their effective usage.

3.8.2.5   Microgravity surveys
Gravimeters can detect tiny variations in the Earth's gravitational field as a result of changes in the ground density. Positive gravity anomalies occur where dense materials

are present, and negative anomalies can be detected where low-density materials or underground voids are present. The location of natural dissolution cavities in limestone and chalk, and of mining-related features, are common applications of this technique.

3.9. Reporting on ground investigations

Good-quality reporting is a fundamental requirement of any ground investigation. Poor reporting can render useless an otherwise comprehensive investigation. When reporting on the ground investigation, the key requirement is that anyone reviewing the factual data should be able to gain the same knowledge and understanding of the data gathered as those who undertook the investigation.

In order to enable the road designers to apply the factual investigation data to the design and construction of the road, an interpretative report must also be prepared that includes an updated assessment of the ground conditions and associated hazards established during the desk study. While the factual and interpretative reports are sometimes combined, in all cases a clear presentation of the factual data should be made before any interpretations (or corrections) are applied.

3.9.1 Factual reporting

As its name implies, the factual report presents an unambiguous record of the ground investigation works undertaken and the ground conditions encountered.

In situ and laboratory testing specifications and methodologies are described within British and/or European standards. In the case of laboratory reports, the laboratory undertaking the tests should be formally accredited: this is often an employer's requirement, as it helps ensure a consistent quality of testing and reporting.

The factual report should include full details of the investigation equipment used and the names of operatives; calibration records for the equipment used should also be available for inspection, if required. The sampling techniques used and the methods of in situ and laboratory testing undertaken should be clearly stated, and (where applicable) geotechnical and environmental test results presented with their detection limits. Issues that may affect the validity of the test results should also be recorded.

Exploratory-hole records should be prepared in accordance with defined description criteria, and, as appropriate, these descriptions should be accompanied by good-quality photographs or digital images. The locations, orientations and elevations of all such holes should be referenced on a site plan, and the dates and times of commencement and completion of all holes, tests and monitoring visits should also be recorded.

The factual report does not generally include an interpretation of the data, although, in some instances, interpretation will be necessary to render the data meaningful: for example, in the case of geophysical surveys and in situ tests such as pressuremeter tests or CPTs. In these instances, the raw data should also be reported, to allow re-analysis if it is required at a later date.

The issuing of factual reports in digital format has now become a standard requirement that allows more efficient transfer of geotechnical data, and reduces cost and time as well as the potential for data transfer errors. The format used (AGS, 2010) was developed for application in the UK, but is now used throughout the world.

AGS data aids the effective management of information during the preparation of subsequent interpretative reports and the design, construction and maintenance of road schemes. Numerous software packages are available for the management of AGS data that allow for the preparation of exploratory hole logs, parameter plots, data tables, site plans, and cross-sections and long sections The use of GISs is now also relatively commonplace, and is growing in importance for the presentation and communication of data sets. The systematic use of such systems is becoming more important with the growing importance of the implementation of building information management – see Chapter 1.

3.9.2 Interpretative reporting

While the contents of the interpretative report will vary according to the scale and complexity of the scheme and the particular requirements of the client, the essential require-ment is that it presents and evaluates the geotechnical data. Typically this should include

- a description of the scheme proposals
- an updating of the information presented in the desk study report, taking into account any new data obtained
- the presentation of the factual investigation data (which may be by reference to the factual report)
- an interpretation of the geology and stratigraphy and the updating of the ground model, based on all available desk study and ground investigation data (this interpretation should be illustrated with cross-sections and long sections, and plans showing the locations of boreholes and trial pits, and of identified geotechnical features in relation to the road scheme proposal(s))
- a particular assessment of the ground hazards and of the risks that these may present to the scheme(s)
- detailed descriptions of all identified or grouped strata, including their physical and chemical properties
- plots of the in situ and laboratory test data for all strata and interpretations of these data; also, upper and lower-limit trend lines should be established, where applicable, to illustrate variations in properties with depth
- the derivation of upper and lower-limit characteristic engineering parameters and coefficients for identified strata where these are required for design purposes, including full details of how the values were derived
- an interpretation of the groundwater and surface water features, including the elevation of the groundwater table and anticipated seasonal fluctuations
- assessments of the soil and groundwater chemistry and contamination, including risks to human health, construction materials and the environment
- recommendations for further investigation
- an updating of the geotechnical hazard and risk register.

When preparing the interpretative report, it is important to be aware of the limitations of the factual data. In some cases, results may be adversely influenced by the quality of the sampling, the manner in which samples were stored or transported, or the limitations of the in situ testing techniques. In certain locations, adverse test results should be considered carefully in order to determine if they are misleading or whether they represent real conditions that must be taken into account in the design.

The *ICE Manual of Geotechnical Engineering* (ICE, 2012) provides an excellent source of additional information to supplement all aspects of ground investigation and reporting discussed here, as well as information relating to the detailed geotechnical design and reporting considerations for road schemes that have not been discussed here.

REFERENCES

AFBI (2015) *The Soils & Environment of Northern Ireland*. See http://www.afbini.gov.uk/index/services/services-specialist-advice/soils-environment.htm (accessed 05/16/2015).

AGS (Association of Geotechnical Specialists) (2010) *Electronic Transfer of Geotechnical and Geoenvironmental Data (Edition 4.0)*. AGS, London, UK.

AGS (2015) http://www.ags.org.uk (accessed 05/16/2015).

BDA (British Drilling Association) (2008) *Guidance for Safe Intrusive Activities on Contaminated or Potentially Contaminated Land*. BDA, Upper Boddington, UK.

BGS (British Geological Survey) (2015) http://www.bgs.ac.uk (accessed 05/16/2015).

BSI (British Standards Institution) (2003) BS EN ISO 14689-1:2003: Geotechnical investigation and testing. Identification and classification of rock. Identification and description. BSI, London, UK.

BSI (2006) BS EN ISO 22475-1:2006: Geotechnical investigation and testing. Sampling methods and groundwater measurements. Technical principles for execution. BSI, London, UK.

BSI (2007) BS EN 1997-2:2007: Eurocode 7. Ground investigation and testing. Geotechnical design. BSI, London, UK.

BSI (2009) NA to BS EN 1997-2:2007: UK National Annex to Eurocode 7. Geotechnical design. Ground investigation and testing. BSI, London, UK.

BSI (2010) BS 5930:1999 + A2:2010: Code of practice for site investigations. BSI, London, UK.

BSI (2013a) BS EN 1997-1:2004 + A1:2013: Eurocode 7. Geotechnical design. General rules. BSI, London, UK.

BSI (2013b) BS EN ISO 14688-1:2002 + A1:2013: Geotechnical investigation and testing. Identification and classification of soil. Identification and description. BSI, London, UK.

BSI (2013c) BS EN ISO 14688-2:2004 + A1:2013: Technical investigation and testing. Identification and classification of soil. Principles for a classification. BSI, London, UK.

Clayton (1995) *The Standard Penetration Test (SPT): Methods and Use*. Construction Industry Research Information Association, London, UK, Report 143.

Clayton (2001) *Managing Geotechnical Risk*. Thomas Telford, London, UK.

Coal Authority (2015) http://coal.decc.gov.uk (accessed 05/16/2015).

Cranfield University (2015) *LandIS: Land Information System*. See http://www.landis.org.uk (accessed 05/16/2015).

EA (Environment Agency) (2015) http://www.gov.uk/government/organisations/environment-agency (accessed 05/16/2015).

EC (European Community) (2000) *Water Framework Directive 2000/60/EC*. EC, Brussels, Belgium.

HA (Highways Agency) (2015) *HA GDMS: Geotechnical Data Management System v5.6.0*. See http://www.hagdms.co.uk (accessed 05/16/2015).

HA, Transport Scotland, Welsh Government and Department for Regional Development Northern Ireland (1993) Section 1: assessment of road schemes. Part 2: scheme assessment reporting. In *Design Manual for Roads and Bridges*, vol. 5. *Assessment and Preparation of Road Schemes*. Stationery Office, London, UK, TD 37/93.

HA, Transport Scotland, Welsh Government and Department for Regional Development Northern Ireland (1997) Section 3: ground investigation. Part 4: specification. In *Manual of Contract Documents for Highway Works*, vol. 5. *Contract Documents for Specialist Activities*. Stationery Office, London, UK.

HA, Transport Scotland, Welsh Government and Department for Regional Development Northern Ireland (2005a) Section 3: ground investigation. Part 1: documentation requirements for ground investigation contracts. In *Manual of Contract Documents for Highway Works*, vol. 5. *Contract Documents for Specialist Activities*. Stationery Office, London, UK, SD 13/97 (amendment No. 1).

HA, Transport Scotland, Welsh Government and Department for Regional Development Northern Ireland (2005b) Section 3: ground investigation. Part 2: ground investigation procedure. In *Manual of Contract Documents for Highway Works*, vol. 5. *Contract Documents for Specialist Activities*. Stationery Office, London, UK SA 9/97 (amendment No. 1).

HA, Transport Scotland, Welsh Government and Department for Regional Development Northern Ireland (2008) Section 1: earthworks. Part 2: managing geotechnical risk. In *Design Manual for Roads and Bridges*, vol. 4. *Geotechnics and Drainage*. Stationery Office, London, UK, HD 22/08.

HSE (2015) http://www.hse.gov.uk (accessed 05/16/2015).

HMG (Her Majesty's Government) (1990) *Environmental Protection Act 1990, Part IIA*. Stationery Office, London, UK.

ICE (Institution of Civil Engineers) (1993) *Site Investigation in Construction Series*. Thomas Telford, London, UK.

ICE (1994) *Manual of Applied Geology for Engineers*. Thomas Telford, London, UK.

ICE (2012) *ICE Manual of Geotechnical Engineering*, vol. 1. *Geotechnical Principles, Problematic Soils and Site Investigation*. ICE Publishing, London, UK.

Lazarus D and Clifton R (2001) *Managing Project Change: A Best Practice Guide*. Construction Industry Research Information Association, London, UK, Report C556.

McDowell PW and working group members (2002) *Geophysics in Engineering Investigations*. Construction Industry Research Information Association and the Geological Society, London, UK, CIRIA Report C562.

National Audit Office (1989) *Quality Control of Road and Bridge Construction*. Stationery Office, London, UK.

Ove Arup (1986) *Mining Subsidence: South Wales Mining Desk Study*. Welsh Office and Department of the Environment, Cardiff, UK.

Perry J and West G (1996) *Sources of Information for Site Investigations in Britain, TRL Report 192*. Transport Research Laboratory, Crowthorne, UK.

Scottish Government (2015) *Scotland's Soils*. See http://www.soils-scotland.gov.uk (accessed 05/16/2015).

SISG (Site Investigation Steering Group) (2011) *UK Specification for Ground Investigation*, 2nd edn. ICE Publishing, London, UK.

Stewart M and Beaven PJ (1980) *Seismic Refraction Surveys for Highway Engineering Purposes, TRRL Report LR 920*. Transport and Road Research Laboratory, Crowthorne, UK.

Stone K, Murray A, Cooke S, Foran J and Gooderham L (2009) *Unexploded Ordnance (UXO): A Guide for the Construction Industry*. Construction Industry Research Information Association, London, UK, Report C681.

TRL (Transport Research Laboratory) (1994) *Study of Efficiency of Site Investigation Practices*. Transport Research Laboratory, Crowthorne, UK, TRL Project Report 60.

Waltham AC (2009) *Foundations of Engineering Geology*, 3rd edn. Spon, London, UK.

Highways
ISBN 978-0-7277-5993-1

ICE Publishing: All rights reserved
http://dx.doi.org/10.1680/h5e.59931.075

Chapter 4
Soils and soil testing for roadworks

Michael Maher   Principal, Golder Associates Ltd, Canada

4.1.   Soil formation and types of soil

The Earth's crust is composed of in-place rock and weathered unconsolidated sediments. The sediments were derived from rock as a result of physical disintegration and chemical decomposition processes, and deposited by gravity, or through ice, water or wind action. Given that the mechanical forces of nature have been instrumental in moving most soils from place to place, it is convenient to describe them according to the primary means by which they were naturally transported to, and/or deposited in, their present locations.

4.1.1   Residual soils

Residual soils are inorganic soils formed by the in situ weathering of the underlying bedrock and, hence, were never transported. The climate (e.g. temperature and rainfall) mainly determined the type of residual soil formed. Mechanical weathering (i.e. disintegration) dominated the soil-forming process in northern cold climates and in arid regions, while chemical weathering dominated the process in tropical regions with high temperatures and high rainfalls. Residual soils are rarely found in glaciated areas: they have either been moved or they are buried beneath other glaciated soils. By contrast, in situ lateritic soils are found to great depths in warm humid areas. The behaviour of laterite soils, which are rich in iron oxides, can change significantly with changes in moisture content, and their compaction and strength characteristics are not as might be expected from their classification.

4.1.2   Gravity-transported soils

Colluvial soils were formed from accumulations of rock debris, scree or talus that became detached from the heights above and were carried down slopes by gravity. These materials are usually composed of coarse angular particles that are poorly sorted, and the upper and lower surfaces of deposits are rarely horizontal. Some deposits are very thin and can be mistaken for topsoil; however, thicknesses of >15 m may be found in non-glaciated valley bottoms and in concave parts of hill slopes. One of the most widespread colluvial soils in the UK is the clay-with-flint overburden that caps many of the chalk areas of south-east England.

Organic soils (e.g. peats and mucks) that were formed in place as a result of the accumulation of chemically decomposed plant residues in shallow ponded areas are termed cumulose soils. Both peat and muck are highly organic: the fibres of unoxidised plant remains are still visible in peat whereas they are mostly oxidised in the older muck soils.

While easily identified on the surface, these soils are often encountered at depths below the surface in glaciated areas.

4.1.3 Glacial soils

Glacial soils were formed from materials transported and deposited by the great glaciers of the ice ages (spanning about 2 million years), which ended about 10 000 years ago. These glaciers, often several kilometres thick, covered vast areas of the northern hemisphere and carried immense quantities of boulders, rock fragments, gravel and sand, as well as finer ground-up debris. When the ice sheet melted and retreated from an area, the carried material, termed glacial drift, was left behind. The drift deposits of most interest to the road engineer can be divided into two groups: (a) those laid down directly by the retreating ice (e.g. glacial tills – especially boulder clays – and drumlins) and (b) glacial melt-water deposits (e.g. moraines, kames, eskers, glacial outwash fans, and varved clays).

The term boulder clay is used to describe the unstratified glacial till that was formed underneath the ice sheet and deposited as an irregular layer of varying depth over the ground surface. This material can be quite variable, and ranges from being entirely of clay to having pockets of sand and gravel and/or concentrations of stones and boulders. The 'clay' is mainly ground rock-flour resulting from abrasion of the rocks over which the glaciers passed. The engineering properties of boulder clays vary considerably and, generally, they require careful investigation before being used in roads.

Drumlins are smooth oval-shaped hills that are rich in boulder clay and oriented with their long axes parallel to the direction of ice movement. It is likely that they were formed as a result of irregular accumulations of till being overridden by moving ice. Drumlins usually occur in groups, and can be up to 2 km long, 400 m wide and 100 m high (albeit 30–35 m is more common).

Moraines are irregular hummocky hills that contain assorted glacial materials. Four types of moraines are generally recognised, namely terminal, lateral, recessional and ground moraines. Found roughly perpendicular to the direction of glacial movement, terminal moraines are long low hills, perhaps 1.5–3 km wide and up to 30 m high, that mark the limit of the glacial advance. As the front of the ice sheet encountered a warmer climate, the material carried was dropped by the melting ice while, at the same time, the ice front was continually forced forward by the pressures from colder regions: hence, the hummocks became greater as more dropped material was added.

Lateral moraines are similar to terminal ones except that they were formed along the lateral edges of the melting ice sheet and are roughly parallel to the direction of ice move-ment. These moraines are most clearly defined along the edges of valleys in mountainous terrain.

As the climate changes of the era caused the planet to become warmer, the ice front melted at a faster rate than it advanced, and the ice mass began to recede. The recession was constantly interrupted and, hence, minor crescent-shaped 'terminal' moraines were formed at irregular intervals: these are known as recessional moraines. If a landform is characterised by recessional moraines interspersed with depressions with small lakes, it is often termed a 'knob-and-kettle' topography.

When the ice sheet melted rapidly, material was continually dropped at the edges of the receding ice sheet. This material is referred to as ground moraine.

During the ice ages, the weather was warmer at particular times of the year, and melt-water flowed out from the ice mass, carrying both coarse and fine particles that were spread over the uncovered land, with finer materials being carried further. In narrow valleys, the waters flowed quickly and freely, and the materials deposited in those valleys tended to be coarse grained, while those left on plains were often extremely varied. Sometimes the outwash waters fed into enclosed valleys and formed lakes within which particle settlement took place. Streams of water flowed in tunnels at the base of the ice, often with great force, especially during periods of rapid ice melting. Kames, eskers, outwash fans and varves are distinctive landforms deposited at these times that are of particular interest to the road engineer.

Kames are hummocky shaped mounds that were formed parallel to the ice front by glacial streams that emerged during pauses in the ice retreat. They mostly contain poorly stratified accumulations of sand and gravel, usually with some clay, that are generally suitable for use in road construction.

Eskers are relatively low (typically <18 m high) sinusoidal ridges, often several kilometres long, that contain highly stratified gravel and sand that is eminently suitable for use in road construction. Usually with flat tops and steep sides, eskers are the end-products of accumulations of glacial material that were deposited by sub-glacial streams at roughly right angles to the retreating ice front. The relative absence of fine-grained particles in eskers is due to them being washed away by streams that flowed rapidly from the melting ice.

Good construction material is also found in outwash fans. These were formed where the streams emerged from the receding ice front and their speeds were checked so that the transported gravel and sand was dropped in a fan shape.

Varves are fine-grained laminated clays that were deposited in glacial lakes or in still waters impounded in front of a retreating glacier. The different rates of settlement of the silty and muddy materials meant that the quicker-settling materials lie below the finer ones, and each pair of coarse and fine layers represents a season's melting of ice (Figure 4.1). The thickness of each varve can vary from being infinitesimally small to perhaps 10 mm thick, and deep accumulations have been found (e.g. 24 m in an engineering excavation in a Swedish lake). Varved soils give rise to troublesome foundations and should be avoided when locating a road.

4.1.4 Alluvial soils

An alluvial soil is a general term used to describe material formed from sediments deposited by river waters that were no longer moving fast enough to keep their loads in suspension or to trundle larger particles along the bottom of the stream.

Figure 4.1 Varved clay exposed at the side of an excavation

An alluvial cone is often the source of good construction materials. This landform is found at a change in gradient where a stream emerged suddenly from a mountain valley and spread out onto a plain. As the water had a higher velocity as it entered the plain, cone material is generally coarse grained and well drained, and contains few fines.

When a river carried an excess of water (e.g. after a heavy rainfall), it often broke its banks and spread out over the adjacent land, dropping coarser material adjacent to the stream and finer material further out. Thus, natural levees of gravel and sand may be found beside the upper reaches of rivers in flood plains. These levee accumulations are often usable for road construction purposes, albeit they normally have to be processed beforehand to remove excess fines. In the lower reaches of mature streams the levee materials are finer, and poor-draining swampy land is found behind them.

Mature rivers tend to be on gently inclined beds, and they use up their energies in horizontal meandering rather than in cutting deep channels. During these meanders, coarse-grained material was often deposited at the inside of the curves while the moving waters cut into the opposite bank. The deposits at the insides of the bends can be very valuable construction materials, and may influence the location of a road as well as its construction.

Over time, mature rivers continually changed their path and old meanders were cut off, leaving behind half-moon-shaped depressions that are termed oxbows. With the passage of time, run-off waters carrying fine-grained materials flowed into the oxbows so that, eventually, they became filled with silty and clayey sediments. As they are not continuous but rather occur as pockets of clay-like materials, oxbows should be avoided when locating a road as they can be associated with differential heaving and/or settlement of a pavement. In cold climates, the removal and replacement of highly frost-susceptible silty sediments to the frost penetration depth may be needed.

For geological reasons, an old meandering river may suddenly have become 'rejuvenated' so that, after excising a new channel in the old alluvium, raised terraces were left on either side of the rejuvenated river. Valuable deposits of gravel and sand are often found in these terraces. Also, terraces often provide good routes for roads through valleys, as they reduce the need for costly excavations into valley sides, and the pavement is usually located above the floodwaters of today's river.

While much sediment carried by rivers was deposited over their floodplains, a considerable amount of finer-grained material still reached the river mouths where it was discharged into an ocean or lake. Over time, roughly triangular-shaped accumulations of these materials, known as deltas, were developed forward into the water, and some eventually became grassed and formed small promontories. Delta deposits can be very varied, and investigation is normally required to determine if they can be used in road construction.

4.1.5   Marine soils

In addition to the material carried to the sea by rivers, the ocean itself is continually eroding the shoreline at one location and depositing eroded material at another. As oceans receded from old land areas, great areas of new land surface composed of a variety of deposited materials became exposed to the weathering forces of nature. These processes have been going on for millions of years.

It is difficult, in the limited space available here, to make definitive comments about marine soils, except that their usage in roads should be treated with very great care, especially if they contain fine-grained material.

4.1.6   Aeolian soils

Aeolian soils were formed from material that was transported and laid down by the wind. Of particular interest are dune sands and loess soils.

Dune sands are found adjacent to large bodies of water where large quantities of beach sand were deposited. When the wind blows consistently from one direction, the coarse sand particles move by 'saltation' across open ground with little vegetative cover, until eventually they lodge together and dunes begin to form. Saltation is the skipping action resulting from sand particles being lifted by the impacts of other wind-blown grains. The dunes formed have fairly gentle slopes on their windward side, while the leeward slopes are at steeper angles of repose.

Because of their tendency to move, sand dunes can be a problem. When a dune is formed so that the amount of sand deposited on it by the wind balances the amount taken from its leeward side by the same wind, the dune can migrate as sand particles move up and over the top. Roads that pass through migrating dunes that are not stabilised by vegetation can be covered by sands and – especially when the roads are low cost and poorly maintained – rendered impassable.

Loesses are porous, low-density aeolian silts of uniform grading that are found away from the beaches from which they draw resources. They mostly consist of silt-sized particles that were picked up by the wind and carried through the air to their final destinations. Deposits at particular locations are typically composed of uniform-sized particles; the particle sizes get smaller the further the distance from the beach source. Loesses are relatively unknown in the UK, albeit they occur extensively in Eastern Europe, China and the USA. Because deposits are both free draining and rich in calcium carbonate ($CaCO_3$), it is possible to excavate road cuttings with near-vertical sides through loess. However, the natural cementing action of the $CaCO_3$ is lost with manipulation of the loess, so that embankments formed with this material must have side-slopes similar to those formed with other soils.

4.2.   Soil profiles

As the sediments described above were moved and reworked, more weathering, abrasion, mixing with organic materials and soluble minerals, and leaching took place until, eventually, soil profiles were formed. The soil profile found in any given locale is mainly, therefore, the historic product of five major soil-forming factors, namely climate, vegetation, parent material, time and topography.

A soil profile (Figure 4.2) comprises a natural succession of soil layers that represent alterations to the original sediment that were brought about by the original soil-forming process. In easily drained soils with well-developed profiles there are normally three distinct layers, known as the A, B and C horizons.

The A horizon is often called the zone of eluviation, as much of the original ultrafine colloidal material and the soluble mineral salts have been leached out by downward-percolating water. This horizon is normally rich in the humus and organic plant residues that are vital to good crop growth, but bad for road construction purposes: this is because the vegetative matter eventually rots and leaves voids, while the layer as a whole often exhibits high compressibility.

The B horizon is also known as the zone of illuviation, as material washed down from the A horizon accumulates in this layer. The fact that this horizon is usually more compact than those above and below it, contains more fine-grained particles, is less permeable and, usually, is more chemically active and unstable makes the B horizon important in road construction. For example, the extra accumulation of fine particles in the B horizon of an active fine-grained soil on a gentle slope may make the layer so unsuitable as a pavement subgrade that it has to be removed and wasted.

The layer in which the road engineer is normally most interested is the C horizon. This contains the unchanged material from which the A and B horizons were originally

Figure 4.2 Hypothetical soil profile

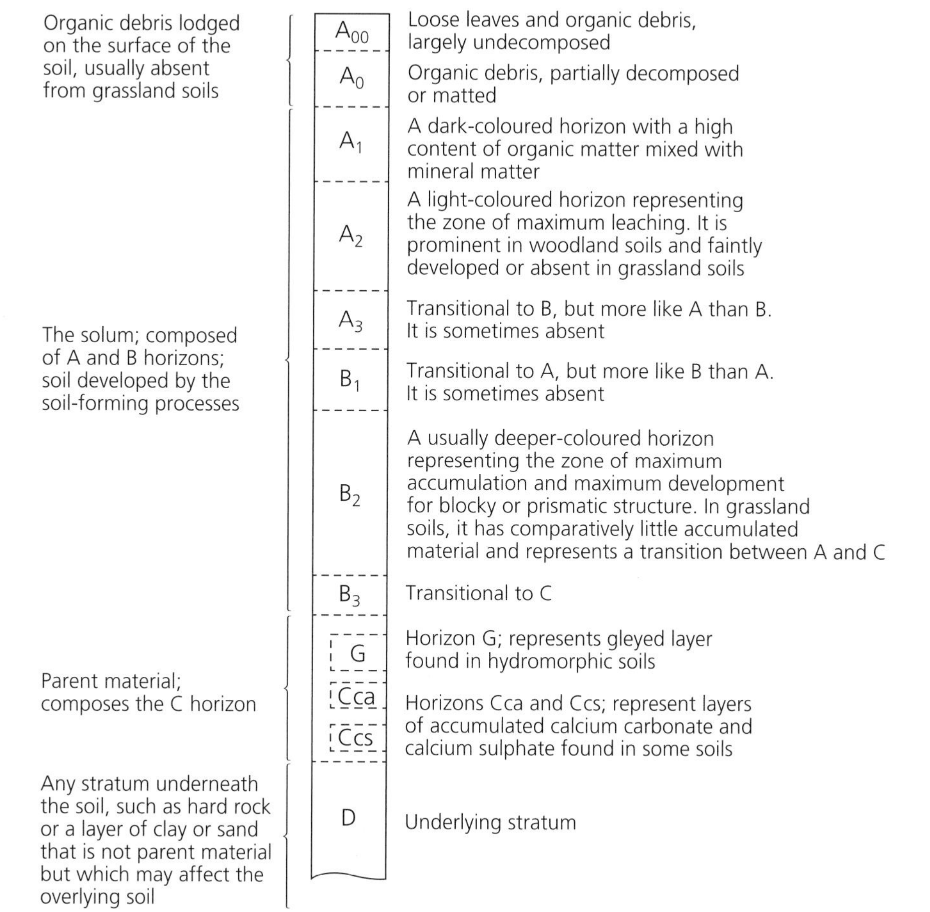

Organic debris lodged on the surface of the soil, usually absent from grassland soils	A_{00}	Loose leaves and organic debris, largely undecomposed
	A_0	Organic debris, partially decomposed or matted
The solum; composed of A and B horizons; soil developed by the soil-forming processes	A_1	A dark-coloured horizon with a high content of organic matter mixed with mineral matter
	A_2	A light-coloured horizon representing the zone of maximum leaching. It is prominent in woodland soils and faintly developed or absent in grassland soils
	A_3	Transitional to B, but more like A than B. It is sometimes absent
	B_1	Transitional to A, but more like B than A. It is sometimes absent
	B_2	A usually deeper-coloured horizon representing the zone of maximum accumulation and maximum development for blocky or prismatic structure. In grassland soils, it has comparatively little accumulated material and represents a transition between A and C
	B_3	Transitional to C
Parent material; composes the C horizon	G	Horizon G; represents gleyed layer found in hydromorphic soils
	Cca / Ccs	Horizons Cca and Ccs; represent layers of accumulated calcium carbonate and calcium sulphate found in some soils
Any stratum underneath the soil, such as hard rock or a layer of clay or sand that is not parent material but which may affect the overlying soil	D	Underlying stratum

developed, and it is in exactly the same physical and chemical state as when it was deposited in the geological cycle. It is this material that, if it is good enough, is normally used as fill when building an embankment or as a subgrade upon which to found a pavement.

Although not a normal occurrence, there may be a further horizon (i.e. the D horizon) beneath the C horizon. This underlying stratum can have a significant effect upon the development of the characteristics of the overlying soil profile if it is within, say, 1– 1.5 m of the surface. (While not shown in Figure 4.2, bedrock is sometimes referred to as the R horizon.)

As shown in Figure 4.2, the A, B and C horizons may also be divided into a number of sub-horizons. More often than not, however, many of these sub-horizons, and sometimes full horizons, are missing in situ, depending upon the soil-forming factors and the erosional features.

4.3.   Soil particles

Depending upon its location in the profile, a soil may be described as a porous mixture of inorganic solids and (mainly in the A horizon) decaying organic matter, interspersed with pore spaces that may or may not contain water. The solids component of this assemblage consists of particles of varying size, ranging from boulders to colloids. The coarseness or fineness of a soil is reflected in terms of the sizes of the component particles that are present: those most commonly considered are gravel, sand, silt and clay.

Before discussing these solids in detail, it is worth noting that while all international organisations now accept that the most convenient way to define gravel, sand, silt and clay is on the basis of particle size, various organisations have assigned different values to the particle sizes. It is especially important to be aware of this when reading and interpreting the technical literature on soils.

In 1908 the Swedish scientist Albert Atterberg published the most important early attempt to put on a scientific basis the limiting sizes of various soil fractions. Atterberg classified gravel particles as being between 20 and 2 mm in size: and said that these were the limits within which no water is held in the pore spaces between particles and where water is weakly held in the pores. Sand was described as being between 2 and 0.2 mm in size: the lower limit was set at the point where water is held in the pores by capillary action. Atterberg visualised silt as being the soil component that ranges in size from where sand begins to assume clay-like features to the upper limit of clay itself; that is, between 0.2 and 0.002 mm. The choice of 0.002 mm as the upper limit for clay was based on the premise that particles smaller than this exhibited Brownian movement when in aqueous suspension.

The current British standard on soil classification – EN ISO 14688 (BSI, 2002, 2004a) – defines particle size limits that are different from, and more comprehensive than, those devised by Atterberg: these are listed in Table 4.1. For completeness, it might also be noted that the term 'soil fines' is also used in the UK to describe soil material passing a 63 μm (0.063 mm) sieve, and that the classification of fine soils is based on grading and plasticity (see Section 4.8).

4.3.1   Sand

Sand particles are quite inactive chemically. They are generally bulky in shape, albeit individual grains may be described as angular, sub-angular, rounded or sub-rounded, depending upon the degree of abrasion received prior to their final deposition. Residual sands are usually angular while river and beach sands are generally rounded. Wind-blown sands are usually very fine and well-rounded whereas ice-worn sand particles can have flat faces.

Clean sand particles do not exhibit any cohesive properties. The pores between the particles are relatively large; hence, sandy soils are very permeable and drain well, and

Table 4.1 Particle size fractions and symbols (from BSI, 2002, 2004)

Soil fraction	Subfraction	Symbol	Particle sizes: mm
Very coarse soil	Large boulder	LBo	>630
	Boulder	Bo	>200–630
	Cobble	Co	>63–200
Coarse soil	Gravel	Gr	>2.0–63
	Coarse gravel	CGr	>20–63
	Medium gravel	MGr	>6.3–20
	Fine gravel	FGr	>2.0–6.3
	Sand	Sa	>0.063–2.0
	Coarse sand	CSa	>0.63–2.0
	Medium sand	MSa	>0.2–0.63
	Fine sand	FSa	>0.063–0.2
Fine soil	Silt	Si	>0.002–0.063
	Coarse silt	CSi	>0.02–0.063
	Medium silt	MSi	>0.0063–0.02
	Fine silt	FSi	>0.002–0.0063
	Clay	Cl	≤0.002

consolidation effects are small. The stability of a sandy soil is greatly influenced by its confinement, degree of compaction, gradation and particle shape. The shearing resistance of a sand increases with compaction, and with particle angularity and size.

4.3.2   Silt
Silts are generally similar to sands in that they derive much of their stability from mechanical interaction between particles. Coarse silt particles are essentially miniature sand particles, and thus they tend to have similar bulky shapes and the same dominant (quartz) mineral.

Unlike sands, however, silts also possess a limited amount of cohesion due to inter-particle water films. While silts are classed as permeable, water can only move through the (small) pore spaces relatively slowly. When the smaller-sized particles predominate, silts exhibit clay-like tendencies, and may undergo shrinkage and expansion when exposed to conditions that cause variations in moisture content. Silts can be problematic in construction, especially where they are excavated below the water table without proper groundwater control.

4.3.3   Clay
Clays differ from sand and silt in respect of both their physical properties and chemical make-up. It is very important for the road engineer to understand what constitutes clay particles.

Physically, clay particles are lamellar (i.e. flat and elongated), and thus have a much larger surface area per unit mass than the bulky shaped silts and sands. A measure of

Table 4.2 Some physical characteristics of soil separates (Millar *et al.*, 1962)

Name	Diameter: mm	Number of particles/g	Surface area of 1 g of each separate: cm^2
Coarse sand	1.0–2.0	90	11.3
Medium and coarse sand	0.5–1.0	722	22.7
Medium sand	0.25–0.50	5777	45.2
Fine and medium sand	0.10–0.25	46 213	90.7
Fine sand and coarse silt	0.05–0.10	722 074	226.9
Silt	0.002–0.05	5 776 674	453.7
Clay	≤0.002	90 260 853 860	11 343.5

Each particle is assumed to be a sphere having the maximum diameter of each group

the differences in surface area of various soil fractions can be gained by assuming that the particles are spherical in shape (Table 4.2). As the intensity of the physicochemical phenomena associated with a soil fraction is a function of its exposed surface area, this table suggests why the clay fraction has an influence on soil behaviour that can be out of proportion to its mass or volume in a soil.

Any analysis of the clay fraction is to a large extent a study of its colloidal component. In theory, a colloid is any particle that exhibits Brownian movement in an aqueous solution; in practice, however, the term is normally applied to particles smaller than 1 μm in size. Clay colloids are primarily responsible for the cohesiveness of a plastic soil, its shrinking and swelling characteristics, and its ability to solidify into a hard mass upon drying. Also, the drainage characteristics of a soil are considerably influenced by the amount and form of its colloid content.

The importance attached to the colloidal fraction is associated with the electrical charges that the colloids carry on their surface. Figure 4.3, which illustrates the Helmholtz double-layer concept of the make-up of a colloidal particle, shows that the inner part comprises an insoluble nucleus or micelle surrounded by a swarm of positively charged cations: the inner sheath of negative charges is part of the wall of the nucleus. These positively charged cations are in equilibrium at different but infinitesimally small distances from the colloid surface. A clay with sodium as the main adsorbed ion is called a sodium clay, and a calcium clay mostly has adsorbed calcium ions.

Figure 4.3 Schematic representation of a colloidal particle

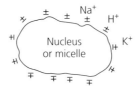

If a clay with adsorbed ions of a particular type is brought into contact with ions of a different type, some of the first type of ions may be released and some of the second ones adsorbed in their place. This exchange of positively charged ions, termed cation exchange, forms the basis of the modification/stabilisation of soil with lime (see Chapter 7).

The manner in which cations are exchanged is most easily explained by considering colloids that are in a solution. Due to heat movement and Brownian motion, the adsorbed ions continually move back and forth (within a limited range) from the surfaces of the particles. If electrolytes are added to the solution, cations are set in random motion because of the Brownian effect, and some of them slip between the negative wall of the nucleus and the adsorbed/oscillating ions. These electrolytic cations then become preferentially adsorbed, and some previously oscillating surface ions are released and remain in the solution as exchanged ions. Obviously, the more loosely the surface ions are held the greater is the average distance of oscillation and, hence, the greater is the likelihood of ion adsorption and/or replacement.

Overall, the efficiency with which ions can replace each other in a clay soil is dependent upon the following factors, namely (a) the relative number or concentration of ions, (b) the number of charges on the ions, (c) the speed of movement or activity of the different ions and (d) the type of clay mineral present. The ease with which cations can be exchanged and adsorbed is expressed in terms of the cation exchange capacity (CEC) of the soil: this is the number of milli-equivalents (meq) of ions that 100 g of soil can adsorb. By definition, 1 meq is 1 mg of hydrogen or the amount of any other ion that will combine with or displace it: thus, if the CEC of a soil is 1 meq it means that every 100 g of dry soil is capable of adsorbing or holding 1 mg of hydrogen or its equivalent.

The affinity of a soil for water is very much influenced by the fineness of its clay particles and upon the clay mineral(s) present. In this respect it should be appreciated that clay particles are essentially composed of minute flakes, and in these flakes, as with all crystalline substances, the atoms are arranged in a series of units to form clay minerals. The atomic structure of most clay minerals consists of two fundamental building blocks composed of tetrahedrons of silica and octahedrons of alumina. Each of the main clay mineral groups was formed from the bonding together of two or more of these molecular sheets. The clay minerals that are of greatest interest in relation to roadworks are kaolinite, montmorillonite and illite.

The kaolinite structural unit is composed of an aluminium octahedral layer with a parallel superimposed silica tetrahedral layer that is inter-grown in such a way that the tips of the silica layer and one of the layers of the alumina unit form a common sheet. The kaolinite mineral is composed of a stacking of these units (Figure 4.4(a)): its structure can be considered akin to a book in which each leaf is 0.7 nm thick. Successive sheets are held together by hydrogen bonds that allow the mineral to cleave into very thin platelets.

Kaolinitic clays are very stable. The hydrogen bonds between the elemental sheets are sufficiently strong to prevent water molecules and other ions from penetrating: hence, the

Figure 4.4 Schematic diagrams of the structure of typical (a) kaolin and (b) montmorillonite crystals

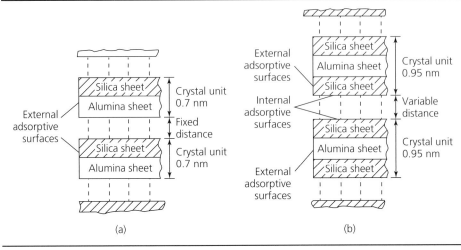

(a)   (b)

lattice is considered to be non-expanding. The kaolinite platelets carry negative electrical charges on their surfaces that attract thick layers of adsorbed water: however, as the lattice is non-expanding, the effective surface area to which water molecules can be attracted is limited to the outer faces, and, consequently, the plasticity, cohesion, and shrinkage and swelling properties of kaolinitic clays are very low compared with other silicate clays.

Figure 4.4(b) shows that a montmorillonite crystal unit comprises two tetrahedral sheets separated by an octahedral sheet (with the tips of each tetrahedral sheet and a hydroxyl layer of the octahedral sheet inter-grown to form a single layer). The minimum thickness of each crystal unit is about 0.95 nm, and the dimensions in the other two directions are indefinite. These sheets are stacked one above the other, like leaves in a book: however, there is very little bonding between successive crystal units, and, consequently, water molecules and cations can readily enter between the sheets. In the presence of an abundance of water, the clay mineral can be split into individual unit layers.

The ease with which water can enter between the crystal units makes it very difficult to work with montmorillonitic clays. The very large areas of charged surface that are exposed mean that hydrated ions/water are easily attracted to them. Each thin platelet of montmorillonite has the power to attract a layer of adsorbed water of up to 20 nm thick to each flat surface so that, assuming zero pressure between the surfaces, the platelets can be separated by 40 nm and still be 'joined'. This capability gives these clays their high plasticity and cohesion, marked shrinkage on drying, and a ready dispersion of their fine flaky particles: for example, typical plastic index (I_p) values for sodium and calcium montmorillonitic clays (CEC = 80–100 meq/100 g) are 603 and 114, respectively, compared with 26 and 37 for sodium and calcium kaolinitic clays (CEC = 5–15 meq/100 g), respectively (ICE, 1976).

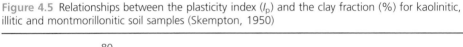

Figure 4.5 Relationships between the plasticity index (I_p) and the clay fraction (%) for kaolinitic, illitic and montmorillonitic soil samples (Skempton, 1950)

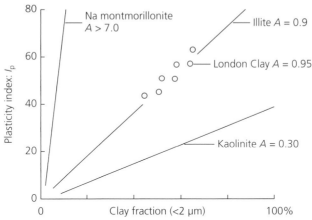

The thickness of each illite crystal unit is 1 nm. Like montmorillonite, illite has a 2:1 lattice structure. It differs from montmorillonite, however, in that there is always a substantial (~20%) replacement of silicon ions by aluminium ions in the tetrahedral layers. The valencies vacated by this substitution are satisfied by positively charged potassium ions that lie between the structural units and tie them together. The strength of this potassium bond is intermediate between the hydrogen bond of kaolinite and the water linkage of montmorillonite, and the net result is that illites have properties that are intermediate between those of the other two clay minerals.

The predominant clay mineral present in a soil can be easily and inexpensively indicated (Figure 4.5) from its activity (A). The activity of a soil is the ratio of its plasticity index (I_p) to the percentage by mass of soil particles within it that are less than 2 μm (i.e. $A = I_p/(\% \text{ clay})$). Inactive (good) and active (difficult) soils have A values of less than 0.75 and more than 1.25, respectively.

4.4.   Soil identification and description in the field

Consistent and standardised soil descriptions are important aspects of a ground investigation. A full description provides detailed information about a soil as it occurs in situ, and very few soils have the same descriptions. Soils are commonly identified and described in the field on the basis of data recovered from boreholes and trial pits, and from undisturbed materials seen in excavations and cuttings.

The description of a soil is based on the particle size grading of the coarser particles and on the plasticity of the finer particles: this is because of the roles that these characteristics play in determining the engineering properties of the soil. The British standard BS EN ISO 14688-1:2002 (BSI, 2002), which details the process for describing undisturbed soils, points out that this is a flexible system for field use by experienced persons: it covers both material and mass characteristics by visual and manual techniques, and does not

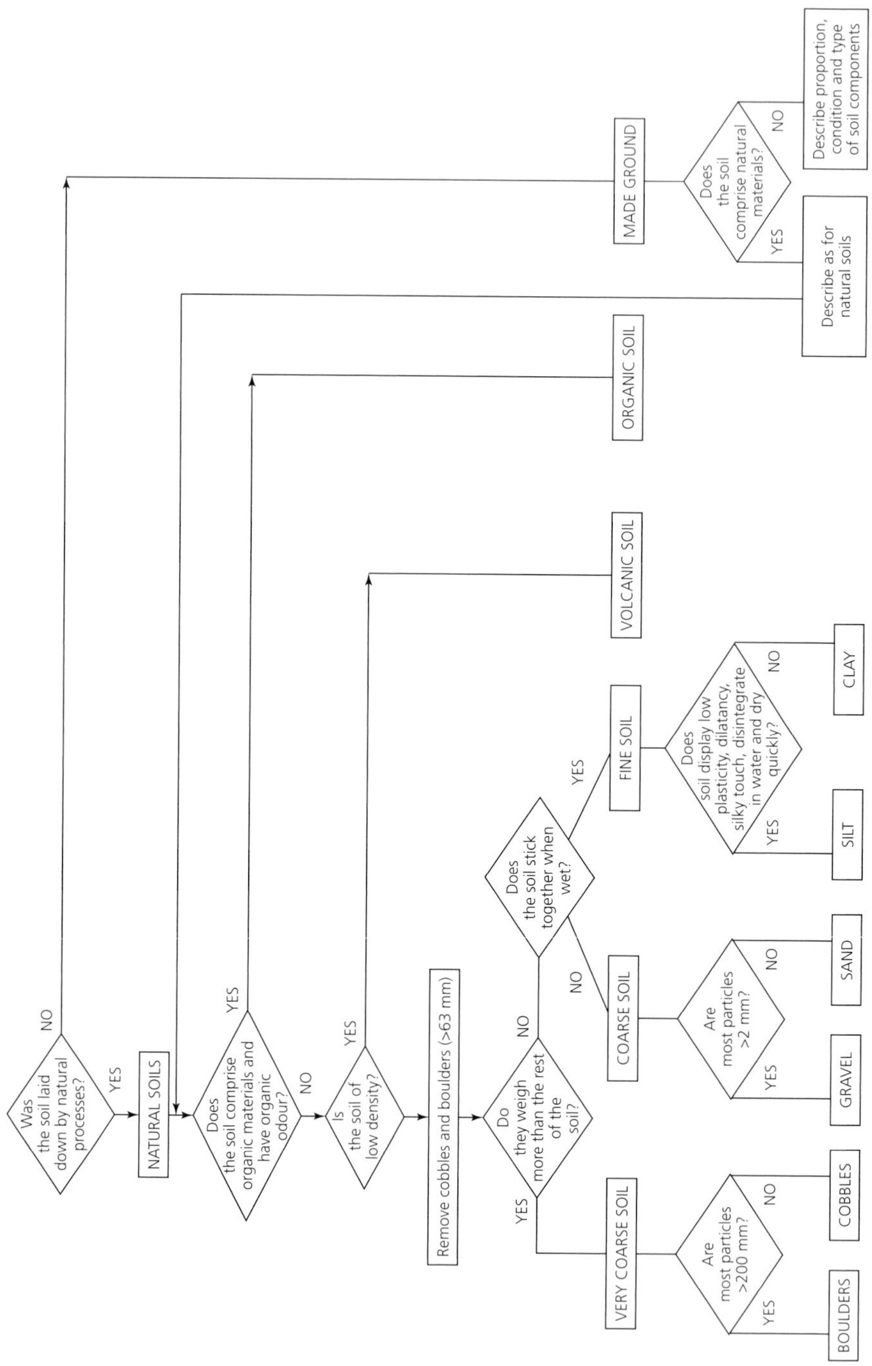

Figure 4.6 Flow chart of the process used in the identification and description of in situ soils (BSI, 2002, 2004a)

require laboratory test results to be obtained. Figure 4.6 summarises, in flow chart form, the descriptive process used, and Table 4.3 is a useful guide for describing soils in the field, based on guidance from BS 5930 (BSI, 2010) and BS 14688 (BSI, 2004b). The criteria used to describe the main characteristics of an in situ soil are as follows:

- particle size distribution
- particle shape
- mineral composition
- fines content
- soil colours
- dry strength
- dilatancy
- plasticity
- sand, silt and clay content
- carbonate content
- organic content.

In relation to this list, note that dilatancy is a behaviour characteristic of a fine-grained soil that allows the content of silt and clay to be evaluated visually. To do so, a moistened sample of the soil is placed in the palm of the hand and shaken (Figure 4.7). The sample becomes shiny by the appearance of water on its surface as it is shaken and then, when the moist sample is pressed with the fingers, the water disappears. For a silt, the water appears and disappears rapidly whereas the reaction is much slower with a predominantly clay soil.

A simple test can also be undertaken to determine the plasticity (toughness) of a fine-grained soil. First, a soil sample is moulded in the hand until it is at the consistency of putty, adding water or drying as needed. The sample is then rolled between the hands to form a 3 mm diameter thread (Figure 4.8): refolding and rolling of the thread is continued until the water content is reduced to the point that the thread just crumbles – which indicates that the sample is at its plastic limit. For a low-plasticity sample, only slight pressure is required to roll the thread near the plastic limit, and the thread is weak and soft, whereas, for a high-plasticity sample, considerable pressure is required to roll the thread to near the plastic limit, and the thread has high stiffness.

The term mass characteristics is used to refer to descriptions of those soil characteristics that depend on structure and, therefore, can only be observed in the field or in some undisturbed samples. The term material characteristics is used to refer to those characteristics that can be described from a visual and manual examination of either disturbed or undisturbed samples in the field. Thus, a fine-grained soil might be described as a firm closely fissured yellowish-brown clay (London Clay formation). A coarse-grained soil might be designated a loose brown sub-angular fine and medium flint gravel (terrace gravels).

In the above examples, additional minor information is given at the end of the main description after a full stop, so as to keep the standard main description concise. As

Table 4.3 Field identification and description of soils (Norbury, 2010)

Soil group	Principal soil type	Particle size: mm		Visual identification	Density/consistency Term	Field test	Discontinuities	
Very coarse soils	BOULDERS	Large boulder	630	Only seen complete in pits or exposures	None defined	Qualitative description of packing by inspection and ease of excavation	Describe spacing of features such as fissures, shears, partings, isolated beds or laminae, desiccation cracks, rootlets, etc.	
		Boulder	200	Often difficult to recover whole from boreholes				
	COBBLES	Cobble	63					
Coarse soils (over about 65% sand and gravel sizes)	GRAVEL	Coarse	20	Easily visible to naked eye; particle shape can be described; grading can be described	Borehole with SPT N-value:			
		Medium	6.3		Very loose	0–4		
		Fine	2.0		Loose	4–10	Fissured	Breaks into blocks along unpolished discontinuity
	SAND	Coarse	0.63	Visible to naked eye; no cohesion when dry; grading can be described	Medium dense	10–30		
		Medium	0.2		Dense	30–50	Sheared	Breaks into blocks along polished discontinuity
		Fine	0.063		Very dense	>50		

Fine soils (over about 35% silt and clay sizes)	SILT	Coarse	0.063	Only coarse silt visible with hand lens; exhibits little plasticity and marked dilatancy; slightly granular or silky to the touch;
		Medium	0.02	disintegrates in water; lumps dry quickly; possesses cohesion but can be powdered easily between
		Fine	0.0063	fingers
	CLAY			Dry lumps can be broken but not powdered between the fingers; they also disintegrate under water but more slowly than silt; smooth to the touch; exhibits plasticity but no dilatancy; sticks to the fingers and dries slowly; shrinks appreciably on drying usually showing cracks

Consistency	Description
Very soft	Finger easily pushed in up to 25mm. Exudes between fingers
Soft	Finger pushed in up to 10 mm. Moulded by light finger pressure
Firm	Thumb makes impression easily. Cannot be moulded by fingers. Rolls to thread
Stiff	Can be indented slightly by thumb. Crumbles in rolling thread. Remoulds
Very stiff	Can be indented by thumbnail. Cannot be moulded, crumbles
Hard	Can be scratched by thumb nail (or extremely weak)

Scale of spacing of discontinuities	
Term	Mean spacing: mm
Very widely	>2000
Widely	2000–600
Medium	600–200
Closely	200–60
Very closely	60–20
Extremely closely	<20

Table 4.3 Continued

Soil group	Condition		Accumulated in situ	
Organic soils	Firm	Fibres compressed together	PEAT	Predominantly plant remains, usually dark brown or black in colour, distinctive smell, low bulk density. Can include disseminated or discrete inorganic particles
	Spongy	Very compressible Open structure	Fibrous peat	Plant remains recognisable and retain some strength. Water and no solids on squeezing
			Pseudo-fibrous peat	Plant remains recognisable and strength lost. Turbid water and <50% solids on squeezing
			Amorphous peat	No recognisable plant remains. Mushy consistency. Paste and >50% solids on squeezing
	Plastic	Can be moulded in hand. Smears fingers	Gyttja	Decomposed plant and animal remains. May contain inorganic particles
			Humus	Remains of plants organisms and excretions with inorganic particles

Figure 4.7 Simple test for dilatancy in a silt where the appearance of water (from (a) to (b)) changes with hand pressure

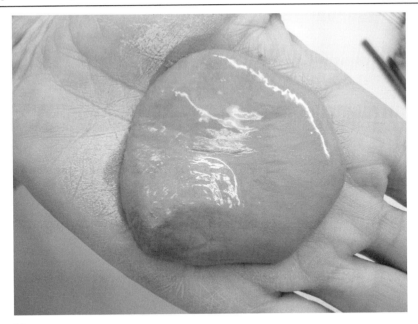

(a)

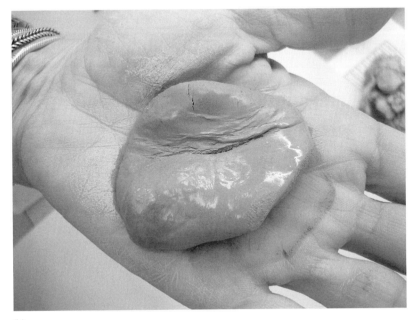

(b)

Figure 4.8 Simple test to evaluate the plasticity of a fine-grained soil sample with (a) low plasticity and (b) medium plasticity

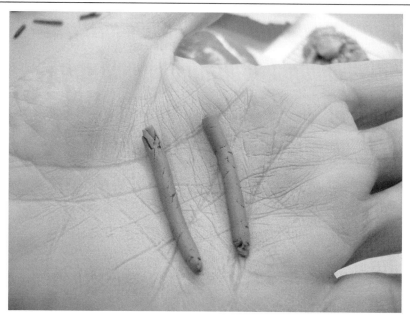

(a)

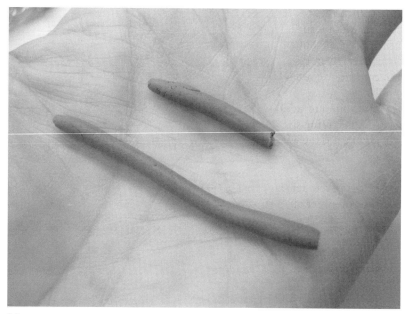

(b)

another example, soil materials in interstratified beds might be described as thinly inter-bedded dense yellow fine sand and soft grey clay (alluvium).

4.5.   Soil water

Water from precipitation, which does not evaporate or flow away in the form of surface runoff, penetrates the ground and percolates downward under the action of gravity until a depth is reached below which the soil pores are completely saturated. The upper surface of this saturated zone, which is most easily determined in coarser-grained soils, is variously termed the water table, the groundwater surface or the phreatic surface. The zone between the ground and the groundwater surface is called the zone of aeration: its thickness can range from >100 m (exceptionally, in very dry climates) to <1 m (commonly, in the UK).

The more coarse-grained the soil in the zone of aeration the more quickly the water filters down. While fine-grained soils have considerable porosity, gravitational water moves more slowly through them due to the resistance to flow (associated with the narrow percolating passages and high sidewall frictions) encountered by the water as it attempts to filter downward. Sometimes the zone of aeration contains a coarse-grained soil that lies on top of an inclined layer of clayey soil (e.g. in glacial till), so that the water is able to infiltrate more quickly laterally than vertically: then, following rain, the percolating water may appear as seepage water in, for example, road cuttings, and special drainage may have to be provided to cope with this.

4.5.1   Groundwater

Below the groundwater surface the soil pores are completely saturated. The water table is not a horizontal plane but changes as constantly as the topography above it. Thus, the groundwater is rarely at rest, as elevation differences in its surface cause it to flow later-ally until it emerges as ground seepage water or to feed lakes, streams or swamps. Hence, in dry weather the water table can be considerably lowered, whereas in wet weather it will rise as precipitation adds to the soil moisture and ponds, lakes and streams are replenished. Thus, during site investigation work, it is important to note the date of any measurement taken to the water table, so that a seasonal correction can be applied later, if appropriate.

By definition, the porewater pressure (u_w) at any depth in the saturated soil is the excess pressure in the porewater above that at the groundwater surface, where it is equal to the atmospheric pressure. Thus, if $u_w = 0$ kPa at the groundwater surface in, say, a gravelly soil, and it increases linearly with depth, then at a depth z m (assuming ρ_w is the density of water, Mg/m^3)

$$u_w = + \rho_w z \tag{4.1}$$

4.5.2   Held water

If gravity was the only force acting on the water in the aeration zone, the soil pores above the water table would be dry at all times except when precipitation water is filtering downward. However, if a cross-section is taken through the zone, it will be found that there is a layer above the groundwater surface within which the pores are wholly or partially filled with moisture. This layer is termed the capillary fringe. It is typically 20–50 mm thick in coarse soils, 120–350 mm in medium sands, 350–700 mm in fine sands, 700–1500 mm in silts and 2000–4000 mm or more (after long periods of time) in clays.

Water that is continuous through the pores within the capillary fringe is held by surface tension and adsorption forces, which impart to it a negative pressure or suction with respect to atmospheric pressure.

With the coarser types of fine-grained soils (e.g. fine sand and coarse silts), the greater proportion of the held water can be attributed to capillarity in the pores between the particles, if it is assumed that the pores in such soils form an interrelated mass of irregular 'capillary tubes'. If the lower ends of a number of different-sized capillary tubes are immersed in water it will be found that (a) the water level in each tube will rise above the free water level until the mass of the water column is just supported by the surface tension force in operation at the interface of the water meniscus at the top of the column and the inside surface of the tube, and (b) the height of this capillary rise will increase as the radius of the tube decreases. Similarly, the height of the capillary fringe will increase as the sizes of the pores in the sands and silts are decreased; that is, the more finely grained the soil the greater the potential for water to rise through capillary action. The porewater pressure (u_w) at any height z above the groundwater surface within the fringe is less than atmospheric pressure, and is equal to $-\rho_w z$.

The mass of water supported by surface tension tends to pull the particles together and compress the soil. Fine sand and coarse silt particles are already in physical contact with each other, and, hence, little volume change occurs as a result of this capillary rise. Below the coarse silt size, however, adsorbed moisture becomes more important in terms of its contribution to the amount and distribution of water within the capillary fringe. This is very important when the soil is an active (e.g. montmorillonitic) clay, as changes in the moisture content within the fringe will affect the density, volume, compressibility and stability of the soil.

The moisture content and its distribution with depth within a clayey soil at any given time are in equilibrium and reflect the environmental conditions, including seasonal precipitation and evaporation. If the ground is covered with a road pavement so that it is protected from precipitation and evaporation, a new moisture equilibrium is eventually reached with respect to the position of the groundwater surface beneath the pavement. In the process, pressure from the overburden and gravity seek to expel the moisture while the suction of the soil seeks to retain it, thus

$$s = \alpha\sigma - u_w \tag{4.2}$$

where s is the soil suction, u_w is the porewater pressure, σ is the vertical pressure from the overburden and α is the compressibility factor (i.e. the proportion of the overburden pressure acting on the porewater). While in coarse-grained soils the overburden pressure is carried by the inter-granular contacts and $\sigma = 0$, it is fully carried by the porewater in active clay soils, and $\sigma = 1$. With intermediate soils the extent to which the overburden pressure is carried by the porewater depends upon the plasticity characteristics of the soil in question: thus the compressibility factor is given by

$$\alpha = 0.027I_{\mathrm{p}} - 0.12 \tag{4.3}$$

where I_{p} is the plasticity index of the soil (between 5 and 40).

A test is available (Croney and Croney, 1998) to measure the soil suction, which is the pressure that has to be applied to the porewater to overcome the capacity of the soil to retain the water. For a given soil there is an increase in soil suction with decreasing moisture content, and this relationship is continuous over its entire moisture range (e.g. the suction value can range from up to $10^7 \, \mathrm{N/m^2}$ for oven-dry soils to zero for saturated soils that will take up no more water). Because of this large variation, the soil suction is usually expressed as the common logarithm of the length in centimetres of an equivalent suspended water column: this is termed the pF value of the soil moisture. Thus, for example, a soil with a pF of 1 equals 10^1 cm of water ($97.9 \times 10^1 \, \mathrm{N/m^2}$), pF $3 = 10^3$ cm ($97.9 \times 10^3 \, \mathrm{N/m^2}$), pF $5 = 10^5$ cm ($97.9 \times 10^5 \, \mathrm{N/m^2}$) and pF $7 = 10^7$ cm ($97.9 \times 10^7 \, \mathrm{N/m^2}$). Because of the logarithmic nature of the scale, pF $= 0$ does not exactly relate to zero suction: oven-dryness is close to pF 7.

The density of wet soil is close to twice that of water, so that if u_{w} and s are also expressed in centimetres of water, Equation 4.2 can be used with an appropriate soil suction curve to deduce the equilibrium moisture content above the groundwater surface (see Croney and Croney (1998) for an excellent treatise on this).

4.6. Soil phase relationships

Soil is an assemblage of mineral particles interspersed with pore spaces that may or may not be filled with water. It is convenient to visualise the different soil phases and the relationships between them by representing, in graphical form, a soil sample in which the solid, liquid and gaseous phases are segregated. Figure 4.9 shows such a sample that is placed in an equi-volume cylinder of unit cross-section: this makes it possible to consider the component volumes as represented by their heights, so that it is easy to develop some important relationships.

Figure 4.9 The three phases of a soil

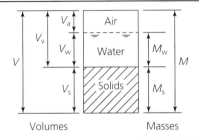

97

4.6.1 Moisture content

The moisture content of a soil is its mass of water expressed as a percentage of the mass of dry solids in the soil. Thus,

$$w = M_w/M_s \times 100 \qquad (4.4)$$

where w is the moisture content (%), M_w is the mass of water and M_s is the mass of solids.

4.6.2 Voids ratio

The voids ratio is the ratio of the volume of voids to the volume of solids in the soil. It is independent of whether the pore spaces contain water, air or other gases. Thus,

$$e = V_v/V_s \qquad (4.5)$$

where e is the voids ratio, V_v is the volume of voids and V_s is the volume of solids. Also,

$$V_v = e/(1 + e) \qquad (4.6)$$

and

$$V_s = 1/(1 + e) \qquad (4.7)$$

In a saturated soil, the voids ratio is directly proportional to the moisture content. In this case, the volume of voids, V_v, is equal to the volume of water, V_w. Then,

$$e_s = V_w/V_s = (M_w/M_s)(\rho_s/\rho_w) = w(\rho_s/\rho_w) \qquad (4.8)$$

where e_s is the voids ratio when the soil is saturated, ρ_s is the density of solid particles and ρ_w is the density of water.

4.6.3 Porosity and percentage of air and water voids

The porosity of a soil is the ratio of the volume of the voids to the total volume of the soil, expressed as a percentage. As with the voids ratio, porosity does not take account of whether or not the pore spaces contain water. Thus,

$$n = V_v/V \times 100 \qquad (4.9)$$

where n is the porosity (%) and V is the total volume of soil mass. Porosity is also expressed by

$$n = e/(1 + e) \times 100 \qquad (4.10)$$

The percentages of the total volume of soil that are occupied by air in the voids and by water in the voids are referred to as the percentage air voids and percentage water voids, respectively. Thus,

$$V_a = V_v - V_w \qquad (4.11)$$

$$n_a = V_a/V \times 100 \tag{4.12}$$

$$n_w = V_w/V \times 100 \tag{4.13}$$

where n_a is the air voids (%), n_w is the water voids (%), V_a is the volume of air voids, V_w is the volume of water voids and V is the total volume of the soil mass. Also, the sum of the percentages of air and water voids equals the porosity:

$$n = n_a + n_w \tag{4.14}$$

4.6.4   Degree of saturation
The extent to which the voids in a soil are filled with water is termed the degree of saturation, S_r (%): this is the ratio of the volume of water to the volume of voids, expressed as a percentage. Thus,

$$S_r = (V_w/V_v) \times 100 \tag{4.15}$$

4.6.5   Bulk density and dry density
The mass of the wet solid particles plus the water contained in the pore spaces of a soil per unit volume is called the bulk (or wet) density. Thus,

$$\rho = M/V = (M_s + M_w)/(V_s + V_w + V_a) \tag{4.16}$$

where ρ is the bulk (or wet) density of the soil (Mg/m^3).

The mass of the dry solids per unit volume of soil is the dry density. Thus,

$$\rho_d = M_s/V = M_s/(V_s + V_w + V_a) = \rho/(1 + w) \tag{4.17}$$

where ρ_d is the dry density of the soil (Mg/m^3).

4.6.6   Relationships between dry density, bulk density, moisture content and percentage air voids
The relationship between the bulk, ρ, and dry, ρ_d, densities of a soil follows from their definitions. Thus,

$$\rho/\rho_d = (M/V)(V/M_s) = (M_s + M_w)/M_s = 1 + w/100 \tag{4.18}$$

and

$$\rho_d = 100\rho/(100 + w) \tag{4.19}$$

The relationship between dry density, ρ_d, moisture content, w, and the percentage air voids, n_a, is deduced as follows:

$$\rho_d = M_s/(V_s + V_w + V_a) \tag{4.20}$$

and

$$1/\rho_d = V_s/M_s + V_w/M_s + V_a/M_s = 1/\rho_s + w/100\rho_w + V_a/V\rho_d \qquad (4.21)$$

Hence,

$$(1/\rho_d)(100 - 100V_a/V) = 100/\rho_s + w/\rho_w \qquad (4.22)$$

$$(1/\rho_d)(100 - n_a) = 100/\rho_s + w/\rho_w \qquad (4.23)$$

and

$$\rho_d = \rho_w(1 - n_a/100)/(\rho_w/\rho_s + w/100) = \rho_w(1 - n_a/100)/(1/r_s + w/100) \qquad (4.24)$$

where r_s is the specific gravity of the soil particles and ρ_w is the density of water.

Equation 4.24 shows that, if any two of dry density, moisture content and percentage air voids are known for a soil of a particular specific gravity, then the third can be easily established.

4.6.7 Specific gravity

Also termed relative density, the specific gravity is the ratio of the mass of the soil particles to the mass of the same (absolute) volume of water. Thus,

$$r_s = M_s/V_s\rho_w = \rho_s/\rho_w \qquad (4.25)$$

where r_s is the specific gravity, ρ_w is the density of water (Mg/m^3) and ρ_s is the particle density (Mg/m^3). Note that particle density, whose units are Mg/m^3, is the average mass per unit volume of the solid particles (where the volume includes any sealed voids contained within the solid particles): it is numerically equal to the dimensionless specific gravity.

4.7. Laboratory testing of soils

This section provides summary descriptions of some of the more common laboratory tests carried out on soils for roadworks. The emphasis in each discussion is upon the purpose for which each test is carried out. Detailed procedures for carrying out these tests are available in the relevant British and European test standards.

4.7.1 Soil sampling

Before discussing individual laboratory tests, it must be emphasised that samples sent to the laboratory for testing must be representative of the field situation being evaluated, enough soil must be sent to enable all of the tests to be carried out that are required by the testing programme, and the conditions in which the soil samples arrive at the laboratory, and their subsequent storage, must meet the requirements of the intended testing programme. In this context, soil specimens sent for testing can be described as disturbed, undistributed, re-compacted, remoulded or reconstituted.

Basic characterisation tests can be carried out on disturbed (i.e. bulk) soil samples, as the results are not affected by sample disturbance. However, it is critical that the sample size be sufficient to yield a test result that is representative of the in situ material: for example, soil samples for water content testing may be disturbed but they should be placed in sealed sample bags or jars so that moisture is not added or lost from the sample during storage or transport to the laboratory. Soil samples that are to be used to evaluate the behaviour of material to be used as compacted fill are not affected by disturbance during transport, as they will be remoulded or re-compacted in the laboratory prior to testing.

If it is necessary to gauge the in situ strength of a soil (e.g. if it is to be used as the subgrade or to form a side slope in a road cutting), then undisturbed samples – such as those obtained using thin-walled samplers (see Chapter 3) – will need to be obtained. These samples should be properly sealed and not exposed to excessive vibration or physical shocks prior to testing.

Every soil test requires a minimum sample size. However, the sample submitted to the laboratory for testing should be significantly larger than the minimum specimen to be used for testing, so that a representative specimen can be prepared from the sample submitted.

Having an adequate sample size is particularly important when testing the grain size distribution of a soil. Eurocode 7 (BSI, 2007a) defines a minimum mass for sieving (MMS) that is based on the maximum particle diameter within the sample: the MMS values range widely from 120 kg for soils with a maximum particle size of 75 mm to a 2 kg sample for a 20 mm maximum particle size to only 100 g for very fine material of less than 2 mm size. Minimum sample masses required for various tests are listed in Eurocode 7, and selected examples of these are shown in Table 4.4.

When the approximate number and types of tests are clear, it is a simple matter to estimate the soil sample size that has to be obtained from the field. However, if the programme of laboratory tests is uncertain, Table 4.5 can be used to provide guidance on desirable amounts of soil for various series of tests.

Before soil samples are sent to the laboratory, the possibility that they may be contaminated with harmful substances should be assessed, so that laboratory personnel can be advised of the need for any special handling procedures that may be necessary to ensure their safety. This is very important for samples taken from man-made ground.

Laboratory tests should be carried out at a constant temperature, ideally in an air-conditioned laboratory, and the test conditions used should be representative of those in the field for the service conditions being considered in the pavement design. Much practical experience underlies soil testing and, when the tests are properly carried out in accordance with the test standards, reliable predictions usually result if the derived data are used with skill and experience.

Table 4.4 Minimum sample sizes for various soil tests (BSI, 2007)

Test	Initial mass required	Minimum mass of prepared test specimen		
		Clay and silt	Sand	Gravelly soil
Water content	At least twice specimen mass	30 g	100 g	$D = 2$–10 mm: MMS $D > 10$ mm: 0.3 × MMS, minimum 500 g
Particle density	100 g	10 g (particle size <4 mm)		
Grain size				
Sieving	2 × MMS	MMS		
Hydrometer	250 g	50 g	100 g	Not applicable
Consistency limits	500 g	300 g (particle size <0.4 mm)		
Density index	8 kg	Mass of specimen depends on soil behaviour during test		
Proctor compaction	S[a] NS[b] 25 kg 10 kg			
California bearing ratio	6 kg			

[a] Soil particles susceptible to crushing during compaction
[b] Soil particles not susceptible to crushing during compaction

Table 4.5 General guidance for masses of soil required for various laboratory tests

Purpose of sample	Soil types	Mass of sample required: kg
Soil identification, including Atterberg limits; sieve analysis; moisture content and sulphate content tests	Clay, silt, sand	1
	Fine and medium gravel	5
	Coarse gravel	30
Compaction tests	All	25–60
Comprehensive examination of construction materials, including soil stabilisation	Clay, silt, sand	100
	Fine and medium gravel	130
	Coarse gravel	160

4.7.2   Basic soils characterisation testing

It is vital that the laboratory testing of soils be carried out in accordance with defined test standards. This allows the results to be interpreted and relied on by others since they are related to a precise test method. Laboratory testing standards can be local/agency-specific, national (e.g. the British Standards Institution (BSI)), regional (e.g. the Comité

Européen de Normalisation (CEN)) or international (e.g. ASTM). Increasingly, the trends are towards greater standardisation between different jurisdictions so as to facilitate international trade and to reduce barriers to competition.

In Europe, the transition from national standards to European standards has been a slow and complex process that has necessitated a coexistence period during which dual test standards were/are used until the old standards were/are withdrawn. In addition, apart from reasons of harmonisation, test standards are updated or modified from time to time to reflect improvements in test equipment, or to incorporate new research information. For these reasons, when referencing test standards in laboratory test reports, the edition year of the standard should be recorded. It should also be noted that European standards have to be applied to a very diverse range of regional needs, and, hence, in addition to the actual standard, related guidance documents may need to be consulted to establish specific national interpretations.

Since a primary reason for testing soils is as part of a site investigation, codes of practice for site investigations provide the framework for such testing. The BSI first published a code on site investigations as CP 2001 in 1957 (BSI, 1957): this was updated as BS 5930 in 1981 (BSI, 1981), revised in 1999, also in 2007, and again in 2010 (BSI, 2010). Section 5 of the current version of the code (BSI, 2010) also provides some general guidance on the laboratory testing of soils. More detail about laboratory test requirements and the evaluation of the results is provided in Section 5 of Eurocode 7 – Geotechnical design (BSI, 2007a). The National Annex to Eurocode 7 (BS1, 2007b) provides a useful checklist for soils classification testing as part of ground investigations.

Specific laboratory test procedures for soils are described in detail in BS 1377, which has eight parts (BSI, 1990). There is also a series of European standards for common soil classification tests (BSI, 2004b): however, at the time of writing, this series has not been adopted in the UK.

4.7.2.1 Moisture content test

Moisture is present in most naturally occurring soils. The proportion of moisture in a soil has a significant effect on its behaviour, and has implications for the reuse of the soil in roadworks.

The test procedure for moisture content involves taking a representative specimen of the soil to be tested and determining its mass. The specimen is then dried to a constant mass in an oven at a temperature of between 105 and 110°C, and its dry mass determined. The moisture content is calculated as the mass of water lost from the sample expressed as a percentage of the dry mass of the sample. The minimum specimen size for the test ranges from 30 g for a fine-grained soil to 3 kg for a coarse-grained soil.

The moisture content aids in the identification and characterisation of a soil, and fine-grained soils (e.g. silts and clays) have higher natural moisture contents than sands and gravels. By plotting moisture contents versus sample depth in the profile of a soil, changes in strata can sometimes be identified. Anomalously, high moisture contents can sometimes be an indication of man-made ground or the presence of organic material in a sample. Natural moisture contents can exceed 100% for highly organic soils (e.g. peats).

4.7.2.2   Consistency (Atterberg) tests

Consistency is that property of a soil that is manifested by its resistance to flow. As such, it is a reflection of the cohesive resistance properties of the finer fractions of a soil rather than of its coarser fractions.

The properties of the finer fractions of a soil are considerably affected by moisture content: for example, as water is continuously added to a dry clay it changes from a solid state to a semi-solid crumbly state to a plastic state to a viscous-liquid state. The moisture contents, expressed as a percentage by mass of the oven-dry soil samples at the empirically determined boundaries between these states, are termed the Atterberg limits (named after the Swedish scientist who was the first to attempt to define them, although Terzaghi (1925) is credited with recognising their full importance in soil classification). The Atterberg limits are determined for particles less than 425 μm.

The shrinkage limit (denoted w_s or SL) is the moisture content at which a soil on being dried ceases to shrink; that is, it is the moisture content at which an initially dry soil sample is just saturated, without any change in its bulk volume. The test is carried out by immersing a cylindrical sample of soil in a mercury bath and measuring its volume as its moisture content is decreased from the initial value, plotting the shrinkage curve (i.e. the volume per 100 g of dry soil versus the moisture content (%)) and reading the shrinkage limit from this graph.

The plastic limit (denoted w_p or PL) is the moisture content at which a soil becomes too dry to be in a plastic condition, and becomes friable and crumbly. This boundary was originally defined by Atterberg as the moisture content at which the sample begins to crumble when rolled into a thread on a glass plate under the palm of the hand. This is now standardised for test purposes as the moisture content at which the soil–water mixture is rolled into an unbroken thread 3 mm in diameter and it begins to crumble. The test set-up is shown in (Figure 4.10).

Pure sands cannot be rolled into a thread, and are therefore reported as 'non-plastic'. The PL values for silts and clays normally range from 5% to 30%, with the silty soils having the lower plastic limits.

The moisture content at which a soil passes from the liquid to the plastic state is termed the liquid limit (w_L or LL). Above the liquid limit the soil behaves as a viscous liquid, while below the liquid limit it acts as a plastic solid. The liquid limit test involves measuring the depths of penetration, after 5 s, of an 80 g standardised cone into a soil pat, at various moisture contents. The moisture content corresponding to a cone penetration of 20 mm, as determined from a moisture content versus depth of penetration relationship (Figure 4.11), is taken as the liquid limit.

Figure 4.10 Test set-up for plastic limit testing

Sandy soils do not have liquid limits, and are therefore reported as non-plastic. For silty soils, liquid limit values of 25–50% can be expected. Liquid limits of 40–60% and above are typical for clay soils; montmorillonitic clays have even higher values: for example, Gault clays are reported (Croney and Croney, 1998) as having liquid limit values ranging from 70% to 121% (and plastic limit values of 25–32%).

Figure 4.11 Determining the liquid limit

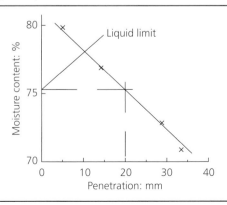

Values commonly used in association with the liquid and plastic limits are the plasticity index (I_p or PI), the liquidity index, I_L, and the consistency index, I_c. The plasticity index is the percentage moisture range over which the soil is in a plastic state, while the liquidity index is the ratio of the difference between the in situ moisture content and the plastic limit to the plasticity index. The consistency index is the numerical difference between the liquid limit and the in situ natural water content percentages expressed as a ratio of the plasticity index. Thus,

$$I_p = w_L - w_p \tag{4.26}$$

$$I_L = (w - w_p)/I_p \tag{4.27}$$

$$I_c = (w_L - w)/I_p \tag{4.28}$$

where w is the in situ moisture content (%), w_p is the plastic limit (%) and w_L is the liquid limit (%).

If I_L is greater than 1, then the in situ moisture content, w, is higher than w_L, which indicates that the soil is extremely weak in its natural state. If I_L is negative (i.e. is less than zero), it indicates that, because the in situ moisture content is less than the plastic limit, the soil is like a brittle solid, and will crumble when remoulded. If the liquidity index is between 1 and zero, it indicates that the in situ moisture content is at or below the liquid limit. If I_c is less than 0.25, it indicates that the soil is very soft, and if I_c is greater than 1, then the soil is very stiff.

The most common use of the plasticity test results is in the classification of fine-grained soils and the fine fractions of mixed soils. Some overall soil characteristics that are directly indicated by the consistency tests are given in Table 4.6.

Both w_L and I_p can be used as a quality-measuring device for pavement materials; that is, to exclude granular materials with too many fine-grained particles. These parameters are also used to evaluate the suitability of silty cohesive material (class 7F) for stabilisation with cement that is to be used as capping (HA *et al.*, 1995). If the plasticity index of a soil is known, it can be used to give a rough approximation of its clay content: for example, soils with I_p values of 13% and 52% can be expected to have clay contents of 20% and

Table 4.6 Some overall characteristics indicated by the consistency tests

Consistency	Comparing soils of equal w_L with I_p increasing	Comparing soils of equal I_p with w_L increasing
Compressibility	About the same	Increases
Permeability	Decreases	Increases
Rate of volume change	Increases	–
Dry strength	Increases	Decreases

60%, respectively (Croney and Croney, 1998). The plasticity index can also be used to estimate the equilibrium subgrade strength of a cohesive soil, as reflected by its California bearing ratio (CBR) value, for use in road capping design (HA *et al.*, 1995): for example, for a heavy clay with an I_p of 70%, the equilibrium CBR would be 2%, and 600 mm of capping would be needed, whereas in the case of a sandy clay with an I_p of 20%, the equilibrium CBR would be 4–5%, and only 350 mm of capping would be needed.

4.7.2.3 Particle density and specific gravity tests

Particle density, which is numerically equal to the dimensionless specific gravity (relative density), is used in computations involving many tests on soils (e.g. determining void ratios and particle size analyses via sedimentation testing). Particle density, denoted by the symbol ρ_s, is expressed in Mg/m^3.

There are three ways (BSI, 1990) of measuring particle density. The first is the gas jar method, which is suitable for most soils, including those containing gravel-sized particles provided that no more than 10% of the material is retained on a 37.5 mm test sieve. The second is the small pyknometer method, which is the definitive method (in the UK) for soils comprising clay, silt and sand-sized particles. The third is the pyknometer method, which is suitable for testing soils containing particles up to medium gravel size: this last test is less accurate than the other two, and is used mainly as a field test.

Both the gas jar and small pyknometer test methods require the soil particles to be oven dried at 105°C and then placed in a container for weighing with and without being topped up with water. The particle density is determined from the following equation:

$$\rho_s = (m_2 - m_1)/[(m_4 - m_1) - (m_3 - m_2)] \tag{4.29}$$

where m_1 is the mass of the container (g), m_2 is the mass of the container and soil (g), m_3 is the mass of the container, soil and water (g), and m_4 is the mass of the container and water (g).

These tests measure the weighted average of the densities of all the mineral particles present in the soil sample. As about 1000 minerals have been identified in rocks (most with densities of between 2 and 7 Mg/m^3), and as soils are derived from rocks, the results obtained can be expected to vary according to the original rock sources and the amount of mixing inherent in the derivation of the soil. However, because of the natural preponderance of quartz and quartz-like minerals, the densities of soil particles in the UK are mostly between 2.55 and 2.75, with the lower values generally indicating organic matter and the higher values metallic material. In practice, a value of 2.65 is commonly used, unless experience suggests otherwise.

4.7.2.4 Particle size distribution (grading)

Gradation testing involves the use of two test processes, sieving and sedimentation, to determine the percentages of the various grain sizes present in a soil. Two methods of sieving are specified: dry sieving and, if the sample contains silt or clay, wet sieving. The relative proportions of silt and clay sizes can be determined only from sedimentation testing.

Dry sieving is appropriate only for soils containing insignificant quantities of silt and clay (i.e. essentially, clean cohesionless soils).

Wet sieving can be used to determine the quantitative distribution of particle sizes down to fine sand size: this first requires the preparation of a measured quantity of soil and the washing of all silt and clay in that sample through a 63 μm sieve. The retained coarser material is then dry sieved through a series of sieves, and the mass retained on each of these sieves, as well as the material passing the 63 μm sieve, is then expressed as a percentage of the total sample mass.

Sedimentation testing is used to determine particle size distributions from coarse sand size (2 mm) down to clay size (0.002 mm). If significant quantities of both coarse- and fine-grained particles are present in a soil, sedimentation testing is only carried out on the material passing the 63 μm sieve, and the results of both the wet sieving and sedimentation tests are combined to give the overall particle size distribution for the soil. In practice, this test is not usually carried out if less than 10% of the material passes the 63 μm sieve.

All sedimentation testing is based on the assumptions that soil particles of different size (a) are spherical in shape, (b) can be dispersed uniformly through a constant-temperature liquid without being close enough to influence each other and (c) have settling velocities in accordance with Stoke's law (which states that the terminal velocity of a spherical particle settling in a liquid is proportional to the square of its diameter). While soil particles, especially clay particles, are not spherical, the results obtained are adequate for practical soil testing purposes.

In a typical sedimentation test, a suspension of a known mass of fine particles is made up in a known volume of water. Appropriate quantities are about 100 g for a sandy soil and 50 g for a silt soil and 30 g for a clay soil. The mixture is shaken thoroughly, and the particles are then allowed to settle under gravity. With the pipette method the mixture is sampled at a fixed depth after predetermined periods of time, and the distribution of particle sizes is determined by mass differences. With the more commonly used hydrometer method, the density of the soil–water mixture is determined at fixed time intervals using a relative density hydrometer of special design, and the distribution of particle sizes is then determined by formula calculation (or a nomograph).

The results of a particle size analysis are presented either (a) in a table that lists the percentage of the total sample that passes each sieve size (i.e. that is smaller than a specified particle diameter) or (b) as a plot of the sieve (i.e. particle) size versus the percentage passing each sieve (i.e. that is smaller than each diameter), with the sieve or particle size on a logarithmic scale and the percentages on an arithmetic scale. Figure 4.12 shows some classic particle size distributions for UK soils.

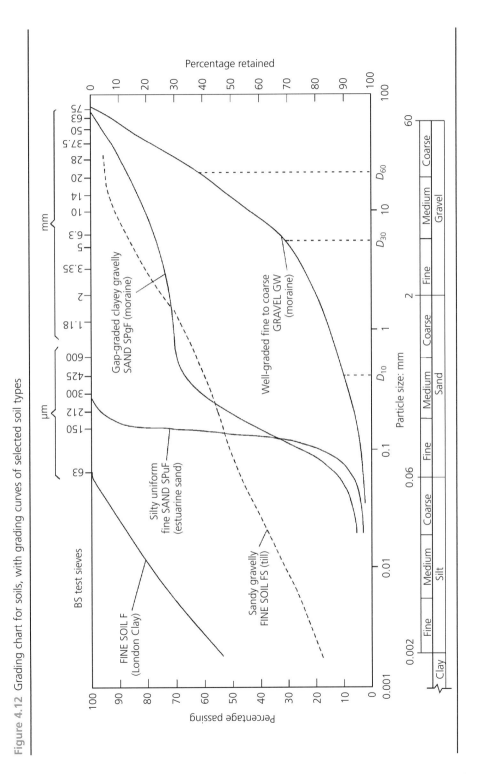

Figure 4.12 Grading chart for soils, with grading curves of selected soil types

Table 4.7 Descriptions of grading curve shapes

Shape of grading curve	C_U	C_C
Multi-graded	>15	$1 < C_C < 3$
Medium graded	6–15	<1
Even graded	<6	<1
Gap graded	Usually high	Any (usually <0.5)
Well graded	>5	$1 < C_C < 3$

A well-graded soil (also referred to as a non-uniform soil) is one that compacts to a dense mass: it has a wide range of particle sizes and a smooth and upward-concave size distribution with no deficient or excess sizes in any size range. A soil is said to be well graded (e.g. the moraine gravel in Figure 4.12) if it has a coefficient of uniformity, C_u, of 5 or more, and a coefficient of curvature, C_c, of between 1 and 3, as determined from

$$C_u = D_{60}/D_{10} \tag{4.30}$$

$$C_c = D_{30}^2/D_{10}D_{60} \tag{4.31}$$

where D_{10}, D_{30} and D_{60} are the particle diameters at which 10%, 30% and 60%, respectively, by mass of the soil are finer. If the range of particle sizes is small, the gradation curve will be steep, and the soil is described as a uniform soil (e.g. the estuarine sand in Figure 4.12): a uniform soil is poorly graded, and has a C_u value of 2 or less. A poorly graded soil may have a near-horizontal 'hump' or 'step' in its gradation curve, indicating that it is missing some intermediate sizes: this is termed a gap-graded soil (e.g. the poorly graded moraine soil in Figure 4.12). Other grading curve shape descriptions are provided in Table 4.7. In practical terms, a soil is often said to be well graded if C_c is close to 1 and to be poorly graded if C_c is much less or much larger than 1.

Particle size distribution tests are of considerable value when used for soil classification purposes. The grading results provide confirmation of soil descriptions for inclusion on borehole and trial pit logs, and they are used in evaluating whether soils will be problematic when encountered during road construction. Use is frequently made of particle size analyses when evaluating reuse options for road cut materials, as well as options for stabilisation by mechanical or chemical means.

The use of particle size distribution analyses to predict engineering behaviour should be treated with care, and limited to situations where other confirming test data are available: for example, where detailed studies of performance or experience have allowed strong empirical relationships to be established.

4.7.3   Soil characterisation to support geotechnical design
The characterisation of soils from the perspective of their suitability to support a pavement and for the purposes of establishing subgrade capping requirements and pavement

thickness designs requires a combination of field and laboratory testing. The field testing aspects are discussed in Chapter 3. The following sections in this chapter describe laboratory tests that complement the basic soil characterisation tests described in the previous section and assist in road and pavement design.

4.7.3.1 Unconfined compressive strength test

The unconfined compressive strength (UCS) test can only be carried out on saturated non-fissured fine-grained soils, or soils that are stabilised with additives that bind the particles together and are sufficiently impermeable to maintain undrained conditions. With fine-grained soils, the UCS test, which is a quick substitute for the undrained triaxial test, is carried out most commonly on 38 mm diameter cylindrical specimens having a height-to-diameter ratio of $2:1$. The undrained shear strength, c_u, is then determined as one-half of the measured UCS. In the UK, stabilised soils are normally tested using 150 mm cubes for coarse-grained and medium-grained materials; for fine-grained soils (i.e. those passing the 5 mm sieve), the use of 100 mm diameter by 200 mm high, or 50 mm diameter by 100 mm high, cylindrical specimens are recommended. Before being tested, stabilised samples are compacted to a predetermined dry density (usually the maximum value obtained during the moisture–density test) either by static (with fine and medium-grained soils) or dynamic (with medium- and coarse-grained soils) compaction. While a load versus deflection curve can be plotted, it is the UCS at failure that is most often used in roadworks, especially if soil stabilisation is involved.

For soil stabilisation work, the UCS test is mostly used to determine the suitability of a soil for treatment with a given additive and to specify the additive content to be used in construction. It is also used as a quality control mechanism in the field. UCS results are considerably influenced by such factors as the amount and type of additive, the method and length of curing of the test specimen, and whether or not the specimens are saturated before testing.

4.7.3.2 One-dimensional consolidation tests

When an embankment is placed on a saturated soil, the immediate tendency is for the particles in the foundation material to be pushed closer together. However, as water in the soil is incompressible, it must initially carry part of the applied load, and this results in the production of porewater pressure, which decreases as water drains from the soil. During the drainage period, which can take many years to finish, the soil particles are continually forced closer, thereby reducing the bulk volume and causing soil settlement. One-dimensional consolidation tests attempt to estimate in an accelerated manner both the rate and total amount of settlement of the soil under an applied load. The results can be used, for example, to establish the safe height of an embankment and its construction staging, and the length of time a pre-loaded embankment might be needed to induce the bulk of the settlement to occur prior to road construction.

Use of the standard (incremental loading) dead-weight oedometer test is restricted to representative samples of saturated clays, fine silts and other soils of low permeability

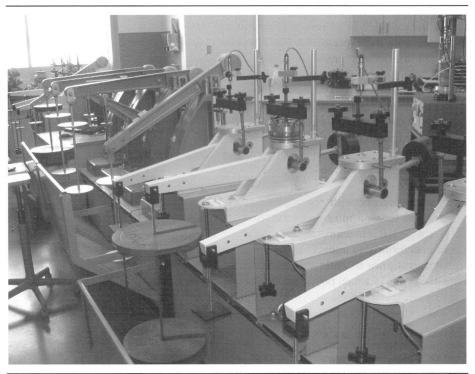

(Figure 4.13). This test involves cutting and trimming a 50, 75 or 100 mm diameter by 20 mm thick soil sample that fits into a special metal ring used in the test, and, after ceramic porous discs have been placed above and below the sample, placing the assembly in a loading unit. In the case of a stiff clay, for example, a careful compressive load–unload–reload sequence is then applied using small increments and decrements, and the changes in thickness of the sample are read at set time intervals. With the continuous-loading oedometer test stresses, strains or pore pressures are continuously varied instead of, as in the standard test, applying the loads in discrete increments.

A number of parameters that are useful in geotechnical design are derived from the consolidation test. The compressibility index, C_c, is used in the analysis of total settlement: it is a dimensionless factor that normally ranges from 0.1 to 0.3 for silty clays, and from 0.2 to 1.0 for clays. The coefficient of consolidation, c_v, ranges from 0.2 to 2.0 cm^2/s for silty clays, and from 0.02 to 0.10 cm^2/s for clays; this coefficient is used to estimate the settlement for a given period of time under a given increment of load. The degree of consolidation, U, is the ratio of the settlement for the time period to the total settlement, expressed as a percentage. In practice, actual settlements can take place more rapidly than may be predicted by the one-dimensional consolidation test. Also, drains can be used in the field to accelerate settlement, by providing readily accessible passageways for the water to escape (see Chapter 5).

4.7.3.3 Shear tests

Laboratory shear tests seek to evaluate representative soil specimens in a way that is similar to that anticipated under field conditions. Hence, test programmes are devised to approximate the expected loading conditions of the soil, so that the results can be translated into reliable predictions of soil responses in terms of the parameters needed for the design analysis contemplated.

Direct shear box tests are carried out only on coarse-grained soils, usually in square boxes with 60 or 100 mm sides. Samples of coarse-grained soils are more easily prepared for shear box testing than for triaxial testing. However, drainage conditions cannot be controlled or pore pressures determined with this form of testing, and the plane of shear is fixed by the nature of the test.

The general practice with respect to laboratory shear testing is to carry out triaxial compression tests on most soils (Figure 4.14). Triaxial tests are normally carried out on 70, 100 or 150 mm diameter samples with a height-to-diameter ratio of 2:1. The tests can be unconsolidated undrained or drained tests or, if the soil sample is consolidated in the apparatus prior to testing, consolidated undrained or drained tests. Any undrained or drained tests in which pore pressures are measured are normally consolidated before carrying out the shear tests.

Figure 4.14 Triaxial test set-up

The tests are designed to derive the angle of internal friction and the cohesion of a soil: these are considered constant for both laboratory and field conditions. The shearing resistance of a soil, which is particularly important in relation to the design of slopes for road cuttings and embankments, is derived from

$$s = c + \sigma_n \tan \phi \tag{4.32}$$

where s is the shear strength (kPa), c is the cohesion (kPa), σ_n is the stress normal to the shear plane (kPa) and ϕ is the angle of shearing resistance (degrees).

Generally, sandy soils develop their shearing resistance through friction, with little or no cohesion. Internal friction is mainly affected by the shape of the coarse particles, and is little affected by the moisture content; however, it increases rapidly with increasing dry density. The bulk of the shear resistance of a clay comes from cohesion associated with water bonds between the particles. Cohesion is greatly influenced by moisture content; that is, it decreases with increasing moisture to reach a low level at the plastic limit and almost zero at the liquid limit. When a UCS test is carried out on a natural clay sample, the internal friction can be assumed equal to zero, and the c value is then one-half of the measured compressive strength. Typical ranges for friction and cohesion are, for sandy soils, $\phi = 28$–$45°$ and $c = 0$–$2.06 \, \text{MN/m}^2$ and, for clay soils, $\phi = 0$–$15°$ and $c = 0.7$–$13.8 \, \text{MN/m}^2$.

4.7.3.4 California bearing ratio test

The CBR test was originally devised by the California State Highways Department, following an extensive study into flexible pavement failures in the 1930s. It was taken up by the US Corps of Engineers during World War II and adapted for airport pavement design purposes. After the war, other road bodies became interested in the test as a means of empirically measuring the soil strength for pavement design purposes, and the original CBR test and design procedures were subsequently adapted to meet particular needs in various countries.

The CBR test is normally carried out in the laboratory. The principle underlying the test involves determining the relationship between force and penetration when a cylindrical plunger of a standard cross-sectional area is made to penetrate a compacted soil sample at a given rate. At certain values of penetration, the ratio of the applied force to a standard force, expressed as a percentage, is defined as the CBR for the soil. The CBR value is used as a basic parameter to establish the supporting strength of a subgrade for flexible pavement design. It is also used to evaluate the potential strengths of unbound base materials for use in road and airfield pavements.

The UK variant of the test is carried out on soil having a maximum particle size not exceeding 20 mm, due to the relatively small size of the mould and the plunger. It involves placing a predetermined mass of soil in a steel mould with an attached collar that is 152 mm in diameter by 178 mm high, and compacting it either statically or dynamically until a 127 mm high specimen is obtained that is at the moisture and dry density conditions required by the design. Following compaction, the specimen, still in

Figure 4.15 Example CBR load–penetration curves

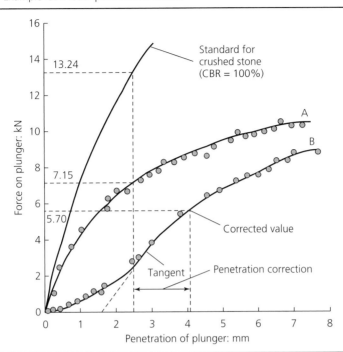

its mould, is covered with annular surcharge weights approximately equivalent to the estimated mass of flexible pavement expected in the field above the soil being tested; it is then placed in a testing machine, and a plunger with an end diameter of 49.6 mm is caused to penetrate the compacted soil at a rate of 1 mm per minute. The plunger penetrations are measured by a dial gauge, and readings of the applied force are read at penetration intervals of 0.25 mm until a total penetration not exceeding 7.5 mm is achieved.

After the test, a load versus penetration curve is drawn for the data obtained: this is (ideally) illustrated by the convex-upward curve A in Figure 4.15. In practice, an initial seating load is applied to the plunger before the loading, and penetration gauges are set to zero, and, sometimes, if the plunger is not perfectly bedded, a load–penetration curve resembling curve B in Figure 4.15 may be obtained. When this happens, the curve has to be corrected by drawing a tangent at the point of greatest slope and then projecting it until it cuts the abscissa: this is the corrected origin, and a new penetration scale is (conceptually) devised to the right of this point. The CBR is then obtained by reading from the corrected load–penetration curve the forces that cause penetrations of 2.5 and 5 mm, and expressing these as percentages of the loads that produce the same penetration in a standard crushed stone mix (Figure 4.15); the forces corresponding to this standard curve are 11.5 kN at 2 mm penetration, 17.6 kN at 4 mm, 22.2 kN at 6 mm, 26.3 kN at 8 mm, 30.3 kN at 10 mm and 33.5 kN at 12 mm. The higher of the two percentages measured at the 2.5 and 5 mm penetrations is then taken as the CBR value for the soil; for example, the CBRs obtained from curves A and B in Figure 4.15 are 54% and 43%, respectively.

The CBR of a soil is an indefinable index of strength that is dependent upon the condition of the soil at the time of testing. This means that the soil needs to be tested in a condition that is critical to its design. At any given moisture content, the CBR of a soil will increase if its dry density increases (i.e. if the voids content is decreased). Thus, a design dry density should be selected for testing that corresponds to the minimum state of compaction expected in the field during service.

At a given dry density, the CBR also varies according to its moisture content. Hence, the design CBR value should be determined for the highest moisture content that the soil is likely to have subsequent to the completion of earthworks. The selection of this moisture content requires an understanding of how moisture fluctuates in a soil.

Before a soil is covered by a road pavement, its moisture content will fluctuate seasonally: for example, a gradient of moisture within the soil profile that is determined during the summer might look like curve A in Figure 4.16, while another taken in the winter might look like curve B. Some time after construction of the pavement, the soil moisture conditions will become stabilised and subject only to changes associated with fluctuations in the level of the water table; when this happens, the moisture gradient might look like curve C. Now, if the pavement is to be constructed during the winter, the soil should be tested at the moisture content at point b; this high content should be used, as the road may be opened to traffic before the moisture content can decrease to that at point c. If the road is to be built during the summer, the soil should be tested at the moisture content at point c, as, ultimately, this will be its stable moisture content.

Figure 4.16 Seasonal variations in the moisture content of a heavy clay soil

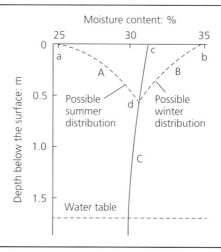

The determination of the equilibrium moisture content (point c) to use in CBR testing is best determined from laboratory suction tests on soil samples taken from the proposed formation level, using a suction corresponding to the most adverse anticipated water table conditions. If facilities for conducting suction tests are not available, this moisture content can be estimated at a depth below the zone of seasonal fluctuation but at least 0.3 m above the water table (i.e. at or below point d). In cohesive soils in the UK, especially those supporting dense vegetation, point d is typically at a depth of 1–1.25 m; in sandy soils, which do not normally have deep-rooted grasses, the moisture content at a depth of 0.3–0.6 m will usually provide a good estimate of the moisture content at point d. If a cohesive soil is to be used in an embankment, the moisture content at a depth of 1–1.25 m should be increased by about 2% to allow for the disturbance to the natural soil structure; if silty and sandy soils are to be used in embankments, the moisture content at point d needs only to be increased by about 1%.

An in situ CBR test can also be carried out in the field, using a suitably rigged vehicle. In situ CBR results can be significantly different from laboratory-derived results (except for heavy clays having an air voids content of 5% or more), and, hence, the in situ CBR test is not normally used as a quality assurance measure during construction.

4.7.3.5   Testing for the resilient modulus

Roads are not static structures: for example, the subsurface moisture regime changes over time (as do the properties of the supporting soils). Also, the main loading from traffic is dynamic in nature, and the response of a soil to repeated loading is quite different to that under static loading. The elastic response of a soil (i.e. its deflection under applied load) changes over time and varies with the magnitude and frequency of loading. The magnitude of these deflections under traffic loading relates directly to pavement performance, and if too large can result in fatigue cracking of the pavement surfacing. Also, the repeated loading can lead to an accumulation of permanent deformation (or creep) that, over time, can contribute to the development of rutting in the wheel paths of a roadway.

In the 1960s and 1970s, as access to digital computers became available, research work was instigated to move from the (then) empirical design methods to more mechanistic ones. Thus, the stresses and strains in multi-layered pavement structures were modelled (as in the design of other types of engineering structures), and these models required the ability to predict the long-term dynamic response of all pavement materials under traffic loading. A team of pioneers in this research at the University of Nottingham (Brown and Pell, 1972; Hyde, 1974) developed repeated load triaxial testing equipment to characterise the load–deflection response of soils and aggregates. With this equipment it became possible to subject soil samples to over a million repetitions of load and to measure the responses. The key material parameter derived from repeated load triaxial testing is the resilient modulus (M_r), which is defined as the ratio of the deviatoric stress to the resilient (elastic) strain experienced by the material under repeated load applications. The term 'resilient' is used to differentiate the elastic or recoverable response of a soil under a dynamic load from the plastic or 'creep' component. M_r is the fundamental subgrade strength parameter needed as input to any rational or mechanistic pavement design

Table 4.8 Subgrade resilient modulus values recommended for use in pavement design in Ontario, Canada

Description	Drainage characteristics	Susceptibility to frost action	M_r: MPa		
			Good	Fair	Poor
Rock, rock fill, shattered rock	Excellent	None	90	80	70
Well-graded gravels and sands	Excellent	Negligible	80	70	50
Poorly graded gravels and sands	Excellent to fair	Negligible to slight	70	50	35
Silty gravels and sands	Fair to semi-impervious	Slight to moderate	50	35	30
Clayey gravels and sands	Practically impervious	Negligible to slight	40	30	25
Silts and sandy silts	Typically poor	Severe	30	25	18
Low-plasticity clays and compressible silts	Practically impervious	Slight to severe	35	20	15
Medium to high-plasticity clays	Semi-impervious to impervious	Negligible to severe	30	20	15

process: but it is a complex parameter, as it is non-linear and depends on the magnitude of the applied stress.

Repeated-load triaxial testing equipment is now commercially available, and is relied upon increasingly for pavement design; as an example, the Ministry of Transportation Ontario, Canada (MTO, 2012a) provides default M_r values (Table 4.8) for use in pavement design in the province in the event that actual test values are not available.

Since it is not easy to obtain M_r values by direct testing, a number of attempts have been made to relate the modulus to other test parameters, in particular to CBR values. This has inherent limitations, since the CBR is a punching test to failure whereas M_r is an elastic parameter derived under relatively low stresses. The first correlations were made by the Shell Research Laboratory in the 1960s (Heukelom and Klomp, 1962), which developed the following simplified relationship based on tests of sand subgrades:

$$M_r = 10 \times CBR \text{ MPa} \tag{4.33}$$

More refined relationships have since been developed to estimate design values for M_r. However, the most reliable results are derived from repeated-load triaxial testing or by iterative back-calculation from the results of falling weight deflectometer testing.

4.7.4   Soil characterisation for construction

One of the objectives of road design on new alignments is to balance the cuts and fills so that soils excavated from cuttings can be used as fill to construct embankments, as there are major cost and efficiency advantages to be gained from maximising the reuse of cut materials for embankment construction. The ability and ease with which excavated soils can be reused depends on the soil type and its water content. The following tests assist in evaluating the suitability of a soil for embankment construction and to determine whether soil stabilisation might be needed to facilitate the reuse of the soil.

4.7.4.1   Moisture–density tests

Compaction is the process whereby the solid particles in a soil are packed more closely together, usually by mechanical means, to increase its dry density and strength. In 1933, Ralph Proctor of the Los Angeles Bureau of Waterworks and Supply devised a laboratory test, now known as the Proctor test, to ensure that dry densities were attained under compaction that gave desired impermeabilities and stabilities to earth dams. After compacting over 200 different soils at various moisture contents, Proctor demonstrated that water acts as a lubricant that enables soil particles to slide over each other during compaction and that, as the moisture content of a soil was increased, its compacted dry density increased to a maximum value, after which it decreased. Only a small amount of air remains trapped in a compacted soil at the point of maximum dry density, and any increase in the moisture content above what Proctor termed its optimum moisture content results in soil particles being replaced by water, and a consequent reduction in dry density. Over the course of time, standardised laboratory procedures were developed to evaluate this moisture–density relationship.

The UK version of this fundamental test is carried out by compacting a prepared air-dried soil sample in a specified way in either a metal cylindrical mould with an internal volume of 1 litre and an internal diameter of 105 mm (for material with particle sizes up to 20 mm) or a CBR mould with an internal diameter of 152 mm and depth of 127 mm (for material with particle sizes up to 37.5 mm). In the standard Proctor test, a 50 mm diameter, 2.5 kg metal rammer is dropped through 300 mm for 27 blows (with the 1 litre mould) or 62 blows (with the CBR mould) onto each of three, approximately equally thick, layers of soil in the mould. In the modified Proctor test, a greater compactive effort is employed, and a 50 mm diameter, 4.5 kg rammer is allowed to fall through 450 mm for the same number of blows as with the standard test, onto each of five layers in the same-sized moulds. Following compaction, surplus compacted soil is struck off flush with the top of the mould, the mass of the soil is determined, and its bulk or wet density is calculated by dividing the mass by the known volume of the mould. A sample is then taken from the soil used in the mould, and its moisture content determined.

After the compacted material has been removed from the mould, a new air-dried sample of soil is prepared, a higher increment of water is added, and a new compacted sample is prepared using the same standard (or modified) Proctor test procedure. The preparation of samples is continued until the mass of a compacted sample is less than that derived from the preceding test. Knowing the bulk density and moisture content of each sample,

Figure 4.17 Moisture–density relationships derived from the standard and modified Proctor tests

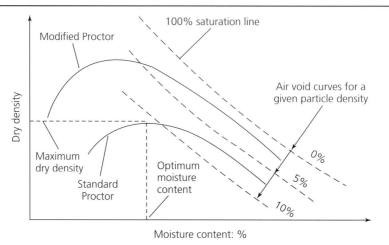

their dry densities are then calculated from

$$\rho_d = 100\rho/(100 + w) \tag{4.34}$$

where ρ_d is the dry density (Mg/m³), ρ is the wet density (Mg/m³) and w is the moisture content (%). If the dry densities are plotted against their corresponding moisture contents, and a smooth curve is drawn through each set of data points, a relationship similar to those shown in Figure 4.17 is obtained, and the optimum moisture content and the maximum dry density can then be read from the moisture–density curve. Also shown in Figure 4.17 are the curves corresponding to 0, 5% and 10% air voids, calculated from

$$\rho_d = (1 - n_a/100)/(1/\rho_s + w/100\,\rho_w) \tag{4.35}$$

where ρ_d is the dry density (Mg/m³), ρ_s is the particle density (Mg/m³), ρ_w is the density of water (Mg/m³; assumed to be 1), n_a is the volume of air voids expressed as a percentage of the total soil voids and w is the moisture content (%).

Table 4.9 compares some typical maximum dry density and optimum moisture content values obtained from standard and modified Proctor tests on various soils. Note that the greater compaction applied in the modified test had the greatest effect upon the heavy and silty clays, with large increases in the maximum dry densities and large decreases in the optimum moisture contents; that is, when the air voids contents are large, the effect of increasing the compaction is significant, whereas it is negligible when the air voids content is small. Table 4.9 also shows that the maximum dry density increases and optimum moisture content decreases as the soil becomes less plastic and more granular. With the uniformly graded fine sand, which has significant voids between its single-size particles, the modified test gives a dry density increase at the optimum moisture content

Table 4.9 Comparison of modified Proctor test results with those derived from the standard Proctor test for various soils

Type of soil	Standard Proctor test		Modified Proctor test	
	Maximum dry density: Mg/m^3	OMC: %	Maximum dry density: Mg/m^3	OMC: %
Heavy clay	1.555	28	+0.320	−10
Silty clay	1.670	21	+0.275	−9
Sandy clay	1.845	14	+0.210	−3
Sand	1.940	11	+0.145	−2
Gravel–sand–clay	2.070	9	+0.130	−1

OMC, optimum moisture content

that is fairly small; further, the optimum moisture content is fairly difficult to determine because of the free-draining characteristics of this soil.

While the Proctor tests provide useful guides to the range of moisture contents suitable for compaction with most soils, they are relatively poor guides for specifications relating to the compaction of permeable granular soils with little silt and clay (e.g. fine-grained clean gravels or uniformly graded coarse clean sands), and so the vibrating hammer test was developed in the UK for use with coarse soils (i.e. a maximum particle size of 37.5 mm). In this test, an electric vibrating hammer is used to compact three layers of moist soil – each layer is vibrated for 60 s – in a CBR mould so that the height of the specimen after compaction lies between 127 and 133 mm. The tamping foot of the hammer has a circular base that nearly covers the area of the mould and produces a flat surface, and the total downward force during compaction is maintained at between 300 and 400 N. The wet or bulk density, ρ, is then calculated from

$$\rho = M/18.1h \tag{4.36}$$

where M is the mass of the wet compacted soil (g) and h is the compacted specimen height (mm). The dry density is then calculated as for the Proctor tests (Equation 4.34). By varying the soil moisture content, a moisture content–dry density relationship is obtained, from which the optimum moisture content and maximum dry density are determined.

Theoretically, the maximum dry density achievable with a soil having a specific gravity of 2.70 is 2.70 Mg/m^3: this could only occur, however, if all the particles could be fitted against each other exactly. As soil particles come in various shapes and sizes, there are always air voids, and, hence, this theoretical density is never achieved. Well-graded dense gravel mixtures can be compacted to maximum dry densities, at their optimum moisture contents, with 5–10% air voids, while uniformly graded sands and gravels may have

10–15% air voids. Clay soils often have in situ dry densities of 50–60% of their theoretical maximum values.

The moisture–density test is designed to assist in the field compaction of earthworks. It assumes that the shear strength of a soil increases with increasing dry density, and (controversially) that the compactive efforts used in the laboratory tests are similar in effect to those achieved by particular construction equipment in the field. Hence, specifications often specify the dry density required for earthworks as a percentage of that achieved in the laboratory when compacted at the optimum moisture content for maximum dry density for the specified compaction effort: for example, soils for earthworks are often required to be compacted to >95% of maximum dry density using the standard Proctor test and with <10% air voids.

Generally, the optimum moisture contents derived from the Proctor tests on a given soil are a useful guide to the moisture content range that is suitable for the compaction of that soil in the field. For UK climatic conditions, well-graded or uniformly graded soils are normally suitable for compaction if their in situ moisture contents are not more than 0.5–1.5% above the optimum moisture content for the maximum dry density as determined from the standard Proctor test. In the case of cohesive soils, the upper in situ moisture content criterion is typically 1.2 times the plastic limit, for compaction in the field.

4.7.4.2   Moisture condition value test

Figure 4.17 clearly demonstrates that for a given soil there is a powerful relationship between the compaction effort, moisture content and dry density. The role of the water is to act as a lubricant, and, at a fixed compaction effort (i.e. a fixed number of rammer blows), there is no increase in the dry (or bulk) density once the optimum moisture content is present in the soil. Table 4.9 and Figure 4.17 show that, as the compaction effort is increased, the optimum moisture content decreases and the maximum dry density increases. Similarly, it can be demonstrated that, if the moisture content of a given soil is held constant, there is a unique compaction effort (i.e. an 'optimum' number of rammer blows) that will produce a maximum dry density, and increasing the number of rammer blows beyond this optimum will not result in any density increase; also, the higher the moisture content of the soil the lower the compaction effort required to achieve the maximum dry density.

The moisture condition value (MCV) test (Parsons and Toombes, 1987) involves testing a soil at a fixed moisture content, and, by incrementally increasing the number of blows of a rammer, determining the compaction effort beyond which no further increase in density occurs. The apparatus (Figure 4.18(a)) used to determine the MCV for a soil sample comprises a 100 mm diameter by 200 mm high mould that sits on a detachable heavy base, a free-falling 97 mm diameter rammer with a mass of 7 kg, and an automatic release mechanism that is adjustable to maintain a constant drop height of 250 mm onto the surface of the sample. The base has a mass of 31 kg, to allow the apparatus to be used in the field on surfaces of varying stiffness (including soft soil). When testing, 1.5 kg of soil passing the 20 mm sieve is placed loosely in the mould, covered with a 99 mm

Figure 4.18 (a) MCV test apparatus and (b) graph of MCV test results for a heavy clay (see also Table 4.10)

(a)

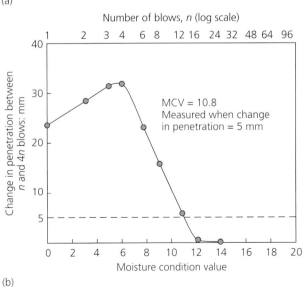

(b)

Table 4.10 Determining the MCV for a heavy clay

Number of blows of rammer, n	Protrusion of rammer above the mould: mm	Change in penetration between n and $4n$ blows of rammer: mm
1	106.3	23.5
2	96.4	29.2
3	89.0	31.8
4	83.8	31.7
6	74.7	23.1
8	67.2	15.6
12	57.2	5.6
16	52.1	0.5
24	51.6	0
32	51.6	
48	51.6	

diameter lightweight rigid fibre disc, and the rammer is then gently lowered to the surface of the disc and allowed to penetrate into the mould, under its own weight, until it comes to rest: an automatic release mechanism is then adjusted to ensure a drop of 250 mm, using the Vernier scales attached to the guide rods and the rammer. The rammer is then raised until it is released by the automatic catch, dropped, and its penetration into the mould (or the length protruding above the mould) is measured using the scale attached to the rammer, and noted (Table 4.10). This process is carried out repeatedly, with the release mechanism being adjusted to ensure a constant drop of 250 mm, with penetration (or protrusion) measurements being taken after selected accumulated numbers of blows, until there is no additional rammer penetration (i.e. no additional densification). A sample of soil is then taken from the mould, and its moisture content determined.

The penetration of the rammer at any given number of blows, n, is then compared with that for $4n$ blows, and the difference, which is a measure of the density change, is calculated and noted. If the change (on a linear scale) is plotted against the initial cumulative number of blows, n (on a logarithmic scale), a curve (Figure 4.18(b)) is obtained in each instance. Note that the steepest possible straight line is drawn through the points immediately before or passing though the 5 mm 'change in penetration' value (which is arbitrarily selected as the point beyond which no significant change in density occurs). The MCV for the sample is then determined as

$$MCV = 10 \log B \tag{4.37}$$

where B is the number of blows at which the change in penetration equals 5 mm, as read from the straight line. For the data in Figure 4.18(b) and Table 4.10, the MCV = 10.8, and the measured moisture content is 31.5%.

The test, which takes about 30 min to complete, is repeated on a number of samples of the same soil, each at a different moisture content, and a different MCV is obtained for each sample. When all the MCVs are plotted against their respective moisture contents, a straight-line relationship is established for the soil that can then be used as a calibration line for subsequent determination of the moisture content. Thus, if the moisture content in the field is required quickly at any time, its MCV can be determined easily using the portable test apparatus, and the relevant moisture content can then be read from the calibration chart.

As a general construction guide, it has been found that for a cohesive soil to be suitable for compaction, its MCV should fall between about 8 and 13 (Driscoll *et al.*, 2009). At values above and below this range, the soil will tend to be too dry or too wet, respectively, to achieve adequate compaction.

4.8.   Soil classification

In contrast to soil identification and description, soil classification is about placing a soil within a limited number of soil groups on the basis of its grading and plasticity characteristics as determined from tests that are carried out on disturbed soil samples (usually in the laboratory). In this case, the characteristics measured are independent of the condition in which the soil occurs, and pay no regard to the influence of the structure, including fabric, of the soil mass.

The soil classification system proposed by Arthur Casagrande (1947) is the basis for the Unified Soil Classification System (ASTM, 2011), which, internationally, is now the most widely used system. Relying on Casagrande's work, the first British system was developed in 1950 as part of the Civil Engineering Code of Practice No. 1 (Child, 1986): this evolved (BSI, 1957, 1981) into the British Soil Classification System described in BS 5930 (BSI, 2010). The current British standard that defines the principles to be used for soil classification is **BS EN ISO 14688-2:2004** (BSI, 2004a): this now partially supersedes BS 5930, albeit this (updated) standard is still broadly used in the UK.

The current preferred approach (BSI, 2010) is that soils should be classified into soil groups on the basis of the grading and plasticity of disturbed samples: it also considers the organic content and genesis of the soil where appropriate. The laboratory-based soil classification of a soil differs from its field description, which records the undisturbed character of a soil. The classification can provide additional useful information as to how the disturbed soil behaves when used as a construction material under various conditions of moisture content.

Using particle size distribution, the particle fractions are used to distinguish the mechanical behaviour of a soil: Table 4.1 shows the terms and symbols used for each soil fraction and its sub-fractions, together with the corresponding range of particle sizes. The plasticity of a soil enables it to be classified as non-plastic or of low, intermediate, high, very high or extremely high plasticity (Figure 4.19). Soils with organic contents are classified as shown in Table 4.11.

While particle size distribution, plasticity and organic content are used to characterise the nature of the soil, other constituents related to the state of the soil in the field are used

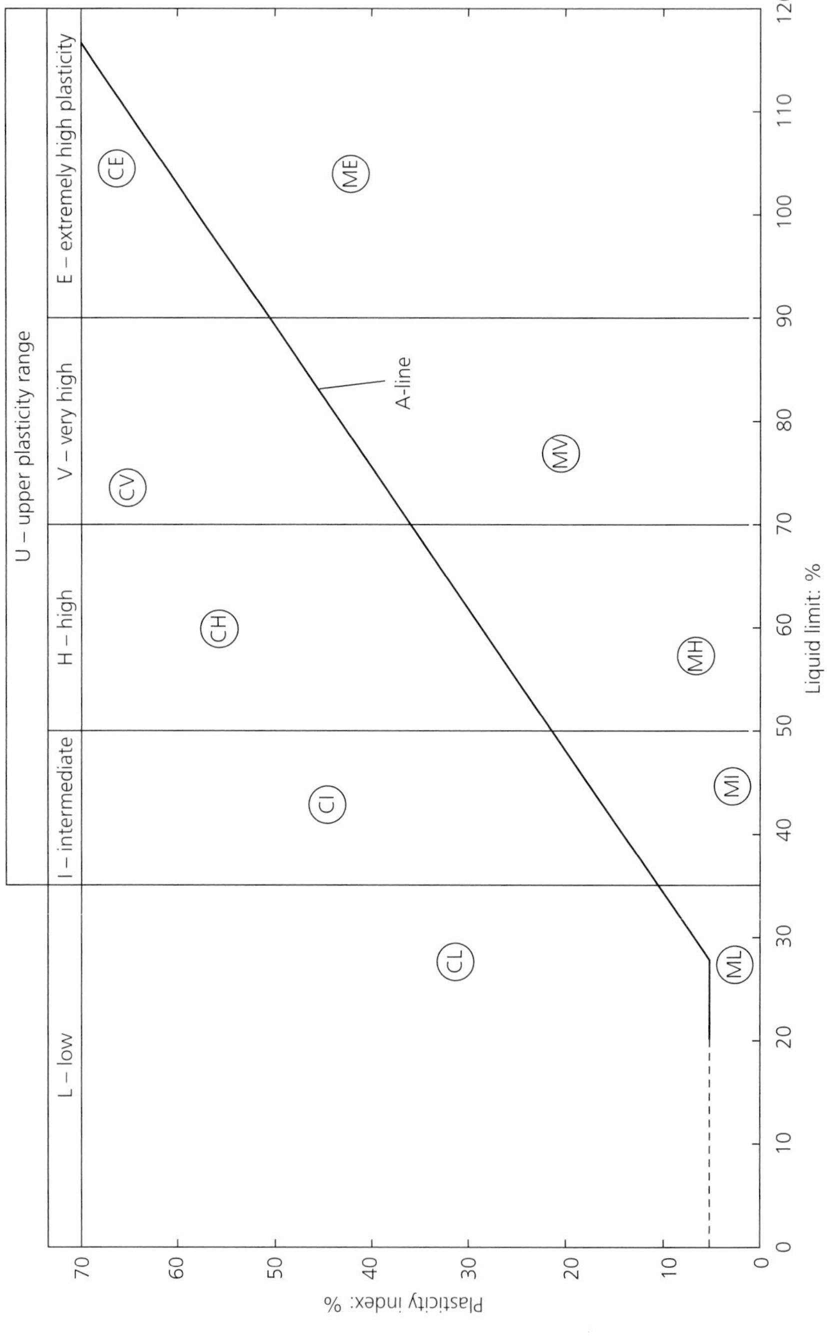

Figure 4.19 Plasticity chart used for the classification of the fine component of soils (BSI, 2010)

Table 4.11 Classification of soils with organic constituents (BSI, 2004a)

Soil	Organic content (<2 mm): % of dry mass
Low organic	2–6
Medium organic	6–20
High organic	>20

Table 4.12 State of soil indicators used in soil classification (BSI, 2004a)

State of soil indicator	Descriptor	Range of values
Density index, I_D (%), for sands and gravels	Very loose	0–15
	Loose	15–35
	Medium dense	35–65
	Dense	65–85
	Very dense	85–100
Undrained shear strength, c_u (kPa), for silts and clays	Extremely low	<10
	Very low	10–20
	Low	20–40
	Medium	40–75
	High	75–150
	Very high	150–300
	Extremely high	>300
Consistency Index, I_C, for silts and clays	Very soft	<0.25
	Soft	0.25–0.50
	Firm	0.50–0.75
	Stiff	0.75–1.00
	Very stiff	>1.00

to refine the classification. BS EN ISO 14688-2:2004 (BSI, 2004a) provides a quantitative basis (Table 4.12) to describe the state of a soil. For sands and gravel soils, it recommends the use of a density index (I_D), typically derived in the field from the standard penetration test or cone penetration test. In the case of silts and clays, the undrained shear strength (c_u) and the consistency index (I_c), derived from laboratory and field testing, can be used. Illustrative examples of soil descriptions are 'Very loose to loose, reddish brown, very silty, sub-angular sand' and 'Firm to stiff, dark brown, slightly gravelly clay'.

4.9. Frost action in soils

Frost action can be defined as any action resulting from freezing and thawing that alters the moisture content, porosity or structure of a soil, or alters its capacity to support loads. The two aspects of frost action that are of major interest are frost heaving and thawing.

4.9.1   Frost heaving

Frost heaving is typically used to describe the process whereby a portion of a frost-susceptible soil, and the road pavement above it, are raised as a result of ice formation in the soil (Figure 4.20(a)). If the heaving occurs non-uniformly, pavement cracking can occur as a result of differential heaving pressures. Differential heaving is likely at sites where there are non-uniform soil conditions (e.g. where subgrades abruptly change from clean sands to silty materials) or where subsurface drainage conditions vary, as can occur between the edges and mid-portion of a road pavement. The extent of the frost depth penetration and any resulting frost-related damage are related to the accumulated duration of below-freezing temperatures during a winter. The frost penetration depth is also affected by soil type: for example, it is less with clays and greater with sandy gravels.

While the volume increase of soil water upon freezing ($\sim$9%) obviously contributes to frost heaving (2.5–5% increase in the soil volume, depending on its void ratio), its major cause is the formation of ice lenses. For ice lenses to develop three conditions must be met: (a) sufficiently cold temperatures to freeze some of the soil water, (b) a supply of water that is available to the freezing zone and (c) a frost-susceptible soil. Ice lensing does not occur when any one of these three conditions is missing. Figure 4.20(b) illustrates the general manner in which layers of segregated ice are formed in a soil. The ice lenses, which normally grow parallel to the ground surface and perpendicular to the direction of the heat flow, can vary from being of hairline thickness to many centimetres thick.

The frost heave mechanism is most simply explained as follows. When the freezing isotherm penetrates into a fine-grained soil, some of the water in the soil pores becomes frozen. As a consequence, the unfrozen moisture content decreases, and the suction associated with its surface tension and adsorption forces rises rapidly and causes moisture from the water source in the unfrozen soil to move to the freezing zone and form ice lenses; this process continues until a suction value is reached that inhibits further freezing at that temperature, and a new state of equilibrium is established. As the temperature drops or the period of freezing is prolonged and the freezing isotherm penetrates further into the soil, the equilibrium is disturbed, and the suction process initiates another cycle of ice lens formation at a different depth in the soil; and so on. The accumulated thickness of the ice lenses then accounts for the frost heave.

Problems from frost heave can be expected when the water table in a frost-susceptible soil is close to the ground surface and just below the freezing zone. The rule of thumb in the UK is that a potential water source for ice lens formation is present when the highest water table level at any time of the year is within, say, 1–1.5 m of the formation or of the bottom of any frost-susceptible material in the road pavement. When the depth to the water source is more than 3 m throughout the year, substantial ice segregation is usually unlikely; however, this does not mean that no ice lenses will be formed but, rather, that those which do occur are likely to be within tolerable limits in most pavements.

Figure 4.20 (a) Differential frost heave where an added lane on the left was constructed with a different material than the original pavement and drainage was blocked, and (b) the mechanism of frost heave in a pavement

(a)

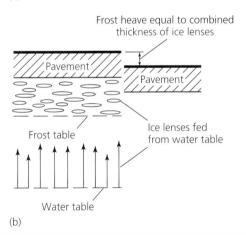

(b)

Investigators of frost-susceptible soils have long found that certain soils are more vulnerable to frost action than others and that particle size distribution provides a practical guide to the identification of soils that are prone to frost heave. Early research (Croney, 1949) identified soils with between 20% and 70% of their particles finer than 0.02 mm as being prone to frost heave; later work (Jones, 1980) concluded that cohesive soils can be regarded as non-frost susceptible when the plasticity index, I_p, is >15% for a well-drained soil, or >20% when the soil is poorly drained (i.e. when the groundwater surface is within about 0.6 m of the bottom of the pavement). The pore sizes in these cohesive soils are generally too small to allow significant migration of water to the freezing front during the relatively short periods of freezing normally associated with UK winters. However, if current global concerns about climate change are fulfilled and associated higher frequencies of severe weather events occur, frost heave could become a greater problem for UK roads in future decades.

Countries with colder climates, where frost penetrates a number of metres (compared with, typically, less than 0.5 m in the Scottish highlands and <200 mm in west and south-west England) each year, have attempted to develop more precise grading criteria so as to minimise the expensive removal of frost-susceptible soils during road construction. In Norway, for example, soils are categorised (Knutson, 1993) into four groups, ranging from T1 (no frost susceptibility) to T4 (high frost susceptibility). T4 soils are typically coarse silts with less than 40% of their particles smaller than 0.002 mm, more than 12% smaller than 0.02 mm and more than 50% smaller than 0.2 mm: these criteria eliminate from concern the clays and clean sands and gravels that are non-frost susceptible.

In Ontario, Canada (MTO, 2012b), frost susceptibility evaluation is based on the percentage of a soil in the 5–75 μm range (Table 4.13). A soil with >55% in this size range is categorised as having high frost susceptibility. A soil with <40% in this size range is considered low risk. When plotted on a grain-size chart, the Canadian and Norwegian frost susceptibility criteria are generally consistent with each other.

Most granular soils are regarded as non-frost susceptible, as the mass of material passing the 75 μm sieve is quite small. However, all chalk soils (and crushed chalk and soft limestone) are considered susceptible to frost heave, and their usage should be avoided in road construction in locales subject to freezing temperatures.

Table 4.13 Frost susceptibility criteria based on the particle size distribution (MTO, 2012b)

Grain size in the range 5–75 μm: %	Susceptibility to frost heaving
0–40	Low
40–55	Moderate
55–100	High

In most UK applications, an evaluation of the particle size distribution and site ground-water and drainage conditions will be sufficient to establish the risks for problematic frost heave. In some high-risk situations, however, testing may be warranted according to a procedure described in Part 5 of BS 1377 (BSI, 1990).

4.9.2 Thawing impacts on a road pavement

When thawing begins, the ice melts primarily from the top down. If the thawing occurs at a faster rate than the melt water can escape into underdrains or into more pervious layers of the pavement system, or be reabsorbed into adjacent drier areas, then heavy traffic loads will result in the generation of excess porewater pressures that decrease the load-carrying capacity of the pavement and result in localised pavement damage (Figure 4.21).

If thawing does not occur at the same rate over all parts of the pavement, non-uniform subsidence of the heaved surface will result. Differential thaw is most commonly associated with (a) different thermal properties of adjacent pavement sections, caused by non-uniformity of subsoil strata and soil conditions, (b) non-uniform exposure to the sun's rays and differing angles of incidence, (c) shading of portions of the pavement due to deep cuts, trees, overpasses or buildings, (d) proximity to the ground surface and surface drainage, and/or (e) differing pavement colours.

Figure 4.21 Pavement damage resulting from local weakening of a pavement following frost heave and subsequent thawing

Another undesirable consequence of thawing is the subsidence of coarse open-graded pavement materials into thaw-weakened silt and clay subgrade soils, as the latter flow up into the large pore spaces in the coarse material. There have been many instances where a pavement has become frost susceptible as a result of the upward impregnation of a silt subgrade into a previously non-frost susceptible base during the thawing process. If the melt water content is high enough and the traffic conditions are sufficiently heavy to cause a reworking of the subsoil, a free-flowing mud may be formed that is forced out at the edges of, or breaks in, the pavement: this action, termed frost boil, is most commonly associated with rigid pavements. Initial thaw damage to flexible pavements is usually reflected in the form of a close network of cracks accompanied by distortion of the surface course.

Acknowledgement

Permission to reproduce extracts from British standards is granted by the BSI. British standards can be obtained in PDF or hard-copy formats from the BSI online shop (www.bsigroup.com/Shop) or by contacting BSI Customer Services for hard copies only (tel. +44 (0)20 8996 9001, email cservices@bsigroup.com).

REFERENCES

ASTM (American Society for Testing and Materials) (2011) ASTM D2487-11: Standard practice for classification of soils for engineering purposes (Unified Soil Classification System). ASTM, Conshohocken, PA, USA.

Brown SF and Pell PS (1972) A fundamental structural design procedure for flexible pavements. *Proceedings of the 3rd International Conference on the Structural Design of Asphalt Pavements, London, UK*, vol 1, pp. 369–381.

BSI (British Standards Institution) (1957) CP2001: Standard code of practice: Site investigation. BSI, London, UK.

BSI (1981) BS 5930:1981: Code of practice for site investigations. BSI, London, UK.

BSI (1990) BS 1377:1990: Methods of test for soils for civil engineering purposes, Parts 1–8. BSI, London, UK.

BSI (2002) EN ISO 14688-1:2002: Identification and classification of soil, Part 1: Identification and description. BSI, London, UK.

BSI (2004a) EN ISO 14688-2:2004: Identification and classification of soil, Part 2: Principles for a classification. BSI, London, UK.

BSI (2004b) DD CEN ISO/TS 17892:2004: Laboratory testing of soil, Parts 1–12. BSI, London, UK.

BSI (2007a) BS EN 1997-2:2007 (incorporating corrigendum June 2010): Eurocode 7. Geotechnical design. BSI, London, UK.

BSI (2007b) NA to BS EN 1997-2:2007: UK National Annex to Eurocode 7. Geotechnical design. BSI, London, UK.

BSI (2010) BS 5930:1999 + A2:2010: Code of practice for site investigations. BSI, London, UK.

Casagrande A (1947) Classification and identification of soil. *Transactions, ASCE* **113**: 901–930.

Child GH (1986) Soil descriptions – *quo vadis*? *Engineering Geology Special Publications* **2**: 73–82.

Croney D (1949) *Some Cases of Frost Damage to Road, DSIR Road Note No. 8*. Road Research Laboratory, Crowthorne, UK.

Croney D and Croney P (1998) *Design and Performance of Road Pavements*, 3rd edn. McGraw-Hill, London, UK.

Driscoll I, Kelly R, Padina S and Foweraker M (2009) Compaction and compactability assessment of difficult soils – an alternative approach. *Australian Geomechanics Journal* **44(3)**: 53–60.

HA (Highways Agency), Transport Scotland, Welsh Government and Department for Regional Development Northern Ireland (1995) Section 1: earthworks. Part 1: design and preparation of contract documents. With Amendment No. 1. In *Design Manual for Roads and Bridges*, vol. 4. *Geotechnics and Drainage*. Stationery Office, London, UK, HA 44/91.

Heukelom W and Klomp AJG (1962) Dynamic testing as a means of controlling pavements during and after construction. *Proceedings of the International Conference on the Structural Design of Asphalt Pavements, Ann Arbor, MI, USA*, pp. 667–685.

Hyde AFL (1974) *Repeated Load Triaxial Testing of Soils*. PhD thesis, University of Nottingham, Nottingham, UK.

ICE (Institution of Civil Engineers) (1976) *Manual of Applied Geology for Engineers*. ICE, London, UK.

Jones RH (1980) Frost heave of roads. *Quarterly Journal of Engineering Geology* **13(2)**: 77–86.

Knutson A (1993) *Frost Action in Soils*. Norwegian Road Research Laboratory, Oslo, Norway.

Millar CE, Turk, LM and Foth, HD (1962) *Fundamentals of Soil Science*. Wiley, New York, NY, USA.

MTO (Ministry of Transportation, Ontario) (2012a) *Ontario's Default Parameters for AASHTOWare Pavement ME Design. Interim Report*. MTO, Downsview, Ontario, Canada.

MTO (2012b) *Pavement Design and Rehabilitation Manual*, 2nd edn. MTO, Downsview, Ontario, Canada.

Norbury D (2010) *Soil and Rock Description in Engineering Practice*. Whittles, Caithness, UK.

Parsons AW and Toombs AF (1987) *The Precision of the Moisture Condition Test: Research Report 90*. Transport and Road Research Laboratory, Crowthorne, UK.

Skempton AW (1950) Soil mechanics in relation to geology. *Proceedings of the Yorkshire Geological Society* **29(1)**: 33–62.

Terzaghi K (1925) Principles of soil mechanics II – compressive strength of clay. *Engineering News Record* **95(20)**: 799.

Highways
ISBN 978-0-7277-5993-1

ICE Publishing: All rights reserved
http://dx.doi.org/10.1680/h5e.59931.135

Chapter 5
Earthworks for roadworks

Paul Nowak   Chief Geotechnical Engineer, Ground Engineering, Atkins Ltd, UK

5.1.   Basic considerations

Earthworks for roadworks mainly involve the preparation and, as appropriate, movement of soil and rock to allow for the safe and economical construction of road pavements on stable foundations (including on embankments) and in cuttings. Fundamental requirements of the completed earthworks are to provide a stable construction platform for the placement and compaction of the overlying pavement layers and to minimise any settlement of the superimposed pavement over its design life.

With respect to a cutting, an additional basic requirement is the stability of its side slopes. If side slopes fail within the design life of the earthworks, the result will be the deposition of debris at the toe of the cutting and/or the failure of material at the base of the cutting. This will likely interrupt the safe movement of traffic and cause damage to the road surface and to verge services (e.g. to drainage and communication cables). With respect to embankments, side slope failures can result in the undercutting of the pavement asset that the embankment supports. Additionally, the gradual settlement of an embankment over its design life, while not necessarily contributing to catastrophic failure, can result in unacceptable differential movement leading to serviceability failure and temporary loss of service.

Some settlement can be tolerated in almost all embankments if it does not result in pavement fracture or excessive roughness. The total amount that can be tolerated after pavement construction is debatable, albeit experience would suggest (NCHRP, 1971) that differential settlements in local areas of >25 mm over a 10 m length would be a cause of much concern whereas 0.15–0.3 m can be tolerated on long embankments if the settlement variations are uniformly distributed along the embankment length.

The most critical area of differential settlement is usually where earthworks adjoin structures. The abutment of a structure is designed to minimise settlement of the overall structure, and this is likely to be less than the settlement of the adjoining embankment. Over the long term, this may lead to differential settlement between the structure and adjacent embankment, with the eventual cracking of the road pavement and the need for appropriate replacement: in the short term, it can lead to a reduction in ride quality.

In order to design economic and sustainable earthworks, the following key considerations need to be addressed:

- minimise soil and rock waste by balancing the volume excavated from cuttings with the volume of material needed for embankment and landscape fill
- maximise the reuse of excavated material so as to minimise the need to import expensive borrow material from offsite
- create opportunities for the optimum use of on-site material
- utilise locally available recycled and waste materials, where economically appropriate
- improve poor-quality on-site materials by treatment (e.g. by cement or lime modification or using geosynthetics).

5.2. Earthworks alignments and quantities

The selection of the horizontal and vertical alignments for a new road pavement normally involves detailed consideration of a number of alternative alignments within a route corridor that allow for the constraints of the site and the application of design standards appropriate to the type of road being built. The end objective is to build a sound road with a 'flowing' alignment that is attractive to and safe for motorists to use, is protective of the environment in which it is located, and is economical to construct. From the earthworks viewpoint, the selection of the preferred alignments should normally result, where practicable, in a route that provides an 'optimum' balance in cut and fill volumes that is based on the topography and the likely nature of the materials to be won from excavations.

Much of the work involved in comparing alternative alignments involves repetitive manipulations of detailed levelling and earthworks data that, in modern practice, is now carried out with the aid of sophisticated computer programs (e.g. Bentley MXROAD and Autodesk AutoCAD Civil 3D) that analyse direct and indirect topographic survey information to create a three-dimensional model in CAD format.

5.2.1 Determining earthworks quantities

The determination of earthworks quantities for each alignment evaluated inside a route corridor is based on cross-section data interpolated from digital ground control models and supplemented, as appropriate, by information gathered in the field. These cross-section data, which are usually established at 20 m chainage intervals and where major ground irregularities and changes occur along the selected alignment, normally indicate the extent of the excavation (i.e. cut) from cuttings and fill for embankments, while taking into account the likely earthwork side slopes.

Ground-level information for use in the quantity modelling of earthworks may be initially determined by carrying out a topographic survey along the route corridor or with the aid of remote sensing techniques such as LiDAR (i.e. Light Detecting And Ranging – a system that measures distance by illuminating a target with a laser and analysing the reflected light) or a GIS (geographic information system). Using both LiDAR and GIS information, a model is established from which earthworks volumes can be calculated geometrically by a computer program, and cross-section drawings can be automatically produced for use in engineering design.

With the aid of geographic positioning systems (GPS) survey data and GPS equipment mounted on earthworks plant, the latitude, longitude, and elevation coordinates produced from the computer model can also be uploaded directly into field instrumentation to aid in the setting out and formation of final earthworks slopes. This method of setting out has largely replaced the use of traditional levelling equipment and 'batter boards' at regular chainage intervals to set out the final cut slopes and embankment profiles.

5.2.2 Balancing earthworks quantities: bulking, shrinkage and haul considerations

The selection of the optimum horizontal and vertical alignment for a road pavement should, ideally, result in the volume of material excavated within the limits of the road scheme being equal to, and within economic haul distances of, the fill required in embankments so that there is no need to waste good on-site excavation material or to import expensive borrow material from off site. In practice, however, the ideal is not always achieved, and suitable excavated material may have to be discarded because it is uneconomical to do otherwise, while material that is unsuitable for use in embankments may have to be conditioned for use. Typically, this type of decision-making arises when earthworks areas are separated by a linear feature such as a road, railway or river that is not scheduled to be bridged until later in the construction programme, after the formation of the adjacent earthworks. Determining the earth-moving costs in such situations, in order to decide what to do, requires a clear understanding of some common earthworks terms, namely bulking, shrinkage and haul.

When calculating cut and fill volumes for a road scheme, and their haul distances along the route corridor, consideration needs to be given to the fact that a unit volume of material excavated from a cutting will occupy a greater volume in its loose state during transport than it did in its natural state prior to excavation, and that the compacted volume of the same material in embankments is also likely to be less than that in its natural state. Table 5.1 shows typical bulking (also called swelling) coefficients applicable, after excavation, to a unit volume of natural soil or rock for its 'loose' transport in a haulage vehicle to an embankment: this bulking must be taken into account when assessing the amount (and cost) of transport required. The term shrinkage is applied to the situation whereby a unit volume of undisturbed natural soil will (except in the case of most rockfill) normally occupy less space when compacted in the embankment (i.e. very few soils are compacted back to their original volumes). Table 5.1 also lists some measured bulking and shrinkage coefficients.

Table 5.1 Some bulking and shrinkage coefficients measured for common materials in their natural state in earthworks

Material	Loose volume: m^3	Compacted volume: m^3
Rockfill	1.75	1.4
Sand and gravel	1.2	0.9
Silt	1.3	0.85
Clay	1.5	0.85

Table 5.1 shows that if 100 m^3 of rock is excavated, it may bulk up to 175 m^3 for transport purposes, and still occupy 140 m^3 even after careful compaction in the embankment. On the other hand, if 100 m^3 of the clay soil is excavated, it may bulk up to 150 m^3 for transport purposes, but will shrink back and occupy 85 m^3 after careful compaction in the embankment.

The unquestioned application of bulking and shrinkage coefficients such as those in Table 5.1 has to be treated with considerable care, however, as earthworks specifications do not always take into account the variable geological conditions under which cut material was originally deposited. For example, with over-consolidated cohesive soils and dense granular materials, the volume placed in embankments can be up to 5% less than the volume won from cuttings, whereas low-to-medium-dense granular soils can often achieve the same compacted embankment density as in the natural state. For rock-fill in embankments, the volume increase can be up to 40% – with the notable exception of unweathered chalk, which can show up to 5% shrinkage in volume from cut to fill (i.e. the chalk shrinkage is due to the natural material having a high voids ratio and open jointing, and these defects are overtaken by the compaction process). Thus, rather than relying on the use of standardised bulking and shrinkage – for no such 'ideal' list exists for soils and rocks throughout the UK – it is best to determine reliable local bulking and shrinkage factors and use these for the detailed volume and cost calculations relating to the project in question.

In earthworks contracts, the contractor may be paid a specified price for excavating, hauling and dumping material, provided that the haulage does not exceed a certain distance: this distance is termed the free haul. The free haul will vary according to the project (e.g. in the UK it might be <150 m for small road schemes and >350 m for motorway projects), and within this free-haul distance the contractor will be paid a fixed amount per cubic metre, irrespective of the actual distance through which the excavated material is moved. When, however, the haulage distance is greater than the free haul, the contractor may be paid at a higher rate for moving the excavated material over the extra distance in excess of the free haul: this extra distance is termed the overhaul.

5.3. Earthworks specifications

A specification for earthworks should adequately describe the design requirements, be easily understood by all parties to the contract, be practicable and capable of enforcement and measurement, and not be unnecessarily costly or time-consuming in its application (BSI, 2009). The specification, which should be developed in collaboration between the client, the designer and the contractor in order to optimise the use of excavated materials for use as engineering fill, and to enable the definition of the properties of material that may have to be imported, should be formulated to address the provision of

■ a well-constructed excavation or formation as a reliable platform for the subsequent construction of engineering fill
■ a sound and durable fill, capable of being handled, placed and compacted to a standard that is appropriate to the requirements of the earthworks as a whole and any structures that they may support

- a means of compaction to enable desired serviceability and performance criteria to be met.

Excellent discussions regarding the various types of earthworks specifications in use at this time are available in the literature (BSI, 2009; Parsons, 1992). Highways England (formerly the Highways Agency) has developed the *Specification for Highway Works* that is used for earthworks in the UK (HA *et al.*, 2014). Other specifications have been developed in many countries (e.g. France, Spain, Russia and the USA) to suit the particular climatic conditions and types of earthworks materials experienced in these locales. The European Committee for Standardization (CEN) is currently endeavouring to develop standard earthworks specifications across Europe through its technical committee CEN/TC 396 (Gilbert and Kidd, 2012).

Three fundamental types of earthworks compaction specifications have been developed for the construction of embankments or engineering fill below foundations, namely method, end-product and performance specifications.

5.3.1   Method specifications
A method specification provides the contractor with a recipe as to how the earthworks should be carried out.

The Highways England's *Specification for Highway Works* is a method specification that was developed from research work carried out by the Transport Research Laboratory in the 1960s and 1970s on UK materials. In Table 6/1 of this specification, the materials that may be used as engineering fill are classified on the basis of their physical properties (e.g. class 1 granular fill, class 2 cohesive fill and class 6 specialist fill), and details regarding the methods of compaction to be used with each material are given in its Table 6/4. The client issuing the specification is required to supply the contractor with information regarding the nature of the fill material that would be acceptable for use in the scheme. Typically, this would include data about proposed fill sources and acceptability levels that would enable the fill materials to be classified, the grading and optimum moisture content for granular soils, and the grading, moisture condition value and/or plastic limit for cohesive soils, and associated requirements relating to the compaction of the fill material used.

The compaction criteria in Table 5.2 specify the types of compaction plant to be used with particular types of soil, the minimum number of passes to be made by each type of roller and the maximum thickness of the compacted layer. These criteria have predominantly been derived from the monitoring of UK materials that were compacted to >95% of the maximum dry density using the standard Proctor test (see Chapter 4) and with <10% air voids.

A method specification is also used in France (LCPC, 2003), but in this case the classification of a soil is based on the soil type and its in situ moisture condition. Weather conditions are also taken into account in the soil classification process, to reflect the climatic conditions encountered in, for example, the more arid conditions found in the south of France.

Table 5.2 Typical compaction characteristics for natural soils, rocks and artificial materials used in earthworks construction (BSI, 2009)

Material	Major divisions	Subgroups	Suitable type of compaction plant	Minimum number of passes for satisfactory compaction	Maximum thickness of the compacted layer: mm	Remarks
Rock-like materials	Natural rocks	All rock fill (except chalk)	Heavy vibratory roller not less than 180 kg per 100 mm of roll. Grid roller not less than 800 kg per 100 mm of roll. Self-propelled rollers	4–12	500–1500[a]	If well graded or easily broken down, then this can be classified as a coarse-grained soil for the purpose of compaction. The maximum diameter of rock fragments should not exceed two-thirds of the layer thickness
		Chalk	See remarks	3	500	This material can be very sensitive to the weight and operation of compacting and spreading plant. Less compactive effort is needed than with other rocks
Artificial	Waste material	Burnt and unburnt colliery shale	Vibratory roller. Smooth-wheeled roller. Self-propelled tamping roller	4–12[a]	300	–
		Pulverised fuel ash	Vibratory roller. Smooth-wheeled roller. Self-propelled tamping roller. Pneumatic-tyred roller	4–12[a]	300	Includes lagoon and furnace bottom ash

		Broken concrete, bricks, steelworks slag, etc.	Heavy vibratory roller Smooth-wheeled roller Self-propelled tamping roller	4–12[a]	300	Non-processed supplied brick slag should be used with caution
Coarse soils	Gravels and gravelly soils	Well-graded gravel and gravel–sand mixture; little or no fines. Well-graded gravel–sand mixtures with excellent clay binder. Uniform gravel. Uniform gravel; little or no fines. Gravel with excess fines, silty gravel, clayey gravel, poorly graded gravel–sand–clay mixtures	Grid roller over 540 kg per 100 mm of roll Pneumatic-tyred roller over 2000 kg per wheel Vibratory plate compactor over 1100 kg/m² of baseplate Smooth-wheeled roller Vibratory roller Vibro-rammer Self-propelled tamping roller	3–12[a]	75–275[a]	—
	Sands and sandy soils	Well-graded sands and gravelly sands; little or no fines. Well-graded sands with excellent clay binder	As above	As above	As above	—
	Uniform sands and gravels	Uniform gravels; little or no fines. Uniform sands; little or no fines. Sands with fines, silty sands, clayey sands, poorly graded sand–clay mixtures	Smooth-wheeled roller below 500 kg per 100 mm of roll Grid roller below 540 kg per 100 mm of roll Pneumatic-tyred roller below 1500 kg per wheel Vibrating roller Vibrating plate compactor Vibro-tamper	3–16[a]	75–300[a]	—

141

Table 5.2 Continued

Material	Major divisions	Subgroups	Suitable type of compaction plant	Minimum number of passes for satisfactory compaction	Maximum thickness of the compacted layer: mm	Remarks
Fine soils	Soils having low plasticity	Silts (inorganic) and very fine sand, rock flour silty or clayey fine sands with low plasticity. Clayey silts (inorganic). Organic silts of low plasticity	Tamping (sheepsfoot) roller Smooth-wheeled roller Pneumatic-tyred roller Vibratory roller over 70 kg per 100 mm of roll Vibratory plate compactor over 1400 kg/m² of baseplate Vibro-tamper Power rammer	4–8[a]	100–450[a]	If the moisture content is low, it may be preferable to use a vibratory roller. Tamping (sheepsfoot) rollers are best suited to soils at a moisture content below their plastic limit
	Soils having medium plasticity	Silty and sandy clays (inorganic) of medium plasticity. Clays (inorganic) of medium plasticity. Organic clays of medium plasticity	As above	As above	As above	– Organic clays are generally unsuitable for earthworks
	Soils having high plasticity	Micaceous or diatomaceous fine sandy and silty soils, plastic silts. Clay (inorganic) of high plasticity, 'fat' clays. Organic clays of high plasticity	As above	As above	As above	Should only be used when circumstances are favourable. Should not be used for earthworks

The information in this table should be taken as a general guide only; when material performance cannot be predicted, it may be established by earthworks trials. This table is only applicable to fill placed and compacted in layers; it is not applicable to deep compaction of materials in situ. The compaction of mixed soils should be based on that subgroup requiring most compactive effort
[a] Depending upon the type of plant

The method approach operates on the basis that if the contractor complies with the exact requirements of the specification, and the earthworks are carried out in the scheme exactly as they were carried out in the full-scale trials that resulted in the development of the specification, the results will meet the requirements of the client. In this case, the onus is on the supervising engineer to ensure that the contractor carries out the work as required by the specification; however, if any deficiencies are subsequently determined, or if the specification is faulty in any way, and the engineer has already agreed that the work was carried out according to the specification, then the contractor can claim that the fault was not theirs, and a dispute can arise.

5.3.2   End-product specifications

With an end-product specification, the earthworks material deemed acceptable for use as engineering fill is determined in a similar way to that for a method specification except that the client defines the end-product desired from the completed work. It is left to the contractor to select the method(s) of compaction, including the layer thicknesses, that will achieve the specified outcome. Desired outcomes are typically specified in terms of relative dry density, air void content and/or some form of strength criterion.

The national specifications used in Spain, for example (Anon., 2003), require the compacted density achieved in a fill to be greater than 95% of the maximum dry density achieved at the optimum moisture content using the standard Proctor test. Note that if a target percentage of maximum dry density only is specified, a stable fill may not always result. For example, compaction that is carried out at a moisture content that is drier than the optimum moisture content may be able to achieve the density target, but if the compacted fill has an air voids content that is too high (>10%) the fill may become unstable if saturated. Hence, to ensure the construction of a stable fill, a maximum air voids criterion should also always be quoted (Charles *et al.*, 1998).

With an end-product specification, it is also not uncommon for the stiffness of a compacted layer to be defined in terms of the modulus derived from plate bearing tests. In this case also, reliance should not be placed on stiffness only as again this criterion will not control the air voids content of the compacted material, so that there could be a significant loss of strength if a significant increase in the moisture content of the fill occurs after compaction.

Contractors generally prefer end-product specifications because they are allowed to supply innovative solutions that can save them money while still meeting the client's requirements. However, regular and rigorous on-site testing must be carried out by the client's supervising engineer to ensure that the end-product requirements are met.

5.3.3   Performance specifications

An earthworks performance specification is normally designed in terms of a required long-term serviceability limit state. For example, the specification may say that the maximum settlement in any one area of a compacted fill should not exceed 25 mm, and that the maximum differential settlement within a defined length of embankment should not exceed 0.05% within, say, 5 years after completion of construction. Another example would be to specify that the compacted fill be able to withstand a bearing pressure of 75 kN/m^2 with a settlement not to exceed 15 mm over the working life of the structure to be placed on the fill.

Use of this latter type of performance specification should be treated with care, as, for example, while the use of an allowable bearing pressure is settlement dependent, it is also dependent on the size of the foundation to which the load is applied. Also, zone loading testing may be required at the time of construction to determine the settlement of the compacted fill under a load similar to that expected to be applied under permanent loading conditions.

Generally, a performance specification may be considered onerous from a contractual viewpoint, as it seeks to place the risk for future events on the contractor and may be very difficult to monitor in practice (McGuire and Filz, 2005). It is also unlikely to take account of future events over which the contractor may have no control (e.g. vegetation planted by the client that may produce volume change and result in greater movement within the earthworks).

Nonetheless, developments currently evolving in respect of continuous compaction control have the potential to mitigate many of the performance risks currently borne by the contractor, and are likely to make the use of performance specifications more attractive in the future (Mooney and White, 2007).

5.4.   Placing and compacting fill materials

In both urban and rural areas, but especially in built-up areas, a problem that has to be dealt with before earthworks can commence is the relocation and/or maintenance of public utility services (i.e. above- and below-ground electrical and telecommunication mains and underground pipes carrying sewage, stormwater, gas, oil and water). The problems of relocation can be compounded by the fact that some underground services are the heritage of bygone days, and information regarding their exact locations is inaccurate or no longer in existence. Also, any new services that lie within the right-of-way of the new scheme should be installed prior to construction instead of after, and any services (whether new or old) that are located within the scheme boundaries should be protected during construction: if they are not, and the services are subsequently interrupted by earthworks, there will normally be significant financial and political reverberations.

Prior to the placement and compaction of engineered fill for embankment construction, some preparation of the formation onto which the fill will be placed will normally be required. This formation preparation will typically include (a) the removal of local irregularities and obstructions so as to provide a level surface, (b) the grubbing up of trees, bushes and hedges, (c) the cleaning out and, possibly, diversion of minor water-courses, and the removal of land drainage, (d) the breaking up and removal of areas of hard standing, (e) the excavation and replacement of local 'soft' soils and, if appropriate, (f) the remediation of any contaminated land. Additionally, if the fill material is to is be placed on a sloping ground surface, benching may be required to key the new fill material into the original sloping ground surface: experience has shown that the steepest gradient upon which a fill can be placed without benching is generally between 12.5% and 20%.

Where embankment material is to be constructed over standing water, it may be possible to effect dewatering measures so that the fill material can be laid in the dry. Guidance of dewatering for both surface water and groundwater is available in the literature (Preene *et al.*, 2000).

5.4.1   Constructing embankments

Achieving effective compaction in the construction of an engineered embankment is a function of the compaction plant used, the type of fill material being compacted, and the drainage conditions existing at the site. With respect to the compaction plant, achieving adequate compaction depends upon (a) the size, weight and operating speed of the compaction plant, (b) the type of material used in the fill, (c) the thickness and moisture content of each fill layer, (d) the number of passes carried out by the compaction plant and (e) the drainage conditions of the underlying ground.

The depth of penetration of the compaction effort is related to the contact bearing pressure exerted by the compaction plant. In practical terms, this means that, for a given width of roller, increasing the weight of the compaction equipment results in the application of greater compaction effort and deeper penetration into the fill layer. Also, the slower the speed at which the plant is operated the greater the compaction effort that will be imparted to the fill layer (particularly in the case of vibratory rollers). Note that the contact area between a compaction roller and the underlying fill is restricted to a narrow strip, the size of which depends upon the mass of the roller and the consistency of the fill (i.e. the contact area is exceedingly small when a light roller is used on a dense fill).

The use of thin layers promotes the more uniform penetration of compaction effort throughout the compacted fill (e.g. if a layer is too thick, then its lower part may not be adequately compacted). The method compaction specifications used in the UK (HA *et al.*, 2014) define the maximum compacted layer thickness for different types of compaction plant.

The achievement of the specified dry density and (low) air voids content is strongly dependent upon the water content of the layer at the time of compaction. The higher the natural water content, the more rapidly the desired air voids content can be achieved, while the lower the moisture content, the more energy will be needed, in terms of plant size and number of passes, to achieve the specified outcome. There are construction difficulties associated with trying to compact materials with too high a water content, especially if they are fine-grained fill soils.

For a given type of compaction plant, layer thickness, and water content, increasing the number of roller passes increases the compaction effort transferred to the fill and reduces the air voids in the compacted layer. However, if the layer thickness is too large for the type of plant used, additional passes will not succeed in compacting the base of the layer, with the result that the top may be compacted to 'refusal', while the lower part remains under-compacted. It is important to remember that once the fill material has been compacted to a condition close to zero air voids, then additional passes with compaction plant will not be beneficial and, in the case of fine-grained materials, will cause problems related to over-compaction.

The class of fill has a significant influence on the effectiveness of different types of compaction plant. Figure 5.1 shows the influence of the class of material on the maximum dry densities and optimum moisture contents achievable with the standard Proctor compaction test. As is suggested in Table 5.2, the class of fill essentially determines what compaction equipment is used to achieve the desired target dry densities and air voids content.

5.4.2 Modification of soils for use in fill

When a fill material has a natural moisture content that is outside the range of acceptability for its use in an embankment, it may need wetting or drying before or during placement. This process, which is termed conditioning, can take a number of forms.

If the fill material is too dry to allow for effective compaction, water may have to be added during the placement process via a spray bar attached to the plant employed to spread each layer prior to its compaction. This process needs, however, to be closely controlled to ensure that too much water is not added, as all or part of a fill layer may then achieve a moisture content that is in excess of the range within which it can be compacted effectively.

If the natural moisture content of a non-cohesive soil is too wet for the material to be immediately compacted, stockpiling can be a suitable way of reducing the moisture content of the material prior to its placement, provided that the prevailing weather conditions are appropriate, a sufficiently large area is available where the fill can be spread, and sufficient time is available in the construction programme to allow the material to dry out.

It is not uncommon for a wet cohesive (plasticity index >10) soil fill to be treated with lime prior to its placement in an embankment. Treatment of a soil with lime (see Chapter 7) can be achieved at two levels: (a) soil modification where a small amount of lime (typically less than 2% by dry weight of soil) is premixed with the wet soil to lower its plasticity index and assist in its compaction, and (b) soil stabilisation where >2% lime may be added to a reactive soil to permanently change its mechanical properties and create a cementitious product for a particular usage (e.g. as a capping layer at the top of the embankment, just beneath the pavement).

Wet granular soils may be treated with cement or a cementitious waste additive such as a reactive pulverised fuel ash (PFA) or granulated blastfurnace slag to a similar level of additive as described for lime.

Figure 5.1 Dry density and moisture content relationships derived for various UK fill materials (Parsons, 1992)

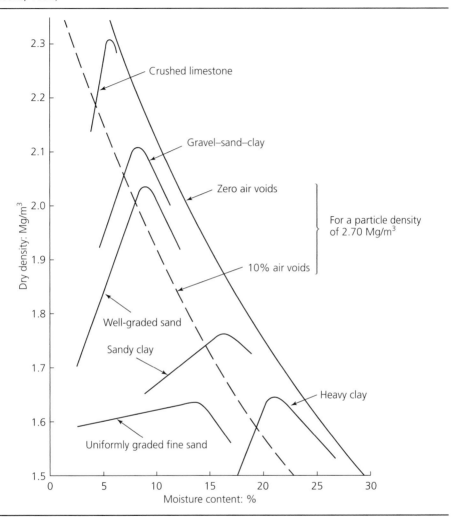

5.4.3 Use of waste materials as fill

The use of waste industrial and construction products for engineering fill has been practised for many years, often with varying success. Legislation in the UK over the last two decades has, however, made their use more attractive for both financial and sustainability reasons.

In 1996, the UK government introduced a landfill tax to discourage the disposal of waste to landfill: as of 2015, this charge amounted to £82.60/t for active waste, and £2.60/t for inactive waste (e.g. concrete, brick, glass and soil). In 2002 the UK government also introduced an aggregate tax on the exploitation of natural aggregates; in 2015 this tax rate stood at £2/t, and recycled waste products were exempt from this tax. (It might be noted that similar taxes also apply in New Zealand and in California, USA.) The benefits underlying these taxes are clear-cut, namely to encourage the demand for recycled materials and reduce the need for 'new' primary materials, reduce the amount of scarce land required for the disposal of waste materials and, generally, to improve the environment.

In 2000, the independent (but government-involved) waste and resources action programme (WRAP) was established in the UK to promote the reduction in use of natural primary resources and the reuse of waste products. In 2004, WRAP launched a quality protocol (QP) that provides a quality assurance and control pathway covering the production of aggregates from inert waste. The QP demonstrates that major criteria to consider when waste has been through an authorised recovery process are the certainty and suitability of its use, and an assurance that the proposed usage does not cause any greater risk to the environment or to human health than the use of equivalent primary aggregates.

The above actions have significantly increased the use of waste materials in earthworks, especially on schemes where the contract earthworks balance is found to be in deficit and an aggregate tax has to be paid on materials obtained from borrow pits. It has also encouraged the use of 'conditioned' on-site materials that otherwise would be disposed off site.

Unfortunately, not all industrial-waste materials are suitable for fill purposes because of their chemical and/or physical properties. Nonetheless, provided that the necessary comprehensive tests can economically be performed within the construction time-scale, industrial materials can offer an attractive solution when used as engineered fill. Commonly used recycled and waste materials are listed in Table 5.3, together with their possible classifications in relation to the (former) Highways Agency's *Specification for Highway Works* (HA *et al.*, 2009).

Table 5.3 Recycled and waste materials commonly used in earthworks in the UK

Material	Possible *SHW*[a] classification	Comments
Unburnt colliery shale	1A, 2C	High sulphate content
Burnt colliery shale	1A	High sulphate content
PFA	2E, 7B	High sulphate content
Crushed concrete	1A, class 6	
Steel and blastfurnace slag	1A	
Glass sand	1B	Limited availability
Spent railway ballast	1A	Limited availability
Road planings	6F1, 6F2	May exhibit ductile behaviour

[a] *Specification for Highway Works* (HA *et al.*, 2009)

5.5. Earthworks equipment

Earthworks plant can be subdivided into two main categories: (a) excavation, haul and placement plant and (b) compaction plant. The space available here does not permit more than a brief overview of the main types of equipment used, and the reader is encouraged to read the many excellent publications available in the literature (e.g. Dumelow, 2012; Nowak and Gilbert, 2015).

5.5.1 Excavation, haul and placement plant

The major types of excavation and earthmoving plant are listed in Table 5.4.

A backacter, also called a backhoe, has a heavy bucket that is connected to the arm of the machine by means of a knuckle joint that allows excavation to take place to depths of 5 m or more, digging toward the machine, below ground level.

Table 5.4 Typical uses of some excavation and earthmoving equipment

Equipment	Typical uses
Backacters (backhoes)	Most commonly used for bulk excavation. Also, excavating below ground level in confined spaces in firm soil (e.g. when digging drainage trenches); as small 'handling cranes' (e.g. when laying drainage pipes)
Bulldozers	Opening up access roads; grubbing and clearing vegetation; stripping top soil; ripping stony and/or non-cohesive soils, and soft and/or stratified rocks with planes of weakness, prior to excavation; shallow excavating; moving earth over short distances (e.g. <100 m); pushing scrapers; spreading and rough-grading earth previously moved by scrapers, trucks or wagons
Dragline crane	Excavation of materials, predominantly granular, including under water
Face shovels	Excavating firm material above ground level (e.g. in cuttings) and previously blasted rock in quarries
Front-end loaders	Miscellaneous localised earthworks operations (e.g. digging and/or filling shallow trenches, excavating for manholes, loading loose materials onto trucks and wagons, stockpiling)
Graders	Accurate finishing work (e.g. trimming the subgrade and establishing the formation, and shaping shoulders, ditches and backslopes); maintaining haul roads. Also, blending materials, including water
Scrapers	Earthmoving operations that involve self-loading, hauling over various distances (e.g. <3 km), dumping and spreading of materials
Trucks and wagons	General haulage operations over long distances (typically 1–10 km). Most commonly used for the transport of earthworks materials. Can be on-site dump trucks with up to 40 t payloads or road-going lorries with up to 20 t payloads

A bulldozer is a rubber-tyred or track-mounted (crawler) tractor that is normally fitted at its front with a straight or angled blade to push earth. While the hardened steel edge of the straight blade is used for general earthmoving purposes, the angled blade is used to push earth sideways. The bulldozer is also used to push scrapers, using a reinforced blade fitted with a stiffened push-plate and shock absorbers. When the ground is very hard, a hardened-steel ripping tool with, usually, up to three teeth can be attached to the rear of the bulldozer to 'plough' the ground up to a depth of 1 m: soil ripping is most effective if carried out downhill in the direction of the slope or, in the case of rocks, in the direction of the dip of the bedding planes.

The most common dragline crane has a crawler-mounted crane and a revolving super-structure, and is fitted with a rope-operated excavation bucket that is cast out from a long boom. As a rule of thumb, the dragline is able to dig to a depth below its track of one-third to one-half of the length of the boom, while the throw of the bucket beyond the radius of the boom can be of a similar length, depending on the skill of the operator. The bucket, which can be perforated for underwater use, has hardened cutting teeth, and is filled by dragging it along or in the ground toward the machine.

The face shovel also has a rotating superstructure mounted on a crawler or wheeled tractor, and has a bucket attached to the boom that can be either rope or hydraulically operated. When in use, the bucket, which is fitted with hardened teeth, is positioned at the bottom of the face being excavated, pressed into it, and pulled upward so that it fills with soil as it rises.

The front-end loader (also called the loader shovel), can do work normally associated with both a face shovel and a small bulldozer. The tracked loader resembles a bulldozer, and its pushed bucket is an effective excavator: the wheeled loader does not have great traction, and is most effectively used to dig in loose soil and/or for stockpiling.

The motor grader is composed of a two- or three-axle frame that is carried by four or six wheels, and has a centrally mounted 3–5 m blade. The blade can be operated horizontally, tilted backward, tilted forward, tilted to the side, or rotated at the side through variable positions, depending on the work to be done.

The scraper is essentially a metal box on wheels with a bottom that can be lowered so that one end digs into the ground when excavating. As the scraper moves forward, a layer of soil (typically 150–300 mm deep, depending on the setting of the cutting edge) 'boils' up into the bowl of the box: when the bowl is full, the bottom is raised, and the scraper transports the soil to where it is to be spread. At the spreading site (e.g. an embankment), the edge of the bowl is lowered to an above-ground height that ensures that the desired thickness of soil is deposited, as an ejector plate pushes from the back of the bowl and the scraper moves forward. Scrapers used on big earthworks schemes are usually motorised and capable of high speeds on local haul roads: depending on the soil conditions, these scrapers may or may not have to be helped by pusher dozers during the digging process.

5.5.2 Compaction plant

Many factors influence the dry densities and air voids contents achievable during earthworks compaction: for example, the soil type and moisture content, the type of compaction machine used and the number of passes made by each machine, and the layer thickness of material being compacted. The main types of compaction plant used in earthworks schemes in the UK are listed in Table 5.5, while Table 5.2 provides overall guidance as to the type of equipment, the number of passes and the layer thicknesses that are commonly used with various earthworks materials.

Table 5.5 Main types of compaction equipment used in earthworks

Compaction plant	Details
Smooth-wheeled rollers	Smooth-wheeled drum rollers can be three-roll or tandem-roll self-propelled units or a single-axle towed roller. Dead weight (i.e. weight increased by adding ballast): 1.7–17 t. Operating speed: 2.5–5 km/h
Vibrating rollers	Similar to smooth-wheeled rollers but with vibration frequencies of 20–35 Hz for large rollers and 45–75 Hz for smaller rollers. Smaller versions are manually guided. Static weight: 0.5–17 t. Speed: 1.5–2.5 and 0.5–1.5 km/h for automatically and manually guided machines, respectively
Tamping rollers	Self-propelled or towed sheepsfoot and pad rollers composed of steel drums with projecting 'feet'. Dead weight: 2.7–27 t. Optimum operating speed: 4–10 km/h
Grid rollers	Generally towed units composed of two cylindrical end-to-end rolls covered with a heavy steel mesh. Roll cores can hold sand or water, and the frame can be ballasted with concrete blocks. Weight: 5.5–15 t when ballasted. Operating speed: 5–24 km/h
Pneumatic-tyred rollers	Can be towed or self-propelled with one or two axles. Using one less wheel on the rear axle of a two-axle roller ensures a smooth, compacted surface. Weight of the roller is generally varied by the addition of ballast to boxes mounted above the axles. A wheel load of up to 3 t is adequate for most soils. Individual wheels can move vertically and maintain a constant load in rough ground. Adjusting the tyre inflation pressure modifies the compaction effort. Optimum operating speed: 5–24 km/h
Vibrating-plate compactors	Manually guided machines suitable for compacting small or awkwardly shaped areas. Weight: 100 kg to 2 t. Plate area: 0.6–1.6 m^2. Typical operating speed: 0.7 km/h
Vibro-tampers	Manually guided machines not suited for heavy compaction but often used in confined spaces. Compaction is induced by vibrations set up in the base plate through a spring activated by an engine-driven reciprocating mechanism. Weight: 50–100 kg

In relation to the value of using self-propelled rollers versus towed rollers, it is important to realise that the use of towed compaction plant is advisable on uniform sands, as self-propelled machines can often dig themselves into the loosely placed sands (Parsons, 1992).

The following are the four main ways by which engineering fill materials are compacted into road embankments:

- Using heavy static weights to press the particles together (e.g. smooth-wheeled rollers). Grid rollers act similarly to smooth-wheeled rollers when compacting coarse materials; however, with fine-grained soils they compact partly by direct contact pressure and partly by a kneading action. Rubber-tyred rollers have an action that is a cross between the static pressure of a smooth-wheeled roller and the kneading action of a tamping roller.
- Kneading the particles while at the same time applying pressure (e.g. tamping rollers).
- Vibrating the materials so that the particles are shaken together into a compact mass (e.g. vibrating rollers).
- Pounding the soil so that the particles are forced together (e.g. impact rollers or tampers).

The main factors controlling the performance of smooth-wheel rollers are the mass per unit width under the rolls and the width and diameter of each roll. The mass/unit width and the roll diameter influence the pressure near the surface of the material being compacted, while the gross mass of each roll affects the rate at which this pressure decreases with depth. Also, the greater the mass of the roll and the smaller its diameter the greater the wave of material pushed ahead of the roller as it advances, and the less even the compacted surface that is left behind. Smooth-wheel rollers are most effectively used to compact gravels, sands and materials that need a crushing action – but they are of no use on rock fill. They also have difficulty in maintaining traction on moist plastic soils, and often cause a 'crust' to be formed at the surface of these soils that prevents proper compaction of the material beneath.

When used to compact coarse materials (e.g. marls or soft sandstones), grid rollers can reduce relatively large boulders obtained during ripping to smaller-sized fragments. They are most effective when it is desired to break and force larger stones below the surface, leaving the top few centimetres of the layer to be made smooth and uniform by other plant. When used with fine-grained soils, these rollers compact with partly a kneading action and partly a direct contact action.

As noted above, rubber-tyred rollers have an action that is partly related to the static pressure of a smooth-wheeled roller and partly to the kneading action of a tamping roller. Some rollers are fitted with wobbly wheels that allow the transmission of additional kneading action to a soil. The main factors affecting the performance of this type of roller are the wheel load and size, ply and, especially, inflation pressure of each tyre. The inflation pressure mainly affects the amount of compaction achieved near the surface while the wheel load primarily influences the compaction depth: for example, research (Lewis, 1959) has shown that there is little value to be gained from using tyre-inflation pressures in excess of 275–345 kN/m^2 with wheel loads of 5 t when compacting clayey soils; however, with granular soils there is advantage in using the highest inflation pressure and the heaviest wheel load practicable, consistent with not overstressing the soil.

The tamping roller, also known as the sheepsfoot roller, is designed to knead the soil as it compacts from the bottom up. During the first pass over a layer of loose soil, the feet penetrate close to the base of the layer so that the bottom material is well compacted: as more passes are made, the feet penetrate to lesser depths, as the density and bearing capacity of the soil increase, and, eventually, the soil becomes so well compacted that the feet of the roller 'walk out' of the layer. In general, the tamping roller is most effective with fine-grained soils; however, it produces more air voids with these soils than either smooth-wheeled or rubber-tyred rollers, and this can be detrimental to soils in which a high air voids content increases their liability to attract moisture. A factor that limits the use of this roller in the UK is that when high dry densities are sought, the optimum moisture contents required by these rollers are often lower than the natural moisture contents of fine-grained soils.

5.6.   Earthworks stability

In the UK, the design of earthworks from the point of view of stability is mainly driven by the possibility of failure of completed earthworks slopes. In relation to this, it should be kept in mind that most new or existing earthworks in the road system have been constructed since the end of World War II, in the period when soil mechanics has been developed as an analytical tool, ground investigations have been applied prior to the commencement of road building, and much-improved compaction equipment has been used in construction, all of which have contributed to the minimisation of slope and embankment failures during construction.

Guidance for the design of earthworks and additional reference sources are provided by Nowak (2012a, 2012b) and Nowak and Gilbert (2015).

5.6.1   A note on factors of safety

Traditionally, the factor of safety against failure of an earthworks slope has been considered as a target 'lump' factor that represents the minimum value that an analysis of failure can achieve. In any analysis of slope failure, the chosen minimum factor of safety can be assumed to depend on (a) a technical assessment of the geotechnical data collected in relation to a potential slip, and (b) engineering judgement regarding the safety, environmental and economic costs of failure (Trenter, 2001). In this respect, the *Code of Practice on Earthworks* (BSI, 2009) recommends the use of minimum factors of safety of 1.3–1.4 for first-time failures, and 1.2 for failures involving ancient shear surfaces. Other traditional target lump factors for specific applications are discussed in the literature (Perry *et al.*, 2003a, 2003b).

It is worth noting that the implementation of Eurocode 7 (BSI, 2004) into UK practice introduced the concept of partial safety factors as input parameters into slope analysis, rather than relying on the use of a final target 'lump' factor for this purpose. Prior to this, the calculation of a satisfactory factor of safety against failure was aimed at achieving a stability state where overall failure of the slope was unlikely to occur: this approach described the ultimate limit state failure of the slope. Conditions can, however, occur where the earthworks do not reach their ultimate failure state, but some movement does occur that adversely affects its performance: this can be considered a serviceability limit state condition that is controlled by the performance of the infrastructure that the earthworks support. Common examples of this condition are (a) the creep of cutting slopes that results in the movement of such services as drainage, electricity or communications cables, and (b) settlement of embankment foundations or creep of embankment slopes resulting in unacceptable deflections of highway surfacings and/or services.

5.6.2   A note on slope failure during operation

Notwithstanding that earthworks slopes in road schemes have historically been designed satisfactorily and have rarely exhibited failure during construction, failure during later operation must still be considered. The failure of a slope during operation can be generally subdivided into two modes: (a) shallow failure by translation or shallow circular failure, and (b) deep failure, usually circular in nature by progressive collapse.

Shallow operational failures in the UK's road and rail network have been extensively studied (Coppin and Richards, 2007; Crabb and Atkinson, 1991; Perry, 1989; Perry et al., 2003a, 2003b; Reid and Clark, 2000) since the late 1980s in order to understand the mechanisms causing failure and, thereby, optimise maintenance requirements over the design life of the earthworks. These studies showed that

- many existing slopes exhibited signs of creep and tension cracking
- while some shallow circular slips can occur, most slope failures identified were translational
- most failures were shallow (usually less than 2.5 m deep) and occurred on slopes >4 m high
- failure incidences were recorded on only a small percentage (<5%) of the total earthworks length of the road cuttings and embankments studied
- generally, failures were associated with slow and insidious movements (e.g. shrink and swell due to seasonal moisture changes, rather than overall instability)
- the most likely failure mechanism is associated with moisture content and porewater pressure increases in the slope surfaces over time due to a lack or non-performance of slope drainage
- any lack of maintenance or remediation that causes prolongation of the potential failure surfaces will lead to more significant volumes of failed material in the long term.

In the light of the above analyses, it is interesting to note that analyses of failures of old London Underground railway slopes built in fissured over-consolidated cohesive London Clay (Ellis and O'Brien, 2007; Potts et al., 1997, 2000) showed that (a) the failures experienced were deep seated, circular and associated with increases in porewater pressure that encouraged the progressive collapse of surfaces emanating from the toes of slopes, and (b) progressive collapse failures could be expected in this clay for up to 125 years after construction if the slopes were constructed at grades steeper than 1:3 (vertical:horizontal) and >8 m in height.

5.7.   Building embankments on soft foundations

When an embankment load is placed on an underlying subsoil, the volume of the voids in the foundation is reduced as it attempts to consolidate. If the soil is saturated, the particles pressing about the pores cause the applied load to be transferred to the incompressible water for a length of time that depends upon the type and compaction state of the foundation material. If the subsoil is granular and has a high coefficient of permeability, the porewater is expelled quickly, and settlement occurs rapidly. Thus, immediate settlement occurs most successfully in granular founding strata (and on weak rock).

As might be expected, the most difficult settlement problems encountered during, and subsequent to, construction are associated with embankments that are underlain by soft, normally consolidated, cohesive soils or by made-ground of variable quality. While the foundation will gain strength as consolidation takes place under the mass of the embankment, a serious stability failure may take place if the rate of strength increase is less than the shearing stresses generated in the foundation during the building of the embankment. Such a failure is typically manifested by a lateral displacement of the foundation soil, an upward bulging of the adjacent soil and excessive settlement of the embankment.

In some circumstances, the risk of subsoil failure during construction can be mitigated by the use of a lightweight fill material (PIARC, 1997) (e.g. polystyrene, PFA or lightweight aggregate) to form the embankment. With or without the use of lightweight fill – and provided that extra land width is available for a wider embankment – the use of flatter side slopes and the building of berms will help prevent the lateral displacement of the soft foundation soil.

An excellent review of the behaviour of embankments on soft clay foundations is available in the literature (Tavenas and Leroueil, 1980).

Various procedures are available to help minimise settlement subsequent to the construction of the embankment. These include the removal and replacement of weak foundation material, stage construction, foundation densification, settlement acceleration, foundation reinforcement at the base of the embankment and the use of columns and piles to directly support the embankment. Before discussing these, however, it should be noted here that the advent of public–private partnership contracts (see Chapter 1), with their emphasis on ongoing long-term maintenance requirements, has required the road designer and the construction contractor to consider more closely the likelihood and consequences of secondary consolidation settlements occurring subsequent to embankment construction.

5.7.1   Removing and replacing the soft foundation material

If an embankment is to be constructed at a site where the weak foundation material exists to a shallow depth, typically <2.0 m, the most reliable, and common, foundation treatment is to remove and replace the poor-quality material with an acceptable embankment fill, usually granular material, prior to the construction of the main embankment. Excavation to a greater depth can be considered if it is economically viable and no problems are likely to emerge with respect to the short-term stability of any excavated slopes: this was the solution adopted during the construction of the N8 Fermoy Bypass in Ireland where 6 m of soft material was removed and replaced with Old Red Sandstone rockfill, prior to the construction of a 9.0 m high embankment.

5.7.2   Stage construction

If, for whatever reason, the length of time spent on construction is not a major priority, a very effective way of avoiding subsoil instability during and after the building period is to stage construct the embankment.

Stage construction is a relatively long-term process whereby the embankment is first built to a height that is limited by the shearing strength of the foundation, and further construction is then delayed until porewater pressure measurements show that the extra stress induced in the subsoil has dissipated. When the extra stress is reduced to a safe value, an additional height of fill is put on top of the embankment material already in place without, again, causing foundation failure, and further construction is then delayed until the porewater pressures are again reduced. The process is then repeated until the full embankment is in place and settlement has ceased.

5.7.3   Settlement acceleration

If, because of time or poor soil thickness constraints, neither stage construction nor removal and replacement are practical solutions to a soft foundation problem, and the stability of the overall embankment is an issue, options available to improve foundation stability include the use of a surcharge with or without vertical sand drains, and foundation densification, to accelerate the rate of consolidation and settlement. Their usage is applicable to two different types of consolidation scenarios.

5.7.3.1   Surcharges and vertical sand drains

A well-established method of accelerating the rate of settlement in relatively shallow (<5 m) soft subsoils is to build the embankment in the normal way to achieve the design elevation and then add a surcharge of an extra height of embankment fill with the intent of accelerating the rate of consolidation so as to attain, during construction, most if not all of the foundation settlement predicted. Then, when the desired foundation settlement has been achieved, the surcharge layer is removed, and the pavement is constructed upon the remaining embankment in the normal way. If the subsoil is of uniform consistency and moisture content, any subsequent (slower) settlement should be relatively uniform: if, however, the foundation material is non-uniform, differential settlement will occur, and a rough pavement will be developed.

Depending upon the foundation conditions, this method of surcharging can lead to foundation instability, as discussed above, if the embankment construction is not carried out very carefully. Also, attention needs to be paid to some clayey subsoils to ensure that they are not over-consolidated by the surcharge load, as swelling could subsequently occur following removal of the surcharge, and this could affect the long-term riding quality of the pavement.

Another method of treating troublesome embankment foundations is to use vertical sand drains with or without a surcharge layer on the embankment. The first step in the sand drain method involves placing a 'working platform' of the first layer of embankment fill on top of the soft soil, and then installing vertical drainage holes or 'shafts' through the layer and down through the subsoil to the desired depth or to a solid stratum. The holes, which are typically 100–150 mm diameter and up to 30 m deep, are spaced about 1.5–2 m apart in a regular pattern and, typically, filled with a clean uniform sand; alternatively, they may take the form of plastic strips encased in geotextile (band) drains that are pushed into the soft soil using a mandrel. The tops of all the vertical shafts and the initial working platform layer are then covered with a 0.5–1 m thick blanket of granular material, and the normal embankment construction is continued to the design height when, if appropriate, a surcharge may be added. If at any time during the building of the embankment the stresses imposed on the subsoil appear likely to exceed its strength, construction is halted until stability is restored.

The purpose of the vertical drains is to speed up the consolidation process by allowing the water to escape quickly from the material beneath the embankment as it is being, and when it has been, loaded. Without the sand drains, the water is only able to escape through the bottom or the top of the compressible material or, if the subsoil is underlain by an impermeable layer, only through the top. With the sand drains, however, as well as being forced upward and downward by the imposition of the embankment load, the water is also forced to move horizontally to the nearby vertical drains, whence it is easily able to move upward through the much more permeable sand, and escape laterally via the horizontal blanket. In addition, by providing the opportunity for radial drainage to take place, this methodology makes use of the fact that many fine-grained soils have a permeability in the horizontal direction that is much greater than in the vertical. Another advantage of this method of accelerated drainage is that the shear strength of the foundation material is more rapidly increased and the likelihood of lateral movement under the embankment load is much lessened.

While the time saved by using vertical drains will obviously vary according to the subsoil conditions encountered, it is not uncommon for the time taken to achieve 95% primary consolidation settlement to be reduced from, say, 5 years without drainage to less than 6 months with vertical drainage.

5.7.3.2 Foundation densification

Dynamic compaction, rapid impact compaction and impact rolling are the methods most commonly used to achieve in situ densification to depths of up to 5 m below the ground surface. These methods are used most effectively with man-made ground or with coarse materials where the groundwater table is not in close proximity to the ground surface. The presence of a high groundwater table or of soft cohesive materials significantly reduces the effectiveness of these forms of treatment.

Dynamic compaction involves the use of special self-propelled compactor machines that use a hoist mechanism to drop a large dead weight, usually a concrete block, from a height in a controlled manner over a regular grid of spaced contact points. This method of compaction causes the immediate densification of the underlying material. More information on dynamic compaction is readily available in the literature (Serridge and Slocombe, 2012).

The rapid impact compactor (Watts and Cooper, 2011) achieves densification of the ground in a similar way to dynamic compaction, but on a smaller scale. While the method was initially developed for the immediate infill of craters on runways resulting from wartime bomb damage, it has been used successfully to stabilise loose granular materials and both cohesive and non-cohesive made-ground materials to depths of up to 4 m. This method is also used to compact granular soil in trenches and close to structures.

Impact rollers are large towed rollers with heavy non-circular rolls that deliver compaction by impact as a result of their large mass and unconventional shape. Various roll shapes have been tried, but the impact rollers seen today are truncated triangular or pentagonal in cross-section that densify through a combination of vibration through the roller and a dynamic effect as the roller rotates, as the extrados of the roller is not circular.

5.7.4 Reinforcing the foundation

Geogrids, also known as geocells, are synthetic thermoplastics formed into an open net-like configuration that are arranged in a horizontal plane on top of the foundation beneath an embankment. Various geotextiles are commercially available for use as geogrids in road construction. Woven multi-filament (i.e. regularly arranged multi-thread) ones have high tensile strength, high modulus, low creep and relatively low water permeability. Non-woven geotextiles are formed by bonding a loose mat of randomly arranged filaments: as the pore sizes and tensile strengths of these materials are generally lower than woven ones, they are most often used for drainage and separation purposes, especially if soil filtration is required.

The role of the woven high-strength geogrid is to provide additional tensile strength along a potential failure plane in a manner similar to its usage in steepened soil slopes. In the case of an embankment constructed on soft ground, its function is to reduce the risk of a slip circle failure of the embankment and foundation soils, and to reduce lateral spreading and cracking of the embankment. It will not significantly reduce embankment settlement associated with the time-dependent consolidation of uniformly weak foundation soils; however, if the foundation has a non-uniform weakness (e.g. if it contains lenses of clay or peat), the reinforcing function of the geogrid will enable it to bridge the weak spots and reduce the risk of localised failure and/or reduce differential shrinkage.

Non-woven drainage geogrids can be placed as separating materials above and below a designed granular drainage layer at the bottom of an embankment on soft ground. In this case, the geogrids will deform without developing high tensile stresses while preventing the embankment soil above and the foundation soil below from passing into the granular drainage layer but allowing the upward-moving water to do so – thereby preventing a build-up of excess porewater pressures and a corresponding loss of foundation shear strength.

The use of geogrids has the advantage that it reduces the lengths of construction time lost due to delays and failures when building embankments on soft ground. Also, when compared with displacement methods, it may be more economical to use geogrids than to import expensive extra fill material to replace removed foundation soil.

The reader is referred to the literature (Manceau *et al.*, 2012) for details of the design of reinforcement geogrids.

5.7.5 Using bearing columns and piles

As noted previously, the greatest potential for differential embankment settlement occurs adjacent to rigid structures such as bridge abutments, and at these interface areas it may be appropriate to support the earthworks approach to the abutment on a structural platform using driven piles (Reid and Buchanan, 1984) or stone or concrete columns. This structural support to the embankment has the advantage of not only minimising settlement between the embankment and the structure but also of reducing the lateral stress transferred onto the abutment foundations (which are also likely to be on piles) from the soft subsoil below the embankment (Leroueil, 1990; Springman and Bolton, 1990).

Stone columns are constructed in a manner similar to vertical sand drains, but, in this case, vibrated coarse gravel is used to form the columns which are up to 600 mm in diameter and up to 10 m deep and, usually, installed in a grid pattern over the footprint of the embankment as it approaches the abutments. As each column is being built, the vibrated gravel particles blend radially into the near-surrounding soil and form a composite gravel–soil 'block' that provides a greater bearing strength (up to 150 kN/m^2) to the embankment than the natural soil. These columns are not very effective in cohesive soils with undrained shear strengths of less than about 25 kN/m^2, as the untreated soil provides insufficient lateral restraint for the coarse gravel as it is vibrated into the vertical shafts.

A different mechanism of load transfer is employed if concrete columns or driven precast piles are used: these methods are used to transfer the embankment load onto a stiffer stratum (usually a granular or rock stratum) below the soft soil. The concrete columns are installed in a manner similar to the stone columns, except that vibrated wet-mix cement concrete is used instead of gravel to form the columns. (Driven precast piling was extensively used as an embankment-support technique on the Channel Tunnel Rail Link in soft-sediment areas in the Thames Estuary.)

5.8. Forming the formation with or without capping layers

After the earthworks forming the subgrade have been completed, the ground surface will still be fairly rough, and it must be graded to the design-formation shape before the pavement is put in place. If the construction programme does not allow for the immediate placement of the pavement, then the formation will need to be protected in the interim period: this usually involves placing a 300 mm protective layer (or more, if the subgrade is a clay soil) of suitable fill material on top of the formation, and this is then removed when the pavement is about to be built.

The formation is normally reflective of the final carriageway shape: this helps to ensure that each subsequent pavement layer is built to its design thickness, and water that gets into the subgrade can drain away. This final shaping, which is usually done with a motor grader, should be delayed until the first layer of the pavement is about to be put in place. When the grading is finished, any formation irregularities should be removed by light rolling. Preferably, no traffic should be allowed onto the formation prior to the place-ment of the first layer of the pavement – which should be carried out as quickly as possible thereafter to minimise the opportunity for rain to fall and soften the subgrade.

If the formation level is at or about the natural ground level, it should be checked to ensure that all topsoil and vegetative matter has been removed from within the roadway width: if left, this material will eventually decay and leave voids that will be the cause of subsequent subgrade or low embankment settlement. If the subgrade comprises a heavy clay soil, then, ideally, the entire plastic B horizon (see Figure 4.2) should be removed so that the C horizon can be used as the foundation soil. Any man-made debris within a metre of the formation should be removed, and all holes left should be properly cleaned out, filled and compacted. If the specification says that a basement need not be removed but it is likely that water may accumulate within it, holes should be made in at least 10% of the slab areas, to permit drainage before backfilling occurs.

If the subgrade soil in a cutting is relatively coarse and loose, it can usually be easily compacted to the design-required depth. If, however, the soil is a heavy clay, a light roller may only be used to smooth the subgrade surface, as such undisturbed clays cannot usually be further compacted beyond their natural state (i.e. the exposed clays will tend to gain water, and rolling with a heavy roller will only result in moisture exudation and the resultant remoulding and weakening of the natural soil state).

If the pavement is to be laid in a rock cutting, a 100–150 mm layer of suitable capping material should be placed between the pavement and the rock, to level any irregularities in the rock surface that were attained during blasting. However, before the capping material is put in place, grooves should be cut in the foundation rock, to allow for the lateral drainage of any water accumulations then or in the future.

When the subgrade passes from the cut to the fill, especially if the change is from rock to soil, there should be no abrupt change in the degree and uniformity of compaction: rather, a capping layer should be installed, to allow any differential effects to occur gradually (e.g. at the 'point' of change the transition layer thickness should be at least 1 m and feathered back to about 150 mm over a distance of, say, 14 m on the rock).

In the case of embankments, the final 600 mm of the engineered fill should be placed and compacted in one continuous operation across the whole fill area, to ensure uniformity of stiffness directly below the pavement to be constructed upon it.

The concept of a capping layer was initially introduced into the UK governmental specifications in 1978 as an economic measure that allowed the use of a capping layer of higher-strength material in the upper 1 m of an embankment, or below a formation in cuttings, so as to enable a reduction in the thickness of the more-expensive granular sub-base material required by the then method of pavement design. This initiative, which was introduced to address the paucity of good-quality pavement material in south-east England at a time of rapid motorway construction, specified that the capping material could be a local granular material with a minimum California bearing ratio (CBR) of 15% when compacted. Subsequently, the use of local gravels of this strength (and equivalent-strength low-cost recycled or waste materials) was encouraged as capping materials on weak subgrades (i.e. on subgrades with a CBR of <5%), in order to reduce the design thickness of higher-quality (and more expensive) sub-base material in the pavement.

With the aim of reducing the thickness of sub-base material required in a pavement, the current pavement design specification (HA, 2009) allows the following thicknesses of class 6F capping material, with a CBR of >15%, to be used on the following subgrades: (a) use 600 mm if the subgrade CBR is <2%, (b) use 450 mm if the subgrade CBR is 2%, (c) use 350 mm if the subgrade CBR is >2% but not >2.5% and (d) use 150 mm if the subgrade CBR is >2.5%.

REFERENCES

Anon. (2003) Construcción de explanaciones, dranajes y cimentaciones. In *Pilego de Pre-scripiones Técnicas Generales para obras de Correteras y Puentes (PG-3)*. Ministerio de Fomento, Madrid, Spain.

BSI (British Standards Institution) (2004) BS EN 1997-1:2004: Eurocode 7. Geotechnical design. BSI, London, UK.

BSI (2009) BS 6031:2009: Code of practice for earthworks. BSI, London, UK.

Charles JA, Skinner HD and Watts KS (1998) The specification of fills to support build-ings on shallow foundations: the '95% fixation'. *Ground Engineering* **31(1)**: 29–33.

Coppin NJ and Richards IG (2007) *Use of Vegetation in Civil Engineering*. Construction Industry Research and Information Association, London, UK, Report C708.

Crabb GI and Atkinson JH (1991) Determination of soil strength parameters for the analysis of highway slope failures. In *Slope Stability Engineering – Developments and Applications: Proceedings of the International Conference on Slope Stability*. Thomas Telford, London, UK, pp. 13–18.

Dumelow PG (2012) Earthworks material specification, compaction and control. In *ICE Manual of Geotechnical Engineering*, vol. 2. *Geotechnical Design, Construction and Verifi-cation*. Institution of Civil Engineers, London, UK, pp. 1115–1141.

Ellis EA and O'Brien AS (2007) Effect of height on delayed collapse of cuttings in stiff clay. *Proceedings of the Institution of Civil Engineers – Geotechnical Engineering* **160(2)**: 73–84.

Gilbert P and Kidd A (2012) An update on new and future earthworks standards in the UK and Europe. *Ground Engineering* **45(2)**: 27–30.

HA (Highways Agency) (2009) *Design Guidance for Road Pavement Foundations (draft HD 25). Interim Advice Note IAN 73/06, Revision 1*. Stationery Office, London, UK.

HA, Transport Scotland, Welsh Government and Department for Regional Development Northern Ireland (2009) Earthworks. In *Manual of Contract Documents for Highway Works*, vol. 1. *Specification for Highway Works*. Stationery Office, London, UK, Series 600.

HA, Transport Scotland, Welsh Government and Department for Regional Development Northern Ireland (2014) *Manual of Contract Documents for Highway Works*, vol. 1. *Specification for Highway Works*. Stationery Office, London, UK.

LCPC (Laboratoire Central de Ponts et Chaussées) (2003) *Practical Manual for the Use of Soils and Rocky Materials in Embankment Construction: Guide Technique*. Laboratoire Central de Ponts et Chaussées, Paris, France.

Leroueil S (1990) *Embankments on Soft Clays*. Ellis Horwood Press, London, UK.

Lewis WA (1959) *Investigation of the Performance of Pneumatic-tyred Rollers in the Compaction of Soils: Road Technical Paper 45*. Stationery Office, London, UK.

McGuire MP and Filz GM (2005) *Specifications for Embankment and Sub-grade Compaction*. Virginia Transportation Research Council, Richmond, VA, USA, Report VTRC 05-CR21.

Manceau S, MacDiarmid C and Horgan G (2012) Design of soil reinforced slopes and structures. In *ICE Manual of Geotechnical Engineering*, vol. 2. *Geotechnical Design, Construction and Verification*. ICE, London, UK, pp. 1093–1107.

Mooney M and White D (2007) *Intelligent Soil Compaction Systems: NCHRP Project 21-09*. Transportation Research Board, Washington, DC, USA.

NCHRP (National Cooperative Highway Research Program) (1971) *Synthesis of Highway Practice No. 8*. Highway Research Board, Washington, DC, USA.

Nowak PA (2012a) Earthworks design principles. In *ICE Manual of Geotechnical Engineering*, vol. 2. *Geotechnical Design, Construction and Verification*. ICE Publishing, London, UK, pp. 1043–1046.

Nowak PA (2012b) Design of new earthworks. In *ICE Manual of Geotechnical Engineering*, vol. 2. *Geotechnical Design, Construction and Verification*. ICE Publishing, London, UK, pp. 1047–1065.

Nowak PA and Gilbert PJ (2015) *Earthworks: A Guide*, 2nd edn. ICE Publishing, London, UK.

Parsons AW (1992) *Compaction of Soils and Granular Materials: A Review of Research Performed at the Transport Research Laboratory*. Transport Research Laboratory, Crowthorne, UK.

Perry J (1989) *A Survey of Slope Condition on Motorway Earthworks in England and Wales*. Transport and Road Research Laboratory, Crowthorne, UK, Report RR199.

Perry J, Pedley M and Brady K (2003a) *Infrastructure Cuttings – Condition Appraisal and Remedial Treatment*. Construction Industry Research and Information Association, London, UK, Report C591.

Perry J, Pedley M and Reid M (2003b) *Infrastructure Embankments – Condition Appraisal and Remedial Treatment*. Construction Industry Research and Information Association, London, UK, Report C592.

PIARC Technical Committee on Earthworks, Drainage and Subgrade (1997) *Lightweight Filling Materials*. PIARC-World Road Association, Paris, France.

Potts DM, Kovacevic N and Vaughan PR (1997) Delayed collapse of cut slopes in stiff clay. *Géotechnique* **47(5)**: 953–982.

Potts DM, Kovacevic N and Vaughan PR (2000) Delayed collapse of cut slopes in stiff clay: discussion by Bromhead and Dixon and authors reply. *Géotechnique* **50(2)**: 203–205.

Preene M, Roberts TOL, Powrie W and Dyer MR (2000) *Groundwater Control – Design and Practice*. Construction Industry Research and Information Association, London, UK, Report C515.

Reid JM and Clark GT (2000) *A Whole Life Cost Model for Earthworks Slopes*. Transport Research Laboratory, Crowthorne, UK, Report 430.

Reid WM and Buchanan NW (1984) Bridge support piling. In *Piling and Ground Treatment: Proceedings of the Conference on Piling and Ground Improvement*. Thomas Telford, London, UK, pp. 267–274.

Serridge CJ and Slocombe B (2012) Ground improvement. In *ICE Manual of Geotechnical Engineering*, vol. 2. *Geotechnical Design, Construction and Verification*. Institution of Civil Engineers, London, UK, pp. 1247–1269.

Springman SM and Bolton MD (1990) *The Effect of Surcharge Loading Adjacent to Piles*. Transport and Road Research Laboratory, Crowthorne, UK, Contractor Report 196.

Tavenas F and Leroueil S (1980) The behavior of embankments on clay foundations. *Canadian Geotechnical Journal* **17(2)**: 236–260.

Trenter NA (2001) *Earthworks: A Guide*. Thomas Telford, London, UK.

Watts KS and Cooper A (2011) Compaction of fills in land reclamation by rapid impact. *Proceedings of the Institution of Civil Engineers – Geotechnical Engineering* **164(GE3)**: 169–193.

Highways
ISBN 978-0-7277-5993-1

ICE Publishing: All rights reserved
http://dx.doi.org/10.1680/h5e.59931.165

Chapter 6
Materials used in road pavements

Andrew Dawson Associate Professor, Nottingham Transportation Centre, UK

Other than soil, the materials most used in road pavements are bitumen, cement, lime, rock, gravel and recycled material aggregates. Each is discussed here.

6.1. Penetration-grade refinery bitumens

Bitumens are viscous liquid or semi-solid materials, consisting essentially of hydrocarbons and their derivatives, which are soluble in trichloroethylene (BSI, 2009a). While bitumens occur naturally (e.g. in lake asphalts containing mineral materials), the vast majority are the penetration-grade products of the fractional distillation of petroleum at refineries. Bitumens that are produced artificially from petroleum crudes (usually naphthenic- and asphaltic-base crudes) are known as refinery bitumens.

The term 'asphalt' is used in the US technical literature to describe what is termed 'bitumen' in the UK and Europe. Following agreement by the European Committee for Standardization (Comité Européen de Normalisation, CEN), in Europe the term 'asphalt' is reserved for materials containing a mixture of bitumen and mineral matter (e.g. lake asphalt or hot rolled asphalt).

Penetration-grade refinery bitumens are designated by the number of 0.1 mm units that a special needle penetrates the bitumen under standard loading conditions; thus, lower-penetration depths are associated with harder bitumens. Twelve grades of bitumen, from 20/30 pen (hardest) to 650/900 pen (softest), are defined for pavements and regulated by the CEN standards. Nine common types are listed in Table 6.1. The harder grades (<30 pen) are used in mastic asphalts, the medium grades (30–70 pen) in stone mastic asphalts, heavy duty macadams and 'enrobé à module élevé' (EME), and the softer grades (>100 pen) in macadams and some thin surface mixes (see Chapter 8). Bitumens are almost invariably black or brown in colour, possess waterproofing and adhesive properties, and soften gradually when heated. Colouring can be added if needed (frequently red or green for cycle and bus lane surfaces).

At 25°C the density of a bitumen typically varies from 1 to 1.04 g/cm^3, while its coefficient of thermal expansion is about 0.00061/°C. Some of the lower-penetration bitumens are 'blown' by passing air through them: this removes some of the volatile fraction and oxidises some of the hydrocarbons, making them less susceptible to temperature change than the other bitumens (BSI, 2009b). Typically, asphalts containing blown binders deform less under traffic; but their mixing, laying and rolling must be

Table 6.1 Penetration-grade bitumens commonly used for road purposes (BSI, 2009a)

Property	Test method	Penetration grade								
		20/30	30/45	35/50	40/60	50/70	70/100	100/150	160/220	250/330
Penetration at 25°C: 0.1 mm units	Penetration needle	20–30	30–45	35–50	40–60	50–70	70–100	100–150	160–220	250–330
Softening point (minimum, maximum): °C	Ring and ball	55, 63	52, 60	50, 58	48, 56	46, 54	43, 51	39, 47	35, 43	30, 38
Resistance to hardening at 163°C	Perform rolling thin-film oven or rotating-flask ageing, then test the bitumen as follows									
Change of mass % (maximum)	Weighing	0.5	0.5	0.5	0.5	0.5	0.8	0.8	1.0	1.0
Retained penetration (minimum)	Penetration needle	55	53	53	50	50	46	43	37	35
Softening point (minimum): °C	Ring and ball	57	54	52	49	48	45	41	37	32
Flash point (minimum): °C	Cleveland open cup	240	240	240	230	230	230	230	220	220
Solubility (minimum): %	Toluene solubility	99.0	99.0	99.0	99.0	99.0	99.0	99.0	99.0	99.0
Kinematic viscosity at 135°C (minimum): mm²/s	Vacuum capillary viscometer	530	400	370	325	295	230	175	135	100

carried out at higher temperatures than are used with conventional bitumens of the same grade.

6.1.1 Bitumen tests and their significance

Specifications with regard to the design and construction of a bituminous pavement are of little value if the properties of the binder used are inadequately controlled. To ensure that the bitumen has the desired qualities, various tests have been devised to measure these properties for particular purposes. As detailed information on these tests is readily available (see the BSI entries in the reference list), they are only briefly described here.

Bitumens are termed 'visco-elastic' because (a) at temperatures above about 100°C they exhibit the properties of a viscous material, (b) at temperatures below about −10°C they behave as an elastic solid and (c) at temperatures in between they behave as a material with viscous and elastic properties, with the predominating property at any given time depending upon the temperature and rate of load application. Viscosity, which is the property of a fluid that retards its ability to flow, is of particular interest at the high temperatures needed to pump bitumen, mix it with aggregate, and lay and compact the mixed materials on site. For example, if the viscosity is too low at mixing, the aggregate will be easily coated but the binder may drain off while being transported; if the viscosity is too high, the mixture may be unworkable by the time it reaches the site. If too low a viscosity is used in a surface dressing, the result may be 'bleeding' (i.e. bitumen flows slowly to the surface of the pavement, especially in hot weather, generating a smooth continuous bitumen layer at the surface) or loss of chippings under traffic.

The absolute viscosity of a bitumen, at a standard temperature, can be measured using a sliding-plate micro-viscometer (Szatkowski, 1967). An increasingly common form is the dynamic shear rheometer (Airey et al., 2002). As illustrated in Figure 6.1(a), the dynamic shear rheometer comprises two circular plates joined by a 1 mm thick film of bitumen. One plate is rotationally oscillated at various torque (or various strain) levels under a range of controlled temperatures. At any particular temperature the viscosity, η, in pascal seconds (Pa s) of the bitumen is the ratio of the shear stress to the rate of strain, defined as

$$\eta = \log[\Delta t / \Delta(\gamma/\tau)] = \log\{\Delta t / \Delta[(\theta_r/d)/(2T/\pi r^3)]\} \tag{6.1}$$

where t is the time of (s) application of the applied torque, T (N m); γ is the shear strain; τ is the shear stress (Pa); r is the radius of each plate (m); θ_r is the angle of rotation (radians); and d is the distance between the plates (m).

Because bitumen is viscous, the peak stress and peak strain do not occur at the same times during cyclic loading, and the difference between their times of occurrence is described by a phase angle, δ. The dynamic shear rheometer therefore routinely measures two properties of the bitumen – the complex shear modulus, G^*, which is the ratio between the maximum stress and the maximum strain, and this phase angle (see also Section 6.1.2). Simpler, rotational, viscometers in which a cylinder is rotated

Figure 6.1 Commonly used viscometers: (a) dynamic shear rheometer; (b) rotational ('Brookfield') viscometer (shown without bitumen bath)

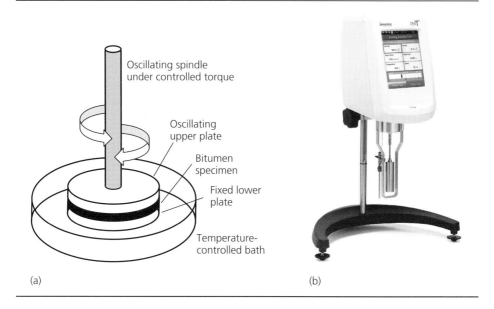

Oscillating spindle under controlled torque

Oscillating upper plate

Bitumen specimen

Fixed lower plate

Temperature-controlled bath

(a)   (b)

about its axisymmetric axis in a bitumen bath, at a fixed temperature, are frequently used for quality assurance and control purposes in bitumen and asphalt production (Figure 6.1(b)).

Although the use of viscometers is increasingly common, older index tests such as the long-established penetration and softening point tests are still used in industrial practice as empirical proxies.

As shown in Figure 6.2(a), the penetration test ((BSI, 2007a) measures the depth to which a standard needle will penetrate a bitumen under standard conditions of temperature (25°C), load (100 g) and time (5 s): the result is expressed in penetration units, where one unit equals 1 dmm (0.1 mm). While the penetration test is simply a classification test, and not directly related to binder quality, the penetration grade of a bitumen is linked to the composition and use of a bituminous material: for example, higher-penetration bitumens are preferred for use in pavements in colder climates (to reduce cracking problems) while lower-penetration ones are preferred in hot climates. When used to bind a well-graded aggregate with high internal friction, a medium-grade bitumen will enhance the workability of the mix. A low-penetration bitumen will provide stability to a mastic type of mix with high sand content, and may be used at locations where traffic stresses are very high (e.g. at bus stops).

The softening point test (BSI, 2007b) determines the temperature at which a bitumen changes from semi-solid to fluid. The softening point is an equi-viscous temperature

Figure 6.2 Penetration and softening point tests for bituminous binders: (a) penetration test; (b) softening point test

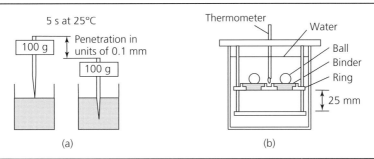

$$(a) \qquad\qquad (b)$$

in that it is the temperature at which all refinery bitumens have the same viscosity (i.e. about 1200 Pa s). There is a rule of thumb to the effect that the mixing temperature of a penetration-grade bitumen is 110°C above its softening point: for example, nominal 50 and 100 pen bitumens are often mixed at about 162 and 156°C, respectively.

As indicated in Figure 6.2(b), the softening point test involves placing a 3.5 g steel ball on a disc of bitumen that is supported by a brass ring and immersed in water. The water is heated uniformly at the rate of 5°C/min until the disc is sufficiently soft for the ball, enveloped in bitumen, to fall 25 mm through the ring onto a base plate. The water temperature at which the binder touches the base plate is recorded as the softening point of the bitumen.

A bituminous binder should never reach its softening point under traffic. Many bitumens, but not all, have a penetration of 800 pen at their softening point. Below the softening point temperature the relationship between the logarithm of the penetration (dmm) and temperature t (°C) is linear for non-blown bitumens; that is,

$$\log(\text{pen}) = \text{constant} + \alpha t \tag{6.2}$$

where, for a bitumen of a given origin, the slope α is a measure of its temperature susceptibility, and is obtained by measuring the penetration at two temperatures. Values of α vary from 0.015 to 0.06; however, other than showing that there are considerable variations in temperature susceptibility, these numbers mean little to the road engineer. Consequently, another measure of temperature susceptibility, the penetration index (PI), was devised (Pfeiffer and van Doormaal, 1936). This assumes that PI = 0 for a 'normal' 200 pen Mexican bitumen, and other road binders are then rated against this standard. The PI is obtained from

$$\text{PI} = [20 \times (1 - 25\alpha)]/(1 + 50\alpha) \tag{6.3}$$

The value of α is derived from penetration measurements at two temperatures, t_1 and t_2, using

$$\alpha = \frac{\log(\text{pen}) \text{ at } t_1 - \log(\text{pen}) \text{ at } t_2}{t_1 - t_2} \tag{6.4}$$

Most road bitumens are considered acceptable if they have a PI of between -1 and $+1$. The PI is also used to estimate the stiffness of a bitumen for analytical pavement design purposes (Heukelom, 1973).

The Fraass breaking point test (BSI, 2007c), which measures the (low) temperature at which a bitumen reaches a critical stiffness and cracks, is used as a control for binders used in very cold climates. The Fraass breaking point is the temperature at which a thin film of bitumen attached to a metal plate cracks as it is slowly flexed and released while being cooled at the rate of $1\,°C/\text{min}$. If the penetration and softening point of a bitumen are known, the breaking point can be predicted, as it is equivalent to the temperature at which the bitumen has a penetration of 1.25.

As the loading times for penetration, softening point and the Fraass breaking point tests are similar for a given penetration-grade bitumen, a process was devised that enables these and viscosity test data to be plotted against temperature on a data chart (Heukelom, 1969, 1973). The temperature scale in this chart (Figure 6.3) is linear, the penetration scale is logarithmic and the viscosity scale has been devised so that low-wax-content bitumens with 'normal' temperature susceptibility or penetration indices give straight line relationships (i.e. curve S, for 'straight'). Hence, only the penetration and softening point of the binder need be known to predict the temperature/viscosity characteristics of a penetration-grade bitumen. Blown (B) bitumens can be represented by two intersecting straight lines, where the slope of the line in the high-temperature

Figure 6.3 Bitumen test chart (RA, rolled asphalt; AC, asphalt concrete; W, waxy; S, unblown; B, blown)

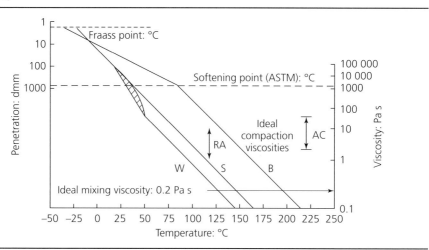

range is approximately the same as that for the unblown (S) bitumen. Four test values are required to characterise such a bitumen. Waxy (W) bitumens also give intersecting lines of nearly equal slopes, but are not aligned. Between the two lines, there is a transition range in which the test data are scattered.

A chart such as Figure 6.3 provides a fingerprint of the rheological behaviour of the binder, and it determines the ideal mixing and compaction temperatures for the bitumen. A viscosity of 0.2 Pa s is generally used for mixing, while that for rolling – which depends on the workability of the mix – ranges from 1–10 Pa s for a rolled asphalt (RA) to 2–20 Pa s for an asphalt concrete (AC) (Nicholls and Daines, 1994). When in service in the pavement, a bitumen is a semi-solid with a viscosity of the order of 10^9 Pa s. (By comparison, water has a viscosity of 0.001 Pa s.)

The loss-on-heating test (BSI, 2009c) is an accelerated volatilisation test designed to ensure that excessive hardening of the bitumen does not occur (through loss of volatile oils) during the storage, transport and application of a binder. A 50 g sample of the binder in a small container is left for 5 h in a revolving-shelf oven that is maintained at 163°C. The percentage loss in mass after the sample has cooled, and the percentage loss in penetration, are then recorded.

The solubility test (BSI, 2014a) ensures that the amounts of impurities picked up during the storage of a bitumen, and the amounts of salt that may not have been removed during the refining process, do not exceed the allowable limits. With this test, a specified amount of binder is dissolved in trichloroethylene, a very toxic solvent; after filtering the solution through a fine-porosity filter, the percentage material that is insoluble is obtained by difference.

With the rolling thin-film oven (RTFO) test (Hveem *et al.*, 1963; BSI, 2014b), eight bottles, each containing 35 g of bitumen, are arranged horizontally in openings within a vertical carriage in an oven that is heated to 163°C. The carriage is rotated at 15 rpm and, every 3 or 4 s, a heated jet of air is blown onto the moving film of bitumen in each bottle. After 75 min, the loss in mass, penetration and softening point of the bitumen are determined. The RTFO test measures hardening or 'ageing' of the bitumen by both oxidation and evaporation, and the results obtained correlate well with the ageing that occurs during mixing with hot aggregate.

The flash point test (ISO, 2002) is carried out by heating a sample of bitumen at a uniform rate while periodically passing a small flame across the material. The temperature at which the vapours first burn with a brief flash is the flash point of the binder, while that at which the vapours continue to burn for at least 5 s is its fire point. The flash point is the more important of the two values, as it indicates the maximum temperature to which the binder can be safely heated. The flash points of most penetration-grade bitumens lie in the range 245–335°C.

6.1.2   Bitumen composition and properties

Bitumen may be considered a colloidal system of high-molecular-weight asphaltene micelles dispersed in a lower-molecular-weight maltene medium. The micelles have an absorbed sheath of high-molecular-weight aromatic resins that act as a stabilising solvating layer (Read *et al.*, 2003). Further away from the centre of the micelles, there is a gradual transition to less-aromatic resins, and these, in turn, extend into the less-aromatic oily maltene dispersion medium.

The asphaltenes are brown to black, highly polar, amorphous solids containing carbon and hydrogen, and some nitrogen, sulphur and oxygen. The asphaltene content signifi-cantly influences the rheological characteristics of a bitumen: for example, increasing the content gives a lower-penetration, harder bitumen with a higher softening point and, consequently, a higher viscosity. The resins are dark-brown semi-solids or solids with a high polarity that makes them strongly adhesive. The aromatics are dark-brown viscous liquids that comprise 40–60% of a bitumen. The saturates are non-polar viscous oils that are straw or white in colour; they comprise 5–20% of a bitumen.

Blowing air through a bitumen during its manufacture increases the asphaltene content and decreases the aromatic content, thereby decreasing the temperature susceptibility of the binder. Mixing (especially), storage, transport, placement and compaction, as well as in-service use, also have the ageing effect of increasing the asphaltene content and decreasing the aromatic content.

The stress–strain relationship for bitumen is more complex than for materials such as steel and cement concrete. At low temperatures and short loading times, bitumen behaves almost as an elastic (brittle) solid, whereas at high temperatures and long loading times, its behaviour is almost that of a viscous fluid. At intermediate tempera-tures and loading times (as experienced at ambient air temperatures and under pulse loadings caused by moving traffic), the response of bitumen is in the visco-elastic range.

The viscous response of bitumen to a load pulse means that peak stress, σ_0, and peak strain, ε_0, are not contemporaneous. Thus, the stiffness response is characterised, mathematically, in terms of a real and an imaginary component. Commonly, the complex shear modulus, G^*, and the visco-elastic phase angle, δ, between the peak stress and peak strain occurrences are defined, where

$$G^* = \sigma_0/[2\varepsilon_0/(1 + v^*)] \qquad (6.5)$$

where v^* is the complex Poisson's ratio.

G^* and δ can be measured directly using a dynamic shear rheometer (Airey and Hunter, 2002). The dissipated energy, $G^* \sin \delta$, is used to characterise the fatigue performance, while the value $G^*/\sin \delta$ is used to characterise the rutting potential.

The stiffness modulus of a bitumen, S_b (MN/m^2), under a given loading time, t (s), at a temperature, T (°C), can be estimated from the van der Poel nomograph (van der Poel, 1954), using the American Society for Testing and Materials (ASTM) softening point (SP) temperature, and the penetration index, PI. For repeated numerical calculations, the following equation, which was derived (Ullidtz, 1979) from the nomograph, can be used to estimate S_b:

$$S_b = 1.157 \times 10^{-7} t^{-0.368} e^{-PI}(SP - T)^5 \tag{6.6}$$

provided that $0.01\text{ s} < t < 0.1\text{ s}$, $-1 < PI < +1$, and $10°C < (SP - T) < 70°C$.

The stiffness modulus of the bitumen can, in turn, be used to estimate the stiffness modulus of the material in which it is used as a binder. Allowing for hardening during mixing and laying to about two-thirds of the original penetration, typical stiffnesses for original 50 and 100 pen bitumens at 15°C are 110 and 55 MPa, respectively, for traffic moving at 50 km/h (i.e. $t = 0.02\text{ s}$).

6.2. Using bitumen at lower temperatures

The consistencies of penetration-grade bitumens vary from very viscous to semi-solid at ambient air temperatures, so they are normally heated to 140–180°C for use in road pavements. Nowadays, there is an increasing interest in finding low-energy means of asphalt production. This has prompted greater take-up of existing, and the development of newer, low-temperature techniques for reducing the viscosity of a bitumen so as to enable efficient mixing with the aggregate.

6.2.1 Cutback bitumens

A cutback bitumen is formed by the addition of a solvent so that it can be applied at ambient temperatures with little or no heating (BSI, 2013a). Various types are in use that can be classified as slow curing, medium curing or rapid curing, depending upon the nature of the volatile solvent used in their preparation.

A medium-curing cutback is mainly used in the UK for surface dressing or maintenance patching purposes, but it may also be used in open-textured bitumen macadams that are sufficiently porous for the solvent to evaporate fairly quickly. This cutback is normally produced by blending kerosene with a 70–300 pen bitumen. After application, the solvent dissipates into the atmosphere, and leaves the bitumen behind as a bonding agent. As full release and evaporation can take many months, or even years, asphalt formed with cutback bitumen will usually remain in a softer condition, especially early in its life, than would be predicted from the original penetration of the binder.

In practice, an adhesion agent is usually included in the formulation of cutback bitumen to assist the 'wetting' of the aggregate and minimise the risk of the early stripping of bitumen from the stone in damp or wet weather, while the solvent is evaporating.

As the viscosity of the cutback bitumen is low, it is specified by the time, in seconds, required for 50 ml to flow through a 10 mm diameter orifice (BSI, 2011a). Its absolute viscosity (in Pa s) can be determined by multiplying the flow time, s, by the density of the cutback bitumen (in g/ml) and by a constant ($=0.400$).

The flash point of a cutback bitumen can be as low as 65°C, due to the solvent. Hence, these binders require care to ensure their safe handling during storage and at working temperatures.

6.2.2   Bitumen emulsions

As bitumen and water are immiscible, another way of reducing the viscosity of a bitumen at low temperature is to add it to water and create an emulsion. An emulsion is a fine dispersion of one fluid (in the form of minute droplets) in another liquid in which it is not soluble. Bitumen emulsions mostly used in the UK are a stable suspension of 70–300 pen bitumen globules, each typically less than 20 μm in diameter, in a continuum of water containing an emulsifying agent; the function of the emulsifier, which is adsorbed on the surfaces of the globules, is to prevent them from coalescing.

Factors affecting the production, storage, use and performance of bitumen emulsions include chemical properties, hardness and quantity of the base bitumen; the bitumen globule size; the type, properties, and concentration of the emulsifying agent; and the conditions under which the manufacturing equipment is used (e.g. temperature, pressure and shear).

Most emulsions are produced by heating the base bitumen to ensure a viscosity of less than 0.2 Pa s, and shredding it in a colloidal mill in the presence of a solution of hot water and emulsifier. The temperature of the emulsifying solution is adjusted so that the temperature of the finished emulsion is less than the boiling point of water, and, for safety reasons, is normally less than 90°C. An emulsifier consists of a long-chain hydro-carbon terminating in a polar molecule that ionises in the water; the non-polar paraffin portion is lipophilic with an affinity for bitumen, while the polar portion is hydrophilic with an affinity for water. A cationic emulsion is produced with the aid of an emulsifying solution prepared by dissolving a long chain of fatty amine in hydrochloric or acetic acid while an anionic emulsion is produced with an emulsifier prepared by dissolving a fatty acid in sodium hydroxide.

When the concurrent streams of hot bitumen and hot water and emulsifier are forced through the minute clearance in the colloidal mill, the shearing stresses created cause the bitumen to disintegrate into globules. Many of the lipophilic chains are adsorbed by these globules, and positively charged ammonium ions are located at the bitumen surface in the case of a cationic emulsion, and negatively charged carboxyl ions at the bitumen surface in the case of an anionic emulsion. In both cases, the like-charged globules repel each other when they come into contact, and this prevents their coalescence. Cationic and anionic types of bitumen emulsion should never be mixed, as breakdown of both emulsions will then occur.

When a bitumen emulsion – which is brown in colour – comes into contact with a mineral aggregate, the emulsifier ions on the globules and in solution are attracted to opposite charges on the aggregate surfaces; this causes a reduction in the charges on the globule surfaces and the initiation of the breaking of the emulsion. The first step in the breaking process, termed 'flocculation', results in electrostatically reduced globules forming a loose network of globule flocs. The greater the amount of emulsifier present, the slower the rate of globule flocculation. Bitumen coalescence takes place rapidly when the charges on the globule surfaces become depleted and the aggregate surfaces become covered with hydrocarbon chains. The breaking process is completed when continuous black bitumen films are formed that adhere to the aggregate particles, and the residual water in the system has evaporated. The water evaporation is fairly rapid in favourable weather, but is slowed by low temperatures, high humidities and low wind velocities. Unlike hot mixes of bitumen and aggregate where the hot liquid bitumen fully coats the hot stones, in emulsion mixes the bitumen globules tend to form spot bonds between stones. This does not mean that the bonding between stones is necessarily any weaker than in a hot mix, provided that the spots are located where the aggregates touch, but care is necessary to achieve durability.

The aggregate grading is important to the breaking process, with finer aggregates (i.e. with large surface areas) breaking more quickly. Dirty aggregates accelerate the breaking process. The more porous and dry the road surface and/or the aggregate being coated, the more quickly water from the emulsion is removed by capillary action, and the more rapidly breaking is completed under the prevailing atmospheric conditions.

Cationic emulsions are generally considered to be more effective than anionic ones, and are more commonly used, under all weather conditions. Most aggregates exhibit negatively charged surfaces so that a positively charged cationic emulsifier attached to a bitumen globule is attracted to the negative charges on the surface of the aggregate, whereas there is no attraction between negatively charged anionic emulsifier and negatively charged aggregate surfaces. (In practice, however, this preference may not be very important, as the negative charge on the aggregate is rapidly neutralised, and further breaking of the emulsion depends on evaporation (Jackson and Dhir, 1997); furthermore, some aggregates exhibit positive charged surfaces.)

6.2.2.1 Emulsion specifications

Cationic bitumen emulsions used in the UK are classified according to Table 6.2 (BSI, 2013c). Using this table, an emulsion designated C 69 B 3 would be a cationic bitumen

Table 6.2 Denomination of the abbreviation terms for cationic bitumen emulsions

Position	Letters	Denomination	Supporting document
1	C	Cationic bituminous emulsion	Particle polarity (BSI, 2009e)
2 and 3	2-digit number	Nominal binder content in % (m/m)	Water content (BSI, 2012a) or recovered binder + oil distillate (BSI, 2009f)
		Indication of the type of binder:	
4 and 5	B	Paving-grade bitumen	Specification for paving grade bitumen (BSI, 2009a)
or	P	Addition of polymers	
4 and 6	F	Addition of more than 2% (m/m) of flux based on emulsion	
5 or 6 or 7	1–7	Relative breaking behaviour	Breaking value (BSI, 2009g)

emulsion with a nominal binder content of 69% produced from paving-grade bitumen and having a class 3 breaking value.

Breaking values of 3 or 4 (the last digit of the designation code) are characterised by a rapid deposition of bitumen on aggregates and road surfaces, with consequent early resistance to rain. While normally fluid enough to be applied at atmospheric temperatures, the high-bitumen-content C grades (typically C 60 to C 69) must be applied hot. While these emulsions are generally unsuitable for mixing with aggregates, they are commonly used for surface dressing, grouting, patching, and formation and sub-base sealing.

With breaking values of 5, the rate of deposition of the bitumen is sufficiently delayed to allow mixing with some clean coarse aggregates. Applied cold, these emulsions are mostly used to prepare coated materials that can be stockpiled (e.g. for remedial patching). With breaking values of 7, the rate of deposition of the bitumen is sufficiently delayed for them to be cold mixed with certain fine aggregates. Cationic slurry seals are also formed with these emulsions.

At the time of writing, anionic emulsions are not regulated by EN classifications. In the UK, a standard (BSI, 2011b) is in use that designates anionic emulsions with a code starting with the initial A. The second part of the code is a number ranging from 1 to 4 that indicates the stability or breaking rate of the emulsion. The higher the number the more stable the emulsion. Class A1, A2 and A3 anionic emulsions are also described as being labile, semi-stable and stable, respectively. A special class A4 emulsion has also been formulated for use in slurry seal work.

The ease with which class A1 emulsions break makes them unsuitable for mixing with aggregates, and they are mostly used for surface dressing, tack coating, patching, formation and sub-base sealing, and concrete curing. Labile emulsions (and semi-stable ones) cannot be stored out of doors in very cold weather, as they will usually break upon freezing. Class A2 emulsions have enough emulsifier to permit mixing with aggregates provided that the percentage passing the 75 μm sieve is well below 5%. Class A3 emulsions can be cold mixed with aggregates, including those with large proportions of fines or chemically active materials such as lime or cement. They can also be stored out of doors in cold climates without breaking and, in warm climates, they can be admixed with soil in soil stabilisation works. When class A4 emulsions are mixed with aggregates specified for use in slurry seals, the mixes form free-flowing slurries whose stability is maintained throughout the laying procedure; also, their setting times can be varied.

The third part of the anionic emulsion classification code is a number that specifies the bitumen content of the emulsion: for example, an A1-70 emulsion is an anionic emulsion, rapid acting, with 70% by mass of residual bitumen.

Emulsion viscosity, which is of considerable importance as most emulsions are applied through nozzles on a spray bar, may be measured using either an Engler viscometer or a Redwood II viscometer (BSI, 2011b). In the case of the Engler apparatus, it is the time (in s) for 200 ml of emulsion at 20°C to flow under gravity through a standard orifice; with the Redwood II test, it is the time (in s) for 50 ml of emulsion to flow from the viscometer at 85°C.

6.2.3 Foamed bitumen

A third technique to encourage bitumen to coat stones without heating them to high temperatures is by injecting hot bitumen (at 170–180°C) into a low-volume stream of cold water that then flash boils to form a foam mousse. As this mousse is mixed with unheated aggregate, the bubbles burst over the stones and leave a coating of bitumen. As the bubbles in the foam are formed of steam, they almost immediately start to shrink, so that mixing with the aggregate must immediately follow foam formation.

Two properties are defined for the foam (Jenkins *et al.*, 1999) (Figure 6.4): (a) the expansion ratio, ER_m, being the volume of foam generated expressed as a ratio of the volume of the same mass of binder once all the bubbles have burst, and (b) the half-life, $\tau_{1/2}$, being the time for the volume of foam to reduce by 50% from its maximum volume. Typically, expansion ratios of around 10–15 s and half-lives of around 20 s are sought. Commonly, foamed bitumen is introduced into aggregate, or road planings from a pavement being rehabilitated, through nozzles placed in an in situ road planer/mixer so that the foam is instantly mixed with the aggregate and placed in position ready for compaction. Plant-based foam mix is also possible.

The foamed bitumen mixing process is very different from that employed in a hot mix and, hence, different properties and durabilities may result. Given that some water is introduced with the foam and the aggregate used is probably not dry, small cement contents – typically 2% or 3% by mass of the total mix – may be added to remove water from the mixture into cement hydration products and deliver some bond strength. The proportion of cement can be critical: too little and there is insufficient to deliver a benefit; too much and the material is so embrittled that it behaves like a weak concrete and loses the flexible nature of an asphalt.

Figure 6.4 Foam characteristics for a typical penetration-grade bitumen showing (a) selection of the percentage water content by mass and (b) the foam volume decay with time (Jenkins *et al.*, 1999)

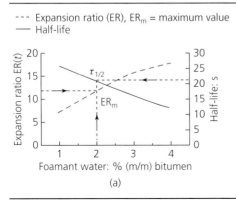
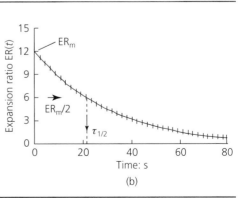

6.2.4   Warm bitumen

An additive can be introduced into bitumen to reduce its viscosity at a medium temperature (typically 90–125°C) or to change the way it interacts with aggregate at these temperatures, or to induce foaming. Above its wax transition temperatures, a wax acts as a very low-viscosity dilutant of the bitumen, so that it attains a sufficiently low viscosity to permit stone coating at a much lower temperature than in a conventional hot mix. On cooling to below the transition temperature the wax changes to a soft solid with very high viscosity, and an almost step change in viscosity is seen, allowing conventionally high viscosities to be achieved at pavement operating temperatures.

Natural or artificial zeolites are compounds that liberate water from their chemical make-up upon heating. If they are mixed in with the aggregate, they will foam when hotter bitumen is added, thereby enhancing the mixing process, even though the bitumen would otherwise be more viscous than desired. Other types of additive work in other ways: for example, by modifying the stone–bitumen wetting so that the stiffer bitumen encountered at medium temperatures coats the stones well. Whatever the additive, the aim is to achieve hot-mix-like properties by having to heat the aggregate much less, thereby saving both energy and carbon emissions, and providing a healthier work environment by reducing volatile hydrocarbon emissions on site.

6.3.   Modified bitumens

As polymer science has advanced, many modifying additives have been developed with the aim of improving the properties of penetration-grade bitumens and emulsions. Some of these modifiers are proprietary products and some are in the public domain. The fact that there has been, and are, a multiplicity of modifiers can create confusion as to their relative roles and values.

When evaluating a modifier, it should be borne in mind that: (a) it is the predominantly viscous response of bitumen at high temperatures (i.e. its ability to flow) that makes it workable for contractors to produce and lay bituminous materials, (b) it is the elasticity of the binder, which predominates at lower pavement temperatures, that gives a bituminous material its structural integrity, and (c) compatibility between bitumen (or bitumen emulsion or foamed bitumen) and aggregate is driven by the physicochemical characteristics of each component and is implicated in the durability of bituminous materials.

At ambient temperatures a small, viscous, strain component still remains after the bitumen has undergone elastic loading and recovery; and it is the accumulation of these viscous responses to millions of axle load applications that contributes to rutting (especially at high ambient temperatures). At low temperatures, bitumen is susceptible to brittleness and cracking. These weaknesses can be alleviated by increasing the overall stiffness of the bitumen used in locales with high ambient road temperatures (to reduce rutting) and by decreasing the viscosity of the binder used in locales with low road temperatures (to reduce cracking). Many additives aim to achieve one of these without changing the other.

Other additives are designed to increase durability, particularly against damage due to loading applied in wet conditions that might induce stripping.

These are the main reasons why small quantities of (relatively expensive) modifiers are added to bitumens used in wearing courses and to emulsions used for surface courses, at difficult road sites, such as at traffic lights, roundabouts and pedestrian crossings, on busy urban streets with large volumes of slow-moving traffic, and in locales subject to temperature extremes. (Modifiers have other advantages also, and it can be expected that their use in bituminous surfacings will grow in future years.)

The main bitumen modifiers in current use can be divided into adhesion agents, thermoplastic crystalline polymers, thermoplastic rubbers, thermosetting polymers and chemical modifiers. The polymer modifiers are classified in a British standard (BSI, 2010a).

6.3.1 Adhesion agents

The coating of aggregates with bitumen is facilitated by using a low-viscosity, high-penetration bitumen in the mix. However, if damp weather conditions prevail during or immediately after the laying, water may be preferentially attracted to the aggregate surfaces so that stripping of the bitumen takes place; however, the more viscous the binder the better is its adhesion to the stone and the less likely it is that stripping will occur. Hence, the choice of bitumen to use in a given circumstance is often a trade-off between ease of mixing (i.e. the workability of the mix) and the desire for better adhesion so as to minimise the risk of stripping.

One role of a conventional filler material in a bituminous mix is to increase the viscosity of the binder, thereby lessening the risk of stripping. It is well established that, if a small amount of hydrated lime or cement (say, 1–2% by mass of the aggregate in the mix) is included as a replacement for some conventional filler material, a chemical action will take place between the additive and the bitumen that results in the formation of compounds that are adsorbed on negatively charged aggregate surfaces, and this has the effect of improving adhesion and rendering the bitumen less vulnerable to stripping.

6.3.2 Thermoplastic crystalline polymers

Also known as thermoplastic plastomers, these polymers include ethylene vinyl acetate (EVA), polyethylene (PE), polypropylene (PP), polystyrene and polyvinylchloride (PVC). Of these, the EVA polymer, which results from the copolymerisation of ethylene and vinyl acetate, is probably most widely used to modify bitumens for roadworks. The properties of an EVA copolymer are controlled by its molecular weight and vinyl acetate content, namely the lower the molecular weight the lower the viscosity and, hence, the stiffness, while the greater the vinyl acetate content the more 'rubbery' (i.e. flexible) the material.

Various copolymer combinations are now available for use with (usually) more workable, lower-viscosity bitumens. All blend well with bitumen and are thermally stable at normal asphalt mixing temperatures. Thus, a 70 pen bitumen with 5% EVA may be used as an alternative to a 50 pen bitumen. Tests have shown that, as well as being more workable, this combination provides increased stiffness and greater rutting resistance at ambient temperatures, while the elastic properties of the bitumen are little affected. The addition of EVA to a surface course also aids compaction during laying at low temperatures by improving the workability of the mix, and by providing more time for the embedment of coated chippings (Nicholls, 1994).

6.3.3 Thermoplastic rubbers

Also known as thermoplastic elastomers, these modifiers include polymers such as natural rubber, vulcanised rubber, styrene–butadiene–styrene (SBS) block copolymer, styrene–ethylene–butadiene–styrene (SEBS) block copolymer, styrene–isoprene–styrene (SIS) block copolymer and polybutadiene. As thermoplastics, they soften on heating and harden when cooled to ambient temperatures.

Experiments on the use of rubber in bitumen have been carried out for over 100 years, and much literature exists on the subject. One review of the literature (Dickinson, 1977), on the use of natural and vulcanised rubber in bituminous mixtures, concluded that rubber additives enhance the elastic responses of bitumen at higher ambient temperatures, and result in materials that have a marked increase in resistance to deformation simultaneously with reduced brittleness at low temperatures. However, when crumb rubber is added to a bitumen, only a proportion of the rubber particles disperse in the binder, and the remainder acts as a soft 'aggregate' or cushion between stones in the total aggregate–binder structure. Later work (Sainton, 1996) showed that if crumb rubber recycled from used tyres is used with a catalyst that makes it soluble in bitumen in dense surfacings, the result is a low-noise-emitting material.

The SBS and SIS styrenic block copolymers are now regarded (Read *et al.*, 2003) as the elastomers with the greatest potential provided that they are admixed with bitumens with which the copolymers are compatible. If they are not, the viscosity (stiffness) of the modified bitumen may not be changed in the desired direction.

A polymer is a long molecule consisting of many small units (i.e. monomers) that are joined end to end, and copolymers are composed of two or more different monomers. By commencing polymerisation with styrene monomer, then changing to a butadiene feedstock, and finally reverting to styrene, it is possible to make a butadiene rubber that is tipped at each end with polystyrene: this product is SBS. The polystyrene end blocks give strength to the polymer while the mid-block butadiene rubber gives it its elasticity. At mixing and compaction temperatures above 100°C, the polystyrene will soften and allow the material to flow; on cooling to ambient temperatures, the copolymer will regain its stiffness and elasticity. At high in situ temperatures of 60°C, SBS-modified bitumens are significantly stiffer and more resistant to permanent deformation than equivalent unmodified bitumens; at very low temperatures, the viscosity is reduced, and the modified binders exhibit greater flexibility and provide greater resistance to cracking (Preston, 1992).

6.3.4 Thermosetting polymers

Polymers such as the acrylic, epoxy or polyurethane resins result from the blending of a liquid resin and a liquid hardener that react chemically with each other. They are described as thermosetting because, unlike thermoplastic polymers, their flow properties are not reversible with a change in temperature, once the material has cooled down to the ambient temperature for the first time.

When the two-component polymers are mixed with bitumen, the resulting binder displays the properties of modified thermosetting resins rather than of conventional thermoplastic bitumens (Sainton, 1996); that is, the binder (a) becomes an elastic material that does not exhibit viscous flow when cured, (b) is more resistant than bitumen to attacks by solvents and (c) is less temperature susceptible than bitumen and is essentially unaffected by in-service pavement temperature changes.

6.3.5 Chemical modifiers

Sulphur and manganese, which are readily available in North America, are two chemical modifiers that have been investigated for use with bitumen.

While sulphur is a yellow solid material at standard conditions of temperature and pressure, it melts at about 119°C and then exists as a low viscosity (in comparison with bitumen) liquid between 120°C and 153°C: at about 154°C, it starts to become, rapidly, very viscous. Above 150°C the molten sulphur reacts vigorously with hot bitumen, to release hydrogen sulphide (H_2S), which is a toxic gas and a health hazard; also, the higher the temperature the more rapid the rate of H_2S evolution.

Depending upon the composition of a bitumen, at least 15–18% by mass of sulphur is normally added to the bitumen prior to admixing with aggregate. Some of the sulphur reacts chemically with the bitumen, while the remainder remains in suspension as a separate phase provided that the temperature does not exceed 150°C. As the mix cools, the excess sulphur in the void spaces slowly recrystallises, keying in the coated particles and increasing inter-particle friction, and imparting high mechanical stability to the mix. Because of the fluidity of sulphur when heated, its addition reduces the viscosity and increases the workability of the binder–aggregate mix; because of the beneficial effect of sulphur on the stiffness modulus after cooling, higher-penetration bitumens can be used, thereby rendering the mix even more workable.

Because of the relative fluidity of the mix, sulphur–bitumen materials are very adaptable to thin-layer work.

Organo-manganese compounds have also been proposed as additives to bitumen on the grounds that they improve the temperature susceptibility of the binder, thereby improving the dynamic stiffness and resistance to deformation of the bituminous material. These compounds must be premixed with a carrier oil if they are to be dispersed quickly in a bitumen.

6.4. Road tar

Tar – a viscous liquid with adhesive properties – was formerly produced as a by-product of 'town' gas production and was widely used as an asphalt binder in the UK until the 1970s. It is now known to produce carcinogenic fumes when hot. So, if encountered during the rehabilitation of pavement layers placed before the 1980s, excavated material should be discarded at an appropriately licensed tip. Hot reworking must never be practised, but cold in situ recycling *may be* legally possible if the material never comes under the EU waste definition.

6.5. Cements

The term 'cement' is usually associated with Portland, slag, pozzolanic and high-alumina cements, all of which are finely ground powders that, in the presence of water, have a chemical reaction (hydration) and produce, after setting and hardening, a very strong and durable binding material.

The term pozzolan indicates materials that, in the presence of lime and water, undergo a chemical reaction that results in the formation of hydrous calcium aluminates and silicates similar to the reaction products of hydrated cement. Setting is the change in the cement paste that occurs when its fluidity begins to disappear; the start of this stiffening process is the 'initial set', and its completion is the 'final set'. Hardening, which is the development of strength, does not begin until setting is complete. The setting time and the rate of setting are considerations of vital importance affecting the use of cement in road pavements. Gypsum (i.e. calcium sulphate) retards setting, and, hence, a small amount is added to cement during its production.

The literature on cement is legion and easily available, as are the methods of testing cement. Hence, the following is only a very brief and general description of the more important cements used in roadworks.

6.5.1 Portland cement

First used in 1824, Portland cement is named after the natural limestone found on the Portland Bill in the English Channel, which it resembles after hydration. The raw materials used in the preparation of this cement are calcium carbonate, which is found in limestone or chalk, and alumina, silica and iron oxide, which are found combined in clay or shale. Marl, a calcareous mudstone (i.e. a mixture of calcareous and argillaceous materials), is also used to make cement.

Cement is usually prepared by grinding and mixing proportioned amounts of the raw materials, and feeding the mix through a high-temperature kiln (traditionally coal fired, but alternative fuels made from organic wastes are being increasingly used). Depending on the hardness of the raw materials used and on their moisture contents, the mixing and grinding process can be done either in water ('wet process') or in a dry condition ('dry process'). The former is technically preferable because of its greater effectiveness in grinding and mixing but it requires much more heat energy in the subsequent kiln stage as the water must be evaporated. In the present energy and emissions-conscious age, most new cement production equipment implements the dry process.

When the temperature of the material in the kiln is approximately 1450°C, incipient fusion takes place, and the components of the lime and the clay combine to form clinkers composed of tricalcium silicate ($3CaO \cdot SiO_2$, abbreviated to C_3S), dicalcium silicate ($2CaO \cdot SiO_2$, or C_2S), tricalcium aluminate ($3CaO \cdot Al_2O_3$, or C_3A) and calcium

Table 6.3 Effects of the main Portland cement compounds

Principal constituent compounds	Rate of chemical reaction and heat generation	Most active period	Contribution towards the final strength
C_3S	Moderate	Days 2–7	Large
C_2S	Slow	Day 7 onwards	Moderate
C_3A	Fast	Day 1	Small

aluminoferrite ($4CaO \cdot Al_2O_3 \cdot Fe_2O_3$, or C_4AF). The burnt clinkers are then allowed to cool, and taken to ball-and-tube mills, where they are ground to a fine powder. Gypsum (typically 1–5%) is added during the grinding process. Table 6.3 compares the contributions to final strength made by the main cement compounds.

In addition to the main compounds noted above, cements contain a number of other minor compounds at low percentages. Two of these, oxides of sodium and potassium (i.e. Na_2O and K_2O), also known as the alkalis, are of interest in that products of their reaction, with some aggregates, can cause a disintegration of the concrete and detrimentally affect the rate of the gain of concrete strength. Thus, it is important that their content is kept low in cement.

The EN standards now used throughout Europe define five types of cement (Table 6.4). While CEM I is the purest form, it is not the most common. CEM II (somewhat incorrectly known as ordinary Portland cement) is the workhorse of the cement industry, and is most commonly used in road pavements. Pulverised fuel ash is commonly included in CEM II in addition to pure Portland cement. Known as 'PFA' in the UK and 'fly ash' in the USA, this waste product (typically obtained from the burning of coal to generate

Table 6.4 Classes of cement according to BSI (2011c)

Designation[a]	Type of cement	Constituents
CEM I	Portland cement	Comprising Portland cement and up to 5% of additional constituents
CEM II	Portland-composite cement	Portland cement and up to 35% of other single constituents
CEM III	Blast furnace cement	Portland cement and higher percentages of blast furnace slag
CEM IV	Pozzolanic cement	Portland cement and up to 55% of pozzolanic constituents
CEM V	Composite cement	Portland cement, blast furnace slag or pulverised fuel ash and pozzolan

[a] Suffixes are added to the basic designators given

electricity) has pozzolanic properties of its own. Powdered limestone may also be included in CEM II. These additives help ensure a medium rate of hardening, and allow economies to be made by the manufacturers.

Rapid-hardening Portland cement is a special material meeting the CEM I requirements but with the additional characteristic that the cement powder is particularly finely ground. There are also CEM II rapid-hardening cements: as well as being finely ground, they typically contain a greater proportion of the C_3A component and less of the pozzolanic 'dilutants' found in standard CEM II. With more fast-setting and less slow-setting components, and with a greater surface area available for hydration reactions, these cements harden more rapidly. Their higher rate of strength development lends itself to usage in situations where an accelerated road-opening to traffic is required (e.g. patching of concrete pavements), or where hardening must be achieved before weather-induced damage is incurred (e.g. by overnight frost action). Rapid-hardening cement is more expensive than ordinary cement because of the extra clinker grinding required to achieve the fine powder.

CEM II/SR, sulphate-resisting Portland cement, is similar to conventional CEM II except for its special capability in resisting chemical attack from sulphur trioxide (SO_3) that is present in seawater, some groundwaters, gypsum-bearing strata, certain clay soils in hot countries and/or some industrial wastes. Damage can occur in concretes if sulphates in external water react with components of the previously hydrated and hardened cement components, because some of the reactions are expansive and cause cracking. Sulphate resistance is achieved by adding tetracalcium trialuminate sulphate to, and reducing the amount of C_3A in, the conventional CEM II cement: the provision of sulphate in the cement ensures that any damaging reactions have taken place before hardening.

CEM III, blast furnace slag cement, is a mixture of Portland cement and up to 65/90/95% (suffixes A/B/C, respectively) of ground granulated blastfurnace slag (GGBS), a by-product of the manufacture of iron. This slag, which is a relatively inert material by itself, is important because it is pozzolanic; that is, the hardening of the pozzolan is characterised by two processes: (a) the cement clinker hydrates when water is added, and (b) as this hydration occurs, there is a release of free calcium hydroxide that enters into secondary reactions with the slag.

GGBS cement concrete hardens more slowly than ordinary CEM II concrete, but there is little difference in the final strength achieved by either. The use of GGBS cement is most justified in locales where it is economically produced and where high early pavement strength is not an essential requirement (e.g. in ground stabilisation works). CEM III is also sulphate resisting due to the higher proportion of non-sensitive hydration products.

6.5.2 Pozzolanic cement

CEM IV, pozzolanic cement, is made by grinding together an intimate mixture of Portland cement clinker and a pozzolanic material such as natural volcanic ash or finely divided PFA. The significant pozzolanic component results in a slow-hardening, low-heat cement with lower water-to-cement ratios for a desired workability, thereby providing a denser concrete of lower permeability and greater durability, and these help to impart a degree of resistance to chemical attack from sulphate and weak acid. Where proximal sources of the pozzolan allow it to be economically sourced, it is particularly suitable for use in soil stabilisation works where large volumes of cement are normally employed and rapid strength gain is less important.

6.6.   Limes

Lime is calcium oxide (CaO) that is produced by calcining (burning) crushed limestone in either shaft (vertical) or rotary (near-horizontal) kilns. If the limestones are pure or near-pure calcium carbonate ($CaCO_3$), the limes produced are termed calcitic or high-calcium quicklimes. If the limestones are dolomitic and contain a high proportion of magnesium carbonate ($MgCO_3$), the lime products are termed dolomitic or magnesium quicklimes.

'Slaking' is the general term used to describe the combining of quicklime with water to produce a hydrated lime. Depending on the amount of water added, the product may be a powder or a slurry of varying degrees of consistency. When just sufficient water is added to a high-calcium quicklime to satisfy its chemical affinity for moisture under the hydration conditions, all of the calcium oxide is converted into calcium hydroxide ($Ca(OH)_2$), with the evolution of heat, and the product is called calcitic or high-calcium hydrated lime. The magnesium oxide (MgO) component of dolomitic quicklime does not hydrate so readily at the temperatures, atmospheric pressure and retention times normally used in the hydration process; in this case the product is termed dolomitic monohydrate lime (i.e. $Ca(OH)_2$ plus MgO).

Calcitic and dolomitic limes are used in road pavements in both quicklime and hydrated lime forms. Hydrated lime powder contains, typically, about 30% water: this makes it more expensive than quicklime to transport, especially if large quantities and long haulage distances are involved. While both hydrated limes and quicklimes release heat upon contact with water (an exothermic reaction), the heat given off by quicklime is much greater, and this makes it dangerous for construction workers to use, especially in windy weather conditions. Thus, hydrated lime powder or slurry is almost invariably used on smaller jobs. The use of quicklime is usually constrained to large projects where specialist plant and appropriate safety measures are economically implementable, and where there are no neighbours close to the site. Where very wet soil needs treatment, the strongly exothermic reaction provided by quicklime can help achieve rapid soil drying, and this may provide sufficient economic impetus to justify the additional safety precautions then necessitated.

6.7.   Conventional aggregates

A little over 200 Mt of aggregate is sold every year in the UK (ONS, 2009; MPA, 2012): this comprises 55 Mt of sand and gravel (80% and 20% land won and sea dredged, respectively), 90 Mt of crushed rock (47%, 43% and 10% limestone/dolomite, igneous and sandstone, respectively), and 60 Mt of recycled and secondary aggregates. Of the total sales, some 90% goes into the construction industry, of which about 37% (mostly gravel) ends up in concrete, 44% in unbound uses, 14% (mainly crushed rock) in asphalt and the rest in other uses.

6.7.1 Natural rock aggregates

If the glacial drift overburden could be removed and the underlying rocks exposed, the simplified geological map of Great Britain and Ireland would appear as in Figure 6.5. Solid rock suitable for the manufacture of road aggregates is almost exclusively quarried from formations of Palaeozoic and pre-Palaeozoic geological age (Blyth and de Freitas, 1984) as few of the later formations have been sufficiently strengthened by heat and/or pressure processes to be strong enough for this application.

Geologists classify rocks into three main groups, based on their method of origin: igneous, sedimentary and metamorphic.

6.7.1.1 Igneous rocks

Igneous rocks were formed at (extrusive rocks) or below (intrusive rocks) the Earth's surface by the cooling of molten material, called magma, which erupted from, or was trapped beneath, the Earth's crust. Extrusive magma cooled rapidly, and the rocks formed are often glassy or vitreous (without crystals) or partly vitreous and partly crystalline with small grain sizes: extrusive rocks may also contain cavities that give them a vesicular texture. Intrusive rocks cooled more slowly, are entirely crystalline, and the crystals are usually large enough to be visible to the naked eye.

The best igneous roadstones normally contain medium grain sizes. Particles with coarse grains (>1.25 mm) are liable to be brittle and to break down under the crushing action of a roller. If the grains are too fine (<0.125 mm), especially if the rock is vesicular, the aggregates are liable to be brittle and splintery.

When an igneous rock contains more than about 66% silica (SiO_2), it is described as acidic, and as basic if it has less than 55%, with rocks containing 55–66% SiO_2 being termed intermediate. Acidic rocks tend to be negatively charged, and aggregates containing large amounts of feldspar and quartz in large crystals do not bind well with bitumen (which also has a slight negative charge), whereas aggregates that are rich in ferro-magnesian minerals (e.g. basalt and gabbro) bind well with bitumen. A summary classification with some common rock names is provided in Table 6.5.

As many rock aggregates differ little in respect of their practical road-making abilities, it is convenient to combine them into groups with common characteristics, namely the important igneous rock aggregates are the basalt, gabbro, granite and porphyry groups. The main rocks in the basalt group (i.e. basalt, dolerite, basic porphyrite and andesite) are mostly basic and intermediate rocks of medium and fine grain size. The gabbro group primarily comprises gabbro, basic diorite and basic gneiss. Members of the granite group are mostly acidic and intermediate rocks of coarse grain size; the heavily used members of this group are granite, granodiorite, acid gneiss and syenite. Porphyry group members are usually acid or intermediate rocks of, essentially, fine grain size (e.g. granophyre and felsite).

Figure 6.5 Simplified geological map of Great Britain and Ireland

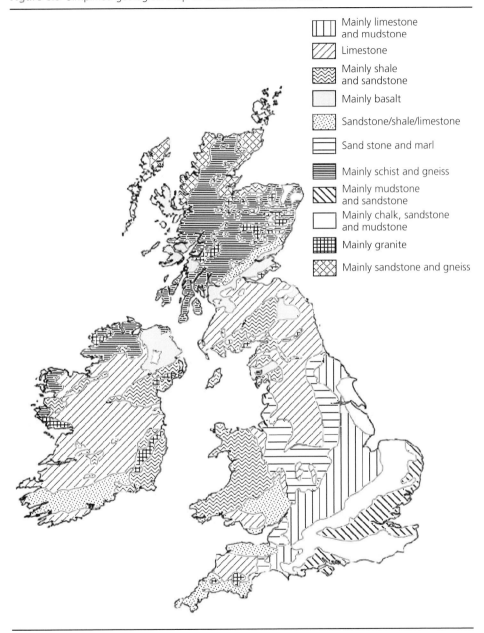

Basalt aggregates are strong but have variable resistance to polishing and variable durability against weathering and climate-induced deterioration. Those containing olivine
that has decomposed to clay have high drying–shrinkage characteristics that lead to
problems in concrete (Hosking, 1992). Granites are strong and their resistance to

Table 6.5 Classification of common igneous and heavily metamorphosed rock types

Origin	Grain size (degree of metamorphism)	Acid	Intermediate		Basic
Intrusive	Coarse Porphyritic[a]	Granite Granophyr	Granodiorite	Diorite	Gabbro
	Medium	Microgranite	Syenite	Microdiorite	Dolerite
Extrusive	Fine[b]	Rhyolite, felsite	Dacite	Andesite	Basalt
Metamorphosed	Coarse (high)	Gneiss, marble, quartzite			
	Fine (moderate)	Schist, marble, quartzite, hornfels			

[a] Some large grains in an otherwise medium/fine-grained matrix
[b] Intrusive rocks are seldom fine grained, and extrusive rocks are seldom other than fine grained, unless porphyritic (e.g. commonly felsite and porphyritic andesite)

polishing is usually good; however, being acidic, more attention may need to be paid to anti-stripping treatment if they are used with bitumen. The porphyries are good all-round roadstones.

6.7.1.2 Sedimentary rocks

These rocks were formed when the products of disintegration and/or decomposition of any older rock were transported by wind or water, re-deposited as sediment, and then consolidated or cemented into a new rock type (e.g. siliceous rocks). Some rocks were also formed as a result of the chemical and/or detrital deposition of organic remains in water (e.g. calcareous rocks).

Argillaceous siliceous rocks were formed when fine-grained particles were deposited as clays or muds and then consolidated by pressure from overlying deposits. These rocks are very fine grained, often highly laminated, and are relatively weak. Hence, they are rarely used as road aggregates and never in asphalt surface courses.

Arenaceous siliceous rocks were formed from deposits of gravel, sand and silt that became lithified as a result of pressure from overlying strata, or by the deposition of cementing material between the grains. Some of these rocks are brittle, while others are quite hard. The predominant mineral is either quartz or chalcedony (both SiO_2), which tends to make good adhesion between these arenaceous aggregates and bitumen more difficult. Depending on the quality of the lithification or cementation, they may suffer from easy abrasion.

Calcareous rocks result from great thicknesses of the remains of marine animals or carbonate precipitation being deposited on the ocean floor. The predominant mineral is calcite ($CaCO_3$), which renders the rocks basic. Some calcareous rocks are too porous to be used as roadstones, and tests (Croney and Jacobs, 1967) have shown that all crushed chalks, and oolitic and magnesian (dolomitic) limestones having an average saturation moisture content greater than 3%, are frost susceptible and, hence, should not be used in pavements in the UK in, for example, a subgrade capping unless their placement is below the maximum depth of frost penetration (typically about 450 mm in the UK).

The important road-making aggregate groups of sedimentary rocks are the gritstones and limestones. Acceptable gritstones are abrasive and highly polishing resistant (e.g. greywackes, tuffs, breccias, fine-grained well-cemented sandstones, siltstones or flagstones). Unacceptable gritstones are mostly coarse-grained sandstones that are deficient in 'cement'. Limestone aggregates are widely used for all construction purposes; however, as they have a high susceptibility to polishing, most are not used in asphalt surface courses.

6.7.1.3 Metamorphic rocks

These are rocks that, as a result of great heat (thermal metamorphism) or heat and pressure (regional metamorphism) were transformed into new rocks by the recrystallisation of their constituents. Thermal metamorphic rocks, which are almost all harder than the rocks from which they were transformed, are in demand as road aggregates. Regional metamorphic rocks are relatively coarse grained, and some are highly foliated, such as schist (from igneous material) and slate (from shale). Aggregates from foliated rocks are less desirable in pavements as they can be quite fissile and may be crushed when compacted with rollers. However, the stronger examples (e.g. some slates) are usable in unbound layers if appropriate compaction plant (e.g. grid-wheeled compactors) is employed.

The metamorphic aggregate groups of importance in road-making are gneiss, hornfels, quartzite and, to a lesser extent, schists. Apart from having poor resistance to polishing, hornfels aggregates are very hard and make excellent pavement materials. While having a tendency to strip because of poor adhesion to bitumen, metamorphic quartzites often make good road aggregates; while they have good resistance to polishing, they are less liked by quarry owners as they induce high wear rates on crushing plant.

6.7.2   Natural rock aggregate production

The production of rock aggregate from intact rock consists of three processes: extraction, crushing and screening.

Extraction is usually carried out using explosives to win stone from the source rock in quarries. Large pieces of won rock have to be reduced in size before being taken to the crusher; this is done by secondary blasting ('pop blasting'), dropping a heavy ball from an overhead jib ('drop balling') and/or using a hydraulic hammer.

Crushing involves continually reducing the size of the stone to the sizes and shapes required by the concrete/asphalt/unbound material specification, using either compression, cone or impact crushers. Compression crushers squeeze the rocks between a fixed plate and a moving member that advances and recedes from the fixed one: the stone moves through the crusher until it is small enough to pass out of the crushing chamber. Cone crushers act in a similar manner, except that the moving part is an eccentrically positioned cone within a conical funnel whereby the annular space opens and closes as the central cone gyrates. Impact crushers repeatedly subject the rock to hammer blows as it passes through the reducing chamber until the particles are small enough to exit the chamber (often through a grid at the outlet).

The predominant size of the multi-sized aggregate produced by a crusher approximates its setting (i.e. the smallest gap between the fixed and moving parts). In practice, the setting required to produce a predominant size of stone is determined by trial and error.

With a compression crusher, the main factor affecting the shape of the aggregate produced is the crusher reduction ratio: this is the ratio of the size of the feed opening to its setting. The lower the ratio the better the aggregate shape; also, the best shapes are obtained with sizes at or close to the setting.

Overall, impact crushers produce better-shaped aggregates than compression or cone ones (Hosking, 1992). However, to minimise wear of crusher parts, impact crushers are mostly used to produce aggregates from more easily crushed rock (e.g. limestone).

The production of aggregates of varying sizes involves a screening process in conjunction with the use of a number of crushing stages. These stages commonly involve a primary crusher, a secondary reducing-crusher, and two or three tertiary reduction-crushers (e.g. cone crushers), which successively reduce the sizes of the stones produced at each stage. Before each crushing stage, material smaller than the next crusher setting is allowed to fall onto vibrating screens: this lessens the amounts of material passing through the following crusher. Usually double- or triple-deck wire mesh screens are used to sort the various aggregate sizes, except at the primary stage when heavy duty perforated plate screens are used to remove waste material (termed 'scalpings').

After the tertiary reduction stage, the complete mix of stones may be stored directly as a 'crusher-run' aggregate that many quarries produce to meet the type 1 grading specification of the UK's highway authorities (HA *et al.*, 2009). For other uses, the crushed stones are fed onto a series of screens to produce commercially saleable single-sized aggregates (typically >40, 40–31.5, 31.5–20, 20–14, 14–10, 10–6.3, 6.3–2 and <2 mm) that are held in storage bins until needed.

A word about terminology: the standardised approach is to define the boundary between coarse and fine aggregates at 2 mm for asphalt, and at 4 mm for all other uses. The term fines is then used to describe the inherent fraction of an aggregate passing a 63 μm sieve while the word filler is applied to the same particles sizes when they are added to influence the properties of a mixture. These definitions have been used in this book but, unfortunately, popular use is not consistent, and the words 'dust' and 'quarry fines' are used somewhat interchangeably: they are also commonly used to refer to the finer fractions of an aggregate after crushing – perhaps >6 or >4 mm, depending on the user (Petavratzi, 2007).

6.7.3   Gravels and sands

Sands and gravels are commonly used in concrete pavements, as their roundness results in good workability and their 'as-dug' gradings often do not require additional processing. Hydrated lime or an anti-stripping agent may be added to some gravels (e.g. those containing flint) if they are to be used in bituminous mixes, to prevent the binder from stripping from the particles. Some gravels are crushed during processing to make them more angular for use in roadworks.

Sands and/or gravels are usually found on land in aeolian deposits such as sand dunes, in alluvial stream deposits such as valley terraces, estuarial margins or alluvial fans, and in glacial deposits such as moraines, eskers and kames (see Chapter 4). They may be won from dry pits using excavation equipment such as drag lines, scrapers, loading shovels and dozers, or from wet pits using suction dredgers, floating cranes and grab or drag line excavators.

Sands and gravels are also extracted offshore, using suction dredgers equipped with hydraulic pumps capable of sucking sand, gravel and water from depths of up to 36 m, through rigid pipes that are dropped to the sea bed. As sand and gravel accumulates in the dredger holds, the sucked water flows overboard. Processing of the materials won from the sea is subsequently carried out on land. Such extraction is deprecated on environmental grounds, as the sea-bed habitat is largely destroyed, and, hence, licensing is limited.

The processing of sand and gravel first involves washing the extracted materials into separate sand and gravel fractions. This washing also removes silt and clay from the fractions and, in the case of marine materials, reduces the chlorides present from the seawater. The gravel fraction (>6.3 mm size material) is then separated into various sizes using vibratory screens: if necessary, crushing is also carried out. The sand fraction is further washed, and, usually, separated into coarse and fine fractions by hydraulic means.

6.7.4   Aggregate properties

Whether used unbound, or bound in a bituminous or cementitious matrix, the most important engineering properties of the aggregates used in road construction and, especially, in pavements are cleanliness; size and gradation; shape and surface texture; hardness and toughness; durability; and relative density.

A clean aggregate is one that has its particles free from adhering silt-size and clay-size material. Aggregate cleanliness is normally ensured by including in the material specification criteria relating to the maximum allowable amounts of adherent deleterious materials in the coarse and fine aggregate fractions, as these reduce the bonding capabilities of cements and bituminous binders.

The aggregate size and gradation (i.e. the maximum particle size and the blend of sizes in an aggregate mix) affect the strength, stiffness, resistance to rutting, density and cost of an unbound layer in a pavement. Similarly, when particles are to be bound together by a cement, pozzolan or bituminous binder, a variation in the gradation will change the amount (and the cost) of binder needed to produce a mix of given stability and quality. Aggregate size and gradation exert a major influence upon the strength and stiffness characteristics of a bituminous mix, as well as on its permeability, workability, deformability under trafficking, and skid resistance. For example, the aggregate grading in a dense bituminous mix – which depends upon being well-graded (i.e. having all sizes) for its denseness and consequent stability – is more critical than the grading used in a bituminous macadam or stone mastic asphalt (in which stability is primarily dependent upon the interlocking of the coarse particles).

The amount of handling and transporting to which well-graded aggregates are subjected should be minimised, as particle-size segregation can easily occur during these operations, and this can be expensive to fix. Well-graded aggregates (see Equation 6.7) have gradation curves resembling a parabola.

The particle shape and surface texture are used to describe individual aggregate particles (Table 6.6). These characteristics influence the ability of the skeleton of an aggregate mix to carry load from one particle to the next, and have a strong influence on the internal friction properties of the aggregate.

Crushed granite, for example, is an excellent pavement aggregate because of the high internal friction associated with the angular shapes and the rough surface texture of the particles. Rounded, smooth gravels have relatively low internal friction, as particle interlock and surface friction are poor; hence, many specifications require gravel aggregate to be crushed to produce jagged edges and rougher surfaces before being used in a pavement. Flat and flaky or long and thin aggregate particles tend to be crushed by rollers, and may result in an aggregate with an excess fine fraction and poor interlock.

While the particle shape affects the physical properties of all asphalt mixes (Benson, 1970), it is critical with respect to those of open-graded mixes. Rough-textured aggregates (both coarse and fine) contribute much more to the stability of a pavement layer than equivalent-sized aggregates with smooth surface textures (Bessa et al., 2014; Powell et al., 2004). Similarly, shape and texture have major effects on the response of aggregate compacted into unbound pavement layers (Lekarp et al., 2000a, 2000b; Pan et al., 2006). Within limits, increased angularity and particle roughness lead to unbound layers that, when compacted, spread load better and resist deformation that would result in pavement rutting. In the lower layers of a pavement these characteristics help to provide a stable platform on which higher layers may be reliably placed. Consequently, and facilitated by modern imaging techniques, there is an increasing drive to measure the shape, angularity and surface texture of aggregate particles (Masad et al., 2005).

Hard aggregates have the ability to resist the abrasive effects of placement (in the short term) and traffic (over the long term). Carriageway macrotexture, which allows water to drain across the road surface while maintaining good frictional interaction between tyres and the aggregate asperities, is dependent for its continuance upon the resistance of its ingredient materials to deformation and abrasion under traffic. The mechanism by which

Table 6.6 Descriptive evaluations of mineral aggregates

(a) Particle shape

Classification	Description	Examples
Rounded	Fully water worn or completely shaped by attrition	River or seashore gravel; desert, seashore and wind-blown sand
Irregular	Naturally irregular, or partly shaped by attrition, and having rounded edges	Other gravels; land or dug flint
Flaky	The thickness is small relative to the other two dimensions	Laminated rock such as slate waste
Angular	Possessing well-defined edges formed at the intersection of roughly planar faces	Crushed rock of all types; talus; crushed slag
Elongated	Usually angular, in which the length is considerably larger than the other two dimensions	Uncommon
Flaky and elongated	The length is considerably larger than the thickness	Uncommon

(b) Surface texture

Surface texture	Characteristics	Examples
Glassy	Conchoidal fracture	Black flint, vitreous slag
Smooth	Water worn, or smooth due to fracture of laminated or fine grained rock	Gravels, chert, slate, marble, some rhyolites
Granular	Fracture showing more or less uniform round grains	Sandstone, oolitic limestone
Rough	Rough fracture of fine or medium grained rock containing no easily visible crystalline constituents	Basalt, felsite, porphyry, limestone
Crystalline	Containing easily visible crystalline constituents	Granite, gabbro, gneiss
Honeycombed and porous	With visible pores and cavities	Brick, pumice, foamed slag, clinker, expanded clay

a road surface becomes slippery is the loss of aggregate microtexture brought about by wearing and smoothing (i.e. polishing) of exposed particle surfaces by vehicle tyres. Also, the abrasion resistance of aggregates in all pavement layers should be sufficient to ensure that the risk of dust irritant being produced by traffic is kept at an acceptable level.

Tough aggregates are those that are better able to resist fracture under applied loads during construction and under traffic. The aggregates in each pavement layer must be tough enough not to break down under the crushing weight of the rollers during construction or the repeated impact and crushing actions of loaded commercial vehicles.

Durable or sound aggregates are those that can resist the disintegrating actions of repeated cycles of wetting and drying, freezing and thawing, or changes in temperature. Aggregates with high water absorptions (i.e. $>2\%$) are often considered to be vulnerable to frost action if placed in a pavement within 450 mm of the road surface. However, aggregates coated with bituminous binder are not normally considered vulnerable to frost problems in the UK (White, 1992).

The relative density on an oven-dry basis (RD_{od}) of an aggregate particle is the ratio of its mass in air to the volume of the particle, where that volume includes all the voids in the particle, whether they are permeable to water intrusion or not. Thus, this measure of relative density (see Equation 6.12) defines the maximum conceivable dry density of an aggregate assembly that has no voids between particles.

Road aggregates are normally proportioned by mass, and, hence, the RD_{od} is of vital importance in determining the proper particle-size blend (e.g. gradation specifications are only valid if the particles in the coarse and fine fractions have approximately the same relative densities). Thus, if the mean relative density on an oven-dry basis of the particles in the fine fraction is much greater than that for the coarse fraction, the result may be a mixture that, because of a lack of fine aggregate, is too 'harsh' (i.e. it has too great a volume of voids), and can potentially lead to instability and, in the case of bituminous mixes, to the possibility of bitumen migration. If the mean relative density of the coarse fraction is the greater, the resultant mix may be too rich in fine aggregate, and the asphalt mixture may not achieve the target resistance to deformation.

The saturated surface-dry relative density (RD_{ssd}) is the ratio of the mass of saturated surface-dry particles to the volume of the same particles, where that volume includes all the pores in the particles. Thus, this (Equation 6.13) is a measure of the maximum conceivable saturated density of an aggregate assembly that has no voids between the particles.

The apparent relative density (RD_{app}) of an aggregate particle is the ratio of its mass in air to the apparent volume of its solid constituents (i.e. excluding the volume of the pores in that particle). This is not the true relative density, as any closed pores entirely within the particle are (in ignorance of their existence) not taken into account. However, as particle sizes get smaller, the apparent relative density (Equation 6.14) becomes very close to the true relative density of the particle, as closed internal voids become almost an impossibility.

By definition, the apparent relative density of an aggregate particle is greater than its relative density on an oven-dry basis, and, consequently, this leads to different estimates of the voids content in bituminous mix design. For this reason, it is preferable to use an effective relative density that reflects the extent to which the binder in the mix penetrates the permeable voids in the aggregate. If it were assumed that the bitumen did not penetrate the permeable voids, the relative density on an oven-dry basis would be appropriate; the alternative assumption, that the bitumen did penetrate the permeable voids, would justify using the apparent relative density.

6.7.5 Aggregate tests

Aggregates are used in all layers of pavement construction, ranging from subgrade improvement (capping layers) to skid-resistant surface courses. Aggregate properties vary from source to source, and the suitability of a given aggregate for a particular usage depends upon the function of the pavement layer and the quality of the material used in that layer. Table 6.7 shows some results derived from roadstone tests of various aggregate groups that illustrate the diversity of engineering characteristics available from different rock types. This diversity has resulted in the development and application of a number of basic tests – which, to be of value, must be carried out on representative aggregate samples (BSI, 1997a, 1999) — to assist in decision-making regarding the suitability of particular aggregates for particular pavement purposes.

The following are summaries of some of the more common tests used in the UK.

Table 6.7 Summary of means and ranges (the latter in brackets) of test values for roadstones in various aggregate groups

Group	Aggregate crushing value	Aggregate abrasion value	Los Angeles value	Micro-Deval value	Polished-stone value	Water absorption	Relative density
Basalt	14 (15–39)	8 (3–15)	17 (8–31)	14	61 (37–64)	0.7 (0.2–1.8)	2.71 (2.6–3.4)
Granite	20 (9–35)	5 (3–9)	24 (10–35)	6 (4–8)	55 (47–72)	0.4 (0.2–2.9)	2.69 (2.6–3.0)
Gritstone	17 (7–29)	7 (2–16)	13 (10–35)	20 (16–25)	74 (62–84)	0.6 (0.1–1.6)	2.69 (2.6–2.9)
Limestone	24 (11–37)	14 (7–26)	26 (20–32)	20 (18–21)	45 (32–77)	1.0 (0.2–2.9)	2.66 (2.5–2.8)
Porphyry	14 (9–29)	4 (2–9)	15	3	58 (45–73)	0.6 (0.4–1.1)	2.73 (2.6–2.9)
Quartzite	16 (9–25)	3 (2–6)	20 (15–26)	3	60 (47–69)	0.7 (0.3–1.3)	2.62 (2.6–2.7)
Gravel	20 (18–25)	7 (5–10)	35 (20–50)	30 (20–40)	50 (45–58)	1.5 (0.9–2.0)	2.65 (2.6–2.9)
Slag	28 (15–39)	8 (3–15)			61 (37–74)	0.7 (0.2–2.6)	2.71 (2.6–3.2)

Gradation tests are used by aggregate producers for quality control purposes and by users for checking compliance with specifications. They provide the quantities, expressed in percentages by mass, of the various particle sizes of which an aggregate sample is composed. In the case of particles >63 µm, these quantities are determined by sieve analysis; that is, by allowing the aggregate sample to fall through a stack of standard sieves of diminishing mesh size that separate the particles into portions retained on each sieve (BSI, 1997b). Although EN standards (BSI, 1996) permit more than one sequence of sieve sizes, in the UK the sequence adopted uses the following sizes: 80, 63, 40, 31.5, 20, (16), 14, (12.5), 10, (8), 6.3, 4, 2, 1, 0.5, 0.25, 0.125 and 0.063 mm (in practice, the sieves in parentheses are infrequently used).

It is not necessary to use all the sieves to define a grading curve, and a particular subset is defined by the standard for a particular application. If the fines content is high, wet sieving is carried out in preference to dry sieving, as otherwise an incorrect grading is likely to be obtained; that is, the amount of the finest material will be underestimated because some fines will stick to the coarser particles.

The particle size distribution by mass is expressed either as the total percentages passing or retained on each sieve of the stack, or as the percentages retained between successive sieves (Table 6.8). The cumulative percentage-passing-by-mass method is very convenient for the graphical representation of a grading, and is used in most aggregate specifications: a sample representation is shown as Figure 6.6. The percentages-retained-on-particular-sieves method is preferred for single-sized aggregate specifications.

An aggregate is said to be 'single sized' if its gradation is such that most particles are retained on a given sieve after having passed the next larger one. A 'gap-graded' aggregate is one where significant components of the particles are retained on a number of larger sieves and then the remainder are mainly retained on a number of smaller sieves, with only

Table 6.8 Laboratory determination of the gradation of an aggregate (sample size = 2000 g)

EN sieve size: mm	Individual mass retained on the sieve		Cumulative mass retained on the sieve		Total mass passing the sieve: %
	g	%	g	%	
31.5	0	0	0	0	100
14	140	7	140	7	93
2.8	440	22	580	29	71
2	200	10	780	39	61
1	400	20	1180	59	41
0.5	280	14	1460	73	27
0.25	180	9	1640	82	18
0.125	140	4	1780	89	11
0.063	80	11	1860	93	7

Figure 6.6 Graphical representation of the grading of the aggregate material presented in Table 6.8

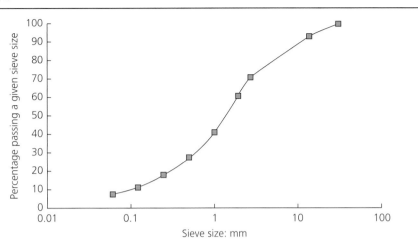

small amounts being retained on the in-between sieves. A 'well-graded' or 'continuously graded' aggregate has sufficient particles retained on each sieve of the nest of sieves to ensure that a dense material is obtained when the aggregate is compacted; the ideal gradation is given by Fuller's curve

$$P = 100(d/D)^n \qquad (6.7)$$

where P is the total percentage by mass passing sieve size d, D is the maximum size of aggregate particles in the sample and n is typically 0.5 (without binder) to 0.45 (with binder). When unbound, a single-sized mixture with a certain principal size will have many open pores: thus, it will be far more permeable than a well-graded mixture that has the same principal size. The well-graded mixture, however, once it is compacted, can be expected to exhibit greater strength and resistance to deformation due to the much denser packing arrangements of its particles.

There is a standard sieving method (BSI, 1997b) for determining the grading of a material as small as 20 μm: if a grading at smaller sizes is needed, a test is available (BSI, 1989) that uses a gravimetric method that applies Stoke's law to the rate of fall of particles in suspension in water.

The cleanliness of an aggregate, which is reflected in the surface-active component of the filler/fines that are <2 μm, is measured by the methylene blue test for the assessment of fines (Brennan et al., 1997; BSI, 2009d). This EN standard stain test, which aims to measure the ionic adsorption potential of at least 200 g of 0–2 mm particle size, involves admixing a series of 5 ml injections of a standard dye solution with a suspension of fines in 500 ml of distilled water. After each dye injection, a drop of the suspension is deposited on a filter paper to form a stain that comprises a generally solid blue-coloured central deposit of material surrounded by a colourless wet zone. The amount of the drop must be such that the deposit diameter is between 8 and 12 mm. The test is deemed to be positive if, in the wet zone, a halo consisting of a persistent light-blue ring of about 1 mm is formed around the central deposit. The methylene blue value (MB, in g/kg) is then recorded from the following equation:

$$MB = 10 \times V_1/M_1 \tag{6.8}$$

where M_1 is the mass of the 0/2 mm test portion (g) and V_1 is the total volume of the dye solution injected (ml).

The measure of particle shape that is usually included in road aggregate specifications is the flakiness index. The flakiness index test (BSI, 2012b), which is carried out on particles of between 63 and 6.3 mm in size, uses a stack of sieves with slot holes that are substantially longer than they are wide, to sieve aggregate in essentially the same way as with a conventional set of sieves. The difference is that flaky particles that would not have passed through standard sieve apertures can pass through the slotted sieves of the same width, so that the resulting grading curve of a flaky aggregate appears to be substantially finer than when evaluated using the standard sieves. Once minute volumes on any sieve have been discounted, the flakiness index is reported as the sum of the masses of aggregate passing through the various slotted sieves, expressed as a percentage of the total mass of the sample that is gauged. The lower the index the more equi-dimensional the aggregate.

Allowable flakiness index levels vary according to purpose: for example, aggregates used in surface dressings usually (but not always) have a maximum allowable index of 20, a maximum of only 15 may be defined for exposed aggregate in concrete surfaces, while aggregates for bituminous mixes may be allowed values up to 50. Unbound sub-bases with even higher flakiness index values have been successfully used (e.g. in north Wales using slate waste (Goulden, 1992; Sherwood, 2001)), but these need careful grading, handling and compaction arrangements if excess fines are to be avoided due to flaky particles crushing under loading.

Many hardness tests have been developed to evaluate the ease (or difficulty) with which aggregate particles wear or break under attrition from traffic. Nowadays, the test most widely used in Europe (including the UK) is the CEN-standardised Los Angeles (LA) test (BSI, 2010b). In this test, 5 kg of aggregate of particle size 10–14 mm, plus 11 47 mm diameter steel balls, are placed in a 711 mm internal diameter by 508 mm long steel drum that is rotated 500 times, with each rotation taking less than 2 s. Figure 6.7(a) shows that the LA drum contains an internal shelf that lifts the aggregate and balls, which then drop to the bottom of the drum as the shelf nears the top: in this way, the aggregate is abraded and repeatedly impacted. At the end of the test, the amount of aggregate passing the 1.6 mm sieve is measured, and the LA value computed as

$$LA \text{ value} = \frac{m_T - m_w}{m_T} \times 100 \tag{6.9}$$

Figure 6.7 Principles of (a) the LA test and (b) the micro-Deval test

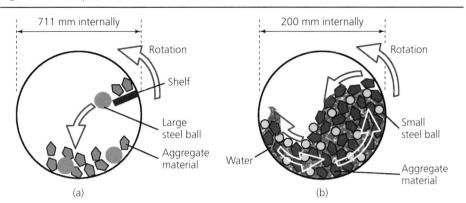

(a)

(b)

where m_T is the total mass of aggregate placed in the drum and m_w is the mass passing the 1.6 mm sieve. An LA value of <30 is considered suitable for many pavement purposes.

Another highly regarded aggregate abrasion assessment is provided by the micro-Deval test (BSI, 2011d). In this test, a 200 mm internal diameter by 154 mm long steel drum is rotated at 100 rpm for 12 000 revolutions (2 h) with a charge of 2 kg of aggregate in the 10–14 mm size range, 5 kg of steel balls of 9.5 mm diameter and 2.5 kg of water (Figure 6.7(b)). The water acts to accelerate any potential degradation caused by mineral softening and decay, and facilitates faster and larger inter-particle tumbling and sliding inside the drum and washes away any potentially protective fines that might act to limit degradation during later rotations of the drum. At the end of the test, the amount of aggregate passing the 1.6 mm sieve is measured, and the micro-Deval value (MDE) is computed in the same manner as in Equation 6.9. While many specifications require the MDE to be reported, few UK ones set a permitted maximum – if they do, a value of MDE of less than 20–30 might be specified, depending on the application (e.g. for surface courses a value <15 might be required).

Both the LA and micro-Deval tests are also widely employed using North American standards. While the LA test specified in the USA is very similar to that used in Europe, this is not true for the micro-Deval test. So, interpretation of test results should proceed with caution where the procedural parentage of the results is not clear.

The aggregate abrasion value (AAV) test (BSI, 2009h), which is still specified for aggregates used in surface courses in the UK, is carried out on two samples of at least 24 aggregate particles, each of which is <14 and >10 mm in size. The particles of each sample are mounted in, but project above, the surface of a resin compound in a small shallow tray, so that, for testing, the aggregate is pressed against a 600 mm diameter steel grinding lap that is rotating at 28–30 rev/min. As the disc rotates, contact is maintained with the aid of a 2 kg load as a standard (Leighton Buzzard) abrasive sand is fed continuously to its surface. On completion of 500 revolutions, the AAV for each test sample is determined from

$$AAV = 3(M_1 - M_2)/D_{ssd} \qquad (6.10)$$

where M_1 and M_2 are the initial and final sample masses, respectively, (in g) and D_{ssd} is their saturated surface-dry particle density (Mg/m^3). The AAV for the aggregate is then taken as the mean of the two values.

The lower the AAV the greater the aggregate resistance to abrasion. Thus, for example, aggregates with AAV values >10 may not be acceptable for chippings used in surface dressing and for micro-surfacing systems when the pavement must carry >1750 cv/ lane/day (cv, commercial vehicles), whereas values as high as 16 may be permitted for thin surface course systems, exposed aggregate concrete surfacing, and coated macadam surface courses carrying <1000 cv/lane/day (BSI, 2009j; HA et al., 2006).

The polished-stone value (PSV) test (BSI, 2009h) is a two-part skid-resistance test that is carried out on aggregate particles subjected to accelerated attrition by a polishing machine in the laboratory. The PSV measures the microtexture of an aggregate – which has been related to traffic skidding and site conditions in the UK (Roe and Hartshorne, 1998).

In part 1 of the PSV test, particles meeting specified size and shape criteria are set in resin and clamped onto the flat periphery of a 406 mm diameter 'road wheel', to form a 46 mm-wide continuous surface of stone particles that is rotated at 320 rpm. A 200 mm diameter solid rubber-tyred wheel with a width of 38 mm is then brought to bear on the aggregate surfaces with a force of 725 N, and corn emery is fed continuously to the interacting tyre–aggregate surfaces for 3 h: at the end of this process, a new wheel is brought to bear on the aggregate for another 3 h, only this time the feed material is emery flour. In part 2 of the test, the PSV of the aggregate sample is determined at the end of the 6 h of attrition by measuring the coefficient of friction between its wet surface and the rubber slider of a standard pendulum-type portable skid-resistance tester. The results are then compared with those of specimens from a control aggregate, and the PSV for the aggregate sample is then determined from

$$PSV = S + 52.5 - C \qquad (6.11)$$

where S and C are the means of the test and control specimen values, respectively.

Aggregates with PSVs of >68 are recommended for use in new asphalt surface courses at difficult sites: for example, at and on approaches to traffic signals, pedestrian crossings, roundabouts, railway level crossings, or on gradients >10% and longer than 50 m. PSVs of 55–65 are used on busy main roads (including motorways), while aggregates with PSVs of <50 are not recommended for use in any surface courses (HA et al., 2006).

There are three main relative density tests for aggregates: (a) the gas jar method for aggregates of between 5 and 40 mm in size (BSI, 2013b), (b) the pyknometer (i.e. density bottle) method for fine aggregates and filler (BSI, 2008), and (c) the test using a wire basket for coarse aggregates (which is not an EN standard test).

The principles underlying relative density determinations are most easily explained in relation to the wire basket method, in which a prepared sample is immersed in water and then weighed in the water after 24 h, following a prescribed procedure. The saturated aggregate is then removed from the water, and the individual particles are surface dried with a towel before being weighed in air: they are next placed in an oven set at 105°C, and, after 24 h, the particles are removed from the oven and weighed again in air. The formulae used to calculate the various relative densities (Mg/m^3) are

- the oven-dry relative density,

$$RD_{od} = D/[A - (B - C)] \qquad (6.12)$$

- the saturated surface-dry relative density,

$$RD_{ssd} = A/[A - (B - C)] \qquad (6.13)$$

- the apparent relative density,

$$RD_{app} = D/[D - (B - C)] \qquad (6.14)$$

where A is the mass of the saturated surface-dry aggregate in air (g), B is the apparent mass in water of the basket plus the saturated aggregate (g), C is the apparent mass in water of the empty basket (g) and D is the mass of the oven-dried aggregate in air (g).

The average relative density of an aggregate composed of fractions of different relative densities can be calculated from the individual values

$$RD_{ave} = 100/[(M_1/R_1) + (M_2/R_2) + (M_3/R_3)] \qquad (6.15)$$

where RD_{ave} is the average relative density of the final 'blended' aggregate, and M_1 and R_1, M_2 and R_2, and M_3 and R_3 are the mass percentages and relative densities of the individual fractions.

The water absorption (WA), as a percentage of the dry mass of the aggregate, is normally determined at the same time as the relative densities

$$WA = 100 \times (A - D)/D \qquad (6.16)$$

where A and D are as defined for relative density. The greater the WA percentage the more likely it is that an aggregate will experience frost susceptibility problems in cold climates.

Once the WA and RD_{ssd} are known, the porosity (%) of an aggregate is determined from

$$porosity = 100 \times (WA)(RD_{ssd})/(WA + 100) \qquad (6.17)$$

6.8. Secondary aggregates
6.8.1 Background

Attitudes to the use of unconventional aggregates has changed dramatically since the 1990s. As a consequence, the demand for conventional rock aggregates is now evaluated against a background of

- a change in public attitude towards a 'greener' agenda
- a desire to reduce energy demands, traffic congestion and vehicle emissions by using local aggregate sources (e.g. urban recycling centres)
- difficulties in obtaining planning consent for natural aggregate extraction
- the use of incentives that discourage the use of natural aggregate sources (e.g. by applying levies for access to virgin aggregates)
- the use of incentives to encourage the reuse of waste as aggregate (e.g. by applying landfill taxation)
- the economic benefits gained from using by-products of materials that have had their initial costs paid for by the primary industry that generated them
- the increasing use of performance-related standards (e.g. BSI, 2002, 2003) and specifications that use 'source-blind' tests in lieu of source-constrained specifications and tests that are biased towards those source materials.

Whereas, historically, crushed hard rocks were primarily considered as suitable road aggregates, the above reasons mean that waste, recycled, by-product and marginal materials are now more likely to be seriously considered as potential aggregates. Nonetheless, there are good reasons to be careful when considering utilising these materials in roadworks, namely

- consistency of supply may be difficult to ensure for some materials (e.g. construction and demolition waste characteristics can vary considerably depending on the source)
- contamination of the supply (e.g. some wastes from industrial processes may contain excessive proportions of leachable heavy metals (e.g. see Sear *et al.*, 2003))
- long-term durability can be uncertain due to lack of historical usage, and durability assessment methods may not be applicable or calibrated to certain materials
- the best use of many candidate materials will often not be as a substitute aggregate but as a wholly new type of material, so that test results need to be obtained and evaluated accordingly.

There is a wide range of secondary materials that, if not usable in a pavement layer, have other roadwork usages, such as drainage materials, mulches for landscaping, and materials for noise attenuation embankments alongside the carriageway. Many are also used in concrete (McNeil and Kang, 2013; Özkan *et al.*, 2007). Permissible applications in the UK national road network of a number of materials are listed in Table 6.9. If other potential aggregates are locally and economically available, their usage may still be permitted, usually on the basis of agreed demonstration and quality-assurance procedures. Local municipalities and private road builders have greater freedom: while they may

Table 6.9 Applications of secondary and recycled aggregates (HA *et al.*, 2004)

Material	Pipe bedding	Embankment	Capping and fill	Unbound granular material for the sub-base	Hydraulically bound mixture for the sub-base and base	Bitumen-bound layers	Pavement quality concrete	Likely waste status
Blastfurnace slag	✓	✓	✓	✓	✓	✓	✓	3
Burnt colliery spoil	✗	✓	✓	✓	✓	✗	✗	3
China clay sand/'stent'	✓	✓	✓	✓	✓	✓	✓	1
Coal fly ash/pulverised fuel ash (CFA/PFA)	✓	✓	✓	✗	✓	✓	✓	2
Demolition and construction waste	✓	✓	✓	✓	✓	✓	✓	2/3
Foundry sand	✓	✓	✓	✓	✓	✓	✓	3
Furnace bottom ash (FBA)	✓	✓	✓	✗	✓	✗	✗	2
Incinerator bottom ash aggregate (IBAA)	✓	✓	✓	✓	✓	✓	✓	3
Phosphoric slag	✓	✓	✓	✓	✓	✓	✓	3
Recycled aggregate	✓	✓	✓	✓	✓	✓	✓	2
Recycled asphalt	✓	✓	✓	✓	✓	✓	✓	2
Recycled concrete	✓	✓	✓	✓	✓	✓	✓	2
Recycled glass (glass cullet)	✓	✓	✓	✓	✓	✓	✗	2
Slate aggregate	✓	✓	✓	✓	✓	✓	✓	1
Spent oil shale/blaise	✗	✓	✓	✓	✓	✗	✗	3
Steel slag	✓	✓	✓	✓	✓	✓	✗	3
Unburnt colliery spoil	✗	✓	✗	✗	✓	✗	✗	1

Waste status: 1, not treated as waste; 2, waste protocol regulates reuse; 3, site-specific reuse requirements apply

adopt national specifications as a basis for their roadworks, their local knowledge of aggregates coming from specific local sources often makes an otherwise rejected material perfectly acceptable.

Advice, although a little dated, on secondary aggregates is available (Coventry *et al.*, 1999; Sherwood, 2001). More recent information is provided by Barritt (2006), and is also available for Scotland (Bateman and Nicholls, 2013).

6.8.2 Some legal and environmental considerations

Waste disposal legislation, which normally requires discarded materials to be tracked and licensed from the point of generation to their ultimate resting place, initially gave rise to concern as to whether roads that used recycled aggregates would need to be licensed as landfills. To overcome this difficulty, a number of quality protocols (i.e. end-of-waste frameworks for waste-derived products) have been produced (e.g. WRAP and EA, 2010, 2013, 2014) that set out a protocol for handling, monitoring and reporting on recycled materials so that the recycler can legitimately claim that the material is no longer a waste but is, instead, a saleable by-product. Where protocols are not available, source-specific practices have to be negotiated with the national regulator, potentially delaying reuse and/or making it uneconomic (e.g. SEPA, 2012).

Due to the variety of sources from which recycled and secondary aggregates come, it is difficult to make general statements concerning their chemical cleanliness. However, many materials (including some natural aggregates) have the capability to leach contaminants at hazard-potential levels, so some environmental assessment of what might take place downstream of the placed material often needs to be undertaken to demonstrate that the leachate is not likely to migrate to any major extent.

Some materials derived from industrial processes contain reasonably consistent and well-known chemicals, and, if there is a risk of these chemicals leaching from the site where the material is to be reused, it may be possible for this problem to be managed by treatment (e.g. by treating the material with cement to change the pH so as to reduce solubility). Demolition wastes can vary considerably, depending from where they were sourced, and these may also be treatment managed (e.g. by washing them to ensure separation of the fines from the coarse particles – because contamination is usually associated with fine particles due to their greater specific surface area) or they may be source managed (e.g. by declining concrete material from the chemistry laboratory at a demolished school but accepting it from other classrooms).

6.8.3　Slag aggregate

Air-cooled blastfurnace slag is a by-product of the steel-making industry. It is skimmed off the top of the molten iron in an iron furnace, and left to cool, before being crushed into aggregate form. The resultant material has good anti-skid properties, and, hence, it is highly regarded as a surface dressing aggregate. It is available in limited volumes (given the small scale of the UK's steel-making industry), so is now little used as a regular aggregate. If found in existing pavements, and if available for recycling, the high angularity and irregular shapes of the slag particles mean that pavements incorporating this aggregate have high internal friction. Asphalt surfaces using this slag are normally very stable; however, additional bitumen is normally required (compared with rock aggregates) to compensate for the binder content absorbed by the slag pores.

Granulated blastfurnace slag is obtained by spraying molten slag into a stream of water, rather than air-cooling it. The resulting material can be ground to a fine cementitious powder (GGBS).

Steel slag produced from the current steel-making process has good polish-resistant properties: it is of high density, and usually comprises vesicular stones. However, it is not normally used as bulk or selected granular fill because, after placing, the presence of free lime (CaO) and magnesia (MgO) can lead to expansion when hydration occurs. Also, when the slag chemistry subsequently causes a reversion from a high-temperature silicate form to a more voluminous low-temperature one, further volume instability can be encountered. The first cause may be addressed by working the material with water (e.g. by intermittently turning heaps exposed to rain), while the latter needs time.

Phosphoric slag, a less common slag, is a by-product from the manufacture of phosphorus. The manner of cooling has a major influence on the porosity of this material. Normally, this slag is spray-cooled in water: this helps to reduce the porosity, and causes shrinkage cracks that break up the slag. The solidified slag is then crushed and screened, typically to a <45 mm graded aggregate. The resulting material is pozzolanic, and gains stiffness over time (e.g. gaining a compressive strength of 4 MPa after 1 year), allowing it to form a high-quality base or sub-base. Phosphoric slag is imported from the Netherlands, so is only likely to be a viable material for large projects, where ship–shore handling can be economically arranged (e.g. into south-east England).

Information on common slags is available on the NSA website (NSA, 2015).

6.8.4 Colliery shale

Colliery shale is the waste product of coal mining that was either removed to gain access to the coalface or was unavoidably brought from the pit with the coal and was separated out at the coal-cleaning plant. It is available in two forms in former coal-mining areas: unburnt shale (or minestone), and burnt shale (i.e. *burnt minestone*) that resulted from the spontaneous combustion of coal particles within the spoil heaps. While an estimated 2000 Mt of the two types may exist within heaps in former mining areas in the UK, most of this has been landscaped and is not now readily recoverable for other uses.

Burnt shales, which are mainly composed of silica, alumina and iron, can differ considerably between and within heaps. Nonetheless, hard, well-burnt, partly vitrified shale (i.e. free from ashy refuse and rubbish that may soften when wet) has a history of successful use as a fill material in embankments and as a capping and sub-base material in pavements that are close to spoil tips. Normally, material of <76 mm maximum size is preferred, for ease of laying and compaction.

Unburnt shale has been less used, mainly because of fears about spontaneous combustion. However, experience has shown that the risk of fire is heavily influenced by the ease with which oxygen can penetrate into and through the material, and that spontaneous ignition does not occur if the unburnt shale is well compacted as it is being laid: hence, unburnt shale is now being as bulk fill material. If layers are thin, heat is lost before spontaneous combustion can occur: thus, stabilisation with cement or lime may allow unburnt shale to form effective capping or sub-base layers.

Excess sulphate present in some shales can cause damage if it leaches into concrete and cement-bound materials. Thus, if a shale is to be used within 0.5 m of a bridge abutment, or a concrete pavement or pipe, the sulphate content of a 1 : 1 shale–water extract should not exceed 2.5 g of sulphate (as SO_3) per litre. Also, unburnt shale should not be used in reinforced-earth structures, as sulphuric acid, which is formed from iron sulphide (FeS_2) reacting with oxygen and water from the atmosphere, could corrode the metal components of the reinforcing straps.

Most unburnt shales, and some burnt ones, are frost susceptible, and are not normally used in pavements in the UK unless they have at least 450 mm of cover or are stabilised with cement.

6.8.5   Spent oil shale

The commercial exploitation of oil-producing shale in the UK has been concentrated in Scotland. At its peak (in 1913), the West Lothian area of Scotland produced about 3.3 Mt of shale oil per annum; this became uneconomic over time, and production was stopped in 1962. During its production, however, various oils were removed from the mined oil shale, and the spent oil shale, and other waste shales brought to the surface, were dumped on land close to the mines and refineries. Spontaneous combustion sometimes occurred in the heaps, causing more changes to them.

This shale, also known as blaise, is most appropriately used as bulk fill in embankments and as a capping material. Its loss on ignition is low, and there is no risk of spontaneous combustion if the material is properly compacted as it is emplaced. The sulphate content can be high, and care needs to be taken when using spent oil shale close to bridge abutments, concrete pipes and so on.

Water absorption tests give an average value of about 15% for spent oil shale. The shale particles are relatively soft and tend to break down under compaction. Tests indicate that spent oil shale normally exhibits frost heaves that are well in excess of 15 mm when subjected to frost–heave testing (BSI, 2009i), and, hence, this material is not normally used within 450 mm of a road surface unless it is cement stabilised.

6.8.6   Pulverised fuel ash and furnace bottom ash

Pulverised fuel ash (PFA) – also termed fly ash – is the solid fine ash carried out in the flue gases of power station boilers that are fired with pulverised coal for electricity generation. The coarser component of the residual ash from the burnt pulverised coal is not carried over with the flue gases but instead falls to the bottom of the furnace as clinker: once crushed, this is termed furnace bottom ash (FBA). The manufacture of environmentally acceptable PFA and FBA is described in the literature (WRAP and EA, 2010).

The main PFA ingredients are finely divided, glassy spheres of silica (SiO_2, 45–51%) and alumina (Al_2O_3, 24–32%). Magnetic and non-magnetic iron compounds, some alkali, and limited amounts (2–3%) of water-soluble materials are also present. The quantity of residual unburnt carbon present in a PFA (determined as the loss on ignition) depends on the efficiency with which the pulverised coal is burnt in the furnace. Thus, old power stations can have carbon contents >10% while modern stations have <2%. A PFA with a high carbon content is normally a lower-quality construction material. As carbon particles are porous, a high carbon content increases the moisture content requirement of any PFA mixture; it also results in lower dry densities, reduces the proportion of reactive surface area available to enter into pozzolanic reactions, and physically limits the contacts of cementitious materials.

The particle size compositions of PFA can vary considerably from power station to power station. Generally, however, ashes with less than 40% of particles passing the 75 μm sieve are unlikely to be frost susceptible in the UK.

As noted previously, PFA is a pozzolan. Because many fly ashes contain some calcium oxide (CaO) or calcium hydroxide ($Ca(OH)_2$) derived from the burning of limestone present in the original coal, they frequently do not require lime activation (unlike other pozzolans) but can be self-cementing. The chemically reactive silica, alumina and haematite materials are mainly found in the finer fractions of the ash (i.e. a reactive PFA tends to have at least 80% by mass of its particles <42 μm).

The main uses of PFA in road construction in the UK are as bulk fills for embankments, in capping layers and in cement-bound and hydraulically bound materials for pavement bases. Sometimes an ash will retain latent electrical charges that cause individual ash particles to repel each other, thereby causing bulking of the massed material: this PFA should be rejected due to the difficulty in achieving adequate compaction. Compacted PFA has a lower dry density (typically 1.28 Mg/m^3) than most other materials used in embankments, and this lightweight property is very advantageous when mass filling is required on compressible soils. The self-hardening properties of some fly ashes are especially useful when selected fill is required behind bridge abutments, as their settlement and any imposed lateral load on the structure are less than those for most conventional fill materials. Overseas, in warm climates, PFA is sometimes used with lime and soil, or cement and soil, to produce stabilised bases and sub-bases (see Chapter 7).

FBA is much coarser than PFA: its particles vary in size from fine sand to coarse gravel. FBA has potential for use as a roadstone, despite its vesicular nature and particle weakness (Dawson, 1989; Dawson and Bullen, 1991; Dawson and Nunes, 1993), and as a concrete aggregate (Bai and Basheer, 2003). It has also been used in the past as an embankment and capping material. Currently, its availability for these purposes is limited, as most FBA is used in lightweight block manufacture.

6.8.7   China clay wastes

The mining of china clay (kaolin), which is concentrated in Devon and Cornwall, involves the use of high-pressure jets of water to extract kaolinised granite from steep-sided open pits. Some 3.7 t of coarse sand, 2 t of waste rock, 2 t of overburden and 0.9 t of micaceous residue result from the production of 1 t of china clay. After removal of the sand and the mica residue, the wastes are normally tipped onto land near the pits, in mounds up to 45 m high.

Except for the micaceous residue, most china clay wastes – especially the sand waste – have potential for use in roads. Many china clay sands are not frost susceptible, and can be used in cement-bound or hydraulically bound bases and pavement-quality concrete, as well as in embankments and capping layers. When used in concrete, they require more cement (15–20% by mass) than river sands to achieve the same strength: this is probably because of the 'harshness' of the waste sand mixes.

A synthetic surface dressing aggregate using china clay sand has been developed that is highly resistant to wear and traffic polishing. Other uses for china clay sand include back-filling for pipe trenches and French drains, and as permeable backings to earth-retaining structures.

A feature of china clay wastes is the presence of mica. All micas have a layered structure in which successive sheets are able to part easily in the plane parallel to their larger surfaces, and to form thin flakes that, when subjected to pressure on their larger surfaces, behave resiliently (like the leaves of a leaf spring). However, studies have shown that the proportion of mica normally present in china clay sand is too low to cause problems in road construction; the main detrimental effect is that its resilience reduces the degree of compaction achievable for a given compaction effort, by about 0.007 and 0.012 Mg/m^3 for each 1% of fine (<75 μm) and coarse (mainly 212–600 μm) mica, respectively.

The waste granite fractions, known as stent, may also be usable in pavements as capping or sub-base materials after appropriate crushing.

6.8.8 Slate waste

For every 1 t of slate produced in the UK, an additional 20 t of waste material is produced as a by-product. Consequently, great quantities of slate waste are available in dumps near quarries in, especially, north Wales. However, its use in roadworks is relatively low, due to the high cost of transporting the waste to where the roads are being built.

A summary of the properties of UK slate-derived aggregates is available in the literature (Goulden, 1992).

While waste slate is a crushed rock, its nature varies with its origin: thus, cherts (which are sometimes inter-bedded with slate) and igneous rocks are often found in waste accumulations. By contrast, 'mill waste' consists mainly of slate blocks and chippings from the dressings of the slate.

Slate waste is used in cement-bound bases and pavement-quality concrete, as well as in capping layers and as a bulk fill material. The flaky nature of the waste particles causes problems in compaction, and grid rollers have been found to be most useful in over-coming these, as they break the longer needle-shaped pieces of slate into short pieces.

6.8.9 Incinerated refuse

As with the burning of coal, there are two waste streams when municipal solid waste (MSW) is burnt in an incinerator.

MSW incinerator fly ash (MSWIFA) is trapped from the chimney flue gases. Due to the nature of the material being burnt and the incineration process, it is likely that the

MSWIFA will contain dioxin – a carcinogen. Therefore, MSWIFA is not recyclable, and must be landfilled in a secure repository.

The second MSW is incinerator bottom ash (MSWIBA). This consists mainly of clinker, glass, ceramics, metal and residual unburnt matter. (The metal content is subsequently separated for its own recycling.) The unburnt residual material may be paper, rag, or putrescible substances, with the amounts depending upon the type of furnace, the fire-bed temperature, and the length of time that the materials are fired – and on the composition of the raw refuse. Unburnt material can be difficult to compact, and is likely, after a period of time, to putrefy, giving rise to odour, leachate and settlement.

While the ash obtained at one incinerator can vary considerably from that obtained at another, there is reasonable consistency in the material obtained at a given source. Provided that the unburnt content is small, the remaining MSWIBA can be beneficially reused in embankments, cappings and sub-bases, with some wastes being better than others. As concrete and cemented products are attacked by sulphates dissolved in water, incinerated refuse in which the soluble sulphate content exceeds 2 g/l should not be used within 0.5 m of a concrete structure, unless the structure is protected through the use of super-sulphated cement or sulphate-resisting Portland cement.

The operators of modern incinerators are usually linked to specialist materials handling and preparation companies that can provide a protocol that users can adopt to ensure satisfactory mechanical and environmental performance.

6.8.10   Recycled asphalt pavement

When flexible pavements are rehabilitated, it is now common for the distressed asphalt in the upper part of the pavement to be removed by a rotating drum equipped with picks. This planer breaks the old asphalt into a coarse granular material, with individual fragments being a collection of stones held together by the old asphaltic mastic. If the old pavement layers are excavated by less specialised equipment (e.g. front-end loaders), then it may be necessary to further crush the asphalt for subsequent reuse. In hot weather, slow crushers tend to get smeared with residual bitumen, so impact crushers are preferred for this task.

Historically, a popular use for recycled asphalt pavement (RAP) material was as a direct replacement for quarried aggregate in the base layers of cycle tracks or farm accesses – until it was recognised that this 'down-cycling' was not making the best use of the capabilities of the material. Nowadays, most RAP is reused in hot-mix, warm-mix or cold-mix asphalts as a partial replacement for the new aggregate and binder fractions.

Because the old binder has usually hardened due to oxidation and volatilisation of some of the bitumen constituents, making new asphalt simply by heating and reworking 100% RAP is normally impossible. While high percentages of RAP have been successfully used in new asphalt mixes, such applications normally necessitate

■   The use of 'rejuvenators'. These are usually proprietary chemicals whose precise actions are not well defined but, in general, they (a) replace lost (volatile) and altered chemical constituents, particularly the aromatics and resins (see Section 6.1.2), and (b) establish similar viscometric properties to those of the original asphalt.

▪ Modification of the grading, usually by adding filler. When compared with conventional crushed aggregate, RAP tends to be low in fine aggregate. When breakage of the old pavement material occurs, it tends to do so through the binder/mastic fraction rather than through the aggregate, and the presence of the old binder tends to keep fragments of rock from the old aggregate adhering to each other. This means that, for 'unbound' uses, the removed RAP needs its grading to be modified by the addition of non-RAP fine aggregate.

In the UK, if more than 25% of a new asphalt comprises old asphalt, then a performance-based framework has to be employed for the resulting asphalt, to ensure that it delivers a durable and correctly functioning mixture (HA *et al.*, 2008).

Aggregates formed of high proportions of RAP can be more difficult to compact than conventional aggregates, particularly in warm weather, due to the viscosity of the old binder, which hinders inter-particle movement under dynamic, vibrating, compaction plant. Conversely, after compaction, further creep movement can take place, as the viscous old binder permits slow, ongoing deformation. Thus, when compared with a conventional aggregate, RAP may provide a stiffer layer that is less resistant to permanent deformation. However, once deformation starts taking place, the old binder in one fragment of RAP slowly comes into contact with adjacent pieces and with any grading modifier. Thus, over time, a layer of compacted RAP slowly becomes less unbound and more like a weakly bonded asphalt.

6.8.11　Recycled concrete aggregate

Concrete obtained for recycling should be homogeneous and not contain >1% foreign material by mass. Unreinforced concrete is relatively easily prepared for recycling by crushing and then grading. If the concrete contains reinforcement, more careful crushing is required to ensure that the reinforcement does not block the crushers: once crushed, however, the steel can be magnetically separated for its subsequent recycling. Prestressed concrete requires specialist handling if the prestress load is not to be released dangerously.

When concrete – whether from an old concrete pavement or from some other demolition source – is crushed to form recycled concrete aggregate (RCA), two important things happen. First, the mortar between the coarse aggregate tends to be liberated, creating excess fines. While standard separation techniques can be used to remove some of these fines, this may be undesirable because of the second occurrence, i.e. unhydrated cement is released from the mortar that will slowly hydrate in the presence of water. Therefore, if the RCA material is reused without being aged in a stockpile, the mix will slowly 'self-cement' and result in a material rather like a lean-mix concrete whose strength and stiffness primarily depends on the quality and make-up of the original recycled concrete.

Up to 20% RCA is commonly used to replace conventional aggregate in new concrete: a higher reuse requires more detailed evaluation and mix design. An important factor in any concrete mix containing RCA is that the recycled aggregate has greater water absorbency, even if its grading is matched with that of the conventional aggregate; that is, the mortar remaining attached to the coarser RCA fragments will more readily absorb water than the old aggregate stones, so that, overall, the RCA absorbs more water from a wet concrete mix than does conventional aggregate. Consequently, concrete mixtures containing RCA normally need an increase in the water:cement ratio to allow for this take-up.

6.8.12   Recycled aggregate

Unbound stone aggregate can be recycled and reused with new aggregate in both asphalt and concrete. The recycled aggregate is normally graded and reblended before reuse, as it may have come from a number of different sources or, if from a single source, may not have been initially emplaced to the specification required for its new use, and/or it may have been damaged (broken) under compaction and usage.

An interesting source of recycled aggregate is railway ballast. Ballast specifications require hard wear-resistant particles, but, due to tamping and the dynamic loading imposed through the track by trains, ballast stones eventually lose their asperities, until they no longer deliver the quality of restraint to track movement required by railway specifications. However, as the original ballast particles are much larger than the maximum particle sizes required for unbound or bound road pavement aggregates, the crushing of these coarser ballast particles can deliver a valuable and high-quality recycled aggregate with new faces of refreshed roughness and angularity.

6.8.13   Glass cullet

About 2.7 Mt of glass is wasted in the UK every year, with >60% being collected for recycling. Clear glass is mostly recycled into new glass, but about 25% of the material sent for recycling cannot be economically reused in this way, so it is crushed and used as an aggregate. With careful handling – to avoid inhalation of glass dust, and cuts – crushed glass can be used as a partial replacement of aggregate in asphalt mixtures: for example, conventional aggregate replacement rates of 10–15% and 10–30% have been reported (Huang *et al.*, 2007) in surface and deeper pavement courses, respectively.

The use of glass in surface courses may require further research, as it is generally considered not to be as resistant to polishing under traffic as many conventional aggregates. Also, there is concern about potential hazards associated with subsequent recycling of pavement courses containing glass (i.e. future recyclers may not be aware of the glass content).

Generally, glass is not a suitable bulk coarse aggregate for concrete, as the glass surfaces are too smooth for good cement adherence. Also, any residual sugars on glass surfaces that remain from the time of consumer usage will hinder cement hydration reactions and introduce further weaknesses into the concrete. While small proportions of glass can be employed as a coarse aggregate, it can be used readily as a fine aggregate, albeit the mix design will need modification because of the different 'flowability' of conventional rock versus glass fines.

The coarser fractions of crushed glass have also been used as road pavement drainage media.

6.8.14 Foundry sand

Metal castings are often formed by pouring molten metal into a former composed of a bed of sand that is held in place by a clay or an industrial resin binding agent. Once the casting has cooled, the sand is removed. The sand may be reused several times before the binding action is lost, or heat damages too many of the sand grains. Then it is discarded.

About 1 Mt of spent sand is discarded each year in the UK, mainly in the Midlands and in south and west Yorkshire. This material has been successfully used as a fine aggregate in asphaltic mixes and in unbound materials as capping or as a grading modifier of coarse recycled aggregate. However, as the sand comes from a metal casting source, its use in an unbound application may create environmental leaching concerns that have to be addressed.

6.8.15 Construction and demolition debris

Demolition wastes are often proposed for roadworks, as they are mostly available in urban areas where there is often a shortage of conventional aggregates. Debris that is relatively free from contamination (e.g. brick and non-reinforced crushed concrete) has potential for road construction, and is sometimes included in specifications as clean rock-like 'hard-core' for use in embankments, capping layers and cement-bound bases. Unfortunately, clean demolition debris is produced spasmodically and in relatively small quantities, so that advanced planning for its use is difficult.

Most debris is variable, and the major components (brick, building stone and concrete) are very often intermixed with plaster, wood, glass and so on, so that the resultant mixture is so heterogeneous that it is unsuitable for use even as a bulk fill. Practically, this means that its most beneficial reuse requires an urban recycling 'hub' operation, which would receive construction and demolition waste from many sites, process them as separate streams and then blend these streams to produce more homogenised and graded aggregate supplies for stockpiling in sufficient quantities for sale. Such centres are now becoming increasingly common.

RCA (see Section 6.8.11) would be the highest quality product of such central operations. In practice, however, RCA materials have properties in excess of those required for most non-road purposes, so that blends of crushed masonry and RCA are typically sold. While these share the slow hydraulic binding properties of RCA, they do not achieve as high a strength: nonetheless, they may be appropriate for some sub-base usages.

6.8.16 Rubber tyre waste

In the USA and EU, approximately 1 and 0.4 tyre/person, respectively, are discarded each year. In the EU, where tyre disposal in landfill is banned, this represents almost 3 Mt of waste tyres per year; less than 0.5 Mt of this is generated in the UK. While the use of whole tyres in roadworks is sometimes feasible (e.g. in retaining walls and crash barriers), breaking of the tyres is almost an invariable requirement to ensuring a potentially useful raw material. A number of techniques are used for this purpose, including (a) shredding, to obtain banknote-sized pieces of old tyre, complete with webbing and steel reinforcement; (b) chipping, to obtain die-sized pieces of rubber; (c) crumbling at low temperatures, to separate the rubber from the fabric and steel elements by making the rubber so brittle that it readily breaks apart; and (d) powdering, by further grinding the rubber crumbs until a fine powder is obtained.

The accepted procedure for manufacturing environmentally acceptable tyre waste products in the UK is described in the literature (WRAP and EA, 2014).

Shredded tyres have been used as embankment fill (Humphrey *et al.*, 1998; Kaliakin *et al.*, 2012): they produce a lightweight bulk material, but are extremely difficult to compact. Chipped rubber has been used as a medium in pavement drainage systems (Humphrey and Swett, 2006): it may also act to improve water quality prior to its release into the environment (Park *et al.*, 1996). Chipped rubber has also been used in unbound aggregate layers, but compaction concerns would appear to limit this application. The most widespread use of waste rubber is as crumb or powdered material in hot asphaltic mixes (Huang *et al.*, 2007; Lo Presti, 2013), where it is used to replace some of the aggregate in the asphalt or, if finely ground, it is dissolved into the bitumen to act as a polymer modifier (see Section 6.3.3).

REFERENCES

Airey GD, Hunter AE and Rahimzadeh B (2002) The influence of geometry and sample preparation on dynamic shear rheometer testing. In *Performance of Bituminous and Hydraulic Materials in Pavements* (Zoorob SE, Collop A and Brown SF (eds)). Balkema, Rotterdam, the Netherlands, pp. 3–12.

Bai Y and Basheer PAM (2003) Influence of furnace bottom ash on properties of concrete. *Proceedings of the ICE: Structures and Buildings* **156(1)**: 85–92.

Barritt J (2006) The evolution of recycled aggregates for concrete. *Concrete Engineering International*, Autumn, pp. 58–60.

Bateman D and Nicholls JC (2013) *Review of Hydraulically-bound Materials for Use in Scotland*. Transport Research Laboratory, Crowthorne, UK, Client Project Report CPR1603.

Benson FJ (1970) *Effects of Aggregate Size, Shape, and Surface Texture on the Properties of Bituminous Mixtures – A Literature Survey*. Highway Research Board, Washington, DC, USA, Highway Research Special Report 109.

Bessa I, Branco V, Soares, J and Neto J (2014) Aggregate shape properties and their influence on the behaviour of hot-mix asphalt. *Journal of Materials in Civil Engineering*, **27**(7): 8pp.

Blyth FGH and de Freitas MH (1984) *A Geology for Engineers*. Arnold, London, UK.

Brennan MJ, Kilmartin T, Lawless A and Mulry B (1997) A comparison of different methylene blue methods for assessing surface activity. In *Mechanical Tests for Bituminous Materials; Proceedings of the 5th International RILEM Symposium* (Di Benedetto H and Francken L (eds)). Balkema, Rotterdam, the Netherlands, pp. 517–523.

BSI (British Standards Institution) (1989) BS 812-103.2:1989. Testing aggregates. Method for determination of particle size distribution. Sedimentation test. BSI, London, UK.

BSI (1996) BS EN 933-2:1996: Tests for geometrical properties of aggregates. Determination of particle size distribution. Test sieves, nominal size of apertures. BSI, London, UK.

BSI (1997a) BS EN 932-1:1997: Tests for general properties of aggregates. Methods for sampling. BSI, London, UK.

BSI (1997b) BS EN 933-1:1997: Tests for geometrical properties of aggregates. Determination of particle size distribution. Sieving method. BSI, London, UK.

BSI (1999) BS EN 932-2:1999: Tests for general properties of aggregates. Methods for reducing laboratory samples. BSI, London, UK.

BSI (2002) BS EN 13242:2002: Aggregates for unbound and hydraulically-bound materials for use in civil engineering work and road construction. BSI, London, UK.

BSI (2003) BS PD 6682-6:2003: Aggregates for unbound and hydraulically-bound materials for use in civil engineering works and road construction. Guidance on the use of BS EN 13242. BSI, London, UK.

BSI (2007a) BS EN 1426:2007: Bitumen and bituminous binders. Determination of needle penetration. BSI, London, UK.

BSI (2007b) BS EN 1427:2007: Bitumen and bituminous binders. Determination of the softening point. Ring and ball method. BSI, London, UK.

BSI (2007c) BS EN 12593:2007: Bitumen and bituminous binders. Determination of the Fraass breaking point. BSI, London, UK.

BSI (2008) BS EN 1097-7:2008: Tests for mechanical and physical properties of aggregates. Determination of the particle density of filler. Pyknometer method. BSI, London, UK.

BSI (2009a) BS EN 12591:2009: Bitumen and bituminous binders. Specifications for paving grade bitumens. BSI, London, UK.

BSI (2009b) BS EN 13304:2009: Bitumen and bituminous binders. Framework for specification of oxidised bitumen. BSI, London, UK.

BSI (2009c) BS EN 13303:2009: Bitumen and bituminous binders. Determination of the loss in mass after heating of industrial bitumen. BSI, London, UK.

BSI (2009d) BS EN 933-9:2009. Tests for geometrical properties of aggregates. Assessment of fines. Methylene blue test. BSI, London, UK.

BSI (2009e) BS EN 1430:2009: Bitumen and bituminous binders. Determination of particle polarity of bituminous emulsions. BSI, London, UK.

BSI (2009f) BS EN 1431:2009: Bitumen and bituminous binders. Determination of residual binder and oil distillate from bitumen emulsions by distillation. BSI, London, UK.

BSI (2009g) BS EN 13075-1:2009: Bitumen and bituminous binders. Determination of breaking behaviour. Determination of breaking value of cationic bituminous emulsions, mineral filler method. BSI, London, UK.

BSI (2009h) BS EN 1097-8:2009: Tests for mechanical and physical properties of aggregates. Determination of the polished stone value – with the aggregate abrasion value (AAV) as an annex. BSI, London, UK.

BSI (2009i) BS 812-124:2009: Testing aggregates. Method for determination of frost heave. BSI, London, UK.

BSI (2009j) BS PD 6682-2:2009: Aggregates. Aggregates for bituminous mixtures and surface treatments for roads, airfields and other trafficked areas. Guidance on the use of BS EN 13043. BSI, London, UK.

BSI (2010a) BS EN 14023:2010: Bitumen and bituminous binders. Specification framework for polymer modified bitumens. BSI, London, UK.

BSI (2010b) BS EN 1097-2:2010: Tests for mechanical and physical properties of aggregates. Methods for the determination of resistance to fragmentation. BSI, London, UK.

BSI (2011a) BS EN 12846-1:2011: Bitumen and bituminous binders. Determination of efflux time by the efflux viscometer. Bituminous emulsions. BSI, London, UK.

BSI (2011b) BS 434-1:2011: Bitumen road emulsions. Specification for anionic bitumen road emulsions. BSI, London, UK.

BSI (2011c) BS EN 197-1:2011: Cement. Composition, specifications and conformity criteria for common cements. BSI, London, UK.

BSI (2011d) BS EN 1097-1:2011: Tests for mechanical and physical properties of aggregates. Determination of the resistance to wear (micro-Deval). BSI, London, UK.

BSI (2012a) BS EN 1428:2012: Bitumen and bituminous binders. Determination of water content in bituminous emulsions. Azeotropic distillation method. BSI, London, UK.

BSI (2012b) BS EN 933-3:2012: Tests for geometrical properties of aggregates. Determination of particle shape. Flakiness index. BSI, London, UK.

BSI (2013a) BS EN 15322:2013: Framework for specifying cut-back and fluxed bituminous binders. BSI, London, UK.

BSI (2013b) BS EN 1097-6:2013: Tests for mechanical and physical properties of aggregates. Determination of particle density and water absorption. BSI, London, UK.

BSI (2013c) BS EN 13808:2013: Bitumen and bituminous binders. Framework for specifying cationic bituminous emulsions (together with National Annexes). BSI, London, UK.

BSI (2014a) BS EN 12592:2014: Bitumen and bituminous binders. Determination of solubility. BSI, London, UK.

BSI (2014b) BS EN 12607-1:2014: Bitumen and bituminous binders. Determination of the resistance to hardening under influence of heat and air. RTFOT method. BSI, London, UK.

Coventry S, Woolveridge C and Hiller S (1999) *The Reclaimed and Recycled Construction Materials Handbook*. CIRIA, London, UK, Report C513.

Croney D and Jacobs J (1967) *The Frost Susceptibility of Soils and Road Materials*. Road Research Laboratory, Crowthorne, UK, Report LR90.

Dawson AR (1989) The degradation of furnace bottom ash under compaction. In *Unbound Aggregates in Roads* (Jones RH and Dawson AR (eds)). Butterworths, London, UK, pp. 169–179.

Dawson AR and Bullen F (1991) Furnace bottom ash – its engineering properties and its use as a sub-base. *Proceedings of the Institution of Civil Engineers* **90**: 993–1009.

Dawson AR and Nunes M (1993) Some British experience of the behaviour of furnace bottom ash and slate waste for pavement foundations. *Proceedings of the Federal Highways Administration Conference on Discarded Materials and By Products for Construction of Highway Facilities*. Federal Highways Administration, Washington, DC, USA, pp. 4-1–4-13.

Dickinson EJ (1977) A critical review of the use of rubbers and polymers in bitumen-bound pavement surfacing materials. *Australian Road Research* **7(2)**: 45–52.

Goulden E (1992) *Slate Waste Aggregates for Unbound Pavement Layers*. MSc dissertation, University of Nottingham, Nottingham, UK.

HA (Highways Agency), Transport Scotland, Welsh Government and Department for Regional Development Northern Ireland (2004) Section 1: preamble. Part 2: conservation and the use of secondary and recycled materials. In *Design Manual for Roads and Bridges*, vol. 7. *Pavement Design and Maintenance*. Stationery Office, London, UK, HD 35/04.

HA, Transport Scotland, Welsh Government and Department for Regional Development Northern Ireland (2006) Section 5: pavement materials. Part 1: surfacing materials for new and maintenance construction. In *Design Manual for Roads and Bridges*, vol. 7. *Pavement Design and Maintenance*. Stationery Office, London, UK, HD 36/06.

HA, Transport Scotland, Welsh Government and Department for Regional Development Northern Ireland (2008) Road pavements – bituminous bound materials. In *Manual of Contract Documents for Highway Works*, vol. 1. *Specification for Highway Works*. Stationery Office, London, UK, Series 900.

HA, Transport Scotland, Welsh Government and Department for Regional Development Northern Ireland (2009) Road pavements – unbound, cement and other hydraulically bound mixtures. In *Manual of Contract Documents for Highway Works*, vol. 1. *Specification for Highway Works*. Stationery Office, London, UK, Series 800.

Heukelom W (1969) A bitumen test data chart for showing the effect of temperature on the mechanical behaviour of asphaltic bitumens, *Journal of the Institute of Petroleum* **55**: 404–417.

Heukelom W (1973) An improved method of characterizing asphaltic bitumens with the aid of their mechanical properties. *Proceedings of the Association of Asphalt Paving Technologists* **42**: 67–98.

Hosking R (1992) *Road Aggregates and Skidding*, *TRL State of the Art Review 4*. Stationery Office, London, UK.

Huang Y, Bird RN and Heidrich O (2007) A review of the use of recycled solid waste materials in asphalt pavements. *Resources, Conservation and Recycling* **52**: 58–73.

Humphrey DN and Swett M (2006) *Literature Review of the Water Quality Effects of Tire-derived Aggregate and Rubber-modified Asphalt Pavement*. Environmental Protection Agency, Washington, DC, USA.

Humphrey DN, Whetten N, Weaver J, Recker K and Cosgrove TA (1998) Tire shreds as lightweight fill for embankments and retaining walls. In *Recycled Materials in Geotechnical Applications. Geotechnical Special Publication No. 79* (Vipulanandan C and Elton DJ (eds)). ASCE, Reston, VA, USA, pp. 51–65.

Hveem FN, Zube E and Skog J (1963) Proposed new test and specifications for paving grade asphalt. *Proceedings of the Association of Asphalt Paving Technologists* **32**: 271–352.

ISO (International Organization for Standardization) (2002) ISO 2719:2002: Determination of flash point. Pensky–Martens closed cup method. ISO, Geneva, Switzerland.

Jackson N and Dhir RK (1997) *Civil Engineering Materials*, 5th edn. Macmillan, London, UK.

Jenkins KJ, van de Ven, MFC and de Groot JLA (1999) Characterisation of foamed bitumen. *Proceedings of the 7th Conference on Asphalt Pavements for South Africa*. See http://asphalt.csir.co.za/FArefs/CAPSA%20%2799%20Jenkins%20100.pdf (accessed 02/06/2015).

Kaliakin V, Meehan C, Attoh-Okine B and Imhoff P (2012) *Long-term Performance Monitoring of a Recycled Tire Embankment in Wilmington, Delaware*. Department of Civil and Environmental Engineering, University of Delaware, Newark, NJ, USA, Report

DCT 232. See http://sites.udel.edu/dct/files/2013/10/long-term-performance-monitoring-133dm1b.pdf (accessed 02/06/2015).

Lekarp F, Isacsson U and Dawson AR (2000a) State of the art. I: resilient response of unbound aggregates. *Journal of Transportation Engineering* **126(1)**: 66–75.

Lekarp F, Isacsson U and Dawson AR (2000b) State of the art. II: permanent strain response of unbound aggregates. *Journal of Transportation Engineering* **126(1)**: 76–84.

Lo Presti D (2013) Recycled tyre rubber modified bitumens for road asphalt mixtures: a literature review. *Construction and Building Materials* **49**: 863–881.

McNeil K and Kang TH-K (2013) Recycled concrete aggregates: a review. *International Journal of Concrete Structures and Materials* **7(1)**: 61–69.

Masad E, Al-Rousan T, Button J, Little D and Tutumluer E (2005) *Test Methods for Characterizing Aggregate Shape, Texture, and Angularity*. Transportation Research Board, Washington, DC, USA, NCHRP Report 555.

MPA (Mineral Products Association) (2012) *Summary Sustainable Development Report 2012*. MPA, London, UK.

Nicholls JC (1994) *EVATECH H Polymer-modified Bitumen*. Transport Research Laboratory, Crowthorne, UK, TRL Report PR109.

Nicholls JC and Daines ME (1994) Laying conditions for bituminous material. In *The Asphalt Yearbook 1994*. Institute of Asphalt Technology, Stanwell, UK, pp. 94–98.

NSA (National Slag Association) (2015) *General Information about NSA, Iron and Steel Slags*. See http://www.nationalslag.org/sites/nationalslag/files/documents/gen_info_sheet.pdf (accessed 02/06/2015).

ONS (Office for National Statistics) (2009) *Annual Minerals Raised Inquiry*. Stationery Office, London, UK.

Özkan Ö, Yüksel I and Muratolu Ö (2007) Strength properties of concrete incorporating coal bottom ash and granulated blast furnace slag. *Waste Management,* **27(2)**: 161–167.

Pan T, Tutumluer E and Anochie J (2006) Aggregate morphology affecting resilient behavior of unbound granular materials. *Transportation Research Record* **1952**: 12–20.

Park JK, Kim JY and Edil TB (1996) Mitigation of organic compound movement in landfills by shredded tires. *Water Environmental Research* **68(1)**: 4–10.

Petavratzi, E (2007) *Sustainable Utilisation of Quarry By-products*. Minerals Industry Research Organisation, English Heritage and the Department for Environment, Food and Rural Affairs, London, UK. See http://www.sustainableaggregates.com/library/docs/mist/l0065_t2c_suqbp.pdf (accessed 02/06/2015).

Pfeiffer JPh and van Doormaal PM (1936) Rheological properties of asphaltic bitumen. *Journal of the Institute of Petroleum Institute* **22(154)**: 414–440.

Powell BD, Zhang J and Brown ER (2004) *Aggregate Properties and the Performance of Superpave-designed Hot-mix Asphalt*. Transportation Research Board, Washington, DC, USA, NCHRP Report 539.

Preston JN (1992) Cariphalte DM – a binder to meet the needs of the future. *Bitumen Review* **66**: 16–19.

Read J, Whiteoak D and Hunter R (eds) (2003) *The Shell Bitumen Handbook*, 5th edn. Thomas Telford, London, UK.

Roe PG and Hartshorne SE (1998) *The Polished Stone Value of Aggregates and In-service Skidding Resistance*. Transport and Road Research Laboratory. Crowthorne, UK, TRL Report 322.

Sainton A (1996) Dense noiseless asphalt concrete with rubber bitumen. *Eurasphalt and Eurobitume Congress, Brussels, Belgium.*

Sear LKA, Weatherley AJ and Dawson AR (2003) The environmental impacts of using fly ash – the UK producers' perspective. *Proceedings of the International Ash Symposium.* Center for Applied Energy Research, University of Kentucky, Lexington, Ky, USA.

SEPA (Scottish Environmental Protection Agency) (2012) *Recycled Aggregates from Inert Waste.* SEPA, Edinburgh, UK, Guidance Note WST-G-033.

Sherwood PT (2001) *Alternative Materials in Road Construction*, 2nd edn. Thomas Telford, London, UK.

Szatkowski W (1967) *Determination of the Elastic Recovery of Binder/polymer Mixtures Using a Modified Sliding Plate Microviscometer.* Road Research Laboratory, Crowthorne, UK.

Ullidtz PA (1979) Fundamental method for prediction of roughness, rutting, and cracking of pavements. *Proceedings of the Association of Asphalt Paving Technologists* **48**: 557–586.

van der Poel C (1954) A general system describing the visco-elastic properties of bitumen and its relation to routine test data. *Journal of Applied Chemistry* **4**: 221–236.

White E (ed.) (1992) *Bituminous Mixes and Flexible Pavements.* British Aggregate Construction Materials Industries, London, UK.

WRAP and EA (Waste and Resources Action Programme and Environment Agency) (2010) *Pulverised Fuel Ash (PFA) and Furnace Bottom Ash (FBA), Quality Protocol.* WRAP, Banbury, UK, Environment Agency Report LIT 8272.

WRAP and EA (2013) *End of Waste Criteria for the Production and Use of Aggregates from Inert Waste, Quality Protocol.* WRAP, Banbury, UK, Environment Agency Report LIT 8709.

WRAP and EA (2014) *End of Waste Criteria for the Production and Use of Tyre-derived Rubber Materials, Quality Protocol.* WRAP, Banbury, UK, Environment Agency Report LIT 8273.

Highways
ISBN 978-0-7277-5993-1

ICE Publishing: All rights reserved
http://dx.doi.org/10.1680/h5e.59931.219

Chapter 7
Soil-stabilised and hydraulically bound mixtures used in road pavements

Kim Jenkins SANRAL Chair in Pavement Engineering, Stellenbosch University, South Africa

7.1. Stabilisation concepts and objectives

The level of service required from a road is a function of the importance of the route and its traffic, and, hence, the materials comprising the pavement infrastructure must be capable of coping with the performance requirements that the traffic demands, within the financial resources available. Typically, pavements that carry in excess of 50 vehicles per day require more than natural gravel in the structural layers in order to perform adequately (Figure 7.1). The improvements required to address this need include

- natural enhancements to the material (i.e. regrading or blending gravels or crushed aggregate)
- treatment of the natural soil (including gravel) or graded crushed aggregate with conventional chemical stabilisers or proprietary products
- converting unpaved roads to surfaced pavements.

As an alternative to the costly importation of off-site high-quality materials from afar to provide high-quality pavements, the treatment of in situ and/or off-site locally available materials with some form of mechanical or chemical stabilisation can often provide an economical solution to the need to provide for greater traffic loads. Such stabilisation treatments are commonly applied to subgrades, capping and sub-base layers; in many countries also – but rarely in the UK – these treatments are also applied to base courses. The selection of the type of treatment depends on the engineering properties desired from the various layers of the pavement structure, and on the prevailing environmental conditions. In practice, however, it must also be acknowledged that differences in philosophy exist between developed and developing countries regarding the use of soil stabilisation techniques to treat subgrades and to enhance the structural layers of pavements.

In general, the application of appropriate soil stabilisation treatments to in situ and/or locally available off-site soils (and some recycled materials) can result in the following:

- enhancement of their compressive and tensile strengths
- increases in their stiffness (i.e. the resilient modulus, M_r)
- improvements to their durability, primarily in terms of their resistance to the effects of moisture

Figure 7.1 Road technology selection and investment, based on the level of service, as generally applied in developing countries (AADT, annual average daily traffic)

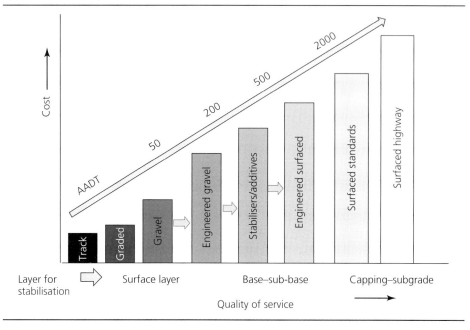

- reductions in their plasticity, which is reflected in a lowering of the plasticity index (PI) and an associated increase in the bearing capacity (e.g. via a higher California bearing ratio (CBR) value)
- improvements in the workability of, especially, clayey soils
- reductions in the in situ moisture content.

In short, soil stabilisation can be used to improve the characteristics of a soil so that its strength, plasticity, durability or other properties can be brought within the specification limits of a pavement. This enables more soil materials from within a road alignment to be utilised in new pavement layers, thereby ensuring more economical haul distances and road construction.

7.2. Types of stabilisation

There are three primary forms of stabilisation, namely mechanical, chemical and bitumen stabilisation.

Mechanical stabilisation involves the compaction and, usually, the blending of two or more soils in order to improve the gradation, reduce the plasticity and improve the bearing capacity.

Chemical stabilisation uses chemical binders, usually lime and/or cement, in a process of soil modification to improve the granular properties and/or cementation of a soil to create a rigid-type bound material. It is not within the scope of this chapter to address the many other proprietary chemical products that are used, for example (with varying degrees of success) to upgrade gravel roads in warm climates. However, when considering the use of these proprietary products, the reader is advised to assess their 'fitness of purpose' for the specific project being treated, by testing for the desired performance properties. (More information on these can be found in appropriate non-marketing publications (e.g. SANRAL, 2014).)

Bitumen stabilisation is a process that is used with cold soil or aggregate to produce a flexible-type bound material by admixing bitumen via either bitumen emulsion or foamed bitumen technology. In cases where the climatic conditions are suitable and good-quality granular materials are locally available, bitumen stabilisation may provide a solution to particular road construction needs. The use of either bitumen emulsion or foamed bitumen binders, often in combination with lime or cement fillers, can be used to provide enhanced flexibility, stiffness and durability without resulting in shrinkage cracking to base, sub-base and capping layers.

7.3. Mechanical stabilisation

The most cost-effective means of mechanical stabilisation is simply to compact a well-graded in situ soil. The removal of oversize material from a soil prior to its compaction is another simple form of this type of stabilisation. If 'as-dug' local materials, including those from road cuttings, do not initially meet the specification requirements for particular pavement layers, it may be found that they will do so when improved by being blended with one or more economically available local materials. As can be gathered, therefore, mechanical stabilisation is a generic term that covers a number of different granular stabilisation activities. For example,

- The term mechanical stabilisation is commonly used to describe the blending of coarse and fine materials with the specific aim of improving gradation and enhancing particle packing, so as to improve the bearing capacity when compacted. Such soil blends are most usually used in structural pavement layers (e.g. bases and sub-bases and capping layers).
- The term soil–aggregate stabilisation is often used to describe a form of mechanical stabilisation whereby a coarser aggregate such as sand (i.e. the fraction between 0.075 and 4.75 mm in size) is admixed with a fine-grained soil in order to 'separate' the clayey particles with plasticity and swell potential. In this case, the main purpose of the stabilisation process is to reduce the effective plasticity of the natural soil when the blended material is compacted and used in a pavement layer.
- The term granular stabilisation tends to be used in warm climatic areas to refer to the treatment of well-graded natural gravels and/or soils with calcium or sodium chlorides for the purpose of maintaining soil moisture – and, thereby, cohesion and durability – in pavement layers that are already mechanically stable. While this well-established form of treatment (for details, see Thornburn and Mura, 1969) is commonly used in the upgrading of low-volume roads in warm climates, its usage would not normally be considered for soil stabilisation purposes in the UK, because of the prevalent inclement weather conditions.

7.3.1 Blending soil materials

The particle size distribution of a granular material dictates the extent to which packing of the particles will occur during compaction – which ultimately impacts on the bearing capacity of the material. If the natural gradation of a soil does not fall within desired specification limits, a second material can be blended in situ or in an off-site plant to improve the gradation, with the aim of enhancing the dry density and bearing strength of the compacted material. Blending can also be used to reduce the plasticity of a soil, primarily through the physical separation of the clay or silt fractions as a result of, say, the supplementation of the sand fraction. In all cases, the success of the stabilisation process typically depends on the economic availability of source materials with soil fractions that can be combined into a composite material with the desired particle size distribution.

The quantities of different soils that need to be blended in order to achieve the desired blend gradation can be determined with the aid of a ternary diagram (Figure 7.2) on which is plotted the actual and desired particle size distributions. The input parameters for the ternary diagram are the three main fractions that comprise each material to be blended, such as the percentages of silt and clay (<0.075 mm), sand (0.075–2.0 mm) and gravel (2.0–37.5 mm).

A simple example of a soil (material A) containing 25% gravel, 25% sand, and 50% silt and clay, is shown in a ternary format in Figure 7.2 so that its particle size distribution is typified by means of a single point or coordinate. If another soil (material B) is to be used for mixing with material A, it is also plotted on the ternary diagram; in addition, the grading of the desired blend (i.e. material C) is also plotted as a coordinate. The blending

Figure 7.2 Ternary diagram for blending road materials (after SANRAL, 2014)

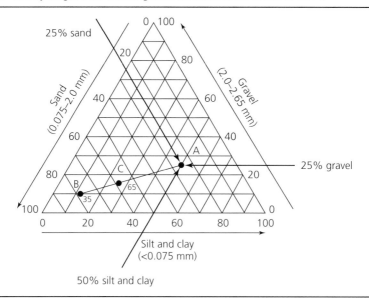

ratio is then determined by the ratio AC : BC to achieve the desired blend. In this example, the ratio of the distances from the plotted point of each material to the coordinate of the desired blend (i.e. 65 : 35) then defines the blending proportions. Thus, 35% of material A plus 65% of material B will create material C with the desired gradation.

Care must be exercised when blending two materials with significantly different specific gravities. The blending ratio will need to be adjusted when significant specific gravity differences are encountered, as packing of particles is a volumetric phenomenon, and is not determined on a mass basis.

Material sources that are in close proximity are best blended in stockpile (e.g. at the borrow pit). This is not always economically possible, however, and it may be necessary for the blending of transported material(s) to take place on site; that is, on the subgrade, sub-base or base of the pavement, as appropriate for the layer being constructed. When employing in situ mixing, the material of greater proportion should first be spread from a windrow into a layer of uniform thickness, and the second material is then placed and preshaped on top of the first. The two materials are then grader mixed to the full depth, cutting the material, turning it over and then spreading it across the width of the pavement. A disc plough is also effective in the mixing process. In situ recyclers may also be used for blending; however, these recyclers blend primarily in the vertical plane, with only nominal horizontal mixing (i.e. the material is pulverised and lifted by the milling drum and placed within 15 cm of its original horizontal position). Hence, only limited longitudinal and lateral blending occurs. Thus, accurate preshaping to provide uniform thicknesses of the two layers being blended is especially important when using this equipment.

In all cases, standard quality control tests (e.g. particle size distribution, Atterberg limits and bearing capacity) should be applied to the blended product, to confirm that the specification requirements are met.

7.4. Chemical stabilisation

In order to create a balanced pavement that distributes stresses as intended to the foundation, the base and sub-base layers must have adequate strength, stiffness and durability. The chemical stabilisation processes discussed in this chapter focus on the use of conventional lime and cement products to achieve these criteria.

If the soil materials intended for use in the pavement require chemical stabilisation, the decision as to whether to modify and/or stabilise them with lime or cement depends on the nature of the material and on the primary function of the layer in which it is to be used. In this context, the term modification is used to describe the use of a chemical to improve the properties of a soil without causing much increase to its elastic modulus or tensile strength, while the term cementation is used to describe the use of a chemical to achieve a soil-stabilised layer with significant strength and stiffness.

Lime is used as a soil treatment for many reasons, ranging from the need to expedite construction on weak clay subgrades to improving the engineering properties of plastic sands and/or gravels, as well as reactive clays. There is normally no value in using lime with low-cohesion sands or gravels (unless a pozzolanic material such as pulverised fuel ash is also added). As a rule of thumb, a small amount of lime – typically between 1% and 3% by mass of the lime–soil mix – is sufficient to modify (improve) a poor-quality clayey subgrade by reducing its PI so that its workability is improved, including during wet weather. If greater amounts of lime are added to the soil, then (dependent upon the clay type and the ambient curing temperature) the additional free-lime content may act as a binding agent and enter into a cementitious reaction with the soil.

Lime-modified soil mixes are most commonly used in the construction of capping and sub-base layers. In countries that are subject to hot climates, it is also not uncommon for lime to be used to bind soil particles in sub-base and soil–aggregate base layers formed from reactive soils with high PIs and/or clay contents in pavements that are not expected to carry major volumes of traffic.

Internationally, conventional cement is the most common cementitious soil-stabilising chemical agent. Factors that ensure its wide usage are that (a) reasonable-quality cement is available in most countries at a relatively low price, (b) the use of cement usually requires less care and control than many other stabilisers, (c) much technical information is easily available on cement-treated soils and (d) most soils – except those with high organic matter or soluble sulphate contents – can be stabilised with cement if enough is added with the right amount of water, and proper compaction and curing is carried out.

The admixing of a small quantity of cement with a poor-quality subgrade will also result in the development of a modified soil that has improved moisture-resisting and stability properties. Unlike with lime, however, in this case the modification results from the formation of weak cementitious bonds that are sufficient to reduce moisture-induced shrinkage and swell (even with reactive soils) without much increase to the elastic modulus and tensile strength of the material. These cement-modified materials have very closely spaced networks of fine cracks, and they fragment under traffic loads and thermal stresses; consequently, they are most effectively used in capping and sub-base layers when formed from soils that already have good particle interlock.

In practice, however, enough cement is commonly added to a soil to enable the creation of a cement-bound hardened material that has significant strength and stiffness when the moisture content of the mixture is adequate for both cement hydration and compaction. This hardened material, often termed soil cement, is most commonly used for subgrade capping and/or sub-base purposes in major road pavements, and in the capping, sub-bases and/or base layers of secondary-type roads.

Cement (or lime) is never used in the surfacings of pavements, as the chemically treated product has poor resistance to abrasion, and needs protection against moisture entry into the cracks that will inevitably form. Shrinkage cracking in a base layer may need to be addressed if cement is used as the stabilising agent in its construction.

Figure 7.3 Varying types and degrees of chemical stabilisation (UCS, unconfined compressive strength) (NITRR, 1986)

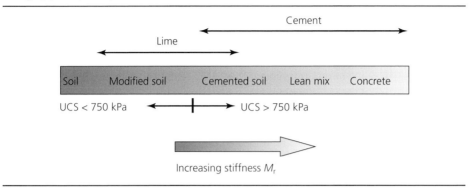

As noted above, the addition of a quantity of lime in excess of that required for soil modification can result in the development of cementation. As is suggested by Figure 7.3, this leads to an overlap between the application of lime and cement in chemical stabilisation processes.

7.4.1 Lime modification and stabilisation: mechanisms and effects

Whether modification or cementation is desired through the stabilisation process, the basic chemistry of the lime–soil reaction needs to be understood in order to appreciate what happens when lime is admixed with a soil. In practice, the term 'lime' is used to describe both the oxides and hydroxides of calcium and calcium magnesium, namely

- quicklime – calcitic (CaO) and dolomitic (CaO plus MgO)
- hydrated lime – calcitic $Ca(OH)_2$ and dolomitic ($Ca(OH)_2$ plus MgO).

The reactions between soil and lime are complex, as they are influenced by the clay mineralogy of the soil, which in itself is multifaceted. In general, the following form part of the soil–lime interaction process: (a) modification, through cation exchange and flocculation; (b) cementation, through pozzolanic reaction; and (c) carbonation, through interaction with the atmosphere.

(a) Modification, through cation exchange and flocculation. When lime is admixed with a moist clayey soil, a cation exchange reaction immediately takes place, and calcium (Ca^{2+}) ions and, to a lesser extent, magnesium (Mg^{2+}) ions from the lime replace hydrogen (H^+), sodium (Na^+) and potassium (K^+) ions that are less firmly attached to the surfaces of the clay particles. This exchange reaction has the immediate positive effect of reducing the amount of adsorbed moisture that is bound at the clay particle surfaces, thereby promoting flocculation of the clay particles and a consequent change in the texture of a soil; that is, the clay particles flocculate and clump together into aggregations that cause the soil to behave like a silt of lower plasticity, thereby modifying and improving the particle size distribution, the permeability and the handling properties of the original soil.

The extent to which ion exchange modifies the properties of a given soil is dependent upon the amount of lime added, the amount of clay in the soil, the reactivity of the clay, and the nature of the dominant cation originally adsorbed on the clay particle. For example, a sodium-dominated montmorillonitic clay, which has a high cation-exchange capacity, will be much more dramatically affected than a kaolinitic clay, which has a low cation-exchange capacity; however, the montmorillonitic clay will require a much larger addition of lime to achieve calcium saturation and the full flocculation effect, whereas a kaolinitic clay requires only a very small amount of lime to achieve its full flocculation potential.

Significant positive effects to note about using lime to modify the properties of a clayey soil are (i) an increase in the plastic limit, and decreases in its liquid limit and PI of the soil; (ii) an increase in the bearing capacity (e.g. as measured by the CBR or the modulus of the subgrade reaction (M_r) tests; (iii) a reduction in shrinkage and swelling; (iv) an improvement in soil consistency and workability; and (v) the 'unbound' lime-modified product does not exhibit any significant compressive or tensile strength.

Both hydrated limes and quicklimes (but not slurry limes) can be used to reduce the moisture content of wet soils. However, quicklime has a faster drying effect, as the chemical reaction between this type of lime and the water in the soil removes free moisture from the soil and the heat produced by the reaction assists in drying (Sherwood, 1995).

(b) Cementation, through the pozzolanic reaction. A pozzolanic material is a siliceous, or a siliceous and aluminous, material that in itself possesses little or no cementitious value but which – when finely fragmented and in the presence of moisture (i.e. the water needs to be available for hydration purposes) – will chemically react with lime at ordinary temperatures to form calcium silicate and aluminate compounds with cementitious properties that bind the soil particles together so that the stabilised material has long-term benefits in terms of strength, volume stability and (in colder climates) resistance to frost action.

Some components of natural soils, notably clay minerals, are pozzolanic materials; however, for a pozzolanic reaction to take place, the amount of lime added to a clayey soil must be in excess of that required by the soil for completion of the cation exchange and flocculation process (i.e. the modification process), as it is the 'free' lime entering into the pozzolanic reaction that results in the slow, long-term, binding together of soil particles at their points of interaction. When enough lime is added, the pH of the soil is raised, typically to about 12.4, and this highly alkaline environment promotes the dissolution of clay particles and the precipitation of hydrous calcium aluminates and silicates – these are broadly similar to the reaction products of hydrated cement – that bind the particles together.

Soil–lime pozzolanic reactions are very temperature dependent. One reason why lime stabilisation is used in warm climates is that high ambient temperatures are very beneficial to the rate of strength development. However, the pozzolanic reaction is curtailed when temperatures drop below about 10°C.

As is suggested in Figure 7.3 and Table 7.1, the boundary between lime–soil modification and cementation is not clearly defined. As noted previously, the amount of lime that causes flocculation/aggregation is usually quite small, and it is the lime content in excess of this amount that becomes involved in the pozzolanic reactions and the cementation of soil particles. Thus, delays in compaction should not be allowed to occur after the design amount of lime has been added to a soil, as, if the lime-treated soil is allowed to gain cementitious strength in the loose state, it will usually result in decreases in the maximum dry density and, consequently, strength, after compaction.

While lime modification occurs rapidly (and is normally completed within 3 days) the cementation process is relatively slow and long term. In general, the early (e.g. first 7 days) gains in the lime–soil strength occur quite rapidly – albeit much more slowly than with soil–cement – and then the strength increases more slowly at a fairly constant rate for many months. For a given curing period, the strength will normally increase as the lime content is increased up to a critical amount beyond which the strength either declines or remains constant; that is, the lime content in excess of that able to react with the pozzolanic material present in the soil during that curing period will not result in any additional strength gain. However, the lime–soil reaction will continue for a longer time, until all the free lime is used up: for example, it may take several years before the pozzolanic reaction is finally completed when plenty of lime is added to a very highly reactive clay soil in the presence of an adequate moisture content.

If a soil has a non-reactive clay mineral and a small clay content then, irrespective of the lime type or content or the length of curing, significant pozzolanic strength development will not take place.

While lime modification helps achieve compliance with granular material specification requirements (e.g. the CBR, PI and particle size distribution), lime cementation enables compliance with compressive or tensile strength specifications.

(c) Carbonation, through interaction with the atmosphere. Carbonation, which is an undesirable reaction, occurs when carbon dioxide from the air or rainwater reacts with free calcium (and magnesium) oxides and hydroxides and converts them back into their respective carbonates (e.g. $Ca(OH)_2 + CO_2 \rightarrow CaCO_3 + H_2O$). Carbonation is especially noticeable in industrial areas, where the carbon dioxide content of the air is much higher than in rural areas.

Carbonation results in a reduction in the soil–lime pH, an increase in the volume of new chemical products, a lower lime–soil dry density and increased permeability. Because of carbonation, some lime that would otherwise take part in the pozzolanic reaction does not do so, and, hence, the soil–lime mix develops lower strength than might otherwise be expected. Thus, if the achievement of high strength is a primary objective of the lime stabilisation process, the carbonation effect means that lime has to be protected while in storage and in transit prior to field use; also, prolonged intensive mixing during construction should be avoided, and compaction should take place as soon as possible after the admixing of the lime with the soil.

Table 7.1 Purpose of lime treatment and implications (Beetham *et al.*, 2014)

Aim of treatment	Physicochemical process	Common terminology	Typical lime application: %	Typical time required
Lower the moisture content of wet/low-strength soil to the optimum moisture content, improving compaction and the bearing capacity	Free-moisture removal by reaction with quicklime. Cation exchange/ clay mineral aggregation effectively increasing the optimum moisture content	Lime improvement/ modification	Low, e.g. 0.5–4% of the dry weight of the soil, dependent on the initial moisture content or the clay content	Immediate and rapid (0–72 h)
Reduce plasticity/potential for volume change	Cation exchange/clay aggregation reduces the clay surface area and the affinity for water. Early pozzolanic reactions restrict the dispersion of aggregations	Lime improvement/ modification	Intermediate	Rapid (0–72 h)
Substantially improve engineering properties, strength, stiffness and durability		Lime stabilisation	High, in excess of the initial consumption of lime, e.g. 2–10% by weight of the dry soil, dependent on mix design results	Continuous improvement beyond 72 h for months/years

Table 7.1, which is based on UK practice, summarises some lime modification and stabilisation effects that are consequent on the lime–soil reaction mechanisms.

7.4.2 Cement stabilisation: mechanism and effects

Cement consists mainly of tricalcium silicate (abbreviated to C_3S), dicalcium silicate (C_2S) and tricalcium aluminate (C_3A). When water is admixed with dry soil and cement, the hydration reactions result in the formation of calcium silicate and calcium aluminate hydrates and the release of calcium hydroxide ($Ca(OH)_2$). As with concrete, the first three of these products, especially the silicates, constitute the major cementitious gel that binds the soil particles together, irrespective of whether they are cohesionless sands, silt or highly reactive clays. The development of the hydration products proceeds rapidly, so that significant strength gains can be achieved within relatively short periods of time.

The released hydrated lime (i.e. $Ca(OH)_2$) enters into secondary pozzolanic reactions with any reactive clays present in the soil – as previously described for lime in Section 7.4.1 – to produce cementitious products similar to those formed during the primary reactions, and this contributes further to inter-particle binding. In addition, there is an exchange reaction whereby calcium ions from the released lime enter into an exchange reaction with cations already absorbed in the clay particles, and this also causes some modification to the plasticity properties of the soil before much stiffening or strength development occurs. The greater the fines content of the soil and the more reactive its clay mineral, the more important is the contribution of the secondary cation-exchange and pozzolanic reactions to the property modification and strength development of cement-stabilised soils.

Ordinary Portland cement is most commonly used to stabilise soils: rapid-hardening cements are normally avoided, as they do not allow the time required for mixing and compaction in the cement stabilisation process. While cement can also be used to modify plastic granular soils when applied in small quantities, with increasing cement application the increase in strength and stiffness rapidly starts to take effect, and the soil stabilisation aspect becomes predominant. There is no clear demarcation between cement-modified and cement-bound (i.e. soil cement) materials, albeit, for well-graded soils, values of 80 kPa for indirect tensile strength (ITS), 0.8 MPa for 7 day moist-cured unconfined compressive strength (UCS), or 700–1500 MPa for resilient modulus, are useful transition guides (NAASRA, 1998). Cement-bound natural gravels and crushed rocks typically have elastic moduli in the range 2000 to 20 000 MPa (vis-à-vis 200 to 500 MPa for unbound materials) and 28 day UCS values >4 MPa, while moduli for cement-bound fine-grained soils are usually less.

Table 7.2 provides a general guide to natural soils that are normally likely to be suitable for stabilisation with cement. The maximum aggregate size is usually 75 mm, but this depends on the mixing plant that is available.

In the UK, grading limits are specified (HA *et al.*, 1998) for the cement-bound materials that are permitted in sub-base and base layers: these are high-quality materials that are

Table 7.2 Guide to soil property limits for effective cement stabilisation (NAASRA, 1998)

Sieve size: mm	Percentage passing	Plasticity limits
4.75	>50	Liquid limit <40%
0.425	>15	Plastic limit <20%
0.075	<50	Plasticity index <20
0.002[a]	<30	

[a] Soils at the upper limit of 2 μm clay may need pretreatment with lime

usually prepared at an off-site central plant from batched amounts of processed crushed gravel or gravel–sand, or rock and/or crushed air-cooled blastfurnace slag. In the case of cement-stabilised capping layers, the main requirement is that, before the addition of cement, the natural soil should have a minimum CBR value of 15% and, in addition, it should meet the following criteria: liquid limit <45%, PI <20%, organic content <2% and total sulphates <1%.

Compared with poorly graded high-void-content soils, well-graded soils only need small amounts of cement to satisfactorily stabilise them.

Strength requirements for cement-bound soils differ internationally and geographically, dependent on research, local material quality and empirical performance data collected over time. The cement content required in any given instance depends upon the design criteria governing the end-product, and these, in turn, are influenced by environmental as well as traffic-load factors: for example, northern US states typically require cement-bound soils to meet strict freeze–thaw criteria while wet–dry durability criteria take precedence in South Africa. In the UK, the practice is to specify the desired stabilities of cement-bound materials in terms of the 7 day compressive strength of 150 mm cubes for the specified layers.

7.5.   Design for chemical stabilisation

The design process for cement and lime-treated soils varies from country to country, and from one environmental location to another within a country. The general practical approach to the design process that is used in South Africa for the chemical stabilisation of natural soils is summarised here as an indication of how to address the issue from first principles. This process involves the following steps: (a) sample the borrow pit or in situ soil, and carry out appropriate laboratory testing; (b) determine the required structural properties or bearing capacity of the modified or stabilised layer; (c) identify the types and quantities of stabiliser available and required in the mixture; and (d) carry out laboratory testing to determine the actual type and quantity of stabiliser required, in accordance with the specifications and/or desired performance outcome.

The flow diagram in Figure 7.4 outlines the design process.

Figure 7.4 The chemical stabilisation design process (SANRAL, 2014) (ICS, initial consumption of stabiliser)

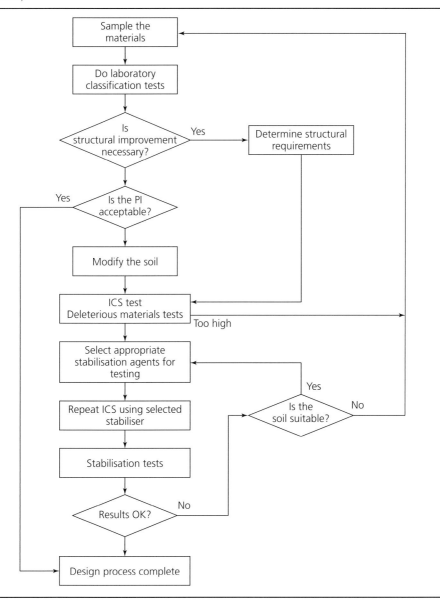

7.5.1   Soil testing

Whether the soil being tested is from the subgrade, or from a cutting, borrow pit or off-site commercial source, it is essential that the samples selected for testing be representative of the material(s) that will be used in construction. The number of samples selected for testing from a given source is therefore a function of the variability of the soil at that source.

Preliminary tests carried out normally include particle distribution, plasticity and bearing capacity evaluations of the raw soil or aggregate. These should provide sufficient information to determine whether the soil requires PI modification and/or structural improvement.

The suitability of modification as the appropriate treatment for the soil being evaluated is ascertained by determining the initial consumption of stabiliser (ICS). When lime is being considered as the modifying agent, the ICL – the percentage of stabiliser (i.e. lime) that must be added to achieve a pH of 12.4 in the soil – is determined in the laboratory: this pH level defines the minimum percentage of lime that is needed to sustain the long-term effects of modification (i.e. to ensure durability of the treatment without a premature reversal of the plasticity reduction). If the ICL content is very high, then this would also suggest that modifying the soil using lime may be economically unviable. Should the soil require structural improvement (i.e. cementitious stabilisation), then tests will need to be carried out to determine whether another form of chemical stabilisation (e.g. cement) would be more cost-effective than adding additional quantities of lime.

7.5.2 Stability requirements

The stability requirements for soil-stabilised layers can be expected to vary considerably, dependent upon the conditions encountered in the field and upon the governing design criteria. In practice, these are normally prescribed in detail in pavement specifications.

The natural soil or aggregate to be treated should be of suitable quality to produce stable pavement layers. In situ and borrow soils often have high silt and clay contents that can be the cause of substandard strengths and bearing capacities. If the source materials have highly variable properties, they may yield modified or stabilised materials that are still variable after treatment, with localised areas of poor stability: this particularly applies to subgrade soils that are treated in situ, so that these may need to be dealt with by dividing them into delineated uniform sections. Materials used in structural layers, namely bases or sub-bases, will need to meet minimum bearing capacity criteria before stabilisation is considered. Following laboratory testing, it should be possible to identify inferior borrow materials containing significant amounts of deleterious components – e.g. organic compounds, secondary minerals (especially those associated with smectite clays), sulphates, soluble salts and sulphuric acids – that would render their usage problematic in pavements. Subgrade materials that have high moisture contents often have to be made drier before their effective compaction can be achieved.

7.5.3 Selecting the stabiliser

Several factors need to be considered when selecting the stabiliser type, namely (a) the physical properties of the natural material (e.g. its liquid limit, PI, clay content, and the ICL value to achieve a pH of 12.4); (b) the purpose of the stabilisation process (i.e. whether it is for modification and/or cementation, and the implications of the appurtenant rapid or slower increase in strength); and (c) the availability and cost of a suitable stabiliser. In practice, however, the choice of an appropriate stabilisation agent depends mainly on the PI of the material (Tables 7.3 and 7.4).

Table 7.3 Recommended stabilisation process in South Africa based on the PI (SANRAL, 2014)

PI of the material: %	Recommended stabilisation process	Comment
<10	Stabilise with cement	The modification reactions from the cement typically reduce the PI to less than 6 for these materials
10–15	Modify and stabilise with cement	Lime, lime–slag, lime–pulverised fuel ash mixtures or, possibly, composite-type cement (<60% clinker, the rest blastfurnace slag) should be evaluated
>15[a]	Modify and improve strength through cementation	These materials would normally not meet the requirements even for upper selected layers, and should only be used if better material cannot be located Lime is the recommended stabiliser, and should always be used to stabilise basic crystalline materials at these PI levels

[a] Stabilisation of materials with a PI of >20% is generally uneconomical

Table 7.4 Suitable chemical stabilisers for soils (after SANRAL, 2014)

Grading	Layer	Requirement	Application	Stabiliser[a]
Fine ($P_{4.75} > 50\%$)	Capping	Reduce plasticity	Modification	Lime
Coarse ($P_{4.75} < 50\%$)	Upper sub-base Lower sub-base	Increase strength	Cementation	Cement Extended cements Lime blends

[a] The effect of the fines content on strength is significant when choosing a stabiliser

Heavy clays or wet materials can be modified for improved workability and compactibility, using lime as the modifying agent. Clayey materials with a predominance of kaolinite are also likely to yield lower PI values with a high liquid limit, and can be treated with lime. Soils derived from basic crystalline rocks should have their PI reduced to non-plastic in order to ensure durability.

Various engineering properties are based on procedures that require only certain soil fractions be tested. It is therefore beneficial to consider the activity (i.e. the plasticity and expansion/shrinkage properties) of the soil in the selection of a stabiliser that takes account of the particle distribution, using the parameter A defined below

$$A = \frac{\text{PI}_{\text{gross}}}{P_{0.002}} \tag{7.1}$$

where PI_{gross} is the weighted PI of the sample (i.e. $PI \times P_{0.425}$, where $P_{0.425}$ is the percentage passing the 0.425 mm sieve) and $P_{0.002}$ is the percentage smaller than 0.002 mm. Thus, when $A > 0.5$, soil modification is appropriate, and the use of lime should be considered, and when $A < 0.5$, modification may not be necessary, and cement or cement–lime blends should be considered.

If the ICL value (i.e. the lime content needed to bring the soil to a pH of 12.4) is greater than, say, 2%, the use of lime and lime blends with ground granulated blastfurnace slag or pulverised fuel ash may be considered. The use of low-clinker slow-set cements may also be considered: this is particularly relevant to soils derived from basic crystalline rocks. Where cement is proposed, the ICS should be determined using the cement type that will be used in the field.

The particle size distribution can be a useful guide to the selection of an appropriate stabiliser for use in laboratory testing: for example, the structural layers of a pavement generally comprise coarser materials for which soil cement is usually more appropriate (see Table 7.4).

Cost should not dictate the use of a potentially inferior or inappropriate stabilising agent over a better, but more expensive, one if early pavement failure is a possible outcome. Thus, when cement stabilisation is being considered (and noting that the availability of specific cement types can be a problem in certain areas of South Africa), only those cement types that are likely to be cost-effective and available on site during construction should be used during the laboratory testing process. Also, as the composition of any cement can change periodically, the final design should always be checked with the materials that are eventually selected for use on the site.

7.5.4   Determining the quantity of stabiliser

The determination of the quantity of stabiliser is primarily based on a strength standard for the soil type being stabilised, as specified by the pavement design. However, there is also a secondary major requirement for stabiliser content, namely the durability of the stabilised layer.

Strength tests are carried out on the material at different stabiliser contents, to determine the optimum amount of cementing agent to add to the material being stabilised in order to achieve the specified strength requirement for each pavement layer. While the strength of cemented soils is commonly measured in terms of the UCS, some authorities prefer ITS testing in order to set specification limits for mix design and quality control, on the grounds that ITS testing is more sensitive to the influence of cementation on the finer components of soils.

An important component of stabilisation mix design is the assurance of long-term durability, as materials that might easily achieve the required UCS or ITS strength can deteriorate in service. If the use of insufficient stabiliser results in marginal strengths, this also increases the likelihood of the pH of the stabilised material dropping over time, and this, in turn, can result in the cementation products of the material becoming unstable, with an eventual consequent loss in strength and stiffness.

It is essential, therefore, that sufficient quantities of the chemical stabiliser be admixed to ensure that adequate durability is achieved over the design life. To ensure this, the conditions that the materials are likely to be exposed to in service may be simulated in the laboratory, to give an indication of likely performance. Typically, laboratory tests may be carried to check on the following:

(a) Chemical durability: ensuring the reliable chemical durability of the stabilised material involves determining compliance with the requirements of the ICS test that the sustained pH of the stabilised-soil material remains at 12.4 and that there is ongoing resistance to carbonation.
(b) Mechanical durability: accelerated carbonation simulation on laboratory-prepared stabilised specimens, followed by UCS or ITS evaluations, provides insight into the sustainability of the engineering properties of the stabilised materials in the various pavement layers over time (Paige-Green et al., 1990).
(c) Erosion resistance: there are several methods for testing the resistance of a stabilised material to erosion after water ingress into a pavement layer, including a wet–dry brushing test (Sampson and Paige-Green, 1990), a rotating cylinder test (van Wijk and Lovell, 1986) and a wheel-tracking test (de Beer, 1989, 1990). (Note: in northern cold climates, a freeze–thaw test would be considered.)

7.5.5 Some construction considerations

Before discussing the construction of cement and lime-treated soil layers, it is important to appreciate that the admixed materials exhibit similar moisture–density relationships as raw soils: that is, for a given compactive effort there is an optimum moisture content (OMC) that will give a maximum dry density. However, for both additives the OMC for the maximum dry density is not necessarily the same as the OMC for maximum strength, and the OMC for strength can vary according to the manner and duration of curing. In general, the OMC for maximum strength of a cement-treated sandy soil tends to be on the dry side of the OMC for maximum dry density; however, in the case of clayey soils treated with either cement or lime, the OMC for maximum strength tends to be on the wet side of the OMC for maximum dry density.

The four basic steps involved in a conventional mix-in-place construction process are (a) the initial preparation of the raw soil being stabilised, (b) the pulverisation of the soil and the admixing of the chemical agent and the water, (c) compaction and finishing, and (d) curing.

If the in situ subgrade soil is to be stabilised, the initial preparation should require the soil to be excavated down to the desired formation level within the C horizon, and the loosened soil is then blade-graded to the specified cross-section and profile. If sub-base borrow material is to be stabilised, it is spread on the formation and graded. In either case, the soil is scarified thoroughly (and pulverised, if specialised mix-in-place equipment is not available), to make it easier to thoroughly admix the cement or lime that is then spread on top of the loose soil. The more uniform the depth of spread the more likely it is that an even distribution of the additive throughout the full depth of the layer being treated will be obtained during the mixing process. If a uniform distribution is not obtained, detrimental cracking may subsequently occur that is associated with localised high and low strengths.

Nowadays, purpose-built travelling plant is normally used to mix the soil, additive and water in one pass of the equipment. A water tanker coupled to the travelling plant can be used to feed the desired amount of water into the mixing hood of the plant during the one-pass mixing process; if a lime slurry is to be admixed, the lime is injected through the mixing hood at the desired moisture content. With some clayey soils, two passes of the admixing plant may be required, to ensure the uniform distribution of the cement or lime; in this case, the water is usually not added until the second pass, and the first pass involves admixing the chemical to the 'dry' soil.

As is suggested by Figure 7.5, compaction should be initiated as soon as possible after the admixing of water to cement-treated soils, and as many roller passes as possible should be carried out in, say, the first 15–20 min of compaction. Ideally, the compaction of cement-treated soils, including trimming, should be completed within 2 h of the beginning of mixing. Trimming is normally initiated before compaction is finished, to ensure good bonding of any corrected shape (see also Bredenhann *et al.*, 2012).

While early compaction is also desirable with lime-treated soils (especially in hot climatic areas), as this will lead to higher strengths, the time factor is far less critical with lime than with cement. In fact, compaction may be deliberately delayed if the soil is difficult

Figure 7.5 Influence of the delay in compaction time on the tensile strength of various cement-treated materials used in road pavements (Bruwer, 2013)

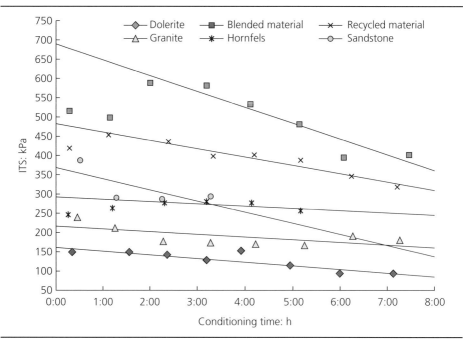

to work: for instance, if it is clayey and its plasticity needs improving, or if it is too wet and needs drying out. In either of these examples, part of the design lime content is first admixed, and the mixture is lightly compacted and allowed to condition for up to 48 h: the rest of the lime is then admixed (with additional water, if necessary), and the final compaction is then carried out.

Table 7.5 is an abbreviation of a method specification that is recommended (Sherwood, 1995) for compacting cement-treated soils in the UK. Note that the type and mass of the

Table 7.5 An abbreviated method specification for compacting cement-stabilised soils in the UK

Compaction plant	Category	Number of passes for a layer thickness of		
		110 mm	150 mm	250 mm
Smooth-wheeled roller	Mass/drum width: kg/m			
	2700–5400	16	US	US
	>5400	8	16	US
Pneumatic-tyred roller	Mass/wheel: kg			
	4000–6000	12	US	US
	6000–8000	12	US	US
	8000–12 000	10	16	US
	>12 000	8	12	US
Vibratory roller	Mass/vibratory drum width: kg/m			
	700–1300	16	US	US
	1300–1800	6	16	US
	1800–2300	4	6	12
	2300–2900	3	5	11
	2900–3600	3	5	10
	3600–4300	2	4	8
	4300–5000	2	4	7
	>5000	2	3	6
Vibratory plate compactor	Base plate mass/area: kg/m^2			
	1400–1800	8	US	US
	1800–2100	5	8	US
	>2100	3	6	12
Vibro-tamper	Mass: kg			
	50–65	4	8	US
	65–75	3	6	12
	>75	2	4	10
Power rammer	Mass: kg			
	100–500	5	8	US
	>500	5	8	14

US, unsuitable

roller, and the number of passes, are influenced very much by the thickness of the layer being compacted. Upon completion of compaction, the final surface of the cement- (or lime-) treated layer should appear smooth, dense, tightly packed, and free from cracks and compaction planes.

The curing of the stabilised material is concerned with promoting strength gain and preventing moisture loss through evaporation for a specified length of time (usually at least 7 days). Its purpose is to allow the cementitious hydration reactions to continue, to reduce shrinkage and shrinkage cracking, and to prevent the development of a variable-strength profile in the stabilised layer. While controlled curing is necessary in all climatic areas, it is especially important in locales with hot ambient temperatures and/or where windy conditions occur. Nowadays, the most common curing methods involve covering the compacted layer with impermeable sheeting or spraying the top of the layer with a bituminous sealing compound or a resin-based aluminous curing compound.

7.6. Stabilisation with bitumen emulsion and foamed bitumen

The mechanisms involved in the stabilisation of a soil, or a recycled pavement material, with bitumen are very different from those involved with cement or lime. With coarse-grained non-plastic soils, the main function of the bitumen is to add cohesive strength; thus, the stabilisation emphasis is upon the thorough admixing of an optimum amount of binder so that soil particles are thinly coated and held together without loss of particle interlock. In the case of a plastic soil that already has cohesion, the function of the bitumen is to waterproof the soil and maintain its cohesive strength; thus, the emphasis is upon impeding the entry of water by adding sufficient bitumen to (a) wrap soil particles or agglomerations of particles in thin bituminous membranes and (b) plug the soil voids. In practice, a combination of these mechanisms occurs in most materials that are stabilised with bitumen.

The main factors that influence the behaviour and design of soils that are stabilised with bitumen (also lime or cement) are summarised in Figure 7.6 (see also NAASRA, 1998). Three types of materials are considered in this diagram, namely natural soils, crushed aggregate and reclaimed asphalt (millings). As can be seen, stabilisation occurs in an environment of uncontrollable variables dictated by climate, as well as by traffic. The selection of an appropriate technology (e.g. in place or in plant treatment) will be guided by economic, practical and performance considerations, as detailed.

Bituminous stabilisation is mostly used in hot climes where there is normally a need for additional fluid to be added to a soil at the time of construction, to ensure adequate mixing and compaction at a uniform fluid content. Its potential for use in the UK is limited because of the regular rainfalls experienced: the moisture contents of most soils would normally be too high throughout the year, and the admixing of additional fluids in the form of bituminous materials could cause loss of strength (Sherwood, 1995).

The materials that are most commonly stabilised with bitumen in warm climatic areas (e.g. in South Africa) include natural granular materials, crushed recycled cement-

Figure 7.6 Factors affecting the design and behaviour of lime, cemented and bitumen-stabilised materials

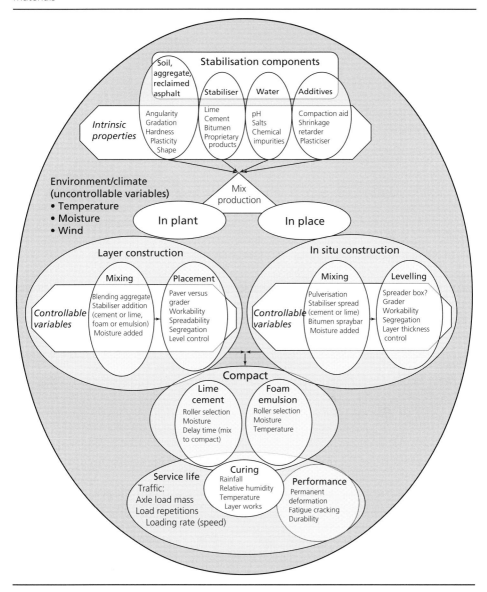

treated pavement materials, and crushed reclaimed asphalts or blends of reclaimed asphalt and raw aggregates. The treatment processes are generally carried out in specialised plants that allow the input materials to be carefully controlled, and the treated materials are usually placed in stockpiles (for limited periods) before being trucked to road sites where they are put in place, using conventional pavers, in the sub-bases and bases of secondary main roads. The ability to stockpile a bitumen-stabilised material

Figure 7.7 (a) Bitumen emulsion and (b) foamed bitumen production

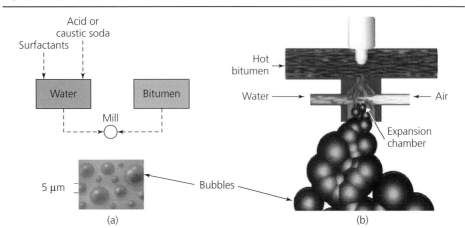

(BSM) – especially when it is treated with foamed bitumen – makes it an ideal material for labour-intensive pavement construction in newly developing countries.

No heating is applied in the treatment of a BSM. The main feature of a successful BSM is that the viscosity of the bitumen must be reduced in order to enable its even dispersion within the material being stabilised, and this is achieved in either of two ways, with bitumen emulsion or with foamed bitumen technology (Figure 7.7). The method of bitumen dispersion differs for bitumen emulsion and foamed bitumen, as follows:

▪ With bitumen emulsion treatment, the emulsion comprises tiny charged bitumen droplets (typically 5 μm) that are suspended in water. The charges on the droplets depend on the type of emulsion; that is, whether anionic (negatively charged) or cationic (positively charged). When admixed with aggregates, the bitumen droplets are attracted to the particles with the greatest surface area and charge concentration. As the droplets are forced together during compaction they flocculate and coalesce, and the contained bitumen adheres to the aggregate particles as the emulsion 'breaks'. Thereafter, excess water is dispelled from the layer during curing (i.e. during 'dry back'), as the bitumen restores its hydrophobic properties.
▪ With foamed bitumen treatment, the 'foamed' bitumen bubbles emerge from a foam spray bar in a colloidal mass that is emitted into the mixing chamber containing the material to be stabilised. The foamed bubbles 'burst' into tiny splinters when they come in contact with the material, and these splinters – which have only sufficient heat energy to warm and adhere to the smallest (<0.075 mm) of the particles of the material – selectively coat the filler. During mixing, the filler–bitumen mastic, in the presence of moisture, is then distributed throughout the mix. Compaction of the admixed material forces the bitumen droplets to adhere to larger aggregate particles, creating viscoelastic 'spot welds' in the mix. After compaction, excess water is dispelled from the layer during the curing/ 'dry-back' process.

The dispersion of bitumen droplets throughout the mixture leads to the definition of BSMs as 'non-continuously-bound' materials that have (Jenkins *et al.*, 2009) low active-filler contents (i.e. not more than 1% cement or lime) and residual bitumen contents ranging from 1.8% to 3%. As such, their behaviour tends to be granular in nature but with significantly improved shear properties that primarily result from an increase in cohesion due to dispersed localised bonding.

BSMs are produced using cold aggregate, and should not be confused with continuously bound materials such as hot-mix asphalts or, indeed, conventional cement-stabilised materials. Note that when the application rates of cement in a BSM exceeds about 2% by weight of dry aggregate, and the residual bitumen content exceeds 3%, the material begins to show bound behaviour, and fatigue damage starts to become evident after trafficking. The matrix of pavement materials based on the cement and bitumen content, and their behaviour, is captured in Figure 7.8. These two binding agents dictate the zones reserved for BSMs, which, in terms of behaviour and distress mechanisms, are different from cement-stabilised and hot-mix asphalt materials. The more expensive bituminous mixes, ranging from half-warm mixes to hot-mixed asphalts, are primarily used in first-world countries with cooler, moist climates and heavy traffic (and large highway budgets), whereas the more economical bitumen stabilisation tends to be used mainly in developing countries, and focuses primarily on lightly bound, granular-type BSMs with a lower bitumen content (1.8–2.4%) and a low active-filler content (1% or less).

Figure 7.8 Pavement material behaviour as a function of the active-filler and bitumen content

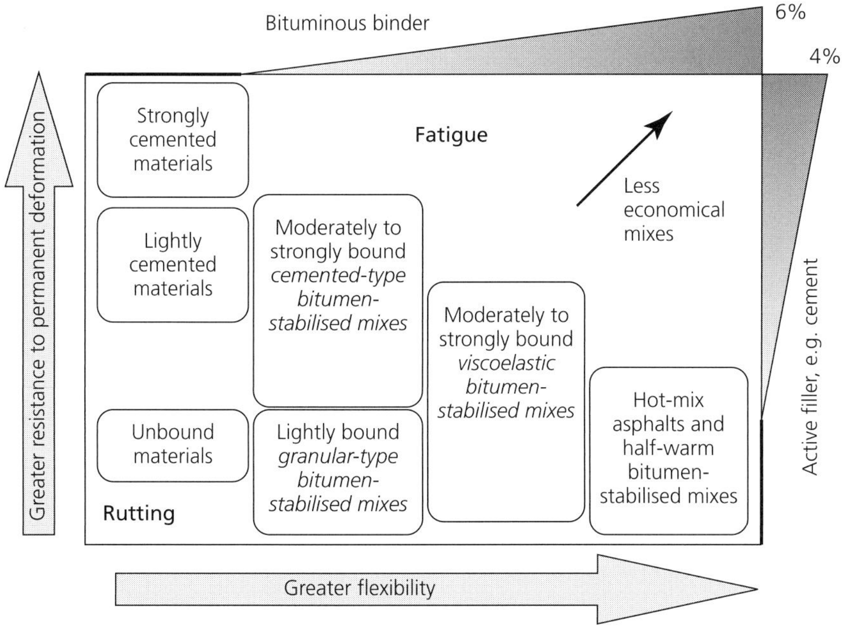

As a consequence of their non-continuous bonding, BSMs are less prone to fatigue cracking; instead, the two basic failure mechanisms are permanent deformation and moisture susceptibility.

By permanent deformation is meant the accumulation of shear deformation as a result of repeated loading, and which is dependent on the densification and shear properties of the stabilised material. Resistance to permanent deformation is enhanced by (a) improved angularity, shape, hardness and roughness of the material being stabilised; (b) increased maximum particle size; (c) improved density through compaction; (d) reduced moisture content; (e) limiting the bitumen content added to, say, <3%, as (depending on the material being stabilised) high bitumen contents encourage instability; and (f) adding active filler up to a maximum of 1%, as high filler contents cause brittleness that encourages shrinkage and traffic-associated cracking.

By moisture susceptibility is meant the damage caused by exposure of a BSM to high moisture contents and the consequent porewater pressures caused by repeated impacts of, especially, commercial vehicle traffic loads. This results in a loss of adhesion between the bitumen and the partially coated particles in the BSMs. Moisture resistance is enhanced by (a) increasing the bitumen content (subject to cost considerations); (b) adding an active filler (up to 1% by mass of dry aggregate); (c) increasing the density through compaction; and (d) achieving a continuous particle size distribution that encourages high compacted densities.

The indicator that is primarily used to guide the selection of the appropriate bitumen content for both foamed BSMs and emulsion-stabilised materials (as well as the selection of the filler type) is the ITS of the material as determined from laboratory-stabilised materials prepared under optimal conditions (i.e. in terms of compaction, moisture content and simulated accelerated curing) to measure the resistance to premature moisture damage. The ITS test does not, however, have a direct link to performance (i.e. to resistance to deformation and moisture damage). For this reason, triaxial testing is also used to provide information on shear properties in order to obtain a more reliable link to the rate of permanent deformation and the risk of moisture damage.

7.6.1 Aggregate selection for BSMs

When considering materials that might be suitable for bitumen stabilisation, the following basic considerations should be kept in mind:

- The minimum inert filler content must be such that 4% passes the 75 μm sieve when stabilising with foamed bitumen, and 2% passes the 75 μm sieve for bitumen emulsion – but if more than 25% reclaimed asphalt (millings) is included, then 2% filler passing the 75 μm sieve is acceptable throughout (excluding the active filler discussed in Section 7.6.3).
- The ideal Nijboer grading, where $P = (d/D_{max})^n$ and $n = 0.45$ (where P is the material passing (%) through a sieve size of d (mm), D_{max} is the maximum grain (sieve) size (mm) and n is a variable), provides the preferred particle size distribution for compaction, and the practice of providing a minimum amount of

voids in the mineral aggregate to allow sufficient space for bitumen (see Chapter 13) does not apply to BSMs.

- To ensure a continuous grading, recycled asphalt needs to be resized through crushing, or blended with crushed aggregate, before being stabilised with bitumen.
- Up to 75% reclaimed asphalt can be incorporated into a BSM, provided that the recovered bitumen is 'aged' so that its hardness is less than 15 pen – but when the recycled asphalt comprises more than 50% of the material being stabilised, then $P_{0.075} > 2\%$ can be applied to BSM foam and BSM emulsion.
- The variability of in situ reclaimed materials needs to be carefully controlled when applying bitumen stabilisation technology to pavement rehabilitation projects.

7.6.2 Bitumen selection for BSMs

Both BSM emulsion and BSM foam require the use of 70/100 pen bitumen in successful stabilisation processes. Harder binders can be used, but they should be carefully evaluated in the mix design.

The following are relevant to the use of bitumen emulsions for stabilisation purposes: (a) either slow-setting stable-grade cationic or anionic emulsions can generally be used; (b) the mixed material must be allowed sufficient time to aerate before compaction so that excess moisture can escape; (c) the breaking rate of the emulsion should be tested with representative aggregate and filler materials, and at water contents and ambient aggregate temperatures applicable to the project, in order to ensure compatibility; and (d) the bitumen emulsion and the material being stabilised must be chemically compatible (e.g. acidic gravel aggregates containing significant quantities of quartzite, granite, sandstone, rhyolite, syenite and felsite may not be stabilised with anionic emulsions, which are basic).

Bitumen selection for use in the production of a high-quality foamed bitumen needs to be characterised in terms of (a) the expansion ratio (i.e. the volume increase ratio from bitumen to foam), which is an indicator of the viscosity reduction needed for good dispersion during mixing with cold aggregates, and (b) the half-life of the foam bitumen (i.e. the time for the foam to collapse from full expansion to 50% expansion), which is a measure of the stability of the foam for mixing.

7.6.3 Use of cement and lime fillers

Active fillers (i.e. cement or lime) are used in BSMs for the following reasons: (a) to supplement the 'natural' fines and thereby improve the dispersion of the bitumen in the mix, (b) to improve the adhesion of the bitumen to the aggregate, (c) to reduce the PI of the in situ materials (e.g. 'dirty' gravels), (d) to increase the stiffness of the mix and the rate of strength gain, (e) to accelerate the curing of the compacted mix, (f) to control the breaking time and to improve the workability of BSM emulsions and (g) to assist in the dispersion of foamed-bitumen droplets.

With BSMs, the amount of hydrated lime and cement additive is normally restricted to a maximum of 1% by weight of dry aggregate, to ensure that the BSM does not become too brittle. Lime, but never cement, is also used with gravels that are relatively plastic (e.g. if the PI is >10): for such soils, the initial consumption of lime (ICL – see Section 7.5.1) is normally adopted as the application rate, and it is added as a pretreatment before adding the bitumen (usually after a minimum delay of 4 h). The application rate of cement should never exceed that for bitumen (i.e. the residual bitumen, in the case of emulsion): if the percentage ratio of bitumen to cement in the mix is <1, the product material will act primarily as a cement-stabilised material.

7.6.4   Construction considerations

The equipment and techniques that are used for the construction of road pavements have changed dramatically since bituminous stabilisation was first used in road construction. Today, the equipment used can be divided into two categories, namely in-place (or in situ) and in-plant (usually off-site) recyclers. In-place, on-site recyclers are used for the bituminous stabilisation of subgrades and pavement structural layers, while in-plant, off-site recyclers are primarily used for the bituminous stabilisation of high-quality bitumen-bound base layers.

There are three types of in-place stabilising and recycling machines that are designed for particular purposes:

- Tyre-mounted recyclers that can admix a bituminous binder to depths of up to 500 mm in subgrade soils or up to 300 mm in existing pavement layers, breaking down the in situ material, and maintaining a pre-set horizon of levels relative to the existing pavement, to produce a quality of mix that is comparable with that obtained with off-site mixing. If the blending of two materials is required in order to improve the particle size distribution, this is achieved by spreading a windrow (a longitudinal heap of material) of the second material to the desired depth on top of the subgrade, before mixing and compacting. In this case, the mixing occurs predominantly in the vertical profile, with only limited (<150 mm) lateral movement and mixing.
- Track-mounted recyclers that can admix to depths of 250 mm, with the ability to mill full-depth in situ asphalt surfacings and bases with helically configured point attack tools (or picks) on the milling drum. Significant lateral mixing of material in the recycled layer occurs over the entire width of the cut.
- Special stabilisation and recycling machines that are fitted with on-board mixing units. These are highly specialised units that are suited to high-productivity bitumen stabilisation (Figure 7.9), and include the mixing of in situ recycled materials in a mobile pugmill mixer.

Figure 7.9 Recycling train used on existing surfaced roads that are being stabilised with foamed bitumen or emulsion

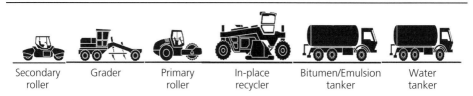

| Secondary roller | Grader | Primary roller | In-place recycler | Bitumen/Emulsion tanker | Water tanker |

Figure 7.10 In-plant stabilisation, then transport of the material to the site and laid by a paver

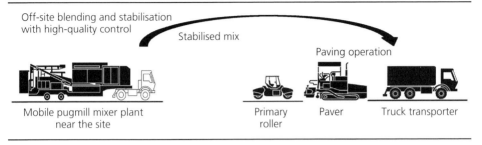

Off-site blending and stabilisation with high-quality control

Stabilised mix

Paving operation

Mobile pugmill mixer plant near the site

Primary roller

Paver

Truck transporter

In-plant off-site mixing is achieved by hauling material recovered from an existing road or a fixed source and treating it with foam or emulsion bitumen in a twin-shaft pugmill, and then transporting the mixed material back to the road. In-plant processing becomes attractive when (a) additional materials need to be added to the structural layers, using stockpiled materials of aggregates from a new source (i.e. both reclaimed asphalt stockpiles or granular sources are eminently suited to bitumen stabilisation using in-plant off-site technology); (b) different materials require blending to achieve the required mix design for BSMs; and (c) material to be recycled from an existing pavement is highly variable and/or, before being stabilised, it requires further processing because it is too

Figure 7.11 Guide to roller selection for stabilised materials in structural pavement layers (Wirtgen, 2012)

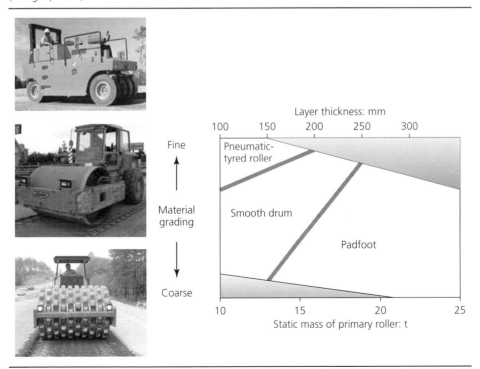

hard and requires ex situ in-plant treatment in a mobile crusher. The in-plant stabilisation process and associated activities are summarised in Figure 7.10 (see also Jenkins *et al.*, 2008, 2009; Wirtgen, 2012).

Achieving the optimum compaction in the field of BSMs requires the judicious selection of rollers. In practice, the roller selection (weight and type) is primarily based on the particle size distribution of the material and the depth of the layer to be compacted. Figure 7.11 provides a useful guideline for primary roller selection for both cement and bitumen-stabilised layers used in structural pavements in South Africa. Note, for example, that a 300 mm thick BSM with a maximum particle size of 25 mm will require a >20 t padfoot roller (bottom photograph) for primary compaction. Most stabilised materials use a pneumatic-tyred roller for finishing of the layer (top photograph).

REFERENCES

Beetham P, Dijkstra T, Dixon N *et al.* (2014) Lime stabilisation for earthworks: a UK perspective. *Proceedings of the Institution of Civil Engineers – Ground Improvement* **168(2)**: 81–95.

Bredenhann SJ, Paige-Green P and Jenkins KJ (2012) Cemented layers: accounting for compaction delays and strength loss with time. *Proceedings of the 7th International Conference on Maintenance and Rehabilitation of Pavements and Technological Control MairePav7*, Auckland, New Zealand, pp. 1–12.

Bruwer JS (2013) *A Laboratory Investigation of the Parameters that Influence the Allowable Working Time for Cement Stabilised Pavement Layers*. MEng thesis, University of Stellenbosch, Stellenbosch, South Africa.

De Beer M (1990) *Aspects of the Design and Behaviour of Road Structures Incorporating Lightly Cementitious Layers*. PhD thesis, University of Pretoria, Pretoria, South Africa.

De Beer M (1989) *Aspects of Erodibility of Lightly Cemented Materials*. CSIR Transportek, Pretoria, South Africa, Research Report DPVT 39.

HA (Highways Agency), Transport Scotland, Welsh Government and Department for Regional Development Northern Ireland (1998) *Manual of Contract Documents for Highway Works*, vol. 1. *Specification for Highway Works*. Stationery Office, London, UK.

Jenkins KJ, Collings DC and Jooste FJ (2008) TG2: the design and use of foamed bitumen treated materials. Shortcomings and imminent revisions. *Proceedings of the Recycling and Stabilisation Conference*, Auckland, New Zealand, pp. 1–15.

Jenkins KJ, Collings DC, Long FM and Jooste FJ (2009) *TG2: A Guideline for the Design and Construction of Bitumen Emulsion and Foamed Bitumen Stabilized Materials*, 2nd edn. Asphalt Academy, Pretoria, South Africa.

NAASRA (National Association of Australian Road Authorities) (1998) *Guide to Stabilisation in Roadworks*. NAASRA, Sydney, Australia.

NITRR (National Institute for Transport and Road Research) (1986) *TRH 13: Cementitious Stabilisers in Road Construction. Technical Recommendations for Highways*. Department of Transport, Pretoria, South Africa.

Paige-Green P and Netterberg F (2004) *Cement Stabilization of Road Pavement Materials: Laboratory Testing Programme Phase 1*. CSIR Transportek, Pretoria, South Africa, Contract Report CR-2003/42.

Sampson LR and Paige-Green P (1990) *Recommendation for Suitable Durability Limits for Lime and Cement Stabilized Materials.* CSIR Transportek, Pretoria, South Africa, Research Report DPVT 130.

SANRAL (South African National Roads Agency) (2014) *SAPEM: South African Pavement Engineering Manual.* SANDRAL, Pretoria, South Africa.

Sherwood PT (1995) *Soil Stabilisation with Cement and Lime.* Stationery Office, London, UK.

Thornburn TH and Mura R (1969) Stabilisation of soils with inorganic salts and bases: a review of the literature. *Highway Research Record* **294**: 1–22.

Van Wijk AJ and Lovell CW (1986) Prediction of subbase erosion caused by pavement pumping. *Transportation Research Record* **1099**: 45–57.

Wirtgen (2012) *Wirtgen Cold Recycling Technology.* Wirtgen, Windhagen, Germany. See http://www.wirtgen.de/en/customer-service/hands-manuals/07_praxisratgeber_1.php (accessed 07/06/2015).

Highways
ISBN 978-0-7277-5993-1

ICE Publishing: All rights reserved
http://dx.doi.org/10.1680/h5e.59931.249

Chapter 8
Standard asphalt mixtures used in pavement layers

Michael Brennan   Consultant (formerly Senior Lecturer, National University of Ireland, Galway), Ireland

Alan Kavanagh   Technical Manager, Atlantic Bitumen Company Ltd, Ireland

All asphalts used in the UK have basic requirements for composition and constituent materials (i.e. aggregate types and gradings, and bitumen types and contents). Where these requirements form the essence of the specifications, they are described as recipe-type specifications. Some asphalts have additional performance-related requirements that require laboratory tests, and these asphalts are referred to as having design-based specifications. However, both the recipe-type and design-based specifications are empirical specifications based on observation and experience.

The discussion in this chapter is focused on describing the specifications of standard mixes. (The concepts relating to the design of bituminous mixes are described in Chapter 13.) Before discussing the standard mixtures, the advantages and disadvantages of recipe mixtures, and some related characteristics of the aggregates used in bituminous mixes are discussed. At the end of the chapter, there are brief discussions on how the mixtures are made, laid and compacted, and on the means by which an asphalt manufacturer is encouraged to comply with specifications.

8.1.   Recipe mixtures: advantages and disadvantages

Over the past 50 years, a substantial body of research work has been built up in relation to the use of asphalt mixtures in road pavements, and this has encouraged the development of laboratory-based mix-design test methods to measure fundamental properties of asphaltic mixtures and ensure that they meet performance-related pavement design specifications. In Europe, this trend has been encouraged by the formation of the EU and the consequent emphasis on performance-based requirements that incorporate theory and technology and are capable of crossing national borders. Towards this end, the concept of a fundamental specification has been contrived. This type of specification relies on a combination of performance-based requirements together with limited requirements for composition and constituent materials, with more degrees of freedom than for a recipe-type specification. It is germane to note that, of all the laboratory tests that have been used hitherto in the UK, only the indirect tensile stiffness modulus test is classified as performance based: the rest are performance related (i.e. they measure properties that are related to performance) (BSI, 2010a).

Worldwide, practice in very many countries is still biased towards the use of particular empirical recipe-type specifications that lay down requirements for the composition, materials and properties of mixtures that, on the basis of proved technical and commercial experiences, correlate with a desired performance. These recipe specifications predominantly prescribe asphalt mixtures in terms of their compositions (i.e. aggregate size and grading, and bitumen type and content) and their placement (i.e. methods of mixing, laying and compacting).

The basic advantages claimed for the recipe-type specification are that (a) they are proved mixtures, and experienced pavement engineers have no difficulty in selecting suitable specifications to meet their design needs; (b) the mixing plant is easily organised to meet 'recipe' needs; and (c) mechanical testing of the finished product is minimised as good quality control is easily ensured (i.e. compliance with the recipe-type specification is easily checked during mixing and construction). The main criticisms of recipe-type specifications are that (a) their proved reliability may be confined to the traffic and climatic conditions within which they were developed, and, consequently, their usage is not necessarily directly transferable across national boundaries; (b) if the quality control checks show that the mixtures supplied do not meet the recipe, the seriousness of the transgression is not easily evaluated without laboratory testing; and (c) construction costs are increased, as the use of non-specified materials is prohibited (e.g. standard aggregate materials that have to be imported to the site preclude the use of locally available materials that might otherwise meet design criteria).

8.2. Aggregates in asphalts

The aggregates most used in asphalts are natural rock aggregates (e.g. basalt, gabbro, granite, gritstone, hornfels, limestone, porphyry, quartzite, flint and gravels) and by-product aggregates (e.g. blastfurnace slag and steel-furnace slag). The aggregates are reduced to usable sizes by crushing, and sorted into different sizes by screening: then, at an asphalt 'factory' plant, they are reconstituted to comply with desired grading curves (Moore, 2000) before being mixed with specified binders to produce the required asphalts.

Aggregates used in asphalt mixtures can be described as comprising three different fractions: coarse aggregate that ranges from some maximum size (but cannot be greater than 32 mm) down to 2 mm, fine aggregate that ranges from 2 mm to 63 μm, and fines and filler materials that pass the 63 μm sieve. Aggregate properties vary considerably, and the suitability of a given crushed aggregate for use in a road built to withstand particular traffic and environmental conditions depends primarily on the pavement layer in which it is to be used and the type and quantity of binder material used in that layer. In general, smaller maximum-sized coarse aggregates are used in the top pavement layers in order to facilitate stringent surface regularity requirements and tighter surface level tolerances. This, in turn, results in the need for increasing bitumen contents in these layers to waterproof and bind together the greater surface areas of aggregate. (Note that, assuming spherical particles, 6.3/10 mm aggregate (i.e. aggregate with minimum and maximum sizes of 6.3 and 10 mm, respectively) has twice the surface area of the same volume of 10/20 mm aggregate.)

The specified aggregate gradings can be either continuous, gap graded or uniformly graded. The use of uniform gradings tends to be limited to precoated chippings that improve skid resistance, and chippings that prevent binder pick-up on vehicle tyres during laying. In addition to the gradation criteria, the quality of the aggregates used in asphalts is controlled by specified requirements for hardness, durability, cleanliness and chemical susceptibility (BSI, 2009a).

In addition to being able to distribute traffic loads, asphalt mixtures must be durable; that is, they must survive for the design life of the pavement and continue to perform to expectations despite the ravages of time, traffic and the elements. Durability is generally improved by the use of high binder contents, low air voids and appropriate maximum aggregate sizes. In the past, many criteria that ensured durability were hidden away in the conservativeness of the recipe approach to material selection, mixing and laying. One consequence of today's trend towards the use of performance requirements is that mixtures are now designed down to minimum binder contents (Nicholls *et al.*, 2008): thus, if care is not taken with the design procedures, they can result in reduced durability.

8.2.1   Use of well-graded asphalt mixtures

The densest aggregate mixes, which minimise the voids in the mineral aggregate (VMA) framework, are obtained with gradations that tend towards those obtained using a Fuller's curve of the form

$$\% \text{ by mass passing sieve size } d = \left(\frac{d}{D}\right)^{n} \times 100 \qquad (8.1)$$

where D is the maximum particle size in the sample, d is a smaller particle size and, for spherical particles such as gravels, n is typically 0.5 but, for rough stone particles, an n value of 0.45 is appropriate (Asphalt Institute, 1997). Gradations must be adjusted within acceptable limits to increase the VMA to accommodate enough bitumen to maximise durability and avoid flushing. In addition, there is the caveat that a Fuller's curve produces a mixture that will be very sensitive to proportioning errors, and that it is best practice to modify the gradation away from the maximum density gradation (DFID, 2002). By definition, continuously graded asphalts have gradations that fairly evenly span all aggregate sizes from the coarsest to the finest. Such mixtures are generally termed 'asphalt concrete' (AC). With well-graded mixtures, the amounts of fine aggregate and filler are relatively low, and the strength and stability of a mix are primarily derived from, and the applied stresses are distributed through, particle-to-particle contact, inter-particle friction and aggregate interlock. With these mixtures, therefore, it is important that the aggregate particles are sufficiently tough that they do not break down under rollers when compacted, or under the subsequent repeated impacts and crushing actions of road traffic.

When added to a well-graded aggregate mix, the initial function of a bituminous binder is to lubricate the aggregate particles and enable them to slide over each other during compaction. The lifelong function is to act as a bonding and waterproofing agent when the road pavement is in service.

The void content, after compaction, of a well-graded recipe-type asphalt can have an impact on its performance: for example, a surface course is likely to be susceptible to fretting if its void content is too high, and to rut under traffic if it is too low. A high void content also makes the binder more vulnerable to accelerated ageing (i.e. hardening due to, mainly, oxidation) and, because of the greater permeability of the resultant asphalt, more vulnerable to loss of adhesion with the stone and, hence, reduced durability. The surface courses of road pavements are the layers that are most vulnerable in this respect: for example, too high a void content can lead to brittle fracture of a surface course at low pavement temperatures.

8.2.2   Use of gap-graded asphalt mixtures

Gap-graded asphalt mixtures consist of aggregate particles that range from the coarsest to the finest, with some intermediate-size particles being deliberately omitted from the gradation. The relevant mixtures are hot rolled asphalt (HRA) and stone mastic asphalt (SMA) (which are very dense mixtures) and porous asphalt (PA) (which is a very open mixture). Gap-graded mixtures have a better workability than well-graded ones, and they are more easily compacted; that is, to obtain high densities in continuously graded mixtures with a full range of particle sizes, the compaction equipment has to overcome greater internal friction resistance between the particles. As HRA is gap graded, and therefore lacks good coarse-aggregate interlock, it derives its stability primarily from the fine aggregates–filler–binder mortar. In this type of mix, the general role of the coarse aggregate is to bulk the material and provide additional stability to the hardened mortar – the coarse particles can be visualised as 'floating' within the mortar. By contrast, in PA and SMA, the fine aggregates–filler–binder mortar content is sufficiently low to allow good coarse-aggregate interlock. Consequently, these mixtures are rut resistant.

8.2.3   Use of hydrated lime as a filler material

Fines are defined as aggregate that passes the 0.063 mm sieve. In practice, hydrated lime is commonly used as a substitute filler material for some of the smallest fines material.

Stripping is the failure, under the actions of water and traffic, of the adhesive bond formed between the bitumen and the aggregate. It has long been known that the addition of 1–2% hydrated lime will reduce or, often, prevent stripping (DSIR, 1962). Hydrated lime combats moisture and frost damage by reducing chemical ageing of the bitumen and by stiffening the mastic more than normal mineral filler (e.g. North American field experience indicates that hydrated lime can increase the durability of asphalt by 20–50%). Hydrated lime is now increasingly used in asphalts in most European countries, in particular Austria, France, the Netherlands, Switzerland and the UK (Lesueur *et al.*, 2012).

8.2.4   Use of reclaimed asphalt pavement material

When asphalt pavements reach the end of their lives, they are either overlaid or replaced. Nowadays, if replaced, the original material is often recycled and reused. While replacement and recycling incur extra costs as a result of scarification, milling and crushing, the ingredients of the old asphalts retain an intrinsic reusable value. Hence, the incorporation of reclaimed asphalt pavement (RAP) materials in new asphalts has benefits associated with the preservation of existing road geometry, reductions in waste, savings in the use of expensive landfill, the conservation of natural aggregate resources and energy, and a reduction in the binder content added to the asphalt in which the RAP is used.

RAP materials may be used in all pavement layers. The RAP material used in AC, HRA and SMA must be in accordance with the PD 6691 guidance document (BSI, 2010a). The basic requirements are that the penetration value of the recovered binder can be measured, the particle size of the RAP aggregate cannot exceed the maximum aggregate size specified for the mixture, and the fresh bitumen added to the mix cannot be more than two grades softer than the nominal grade specified for the mixture (e.g. for a 40/60 nominal grade, a 100/150 grade is the upper limit for the new bitumen added).

For RAP contents of up to 10%, the penetration of the blended bituminous binder can be estimated from the penetrations of the recovered binder in the RAP and the fresh binder as follows:

$$\log P = \frac{A \log P_a + B \log P_b}{100} \tag{8.2}$$

where P is the penetration of the final bitumen blend, P_a is the penetration of the recycled bitumen, P_b is the penetration of the fresh bitumen, A is the percentage of recycled bitumen in the blend and B is the percentage of fresh bitumen in the blend (Hunter et al., 2015). For RAP contents of >10%, it is necessary to monitor the penetration of the bitumen from the plant-mixed blended material to ensure that it lies within specified criteria: for example, for a 40/60 nominal binder grade material, the penetration value should be between 60 pen (i.e. the upper limit for the grade) and 30 pen (i.e. the lower limit of the next harder (30/45 pen) grade of bitumen).

8.2.5 Blending aggregates to achieve target gradations

The first step in blending aggregates from the aggregate stockpiles available at an asphalt plant is to target an aggregate gradation that is about mid-way within the specified gradation envelope for the desired mixture. The optimum stockpile combination can then be obtained by trial and error, using a simple calculator.

When carrying out the stockpile blending process, the usual practice is to start by targeting the percentage retained on a coarse sieve from a coarse aggregate stockpile, followed by the percentage passing a fine sieve from a fine aggregate stockpile, and then to make up the balance from the remaining stockpiles. In the example in Table 8.1, the sieve sizes and the target limits of the specified grading envelope for a 20 mm maximum size high-modulus AC are shown in columns 1 and 2, and the mid-range target grading is in column 3. The gradings of three stockpiles, namely with nominal sizes of 10/20, 4/10 and 0/4, that are available on site are given in columns 4–6. Note that the stockpile designated as 10/20 has 100% passing the 32 mm sieve, 85% passing the 20 mm sieve, 86.9% retained on the 14 mm sieve and 1.3% passing the 10 mm sieve, and its nominal maximum aggregate size is 14 mm. For blending purposes, it is clear that the target of 17.5% (i.e. 100 − 82.5) of aggregate to be retained on the 14 mm sieve must all come

Table 8.1 A trial-and-error solution to blending aggregates from different stockpiles

1	2	3	4	5	6	7	8
Sieve size: mm	Percentage passing						Tolerances[a]
	Limits	Target	10/20	4/10	0/4	Blend	
32	100	100	100	100	100	100	–
20	90–99	94.5	85.0	100	100	97.0	−2/+0
14	70–95	82.5	13.1	100	100	82.6	−8/+5
10	55–90	72.5	1.3	92.9	100	78.1	–
6.3	42–75	58.5	1.1	21.1	99.7	55.6	±7
2	18–35	26.5	1.1	3.1	51.9	26.6	±6
0.25	8–18	13	1.1	2.9	24.0	12.9	±4
0.063	5–9	7	0.9	23	12.1	6.8	±2
			20%	31%	49%		

[a] Based on BSI (2010a)

from the 10/20 stockpile; however, this stockpile on its own cannot meet the remainder of the target specification requirements. Accordingly, the percentage, A, of the blend that must come from the 10/20 stockpile is obtained by solving $86.9 \times (A/100) = 17.5$. The solution requires that 20% of the blend come from the 10/20 stockpile.

The target grading must contain 26.5% passing the 2 mm sieve. There is 51.9% passing the 2 mm sieve in the 0/4 stockpile, but a small contribution will also come from the 10/20 and 4/10 stockpiles. Assuming that the contributions of aggregate passing the 2 mm sieve from the 10/20 and 4/10 stockpiles will amount to 1%, then 25.5% (i.e. $26.5 - 1$) will be required from the 0/4 stockpile. Accordingly, the percentage, B, of the blend that must come from the 0/4 stockpile is obtained by solving $51.9 \times B/100 = 25.5$: thus, 49% of the blend must come from the 0/4 stockpile. The balance of the blend (i.e. $100 - 20 - 49 = 31\%$) must therefore come from the 4/10 stockpile. The blend is shown in the bottom row of Table 8.1: note that the resulting grading of the blend shown in column 7 complies well with the specified limits in column 2. This blend constitutes a declared target grading that the asphalt producer should meet.

Compliance of individual samples of asphalt with the declared grading is required in the UK. Note also that the allowable tolerances (BSI, 2010a) for this mixture from BS EN 13108-21 are given in the final column of Table 8.1.

8.3. Specifications currently used in the UK

The European Committee for Standardization (CEN, Comité Européen de Normalisation) is the national standards organisation of the EU and the European Free Trade Association (EFTA). The CEN, via its technical committees, has the responsibility for the development of common European (EN) standards so as to facilitate trade and communication between member states: its national members are committed to conferring the status of

Figure 8.1 Volumetric models showing typical compositions of AC, PA, SMA, gussasphalt, HRA and mastic asphalt

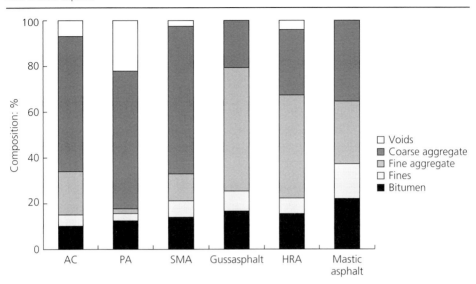

a national standard on approved ENs and to withdrawing existing standards that conflict with them. This process has affected the recipe specifications historically developed in most EU countries, and, consequently, there has been a rationalisation and harmonisation of 'national' specifications.

Currently, the harmonised European standards or 'norms' (i.e. the EN specifications) recognise seven distinctly different asphalt types for use in pavement construction. Their allowable limits for composition and performance encompass the wide range of different mixtures that are used within the EU member states. The UK national guidance document PD 6691 (BSI, 2010a) specifies requirements for these mixtures that make them more specific and applicable to UK climatic, traffic loading and safety conditions including skidding resistance (Bradshaw-Bullock, 2007). Models of six asphalt types, showing typical compositions of coarse aggregates, fine aggregates, fines and bitumen, are shown in Figure 8.1. Of these, AC, HRA and SMA are particularly relevant for pavement construction in the UK. There has been a tendency for a fourth material, thin asphalt surfacing (not shown in this diagram), to replace PA because of the dwindling use of this asphalt as a low-noise surfacing (BSI, 2010a). Thin asphalt surfacings are considered in Chapter 18.

Before discussing the recipe mixtures used in the UK, it should be noted that the BS EN reference designation for an asphalt material has the following five basic components:

■ mixture
■ *D* and %/*D*
■ base/bin/surf

- binder
- rec/des.

The term 'mixture' is used to designate the material being specified (e.g. AC, HRA, SMA or PA). For AC, SMA and PA, D is the upper sieve size of the aggregate (i.e. the maximum nominal aggregate size). For HRA, $\%/D$ is the percentage of coarse aggregate and its maximum size D. The terms 'base', 'bin' and 'surf' refer to where the material should be used in the pavement (i.e. in the base, binder or surface courses, respectively). By 'binder' is meant the grade to be used in the mixture (e.g. 40/60 means 40/60 pen grade bitumen). The terms 'rec' and 'des' refer to recipe-type asphalts and design asphalts, respectively.

8.3.1 Asphalt concrete

AC, which is widely used in pavement bases and binder and surface courses, is covered by BS EN 13108-1 (BSI, 2006a). A list of standard types of AC in current use is given in Table 8.2. Bitumens of various viscosities, ranging from 10/20 to 250/330 pen are used in these mixes, with the softer binders tending to be used in surface and binder courses. Proprietary polymer-modified bitumen may also be used as agreed between the supplier and the user.

Table 8.2 AC mixtures used in the UK

Material description	Designation	Bitumen options XX/YY
32 mm dense base and binder course AC (recipe mixture)	AC 32 dense base/bin XX/YY rec	40/60, 100/150, 160/220
32 mm heavy duty macadam base and binder course (recipe mixtures)	AC 32 HDM base/bin XX/YY rec	40/60
20 mm dense binder course AC (recipe mixture)	AC 20 dense bin XX/YY rec	40/60, 100/150, 160/220
20 mm heavy duty macadam binder course (recipe mixture)	AC 20 HDM bin XX/YY rec	40/60
6 mm dense AC surface course (recipe mixture)	AC 6 dense surf XX/YY rec	70/100, 100/150 (preferred)
10 mm close-graded AC surface course (recipe mixture)	AC 10 close surf XX/YY rec	70/100, 100/150 (preferred)
14 mm close-graded AC surface course (recipe mixture)	AC 14 close surf XX/YY rec	70/100, 100/150
4 mm fine-graded AC surface course (recipe mixture)	AC 4 fine surf XX/YY rec	160/220 (preferred), 250/330
10 mm open-graded AC surface course (recipe mixture)	AC 10 open surf XX/YY rec	70/100, 100/150, 160/220

Table 8.2 Continued

Material description	Designation	Bitumen options XX/YY
14 mm open-graded AC surface course (recipe mixture)	AC 14 open surf XX/YY rec	70/100, 100/150, 160/220
32 mm dense base and binder course AC (design mixture)	AC 32 dense base/bin XX/YY des	40/60
32 mm heavy duty macadam base and binder course (design mixture)	AC 32 HDM base/bin XX/YY des	40/60
32 mm high modulus base and binder course AC (design mixture)	AC 32 HMB base/bin XX/YY des	30/45
20 mm dense binder course AC (design mixture)	AC 20 dense bin XX/YY des	40/60
20 mm heavy duty macadam binder course AC (design mixture)	AC 20 HDM bin XX/YY des	40/60
20 mm high modulus binder course AC (design mixture)	AC 20 HMB bin XX/YY des	30/45
10 mm EME2 base and binder course AC (design mixture)	AC 10 EME2 base/bin XX/YY des	10/20, 15/25
14 mm EME2 base and binder course AC (design mixture)	AC 14 EME2 base/bin XX/YY des	10/20, 15/25
20 mm EME2 base and binder course AC (design mixture)	AC 20 EME2 base/bin XX/YY des	10/20, 15/25
6 mm medium graded surface course*	AC 6 med surf rec	160/220 (preferred), 250/330
4 mm fine graded surface course*	AC 4 fine surf rec	160/220 (preferred), 250/330
20 mm open graded binder course – crushed rock or slag	AC 20 open bin rec	160/220, 250/330
20 mm open graded binder course – gravel	AC 20 open bin grav rec	160/220, 250/330

EME2, *enrobé à module élevé* class 2 asphalt
* AC 4 fine surf and AC 6 med surf are unlikely to be suitable for use on carriageways

Table 8.3 Grading target limits for AC 32 recipe and design mixtures

Test sieve: mm	AC 32 dense/HDM base/bin XX/YY rec		AC 32 dense/HMB/HDM base/bin XX/YY des		Fuller's curve, $D = 31.5$ mm
	Passing limits: %	Mid-range: %	Passing limits: %	Mid-range: %	Passing: %
40	100	100	100	100	–
31.5	99–100	99.5	90–100	95	100
20	80–86	83	71–95	83	82
14	–	–	–	–	69
10	–	–	–	–	60
6.3	52	52	44–60	52	48
2	27–33	30	20–40	30	29
0.25	11–15	13	6–20	13	11
0.063	6 (dense)	6	2–9 (dense and HMB)	6	6
0.063	8 (HDM)	8	7–11 (HDM)	9	

The specified aggregate grading limits for both recipe and design mixtures of AC 32 dense base and AC 32 dense binder course are given in Table 8.3. Where there are single values in the target composition (e.g. for the percentage passing the 6.3 mm sieve in the recipe grading), only this value can be used. While the mid-range values are the same on all the sieves, excepting the D sieve, the limits for the design mixture are more generous, as this material must satisfy additional performance requirements. There is close agreement between the mid-range grading and the Fuller's curve grading (see Table 8.3). For the recipe mixtures, the grading must simply fall within the limits (BSI, 2010a), whereas for the design mixtures the chosen grading must be smooth and continuous, and not vary from the low limit on one size of sieve to the high limit on the adjacent sieve, or vice versa.

For the recipe mixtures, there is a specified binder content that is both the minimum and maximum target binder content; for the design mixtures, there is a specified minimum target binder content. It is standard practice in the UK to specify different binder contents depending on the type of aggregate used (i.e. limestone, basalt, other crushed rock, steel slag, blastfurnace slag or gravel). Thus, for example, the specified binder contents for AC 32 dense base vary little from one stone type to another (3.9–4.4%), but they vary significantly (increasing from 4.4% to 6.2%) for blastfurnace slags with bulk densities decreasing from 1.44 to 1.12 Mg/m^3. While the same aggregate grading covers both AC 32 dense base and the AC 32 dense binder course, the specified bitumen content for the recipe binder course is higher (4.6%) than that for the base course (4.0%) containing, in this example, limestone aggregate.

For AC mixtures, the specified flakiness category of the coarse aggregate is FI_{35} (i.e. the flakiness index of the aggregate cannot exceed 35%); the fines category for gravel aggregate is f_1 (i.e. the percentage passing the 0.063 mm sieve cannot exceed 1%); and the fines content category for sand, when it is the fine aggregate, is f_{10} (i.e. it cannot exceed 10%).

When RAP is used, it is required to conform to category F_5 for foreign matter (i.e. the percentage of foreign matter – cement concrete, bricks, sub-base material excluding natural aggregate, cement mortar and metal – should not exceed 5%), and its binder penetration to category P_{15} (i.e. the average penetration of the recovered binder is at least 15). The amount of RAP in surface courses is limited to 10%, and, in other layers, to 50%.

Heavy duty macadam (HDM) is a particular type of AC that was introduced into the UK specifications to emulate the performance of the French dense AC *grave-bitume* (Hingley *et al.*, 1976). It can be specified as a recipe or a design mixture, in accordance with PD 6691. The recipe HDM contains more fines and a harder grade of bitumen than dense AC: accordingly, except for more fines (e.g. 8% instead of 6%), the same aggregate grading as in Table 8.3 applies. High-modulus base (HMB) is a generic name for another dense base course AC made with a 30/45 grade bitumen: this is a design mixture. HDM and HMB are used in long-life (40 years or 80 msa) road pavements (Nunn *et al.*, 1987, 1997).

Design base and binder course AC mixtures are subjected to durability tests unless they contain 2% hydrated lime. The durability test is quite elaborate, requiring comparisons of the indirect tensile stiffness moduli of standard asphalt specimens and conditioned asphalt specimens that have been subjected to accelerated ageing and moisture saturation under pressure; the conditioned specimens are required to retain 80% of the stiffness of the standard specimens.

For designed base and binder course mixtures, there is a maximum allowable void content for 150 mm diameter cores and a minimum allowable void content for cores compacted to refusal. The resistance to permanent deformation is measured with a small wheel-tracking device, using 200 mm diameter cores taken from a full-scale trial strip. In the absence of experience of wheel-track slopes (mm/1000 cycles) and proportional rut depths (expressed as percentages), maximum rut rates and rut depths when tested with BS 598-110 (BSI, 1998) are recommended (e.g. 5 mm/h and 7 mm depth at 60°C for heavily stressed sites that require high rut resistance).

Enrobé à module élevé (EME) is another long-life AC that is covered by BS EN 13108-1. It was developed in France in the 1980s for strengthening roads in built-up areas and for the reconstruction of slow lanes on motorways (Brosseaud, 2006–2007). It is a very high-stiffness AC base whose use in France justifies reductions in thickness of 25–40% in base thickness compared with *grave-bitume*, the French dense AC. There are high and low binder content EMEs: EME2 and EME1 are required to have void contents of less than 6% and 10%, respectively, after 100 gyrations in the French gyratory shear compactor (Nunn and Smith, 1994). The higher-binder-content EME2 mixture would be expected to have a longer fatigue life. The aggregate grading limits in PD 6691 for AC 20 EME2 base/bin were used in the blending exercise in Table 8.1. The grading limits are generous, but the material has several performance thresholds to exceed. As a design mixture, the grading that is chosen must be smooth and continuous, and not vary from the low limit on one size of sieve to the high limit on the adjacent sieve, or vice versa. Minimum target binder contents are specified (e.g. 5.1% for AC 20 EME2). Only crushed rock and steel slag are used in EME2.

The void contents of EME2 after testing in the gyratory compactor in accordance with BS EN 12697-31 (BSI, 2007) cannot exceed 6% after 80, 100 and 120 gyrations for EME2 10, 14 and 20, respectively. The average void content of 150 mm diameter core pairs taken from a full-scale trial strip is also controlled.

The water sensitivity of EME2 is assessed using the Duriez test, in accordance with NFP-98 251-1 (AFNOR, 2002); it is recommended that the retained strength should be ≥ 0.75 (Brosseaud, 2006–2007). The deformation resistance of EME2 at the target composition and the target void content of between 3% and 6% in the French large-wheel-tracking test should be category $P_{7.5}$ (i.e. less than 7.5% at 60°C) after 30 000 cycles. The indirect tensile stiffness moduli of 150 mm diameter cores taken from a full-scale trial should be class $S_{min\ 5500}$ (i.e. the minimum stiffness should be 5500 MPa). When the binder richness modulus, K, is ≥ 3.6, resistance to fatigue is deemed to be satisfactory, as the French fatigue test is very lengthy and not generally available in the UK. The binder richness modulus is a surrogate for the thickness of binder surrounding the aggregate, and it is related to the specific surface area, Σ, and the density of the aggregate, ρ_{mc} (Nunn and Smith, 1994). The binder richness modulus, K, is calculated using the target composition from

$$B_{PPC} = Ka\sqrt[5]{\Sigma} \qquad (8.3)$$

where B_{PPC} is the mass of soluble binder expressed as a percentage of the total dry aggregate, including filler, K is the binder richness modulus, a is a correction factor ($=2.650/\rho_{mc}$, where ρ_{mc} is the theoretical aggregate density calculated using the apparent density) and Σ, the specific surface area of the aggregate (m²/kg), is given by

$$\Sigma = 0.25G + 2.3S + 12s + 135f \qquad (8.4)$$

where G is the proportion by mass >6.3 mm, S is the proportion by mass between 6.3 and 0.315 mm, s is the proportion by mass between 0.315 and 0.08 mm, and f is the proportion by mass <0.08 mm. The proportions are expressed as decimal fractions, not percentages: for example, an EME2 aggregate with values of 53%, 28%, 12% and 7% for G, S, s and f, respectively, has a specific surface area of 11.67 m²/kg.

The required stiffening effect of filler used in EME2 is category $\Delta_{R\&B}$ 8/16: that is, there must be an increase of between 8 and 16°C in the ring and ball (R&B) temperature of the binder–filler mixture consisting of 27.5 parts of filler to 62.5 parts of pure binder by volume, when related to the R&B temperature of the pure binder (BSI, 2013).

The workability of the final surface course mixtures listed in Table 8.2 can be enhanced by adding a flux oil (e.g. kerosene), for hand-laying in maintenance work: these fluxed mixtures are identified with the letter F. With a flash point ranging from, say, 37–65°C, it would be dangerous to add kerosene to a hot asphalt mixture, so the kerosene is added to the bitumen; however, this should not result in fluxed bitumen with a penetration value greater than 400 pen. Delayed set macadam, a material that is used in temporary works such as utility or pothole repairs, is identified with the additional designation DS (e.g. AC 6 med surf 160/200 DS).

8.3.2 Hot rolled asphalt

HRA has its origin in experimental trials that were first carried out in London on Kings Road, Chelsea, and Pelham Street, Kensington, in 1895 (DSIR, 1962). The first British standard specification for HRA was devised in 1928, and, since then, the specification has been revised numerous times as a result of experience gained on many different types of road under varying traffic and weather conditions. Covered by BS EN 13108-4 (BSI, 2006b), it is now mainly used in surface courses.

HRA can be produced using recipe, design and performance-related design specifications. The objective underlying the use of HRA mixtures is to provide a dense and impervious pavement layer that, from the time of laying, will not undergo much subsequent compaction under traffic. Thus, a majority of HRA mixtures are dense, low air void content (typically 3–6%), nearly impervious, durable and gap graded (i.e. with a low percentage of medium-sized 2–10 mm aggregate), and are composed mainly of fine aggregate, filler and a high bitumen content that form a mortar within which the coarse aggregate is dispersed.

The minimum amount of mortar that can be used depends on the thickness of the layer: with layers <50 mm thick, it is difficult to obtain imperviousness without 55% mortar, but 45% mortar can be adequate for up to 75 mm thick layers. With mortar contents of this magnitude, the viscosity of the bitumen is critical; thus, hard grades of bitumen are essential to provide mechanical stability.

The binders used in HRA can range from 30/45 to 100/150 pen grade bitumen but the preferred grade is 40/60 pen, and these mixtures must be hot mixed and hot laid. Paving grades 70/100 and 100/150 may be produced by blending in the mixer at the asphalt plant, but the grades for blending must be no harder than 30/45 nor softer than 160/220. While paving grade refinery bitumens are most commonly used, polymer-modified bitumens or a blend of either with natural asphalt, usually Trinidad lake asphalt, are also used. The inclusion of the lake asphalt has the effect of enhancing the skid-resistant properties of a surface course HRA.

For the coarse aggregate, the flakiness category should be FI_{35}, and the fines content should be f_4. When the coarse aggregate is gravel, 2% of either hydrated lime or cement is added in order to improve bitumen adhesion.

The fine aggregate can be either sand or crushed aggregate, or a mixture of both, that substantially passes the 2 mm sieve. Until relatively recently, the fine aggregate used for HRA in the UK was mainly natural sand, and the grading envelopes historically developed for use in various HRA recipes relied on commonly available sands. Sand particles are relatively rounded, and their use provides a more workable mixture on site: by contrast, crushed rock and slag fine aggregate – which are now more commonly used, as the supply of suitable sands has decreased – are more angular and elongated, and give a harsher and more stable mixture, albeit at the expense of some workability.

For an HRA surface course, there are two types of mixture: type F, containing 0/2 mm G_A90 fine aggregate (i.e. an all-in grading with at least 90% passing the 2 mm sieve), usually sand, and type C, containing 0/4 mm G_A85, usually crushed rock (BSI, 2010a). The appropriate fines categories are f_{10} for type F surface course mixtures, f_{16} for type C surface course mixtures, and f_{22} for base and binder course mixtures. Type F mixtures are characterised by a gap grading that is typical of traditional HRA. A list of HRA mixtures in current use (BSI, 2010a; HA et al., 2008) is given in Table 8.4.

When RAP materials are used in HRA mixtures, the RAP must be category F_5 with respect to foreign matter, and category P_{15} for binder penetration. HRA surface courses may contain up to 10% RAP material: all other layers may contain up to 50%, but this must decrease to 20% when either the RAP or the mixture contains modified binder.

For HRA surface courses, there now appears to be a movement away from the use of recipe mixtures towards the use of design mixtures (albeit the same aggregate grading limits are used) to cope with increasing traffic loads and the damage that can be done during hot summers (e.g. as occurred in 1976 in the UK) (Xu et al., 2015). Accordingly, in a situation where a pavement is facing south towards the midday sun and is carrying heavy commercial traffic in a climbing lane, a design mix incorporating a laboratory study of the rutting performance of the HRA might be preferable to a recipe mix. Recipe HRA surface course mixtures are all type F mixtures with fixed specified contents; that is, they have coarse aggregate contents of 0, 15, 30 and 35%, and the coarse aggregate maximum size can be either 10 or 14 mm. In these mixtures, the coarse aggregate is dispersed in the mortar, leaving a smooth surface after compaction. When the coarse aggregate content is $\leq 35\%$, it is standard practice to apply coated chippings to the surface to give it texture and skid resistance: the binder used to coat the chippings must be either 30/45 or 40/60 pen grade, and the binder content must be at least 1.5%. The chippings are applied after initial compaction by the paver and before the first pass of the roller.

The rate of spread of the chippings is normally required to be at least 70% of that needed for shoulder-to-shoulder contact, or 60% when there is evidence that this is sufficient to provide the required initial surface texture depth. The minimum rolling temperature at which this can be done is based on a binder viscosity of 30 Pa; for a 40/60 pen bitumen this is 85°C. Early chip loss of the precoated chippings is often a result of poor chip embedment (e.g. if the material is laid in low ambient temperatures in winter months). In general, for HRA surface course mixtures that must be chipped, coarse aggregate with a minimum polished-stone value (PSV) of 44 is required in the mixture (HA et al., 2008).

Table 8.4 HRA recipe mixtures used in the UK

Material description	Designation	Bitumen options XX/YY
Type F 0/2 HRA surface course (recipe mixture)	HRA 0/2 F surf XX/YY rec	40/60
Type F 15/10 HRA surface course (recipe mixture)	HRA 15/10 F surf XX/YY rec	40/60
Type F 30/10 HRA surface course (recipe mixture)	HRA 30/10 F surf XX/YY rec	40/60
Type F 30/14 HRA surface course (recipe mixture)	HRA 30/14 F surf XX/YY rec	40/60
Type F 35/14 HRA surface course (recipe mixture)	HRA 35/14 F surf XX/YY rec	40/60
Type F 0/2 HRA surface course (design mixture)	HRA 0/2 F surf XX/YY des	40/60
Type F 30/10 HRA surface course (design mixture)	HRA 30/10 F surf XX/YY des	40/60
Type F 35/14 HRA surface course (design mixture)	HRA 35/14F surf XX/YY des	40/60
Type F 55/10 HRA surface course (design mixture)	HRA 55/10 F surf XX/YY des	40/60
Type F 55/14 HRA surface course (design mixture)	HRA 55/14 F surf XX/YY des	40/60
Type C 0/2 HRA surface course (design mixture)	HRA 0/2 C surf XX/YY des	40/60
Type C 30/10 HRA surface course (design mixture)	HRA 30/10 C surf XX/YY des	40/60
Type C 35/14 HRA surface course (design mixture)	HRA 35/14 C surf XX/YY des	40/60
Type C 55/10 HRA surface course (design mixture)	HRA 55/10 C surf XX/YY des	40/60
Type C 55/14 HRA surface course (design mixture)	HRA 55/14 C surf XX/YY des	40/60
60/32 HRA base and binder courses (recipe mixtures)	HRA 60/32 base/bin XX/YY rec	30/45, 40/60
60/20 HRA base and binder courses (recipe mixtures)	HRA 60/20 base/bin XX/YY rec	30/45, 40/60
50/10 HRA base/binder course (recipe mixtures)[a]	HRA 50/10 base/bin rec	40/60
50/14 mm HRA base/binder course (recipe mixtures)[a]	HRA 50/14 base/bin rec	40/60
50/20 HRA base/binder course (recipe mixtures)[a]	HRA 50/20 base/bin rec	40/60

[a] Suitable for a regulating course

The coated chippings are normally spread evenly using a mechanical spreader (except in confined areas that require hand spreading), and are rolled into the surface until they are effectively held. After rolling, the following initial texture depths should be achieved: from 1.5 to 2.0 mm for high-speed roads (≥80 km/h speed limit) and from 1.2 to 1.7 mm for lower-speed roads (≤60 km/h speed limit): for roundabouts with the same speed limits, the specified minimum and maximum values are 1.2 and 1.7 mm, and 1.0 and 1.5 mm, respectively (HA, 2012).

Design surface course mixtures can be either type F or type C mixtures. They have coarse aggregate contents of 0, 30%, 35% and 55%, and the coarse aggregate maximum size can be 10 or 14 mm also. Coated chippings are not required for the mixtures with 55% coarse aggregate. The design binder content is established using the Marshall stability test according to BS 594987. The performance-related surface course is limited to HRA 35/14 F surf des. This specification requires additional tests for the volumetric binder content (15.5% is the minimum required), void content and resistance to permanent deformation of cores extracted from a trial strip, in accordance with BS 594987 (BSI, 2010b). Cores of design HRA mixtures are subject to the same rutting limitations that apply to AC.

The recipe mixtures specified for HRA base and binder courses have gradations with 50% and 60% coarse aggregate, and either 32 or 20 mm maximum-size aggregate (e.g. HRA 60/32 base 40/60).

Seventeen HRA aggregate gradations suitable for different purposes, including design mixtures, are listed in PD 6691 (BSI, 2010a), and a selection of four of these is given in Table 8.5. In this table, HRA 60/32 is a base material, HRA 50/14 is suitable for use

Table 8.5 Selection of gradations (% by mass passing BS sieve sizes) and binder contents (% by mass of the total mix) for four HRA mixtures. (Note: the recommended binder contents are given in the bottom row.)

Sieve size: mm	HRA 60/32	HRA 50/14	HRA 35/14 F	HRA 35/14 C
40	100	–	–	–
31.5	99–100	–	–	–
20	59–71	100	100	100
14	39–56	98–100	95–100	95–100
10	–	72–93	62–81	62–81
6.3	–	–	–	–
2	37	40–50	61	59
0.5	13–30	17–51	44–63	24–41
0.25	10–25	14–31	16–46	16–26
0.063	4	3–6	8	8
Binder: %	5.5 for limestone	6.2 for limestone	7.3 and 7.8 for crushed rock	6.4 (minimum) for limestone

as a regulating course, HRA 35/14 F is a recipe surface-course material and HRA 35/14 C is a design surface course material. For recipe surface courses, two binder contents are specified, namely a lower binder content for most conditions, and a higher one for cold, elevated, wet conditions, or more lightly trafficked roads. Note also that the binder content varies according to the aggregate type: for example, a lower binder content is specified for mixtures that use gravel in order to combat the increased proneness of rounded gravel mixtures to rutting. Also, the required binder content of mixtures made using blastfurnace slag increases as the density of the slag decreases. For design mixtures, different minimum allowable target binder contents are specified for mixtures containing limestone, basalt and other crushed stone aggregates.

The distinctive gap gradings of the four mixtures in Table 8.5 are evident if they are plotted on standard semi-logarithmic paper, with the percentage passing on the linear ordinate and the sieve size on the logarithmic abscissa.

8.3.3 Stone mastic asphalt

SMA was initially developed in Germany in the mid-1960s, to create a road surface with a high abrading resistance to the studded tyres that were then commonly used on vehicles during the snow season. While these tyres were banned in 1975, it was noted that SMA was durable, had a good resistance to rutting under heavy traffic, and was a quiet surface compared with other materials in use at the time. As a consequence, it was further developed as a surfacing material for autobahns and other major roads, and was included in the German specifications. It is now also used in many major roads in northern Europe, including the UK, and is specified by BS EN 13108-5 (BSI, 2006c).

A feature of SMA is that the carriageway surface resembles minute 'plateaus with ravines' that give rise to lower noise emissions. With SMA, small plateaus of the same height are located irregularly next to one another, so that interim spaces (i.e. the ravines) are left between them: this reduces air pumping, as the tyre profile releases trapped air. Notwithstanding the plateaus and ravines, the resultant carriageway surface is sufficiently smooth that the tyre vibration excitement is low when small aggregate sizes are used in the mixture, and this is reflected in low noise emissions (Ripke *et al.*, 2005).

SMA can be described as a gap-graded aggregate (Hüning, 1996) that is bound with a stiff bitumen-rich mortar. Its main feature is a self-supporting interlocking coarse (>2 mm) aggregate 'skeleton' with a high angle of internal friction that distributes the applied traffic loads: the stone skeleton comprises 70–80% of the aggregate for a gradation with a nominal size of 8 mm or more. The relative quantities of fine aggregate, filler and bitumen are chosen so that the spaces in the skeleton are almost filled with mortar to give a void content in the range 2–4%: in this way, the coarse aggregate framework is effectively locked into place by the mortar, and continued good contact between the particles is ensured. The mortar also gives the mix its durability against ageing and weathering (Liljedahl, 1992). When SMA has a low void content, the framework mechanism enables it to be both rut resistant and durable to the elements (EAPA, 1998).

It is important that firm contact be maintained between the coarse aggregate particles during, and after, roller compaction, as it is through this interlocking framework that the wheel-load stresses are mainly distributed. If the mortar content is too high (i.e. if the volume of mastic material is greater than the available voids), the excess mortar will push apart the stone skeleton, with consequent flushing of the bitumen after compaction and the creation of an inferior surface texture.

Because of its gap grading and high binder content, it is essential that either stabilising agents or a polymer-modified bitumen be used with SMA mixes so that neither the bitumen nor the mortar will drain from the aggregate between mixing and compaction, while being handled and transported to the site. Cellulose fibres are the most commonly used stabilising agent, and are added at a minimum rate of 0.3% by mass of the total mixture. Binder drainage can also occur if there is a deficiency of fine aggregate or insufficient fibres to hold the bitumen, or if the bitumen is overheated. If it is allowed to occur, binder drainage leads to fatting up and a reduced surface texture. A deficiency of bitumen in the mortar also produces a porous material that is susceptible to ravelling.

There should also be point-to-point contact between the fine aggregate particles within the mortar. If too much bitumen is present so that the fine aggregate is saturated with binder, firm contact between the fine aggregate will be lost, and this will reduce the shear resistance of the mortar; moreover, when the binder tries to expand in hot weather, there is not enough room for this expansion, and the fine aggregate will spread apart (Liljedahl, 1990).

Recipe-mix trials in the UK led to the development of a draft specification in which the German aggregate grading was modified to suit British standard sieve sizes and surface texture requirements for wet weather conditions (Nunn, 1994). A rational design method, which fixes the volumetric composition of the mix ingredients according to well-founded objectives, can be used to design SMA (Liljedahl, 1992): while this design method seems attractive (Brennan et al., 2000), practitioners tend to prefer to use the performance-proved recipe specifications.

While SMA is used primarily as a surface course, it is also suitable for use as a binder course. The preferred binder is 40/60 pen bitumen, but 70/100 and 100/150 grades are also suitable. Flint aggregates and blastfurnace slag are omitted as suitable coarse aggregates. An Fl_{20} flakiness index and an f_4 fines content are required. Because of its low PSV value, limestone aggregate cannot be used in the surface course. The fines category of the fine aggregate used should be f_{22} (BSI, 2010a). The four SMAs used in the UK are listed in Table 8.6. Due to the required aggregate gap grading, the use of all-in graded RAP is not normally recommended for SMA.

Aggregate grading limits for the four SMAs commonly used in the UK, in accordance with PD 6691 (BSI, 2010a), are given in Table 8.7. The adjustments to the limits that are in accordance with Transport Scotland requirements (TS, 2010) for the 6, 10 and 14 mm nominal sizes are noted in brackets in this table. The Scottish specification, which is intended to be more in harmony with German practice, requires binder contents that are

Table 8.6 SMA materials used in the UK

Material description	Designation	Bitumen options
6 mm SMA surface course	SMA 6 surf	40/60 (preferred), 70/100, 100/150. Polymer-
10 mm SMA surface course	SMA 10 surf	modified bitumen or a blend with natural
14 mm SMA surface course	SMA 14 surf	bitumen (lake asphalt). 70/100 and 100/150
20 mm SMA surface course	SMA 20 surf	may be produced by blending in the mixer

Table 8.7 Aggregate gradations (% by mass passing British standard sieve sizes) and binder contents (% by mass of the total mix) for SMA surface and binder courses used in the UK. (Note: the gradation and binder variations used in Scotland are in brackets.)

Sieve size: mm	SMA 6 surf/bin	SMA 10 surf/bin	SMA 14 surf/bin	SMA 20 surf/bin
31.5	–	–	–	100
20	–	–	100	94–100
14	–	100	93–100	–
10	100	93–100	35–60	25–39
6.3	93–100	28(35)–52	22–36	22–32
4	26(32)–51(45)	–	–	–
2	24(25)–39(35)	20–32	16–30	15–26
0.063	8–14	8–13(12)	6(8)–12	8–11
% binder	6.6(7.1)	6.2(6.7)	5.8(6.3)	5.4

0.5% higher than the PD 6691 binder contents shown in the table: also, a polymer-modified 75/130 grade bitumen with an R&B temperature greater than 75°C is required.

PD 6691 also specifies minimum and maximum air void content requirements of SMA Marshall specimens as $V_{min\ 1.5}$ and $V_{max\ 5}$ for SMA 10 and SMA 14 surface course mixtures. The corresponding Scottish specification (TS, 2010) is much tighter at $V_{min\ 3.0}$ and $V_{max\ 3.5}$. PD 6691 also contains void content and rut resistance requirements for cores taken from a trial strip of SMA binder course. The Schellenberg method (BSI, 2004) is used to measure the resistance of SMA to binder drainage. The limiting average binder drainage is 0.3% (i.e. $D_{0.3}$), and it is part of the specification. Cores of SMA mixtures are subject to the same rutting limitations that apply to AC.

8.3.4 Porous asphalt

PA is an open-graded surface course material that facilitates the rapid drainage of rainwater from the carriageway. It uses a gap-graded aggregate that, when compacted, produces an asphalt matrix with about 20% air voids (Nunn et al., 1997). At this void content, rainwater falling on the carriageway is absorbed in the interconnected open voids and is able to drain away laterally and longitudinally under the influence of the longfall and crossfall of the pavement, for collection in roadside edge drains. As it is highly pervious, a PA surfacing must be laid on an impermeable layer so that the water cannot penetrate further downward into the pavement. Thus, to enable rainwater to be drained from the PA, and to provide protection against rain penetrating into the base, the top of the binder course may need to be sealed with the same binder as the PA at a rate of about 2.5 kg/m^2 (Schäfer, 2004).

While PA was developed to quickly remove rainwater from the carriageway, it has a number of ancillary advantages, namely that (a) aquaplaning is avoided and spraying is reduced in wet weather on high-speed roads, (b) the glare reflected from a wet carriageway is reduced, (c) road markings are more visible and (d) tyre/road noise is reduced. As a consequence, the use of PA is most appropriate on high-speed roads without kerbs or many junctions: for example, on motorways and rural dual carriageways, at accident sites where high-speed skidding accidents are attributable to wet weather spray, and on high-speed bypasses where noise is likely to trouble adjacent householders.

However, there are disadvantages associated with the use of PA that obviate its use at road locations where (a) excessive droppings of detritus and oil from vehicles is expected, as these will block the carriageway drainage, and (b) free drainage cannot be accommodated at the side(s) of the carriageway. Also, (c) because of its open surface structure, the pavement cools more quickly in cold weather, so that the onset of snow and ice requires the early spreading of greater amounts of deicing salts, and (d) due to the high void content of PA, the binder is vulnerable to oxidation ageing (e.g. the R&B temperature increases by about 1.5°C per annum), which leads eventually to brittle failure (Ripke, 2004). To obviate the oxidation problem, a high bitumen content must be used to ensure a thick film of bitumen on the coarse aggregate particles.

While the noise reduction associated with PA is good on high-speed roads, the channels are more likely to get clogged on lower-speed urban roads, and the spray and noise level reductions achieved on these roads are relatively small. A twin-layer PA, in which a finer upper layer acts as a filter while the lower layer forms the resonance space for sound absorption, can mitigate this effect: for example, a 20 mm thick layer of PA 5 on top of a 50 mm thick layer of PA 16 (Nielsen et al., 2004). Where clogging occurs, it increases the noise level by about 0.5 dB(A) per year, and annual high-pressure cleaning may be required to overcome this.

PA is also considered to have two lives, namely an ultimate life (when it has to be replaced due to progressive binder hardening) and a spray-reducing life. A review of factors affecting the lives of PA in UK trials showed that (a) the flow of traffic has only a small effect on both lives, which is beneficial for the ultimate life and detrimental for the spray-reducing life; (b) the use of larger nominal size aggregates benefits both lives, as does the use of softer bitumen; and (c) the inclusion of polymers in the binder can benefit both lives. The average ultimate life for PA 20 is 8 years, and 5 years for PA 10; the spray-reducing lives for PA 20 and PA 10 are 6.5 and 4 years, respectively (Nicholls, 1997). By comparison, the expected ultimate life of PA is 7 years under Dutch climatic conditions (Nielsen et al., 2004). Noise levels from PA surfaces are 3–4 dB(A) lower than those from non-porous surfaces: this is equivalent to halving the traffic flow or doubling the distance from the traffic stream (Nicholls, 1997).

In Germany, PA is mainly used for its noise-reducing quality. To maintain a noise reduction of 5 dB(A) over the lifetime of the pavement, a void content of at least 22% is required by the German specification to avoid the pumping effect under tyres, and a fine-grained aggregate surface (i.e. 8 mm maximum size aggregate) is specified to reduce tyre tread impact and 'snap-out' noises (Renken, 2004). A typical aggregate gradation for a German PA 8 would have 85% >5 mm and 90–95% >2 mm, and it would use 6.2–6.8% polymer-modified bitumen and 0.5% stabilising additive (Schäfer, 2004).

PD 6691 (BSI, 2010a) requires compliance with BS EN 13108-7 (BSI, 2006d) when PA is used in the UK.

8.4. Mixing, laying and compaction of asphalt mixtures

Once the stockpile aggregates have been proportioned and blended to the desired specification, they must be dried and heated and thoroughly mixed with the specified amount of bitumen, to ensure that individual particles are well coated and a homogeneous mixture is obtained for transport to the road site to be laid and compacted. The blending and mixing process takes place at an asphalt mixing plant, of which there are various types: some are batch plants employing a continuous-flow operation (see Figure 13.17), while others prepare individual batches (see Figure 13.18). These plants can also be stationary (i.e. fixed-in-place) or mobile (i.e. the component parts can be dismantled and moved) (Figure 8.2).

The *Specification for Highway Works* (HA *et al.*, 2008) requires asphalts to be produced in mixing plants that are registered to BS EN ISO 9001 (BSI, 2008) and laid by registered contractors operating in accordance with governmental requirements (HA *et al.*, 2008).

With the continuous-flow type of operation shown in Figure 8.2, the aggregates, which are often damp from exposure to the elements, are supplied in different nominal sizes from the cold feed bins, and the filler is usually fed from a silo. They are proportioned and fed directly into a drum mixer that has two zones: the drying and heating of the cold damp aggregate takes place near the flame in the first zone of the drum, after which the

Figure 8.2 Schematic drawing of a mobile hot-mix asphalt plant

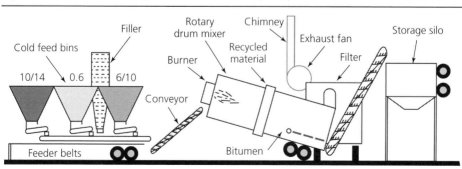

desired amount of hot bitumen is fed in, and hot mixing takes place in the second part of the drum. RAP may also be added to the drum and included in the mixture. In the process of drying the aggregate and achieving the desired mixing temperatures, much unwanted dust is created: this has to be extracted before the bitumen is added, and prevented from polluting the environs. After mixing, the asphalt exits the drum and is carried to a silo, from which it is discharged into insulated trucks for transport to the road site. A mobile drum plant, such as that illustrated in Figure 8.2, can produce asphalt at rates of up to 100 t/h, while high-capacity fixed-drum facilities can produce up to 600 t/h.

To prevent heat loss, the mixed material should be covered while being transported from the mixing plant to where it is tipped into the hopper at the front of the on-site paving machine. Hot asphalts are laid with a self-propelled paving machine (see Figure 13.19). Hand placement is restricted to confined areas, and hand raking to the edges of layers, and to gullies and utility covers.

BS 594987 (BSI, 2010b) and the *Specification of Highway Works* (HA, 2012; HA *et al.*, 2008) gives control criteria regarding the transport and laying of hot-mix asphalt.

The quantity of asphalt that is required from the asphalt plant depends on the surface area to be covered and the superage – the area covered per tonne of laid material (m^2/t). The superage is calculated from

$$\text{superage (m}^2/\text{t)} = \frac{1000 \times \text{volume/unit weight (m}^3/\text{t)}}{\text{thickness (mm)}} \tag{8.5}$$

For any combination of specification and aggregate source, superage can vary by 10%. The volume per unit weight is likely to be in the range 0.40–0.45 m^3/t (Hindley, 2000).

Rolling is started when the spreading and finishing operations have been completed and while the laid mat of asphalt is still hot, using steel-wheeled or pneumatic-tyred rollers, or a combination of both roller types (see Chapter 13). For a vibrating roller in motion, it is standard practice to regulate the spacing between the drum impacts to one imperial inch (~25 mm), to ensure a smooth compacted surface. This control limits the speed of the roller according to the vibration frequency (Hz): a roller with a vibrating frequency of 1500 vib/min (25 Hz) is limited to a speed of 2.3 km/h, whereas the speed can increase to 6.1 km/h at a frequency of 4000 vib/min (67 Hz).

BS 594987 (BSI, 2010b) specifies the requirements for the transport, laying and compaction of AC, HRA and SMA. These include recommended delivery and rolling temperatures, nominal and minimum layer thicknesses, approximate rates of spread, and the number of rollers required. The minimum rolling temperatures depend on the binder grade and mix type: for example, for an AC dense-base material and SMA made with a 40/60 pen bitumen, the target delivery temperature is 130°C and the material must be at 100°C just prior to rolling; for all HRA surface layers made with 40/60 pen bitumen, the specified temperatures are 140°C and 110°C, respectively.

The capacity of a roller is calculated from

$$Q = f \frac{bvh \times 1000}{n} \rho \qquad (8.6)$$

where Q is the output of the roller (t/h), f is the efficiency factor, b is the drum width (m), v is the roller speed (km/h), h is the compacted layer thickness (m), n is the number of passes and ρ is the compacted asphalt density (t/m^3). The efficiency factor f, which is the practical capacity divided by the theoretical capacity, depends on the extent of the overlap and the effective operating time as the roller stops and starts at the end of a run: a value of 0.6 is used for asphalt. This expression can be used with basic information on the asphalt being laid, to calculate the numbers of rollers required. However, in practice, roller outputs are estimated from charts provided by roller manufacturers that are based on experience with commonly laid mixtures.

Maximum and minimum compacted layer thicknesses are specified for all asphalt mixtures. Layers up to a limit of 150 mm thick are allowed for mixtures with a maximum aggregate size of 32 mm, and the allowable maximum thickness decreases with aggregate size. Attempts to compact thicker layers risk having surface undulations left by the compacting rollers.

When layers thicker than the maximum allowable compacted layer thickness are required, two layers should be laid and bound together with either a bond coat or a tack coat. These coatings are also required prior to laying a new asphalt layer on any bound substratum. In BS 594987 (BSI, 2010b), a bond coat is described as a polymer-modified bitumen emulsion, whereas a tack coat is a conventional bitumen emulsion. The bond coat normally provides better adhesion and allows heavier rates of application when this is needed to improve the impermeability of the lower layer (e.g. when placing a PA surface on a binder course), and prevent the ingress of rainwater into the pavement. A bond coat typically provides 0.35 kg/m^2 of residual bitumen, while a tack coat provides 0.20 kg/m^2.

Thin layers (i.e. <60 mm thick) lose heat rapidly during laying and compaction. In general, such a layer should not be thinner than two to three times the upper sieve size of the aggregate gradation if it is to be effectively compacted, and if surface layer aggregates are to be prevented from being crushed under the steel wheels of a roller.

If a bituminous layer other than the surface course is temporarily opened to traffic, it should be surface dressed with chippings that are not less than PSV$_{50}$. Chippings to prevent bond coat or tack coat pick-up should be 2/4 or 2/6 G$_C$85/35.

The compaction of designed AC base and binder courses (including EME2), performance-related HRA surface courses, and SMA binder courses is controlled on site by measuring the air voids contents of extracted cores, for which an upper limit is specified (Table 8.8). For the AC designed base and binder courses, there is the additional requirement to measure the air voids content of extracted cores compacted to refusal, for which

Table 8.8 Maximum and minimum average air voids content for cores from trial strips for design AC, EME2, HRA and SMA

Design material	V_{max}		V_{min}
	2 cores	6 cores	6 cores
AC 32 dense/HDM/HMB base/bin XX/YY des	7.0	–	0.5
EME2	6.0	–	–
HRA 35/14F surf XX/YY des	7.5	5.0	–
SMA bin	6.0	4.0	–

a lower limit is specified (see Table 8.8). The air void content is obtained from

$$\text{air void content (\%)} = \left(1 - \frac{\rho}{\rho_{max}}\right) \times 100 \qquad (8.7)$$

where ρ is the bulk density and ρ_{max} is the maximum density. In general, of the four ways of calculating the bulk density (BSI, 2012), the saturated and surface dried density method is suitable for AC with an air voids content of $\leq 5\%$, and for SMA with an air voids content of $\leq 4\%$, whereas the density in a dried condition method is suitable for HRA, which has relatively fine pores: that is, the objective is to measure those voids that are part of the material but neglect those that occur as surface irregularities. Three procedures are available to calculate the maximum density: volumetric, hydrostatic and mathematical (BSI, 2009b). The volumetric method of measuring the density of crumbed particles of the mixture experimentally, using a pyknometer under a vacuum to evacuate air in the accessible pores, is preferred, as it recognises that the bitumen in an asphalt mixture can permeate the voids on the surface of the aggregate particles.

8.5. Evaluation of conformity

In the production of AC, HRA or SMA for roadworks, asphalt manufacturers must be able to prove that (a) the particular ingredient formulations that they are using comply with the relevant specifications and BS EN product standards, and (b) the products coming from the asphalt-mixing plant are consistently compliant with their own conformity specifications. The accidental intermixing of stockpiles, or segregation of stored aggregate, or even an occasional stone picked up by a front-end loader feeding the cold-feed bins that blocks the gate at the bottom of a bin, will alter the composition. The consequence of such non-conformity will be reduced performance and premature failure. The first requirement is determined by an initial type test (ITT) of the product, and the second requirement is controlled by operating a system of factory production control (FPC) at the asphalt-manufacturing plant.

The requirements for the ITT detailed in BS EN 13108-20 include a list of the constituent materials, a declaration of the target composition of the mixture in terms of aggregate grading, binder content, and test data showing conformity of the mixture with any specified performance requirements (BSI, 2006e). These test data come from laboratory tests or from measurements of the air voids content, stiffness modulus and/or rut resistance on cores taken from a trial section of the paved material. An ITT report is normally valid for 5 years, but must be revised if the target composition is changed or if the source of a constituent material is changed.

BS EN 13108-21 specifies the procedures that the manufacturer must undertake when checking that a mixture is being produced in compliance with its conformity specification. FPC is based on the principle of controlling materials, processes and, especially, mix composition to replicate consistently the formulations that are validated in the ITT. The FPC requirements, which form part of the National Highway Sector Scheme 14 for the quality management of asphalt mix production, specify the minimum frequency at which compliance tests are to be carried out and the tolerances associated with their results: for example, with zero non-conforming tests, just one test is required for each 600 t of surface course asphalt, whereas in the event of more than eight non-conforming results in the previous 32, an immediate and comprehensive review is required in accordance with the FPC system (BSI, 2006f) to determine corrective and preventive action.

In order to display conformity with the standards, a manufacturer must have an FPC system audited by a notified body on an annual basis. In the EU, a notified body is an organisation that has been accredited by a member state to assess whether a product meets certain pre-ordained standards. In addition, since the implementation of the EU's Construction Products Regulation in 2013, manufacturers of asphalt are required to (a) make a declaration of performance (DoP) regarding any product that they place on the market and assume responsibility for this performance, (b) affix a CE mark – this is the mandatory conformity logo for Conformité Européenne – to the product documentation, and (c) retain and stand by the DoP for a period of 10 years (EPC, 2011).

REFERENCES

AFNOR (Association Française de Normalisation) (2002) NF P 98-251-1: Essais relatifs aux chaussées. Essais statiques sur mélanges hydrocarbonées. Essai DURIEZ sur mélanges hydroncarbonées à chaud. AFNOR, Paris, France.

Asphalt Institute (1997) *MS-2: Mix design methods for asphalt concrete and other hot-mix types*. Asphalt Institute, Lexington, KT, USA.

Bradshaw-Bullock J (2007) Operation of the new European asphalt standards in the UK. *Asphalt Professional* **27**: 19–25.

Brennan MJ, Nolan J, Murphy D and Lohan G (2000) Designing stone mastic asphalt. *International Journal of Road Materials and Pavement Design* **1(2)**: 227–243.

Brosseaud Y (2006–2007) Les enrobés à module élevé: bilan de l'expérience de française et transfer de technologie. *Revue Générale des Routes et de l'Aménagement* **854**: 70–76.

BSI (British Standards Institution) (1998) BS 598-110: Sampling and examination of bituminous mixtures for roads and other paved areas. Methods of test for the determination of wheel-tracking rate and depth. BSI, London, UK.

BSI (2004) BS EN 12697-18:2004: Bituminous mixtures. Test methods for hot mix asphalt. Binder drainage. BSI, London, UK.

BSI (2006a) BS EN 13108-1:2006: Bituminous mixtures. Material specifications. Asphalt Concrete. BSI, London, UK.

BSI (2006b) BS EN 13108-4:2006: Bituminous mixtures. Material specifications. Hot rolled asphalt. BSI, London, UK.

BSI (2006c) BS EN 13108-5:2006: Bituminous mixtures. Material specifications. Stone mastic asphalt. BSI, London, UK.

BSI (2006d) BS EN 13108-7:2006: Bituminous mixtures. Material specifications. Porous asphalt. BSI, London, UK.

BSI (2006e) BS EN 13108-20:2006: Bituminous mixtures. Material specifications. Type testing. BSI, London, UK.

BSI (2006f) BS EN 13108-21:2006: Bituminous mixtures. Material specifications. Factory production control. BSI, London, UK.

BSI (2007) BS EN 12697-31:2007: Bituminous mixtures. Test methods for hot asphalt. specimen preparation by gyratory compactor. BSI, London, UK.

BSI (2008) BS EN ISO 9001:2008: Quality management systems. Requirements. BSI, London, UK.

BSI (2009a) PD 6682-2:2009 + A1:2013: Aggregates. Aggregates for bituminous mixtures and surface treatments for roads, airfields and other trafficked areas. Guidance on the use of BS EN 13043. BSI, London, UK.

BSI (2009b) BS EN 12697-5:2009: Bituminous mixtures. Test methods for hot mix asphalt. Determination of the maximum density. BSI, London, UK.

BSI (2010a) PD 6691:2010: Bituminous mixtures. Material specifications. Guidance on the use of BS EN 13108. BSI, London, UK.

BSI (2010b) BS 594987:2010: Asphalt for roads and other paved areas. Specification for transport, laying, compaction and type testing protocols. BSI, London, UK.

BSI (2012) BS EN 12697-6:2012: Bituminous mixtures. Test methods for hot mix asphalt. Determination of bulk density of bituminous specimens. BSI, London, UK.

BSI (2013) BS EN 13179-1:2013: Tests for filler aggregate used in bituminous mixtures. Delta ring and ball test. BSI, London, UK.

DFID (Department for International Development) (2002) *A Guide to the Design of Hot Mix Asphalt in Tropical and Sub-tropical Countries*: *Overseas Road Note 19*. Transport Research Laboratory, Crowthorne, UK.

DSIR (Department of Scientific and Industrial Research) (1962) *Bituminous Materials in Road Construction*. Stationery Office, London, UK.

EAPA (European Asphalt Pavement Association) (1998) *Heavy Duty Surfaces: The Arguments for SMA*. EAPA, Breukelen, the Netherlands.

EPC (European Parliament and Council) (2011) Regulation (EU) No. 305/2011 of the European Parliament and of the Council of 9 March 2011 laying down harmonised conditions for the marketing of construction products and repealing Council Directive 89/106/EEC. *Official Journal of the European Union* **L88/5–43**.

HA (Highways Agency) (2012) *Interim Advice Note 154/12. Revision of SHW Clause 903, Clause 921 and Clause 942*. HA, London, UK.

HA, Transport Scotland, Welsh Government and Department for Regional Development Northern Ireland (2008) Road pavements – bituminous bound materials. In *Manual of Contract Documents for Highway Works*, vol. 1. *Specification for Highway Works*. Stationery Office, London, UK, Series 900.

Hindley G (2000) Good surfacing practice. In *Asphalts in Road Construction* (Hunter RN (ed.)). Thomas Telford, London, UK, pp. 235–262.

Hingley CE, Peattie HR and Powell WD (1976) *French Experience with Grave-bitume, a Dense Bituminous Roadbase*. Transport and Road Research Laboratory, Crowthorne, UK, Report SR 242.

Hüning P (1996) Stone mastic asphalt justified. *European Asphalt Magazine* **3–4**: 9–13.

Hunter RN, Self A and Read J (eds) (2015) *The Shell Bitumen Handbook*, 6th edn., ICE Publishing, London, pp. 744–747.

Lesueur D, Petit J and Ritter HJ (2012) Increasing the durability of asphalt mixtures by hydrated lime addition: what evidence? *Proceedings of the 5th Eurasphalt and Eurobitume Congress, Istanbul, Turkey*. European Asphalt Pavement Association, Brussels, Belgium, paper P5EE-255.

Liljedahl B (1990) Heavy duty asphalt pavements. In *Bituminous Roads in the 1990s, Seminar Papers, Dublin*. Institute of Asphalt Technology, Edinburgh, UK, pp. 151–163.

Liljedahl B (1992) Mix design for heavy duty asphalt pavements. In *The Asphalt Yearbook 1992*. Institute of Asphalt Technology, Stanwell, UK, pp. 18–24.

Moore J (2000) Production: processing raw materials to mixed materials. In *Asphalts in Road Construction* (Hunter RN (ed.)). Thomas Telford, London, UK, pp. 202–234.

Nicholls JC (1997) *Review of UK Porous Asphalt Trial*. Transport Research Laboratory, Crowthorne, UK, Report 264.

Nicholls JC, McHale MJ, Griffiths RD *et al.* (2008) Encouraging durability in the specification and laying of asphalts. *Proceedings of the 4th Eurasphalt and Eurobitume Congress, Copenhagen, Denmark*. European Asphalt Pavement Association, Brussels, Belgium, paper 300-006.

Nielsen CB, Nielsen E, Andersen, JB and Raaberg J (2004) Development of durable porous asphalt mixes from laboratory experiments. *Proceedings of the 3rd Eurasphalt and Eurobitume Congress, Vienna, Austria*. European Asphalt Pavement Association, Brussels, Belgium, paper 90.

Nunn ME (1994) *Evaluation of Stone Mastic Asphalt (SMA): A High Stability Wearing Course Material*. Transport Research Laboratory, Crowthorne, UK, Report PR 65, E111A/HM.

Nunn ME and Smith T (1994) *Evaluation of Enrobé Module Eleve (EME): A French High Modulus Roadbase Material*. Transport Research Laboratory, Crowthorne, UK, Report PR 66.

Nunn ME, Rant CJ and Schoepe B (1987) *Improved Roadbase Macadams: Road Trials and Design Considerations*. Transport and Road Research Laboratory, Crowthorne, UK, Research Report 132.

Nunn ME, Brown A, Weston D and Nicholls JC (eds) (1997) *Design of Long-life Flexible Pavements for Heavy Traffic*. Transport Research Laboratory, Crowthorne, UK, Report 250.

Read J and Whiteoak D (2003) *The Shell Bitumen Handbook*, 5th edn., Thomas Telford, London, UK.

Renken P (2004) Noise reducing asphalt pavements (porous asphlalt) – optimal composition, prediction of material properties and experience with long-term performance. *Proceedings of the 3rd Eurasphalt and Eurobitume Congress, Vienna, Austria*. European Asphalt Pavement Association, Brussels, Belgium, paper 171.

Ripke O (2004) Reducing traffic noise by optimising hot-mix asphalt surface courses. *Proceedings of the 3rd Eurasphalt and Eurobitume Congress, Vienna, Austria*. European Asphalt Pavement Association, Brussels, Belgium, paper 92.

Ripke O, Andersen B, Bendtsen H and Sandberg U (2005) *Report of Promising New Road Surfaces for Testing. European Commission DG Research 6th Framework Programme Priority 6, Sustainable Development, Global Change and Ecosystems Integrated Project – Contract No. 516288.* European Commission, Brussels, Belgium.

Schäfer V (2004) Experiences with porous asphalt of a new generation on the motorway A2 in northern Germany (Lower Saxony). *Proceedings of the 3rd Eurasphalt and Eurobitume Congress, Vienna, Austria.* European Asphalt Pavement Association, Brussels, Belgium, paper 192.

TS (Transport Scotland) (2010) *Surface Course Specification and Guidance.* TS, Glasgow, UK.

Xu S, Wang K, Wayira A and Lu J (2015) Influence of binder properties on the performance of asphalts. In *The Shell Bitumen Handbook*, 6th edn. (Hunter RN, Self A and Read J (eds)). ICE Publishing, London, pp. 503–507.

Highways
ISBN 978-0-7277-5993-1

http://dx.doi.org/10.1680/h5e.59931.277

Chapter 9
Moisture control: surface and subsurface drainage

Andrew Todd   Principal Engineer, Jacobs Ltd, UK

9.1.   Importance of road drainage

Probably the most important factor in good pavement design is drainage, because without good drainage the road construction can deteriorate very rapidly. Excess water, whether it results from precipitation or groundwater, will have a detrimental effect on road foundations and pavements and on traffic movement and safety. Consequently, good drainage design is fundamental to economic road design.

Good road drainage has three functions:

- It prevents water from entering the pavement, its foundation and associated earthworks, and thereby maximises their longevity.
- It removes rainwater efficiently from the pavement surface, thereby making it safer for traffic and minimising the nuisance caused to other road users and adjacent properties.
- It minimises the impact of shed water on the receiving environment in terms of flood risk and water quality.

Good drainage design begins with good route location. Roads that avoid poorly drained areas, unstable and/or weak soils, frequently flooded sites and unnecessary stream crossings greatly reduce the costs and dangers associated with the need for road drainage.

9.2.   Steps in the drainage design process

While the material, construction and maintenance costs of road drainage systems can be high, the benefits derived from their installation more than make their use economically worthwhile. Aspects of drainage normally undertaken during design, generally in order of process as currently carried out in UK practice, are as follows.

(a) Carry out the water environment component (DfT, 2014) of the environmental impact appraisal, to assess the impact that the planned road may have on the existing water environment within its zone of influence, including on water quality and volume.

(b) Assess the flood risk and the associated drainage needs of the land area affected by the final road alignment, after determining the effect that constructing the road through the existing landscape may have on watercourses and on the watershed associated with the adjacent topography. In this context, the road design should ensure that the construction is not the cause of an increase in the risk or severity of flooding compared with the previous unpaved 'green-field' situation as a result of, for example, restrictions imposed on the natural drainage systems and/or increases caused by the runoff rate from the carriageway surface. Of particular concern is the risk of flooding resulting from the transfer of water from one catchment to another, or an excessive increase in the natural runoff into one watercourse to the detriment of another. While ditches or underground piped land drains may be used to trap and carry runoff from a natural catchment, these drainage systems will need to be intercepted at some point(s) so that the water can be safely conveyed downstream using, for example, culverts that cross under the pavement.

(c) Assess the pre-earthworks drainage needs (separately from the land drainage needs discussed above) with the aim of minimising the risk of the roadworks becoming flooded during construction. While this is obviously linked to the land drainage in that these works will also form part of the final drainage system that ensures that the road is not inundated after being opened to traffic, this component of the drainage system should, wherever practical, be kept separate from the carriageway drainage system.

(d) Assess the subsurface drainage needs. Subsurface drainage systems are usually installed during road construction to intercept water before it can enter a pavement, and to remove already in-ground water and/or rainwater that may subsequently enter the pavement structure after construction and damage a subgrade or foundation if left unattended. However, protection of the road foundation and subgrade is not always easy because, as is shown in Figure 9.1 there are many ways whereby moisture can enter a pavement structure.

While some water can be expected to enter pavements from groundwater sources, most free water enters through cracks and potholes in wheel tracks in the road surface,

Figure 9.1 The many ways whereby water can enter and leave road pavements, foundations and subgrades

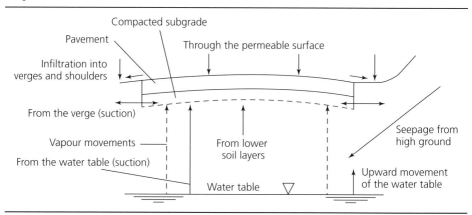

Table 9.1 Typical values of capillary rise for various materials

Material	Particle size range: mm	Capillary rise: mm
Fine sand	0.05–0.25	300–1000
Medium sand	0.25–0.5	150–300
Coarse sand	0.5–2	100–150
Well-graded sand	0.25–2	150–1000
Fine gravel	2–6	20–100
Coarse gravel	6–20	5–20
One-sized aggregate	>5	<5

through surface joints and at the edges of carriageways, and through pores in the surface course material. Seepage from high ground, which is common in hilly topography, is especially noticeable in road cuttings that extend below the natural water table where inflows from springs, and seepage from water-bearing permeable strata, have easy access to the pavement foundations and subgrade. (Note that seepage through the face of a non-rock cutting may also erode the slope of the cutting if left untreated.) A road that is located in flat terrain may be built on the natural subgrade or on a low embankment, so that the formation is relatively close to the natural water table, and, if the subgrade soil is fine grained, the water may be drawn up through capillary (suction) action. The measured capillary height rises shown in Table 9.1 for various soil materials (NAASRA, 1983) illustrate the potential danger from this source.

Moisture movement in the vapour phase is associated with differences in vapour pressure arising from temperature and/or moisture content differences in vertical positions in coarse-grained (rather than fine-grained) subgrade soils. Vapour movements due to temperature differences can be significant in climatic areas where there are substantial fluctuations in the daily temperature – a state of affairs that is not normally experienced in the UK. Vapour movement due to moisture differences requires the soil to be relatively dry, and this is a moisture condition that is also not very common in the UK's temperate climate.

(e) Assess the surface drainage requirements of the carriageway. As a basic design principle, precipitation – most usually rainwater in the UK – should be removed from the carriageway as quickly as possible and conveyed to a suitable outfall, preferably via a drainage system that is separate from any land or pre-earthworks drainage.

(f) Assess the extent to which sustainable drainage techniques must be or can be employed in the design of the surface water drainage system, so as to ensure that receiving waters, groundwater and adjacent habitats are protected at points of drainage discharge.

(g) Check that the maintenance needs of the surface and subsurface drainage facilities have been adequately incorporated into the drainage design.

9.3.   Environmental impact appraisal

At the initial stage in the design process (i.e. when choosing the preferred route option and carrying out the feasibility design), an assessment of the impact that the proposed route may have on the surrounding water environment should always be undertaken. As part of this exercise the significant water resource features within the zone of influence of the proposed route will need to be identified, and parameters with respect to flow rates and water quality criteria established. In most countries, much information regarding these matters is normally available from the relevant environment protection agency.

Typical water resource features identified include those that impact on water supply, transport and the dilution of waste products, aesthetics, biodiversity, cultural heritage, recreation, value to the economy, and the conveyance of flood flows. Subjective judgements as to the importance of each feature will need to be made under the parameters of quality, scale, rarity and substitutability. The quality criterion provides a measure of the physical condition of the feature; the scale criterion classifies its importance in terms of, for example, whether it is of global, national or regional significance; the rarity criterion allows consideration of whether the water attribute being evaluated (e.g. its potability) is commonplace or scarce; the term substitutability refers, in this case, to whether the water feature is sustainable or whether it is feasible for it to be replaced over a given time-frame and within the funds available to the scheme.

Using the above criteria, the importance value of each of the above parameters is then assessed on the basis of it being very high, high, medium or low. For example, a resource feature such as an aquifer that provides potable water to a large population, or a floodplain that provides protection against flooding to more than, say, 100 residential properties, might be valued as being of very high importance, while an unproductive or unattractive land feature, or a floodplain with limited existing development, might be rated as being of low importance.

The magnitude of the potential impact of the road scheme upon each water feature is then appraised and classified as to whether it is largely adverse, moderately adverse, slightly adverse, negligible, slightly beneficial, moderately beneficial or largely beneficial. Finally, the significance of each impact is then determined by comparing the importance of the feature against the magnitude of the potential impact using the criteria described in Table 9.2, and, from this, an overall water environment assessment score can be determined for the proposed scheme.

In practice, it is not common for a road scheme to result in a score in the beneficial range. Consequently, assuming that the road scheme is proceeded with, the detailed road design should seek to address measures that aim to mitigate the adverse impacts measured in the environmental impact appraisal and to improve the overall score.

9.4.   Assessing land drainage needs

A new road inevitably cuts across natural drainage paths in the landscape, and the function of the land drainage design therefore is to ensure that its design will not (a) cause flooding that will result in damage to the new pavement construction and/or hinder

Table 9.2 Definitions of final assessment scores used in the water environment component of the environmental impact appraisal (DfT, 2014)

Score	Comment
Large beneficial impact	It is extremely unlikely that any scheme incorporating the construction of a new transport route (road or rail) would fit into this category. However, a scheme could have a large positive impact if it is predicted that it will result in a 'very' or 'highly' significant improvement to a water feature(s), with insignificant adverse impacts on other water features
Moderate beneficial impact	Where the scheme provides an opportunity to enhance the water environment, because it results in predicted ■ significant improvements for at least one water feature, with insignificant adverse impacts on other features ■ very or highly significant improvements, but with some adverse impacts of a much lower significance The predicted improvements achieved by the scheme should greatly outweigh any potential negative impacts
Slight beneficial impact	Where the scheme provides an opportunity to enhance the water environment, because it provides improvements in water features that are of greater significance than the adverse effects
Neutral	Where the net impact of the scheme is neutral, because ■ it has no appreciable effect, either positive or negative, on the identified features ■ the scheme would result in a combination of effects, some positive and some negative, which balance to give an overall neutral impact – in most cases these will be slight or moderate positive and negative impacts; it may be possible to balance impacts of greater significance, but, in these cases, great care will be required to ensure that the impacts are comparable in terms of their potential environmental impacts and the perception of these impacts
Slight adverse impact	Where the scheme may result in a degradation of the water environment, because the predicted adverse impacts are of greater significance than the predicted improvements
Moderate adverse impact	Where the scheme may result in a degradation of the water environment, because it results in predicted ■ significant adverse impacts on at least one feature, with insignificant predicted improvements to other features ■ very or highly significant adverse impacts, but with some improvements that are of a much lower significance and are insufficient positive impacts to offset the negative impacts of the scheme

Table 9.2 Continued

Score	Comment
Large adverse impact	Where the scheme may result in a degradation of the water environment, because it results in predicted ■ highly significant adverse impacts on a water feature ■ significant adverse impacts on several water features
Very large adverse impact	Where the scheme may result in a degradation of the water environment because it results in predicted ■ very significant adverse impacts on at least one water feature ■ highly significant adverse impacts on several water features

traffic by flooding the carriageway, and/or (b) inhibit water flows in the existing drainage channels so that unwanted redirection and redistribution of flows result, and local catchment areas and boundaries are altered. Ideally, road drainage and land drainage should be kept entirely separate but, in practice, due to (say) land constraints the land drainage design may require smaller watercourses to be locally diverted into toe drains alongside embankments and directed into culverts that cross beneath the new road, while larger watercourse crossings such as streams may require multiple-barrel culverts or, in the case of significant rivers, the building of bridges over the waterways.

Irrespective of whether culverts or bridges are required, the presence of watercourse crossings may impact on the vertical alignment of the road, so it is important that the dimensions and levels of the crossings be conveyed to the designer responsible for the alignment design early in the design process.

Flood flows from natural catchments can last for many hours after heavy rain, so that the potential for traffic to be disrupted as a result of a road being flooded is much greater from these sources than from the runoffs from carriageway surfaces, which, typically, last for the duration of the rainstorm. Thus, while road surface drainage systems are designed to intercept and remove water resulting from short-duration, high-intensity rainfalls with return periods of 1 year (for normal, longitudinal, edge-of-the-pavement surface channels) or of 5 years (so that there is no flooding of the carriageway), storms with a return period of 75 or 100 years are commonly assessed when designing culverts to handle intercepted flood flows from natural catchments. (Note that, in practice, some culverts that have to carry the flows from permanent watercourses across roads may cater for flow rates based on return periods varying from 25 to 100 years, depending on the location of the site and the significance of the detrimental effects that could arise from flooding.)

9.4.1 Culverts

Culverts can be circular, rectangular box, piped arch, arch or complex structures. Their design is a subject unto itself, and space does not permit any detailed design discussion here: rather, the reader is referred to an authoritative guidance that is readily available in the literature (Balkham *et al.*, 2010) for information on the hydraulic design of culverts.

Figure 9.2 Culvert alignment options relative to road crossings (after HA *et al.*, 2004a): (a) online; (b) partial diversion, 90° crossing; (c) full diversion; (d) full diversion, 90° crossing

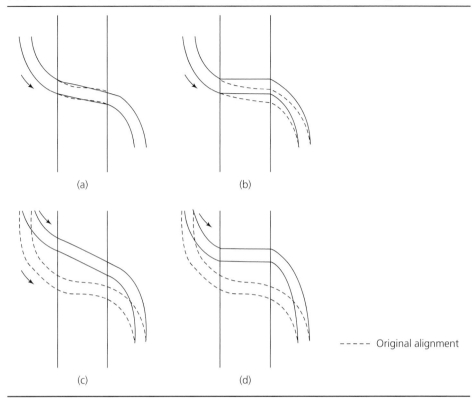

(It might be noted here, however, that if the cross-section of a culvert has a dimension greater than 900 mm, experience has shown that the culvert will also need to be structurally designed, even if it is simply a pipe.)

Ideally, the alignment of a culvert crossing should be as direct as possible so as to minimise its length – for hydraulics, cost, habitat migration, and ease of maintenance reasons. On the other hand, a culvert that is laid along the line of an existing watercourse (e.g. Figure 9.2(a)) is normally the best option with respect to maintaining the existing hydraulic conditions. The slope of the culvert should conform closely to the slope of the natural watercourse: if the speed of the water is reduced by flattening the slope, sediment will be deposited within the culvert, and if the slope is made steeper, more scouring is likely at the culvert exit. Off-line construction (e.g. Figure 9.2(c)) has the considerable advantage of being able to maintain the existing channel flow while building the culvert in the dry; note, however, that the flow is returned to its natural watercourse as soon as possible after passing through the culvert.

Further consideration of Figure 9.2 also suggests that the drainage engineer normally needs to incorporate erosion protection at transition points into and out of culverts and where the watercourse is diverted. The introduction of bends into a watercourse usually results in the deposition of silt on the inside of the bend, and the scouring of the bank on the outside. Scour erosion of a downstream bed at the outlet of a new culvert is also very likely if no energy dissipation device is incorporated, especially if the maximum discharge velocity under the design flood-flow conditions exceeds about 1.2 m/s.

Where a natural watercourse is intercepted or diverted, the drainage designer should ensure that the captured water remains within a channel that drains the original catchment, so that there is no cross-catchment transfer that might lead to an increased flow volume that could have a subsequent adverse impact somewhere downstream. As part of the environmental impact appraisal, the catchment for each individual watercourse should have been defined; however, if this is not available a review of topographical mapping, especially of the contouring, should enable the catchment area of the watercourse at each culvert crossing point to be determined for design purposes.

Normally, the overland runoff towards the road is calculated using either of two approaches (Faulkner, 1999; Marshall and Bayliss, 1994): governmental guidance on their usage is available (HA *et al.*, 2004a).

9.5. Pre-earthworks drainage works

The pre-earthworks drainage work is intended to ensure that overland surface flows do not interfere with the roadworks during their construction and that, post-construction, there is no subsequent runoff inundation from each catchment along the line of the road. As with the previously discussed land drainage design, when carrying out this protective work, care has to be taken to ensure that intercepted overland flows are normally redirected to a watercourse within the same catchment – the intent being to avoid cross-catchment transfer of flows that may be the cause of subsequent flooding from the downstream watercourse(s).

Any connection of land drainage and/or pre-earthworks runoff flows to the drainage system of the road surface should also be avoided, if at all possible. If this is not practical, and such a connection is found to be necessary, then it should occur downstream of any attenuation or treatment facilities, so as to avoid overestimating their discharge rate. The reason for this is that, when the land drainage flows based on the greenfield runoff rate are added to the discharge rate determined for the road surface, the result will be an overestimation of the permitted pond outflow rate, as the peak of the land-drainage hydrograph occurs later than that of the peak for the road surface runoff.

Table 9.3 shows that there are very considerable differences in the catchment runoff rates experienced with various types of soil. The 'greenfield' runoff from a rural catchment greater than 0.5 km² (i.e. >50 ha) can be estimated from Equation 9.2

$$\text{SOIL} = (0.1S_1 + 0.3S_2 + 0.37S_3 + 0.47S_4 + 0.53S_5)/(S_1 + S_2 + S_3 + S_4 + S_5) \quad (9.1)$$

$$\text{Qbar}_{\text{rural}} = 0.00108 \times \text{AREA}^{0.89} \times \text{SAAR}^{1.17} \times \text{SOIL}^{2.17} \quad (9.2)$$

Table 9.3 The runoff potentials of different soil classes (HA *et al.*, 2004b)

General soil description	Runoff potential	Soil class
Well drained sandy, loamy or earthy peat soils Less permeable loamy soils over clayey soils on plateaux adjacent to very permeable soils in valleys	Very low	S_1
Very permeable soils (e.g. gravel, sand) with shallow groundwater Permeable soils over rocks Moderately permeable soils, some with slowly permeable subsoils	Low	S_2
Very fine sands, silts and sedimentary clays Permeable soils (e.g. gravel, sand) with shallow groundwater in low-lying areas Mixed areas of permeable and impermeable soils in similar proportions	Moderate	S_3
Clayey or loamy soils	High	S_4
Soils of the wet uplands: bare rocks or cliffs; shallow, permeable rocky soils on steep slopes Peats with impermeable layers at shallow depth	Very high	S_5

Note: chalky soils can have a wide range of permeabilities, and runoff potentials that vary between those of clay loams and those of coarse sands

where Qbar$_{rural}$ is the mean annual flood (m³/s), AREA is the catchment plan area (km²), SAAR is the standard average annual rainfall for the location (mm) and SOIL is a soil index derived from Equation 9.1, in which S_1 to S_5 denote the proportions of the catchment covered by each of the soil classes described in Table 9.3. (Note: the SOIL index is an indication of the overall permeability of the catchment and, consequently, its runoff potential.)

A method that is applicable to the estimation of runoff from catchments smaller than 0.3 km² (<30 ha), which was primarily developed for the sizing of field drainage pipes, is also available in the literature (Young and Prudhoe, 1973).

9.6. Subsurface drainage

While some moisture is required to achieve the desired compaction of unbound pavement layers when building a road, excess moisture in the subgrade and in unbound pavement (including foundation) layers post-construction encourages their rapid deterioration and the eventual failure of the pavement. Where the pavement formation is allowed to become wet, there is a significant risk of the pressure from passing wheel loads forcing water deeper into the subsoil, which in turn forces the fine particulates up into the pavement construction. This is often called 'pumping', and will eventually result in the permanent deformation of the road pavement.

The reason for installing subsurface drainage is to prevent surface and seepage moisture from entering the pavement in both cut and fill situations, or to prevent the upward movement of groundwater from the water table. This is most commonly achieved in the UK by installing longitudinal subsurface drains – typically either a narrow filter drain or a fin drain – at the low edges of the pavement to sufficient depths to enable the shaped foundation layers and formation to be drained of any moisture that may trickle down to them, to prevent the lateral ingress of water from shoulder, verge and central reserve areas and, sometimes, to lower the groundwater level and prevent capillary moisture from rising up. The general (minimum) guidance for UK conditions is that the bottom of an edge-of-the-carriageway subsurface drain should extend to at least 600 mm below the compacted and shaped formation, or that the invert of the drainage pipe should be set at a depth of at least 50 mm plus the nominal pipe diameter below the sub-formation (bottom of capping) level, whichever results in the greater depth of drain. However, deeper drains may be required, depending on the situation encountered in the field.

Cross-sections of four typical subsurface drains used in UK roads are shown in Figure 9.3.

The type 5 fin drain (Figure 9.3(a)) is a thin, high-strength, box-shaped 'sandwich' system, the main features of which are two vertical geotextile fabric walls 25 mm (minimum) apart separated by a core of horizontal polyethylene spacers, that is placed in a narrow trench in contact with the wall abutting the foundation layers. The trench is then backfilled with a suitable granular material, with care being taken to ensure that the filter fabric is not damaged in the process. Any subsurface water is able to enter the filter through the geotextile sidewalls, trickle down to the (no-pipe) enclosed bottom and flow along a conventional slope to outlets located at appropriate intervals along the trench length.

The type 6 fin drain shown in Figure 9.3(b) comprises a thin twin-skin geotextile with a permeable core that also envelops a perforated plastic pipe installed in the bottom of the trench. The geotextile is placed against the pavement-side trench wall in contact with the layers to be drained, so that it acts as a filter for the subsurface water that trickles down the core and enters the trench pipe for its subsequent removal at spaced outlets.

The type 8 narrow filter drain shown in Figure 9.3(c) typically comprises a 200 mm wide (maximum) longitudinal trench that is installed with one of its faces in contact with the pavement construction, a sloping, small-diameter (100 mm maximum) perforated pipe resting in and covered by a granular-surround filter material at the bottom of the trench, and a coarser granular backfill that extends up to the level of the top of the sub-base and is then sealed with non-permeable material. As such trenches are relatively shallow, with vertical walls, they are most usually dug with a ditching machine. The pipe may be fitted with a geotextile filter 'sock' whose function is to prevent the granular surround, and sediment in the water that is not removed by the backfill or surround, from entering and settling in the pipe. Alternatively, instead of placing a sock around the pipe, the contents of the backfilled trench may be completely wrapped in a geotextile fabric (see the type 9

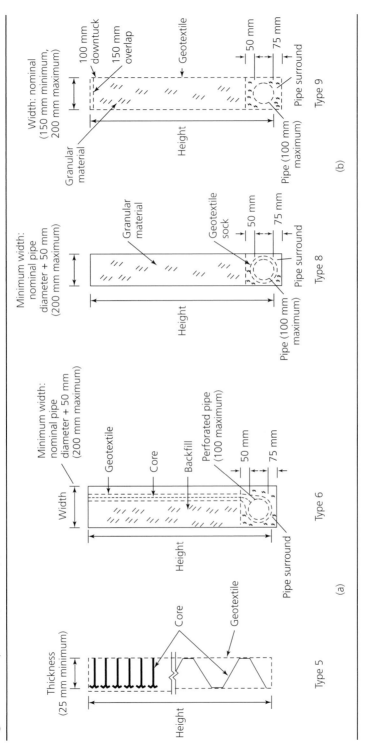

Figure 9.3 Examples of (a) fin and (b) narrow filter drains laid in narrow trenches

287

narrow filter drain in Figure 9.3(d)) whose permeability is designed to be compatible with the gradation of the adjacent subgrade and foundation materials: in this case, the geotextile at the side of the trench carries out the initial filtering of the sediment carried by the water entering the trench. In either case, the slope of the trench pipe in narrow filter drains (and in fin drains) is as steep as practical, but preferably no flatter than $1:200$ (0.5%), and provided with outlets to carrier drain chambers at regular intervals (normally about 100 m, but never more than 200 m, apart), from whence the filtered water may be picked up by a carrier pipe and taken to a suitable outfall point for discharge into a watercourse or stormwater drain.

9.6.1   Combined subsurface and surface drains

Combined surface and subsurface filter drains (i.e. those that drain moisture from both the carriageway and subsurface and carry it away in the same carrier pipe) were once the norm for major roads in the UK. Known as French drains, they fell from favour, however, as reports emerged that they contributed to pavement failures by bringing large quantities of surface water into the ground close to foundations where malfunctioning of the drainage system is not easily detected and the inherent risk to pavement stability is greatest. Subsequent research showed that the causes of many of these failures were associated with poor construction and, especially, poor maintenance practices that failed to ensure that (a) adequate measures were taken to prevent the build-up of vegetation (especially along the unsealed edge of the drain abutting the pavement) that restricted surface water inflow, and/or (b) 'clogged' filter materials were regularly replaced or cleaned and recycled to remove sediment retained in the filter. Combined surface and subsurface drains are not now the preferred systems for new construction but tend to be used for reconstruction work, for situations where significant groundwater flows have to be dealt with in cuttings, and for locations where a road pavement has long lengths of zero longitudinal gradient.

As well as providing for the normal drainage of foundation layers and subgrade, a subsurface system in a cutting may also be required to lower the groundwater to an appropriate depth below the pavement. A conventional surface drain may also be located in the verge in a cutting to collect both runoff from the carriageway and from the slope of the cutting (including from drains installed in the side-slopes). In such instances a combined surface-and-subsurface filter drain can be used advantageously to deal with both functions – including lowering the water table beneath the pavement to a greater depth than would normally be possible with conventional fin or narrow filter drains – because the hydraulic capacity needed to deal with the surface water during heavy storms means that the combined drain system normally has ample capacity to handle the intercepted groundwater.

In combined drains in cuttings, pipes with perforations or slots are laid uppermost within trenches that are situated in verges and/or central reserves adjacent to the low edges of pavements, and water is able to run off the carriageway directly onto the open trench top, and make its way down through the permeable backfill to the pipe at the base of the trench. The pipe joints are sealed so that surface water input at the trench base level is limited. Foundation and capping layers are contiguous with the side of the trench,

Table 9.4 Summary grading requirements for a type B filter drain backfill material used in subsurface drains in the UK (BSI, 2007)

BS sieve size: mm	% by mass passing a given sieve size
80	100
63	98–100
40	80–99
20	0–20
10	0–5
4	–
0.063	–

so that any water that reaches these crossfall-shaped layers from the road surface is also collected by the drain. Depending upon the aims underlying the design of the system the trench bottom may be deepened to, say, 1–1.5 m below the formation (if the objective is to lower the water table), or lined with an impermeable membrane up to the pipe soffit level (to prevent surface water from wetting subsoil that would otherwise be dry).

The prime function of the filter material in a combined surface and subsurface system is to intercept the fine sediment in the runoff water and to reduce the amount that is washed into the pipe at the base of the trench. In the UK, the *Specification for Highway Works* (HA *et al.*, 2009a) requires the use of a type-B-graded natural or recycled aggregate as backfill in these drain trenches (Table 9.4): this grading has a high void ratio that facilitates the movement of water through the medium and also provides an element of short-term flow attenuation. As it is to the filtered fine sediments that polluting contaminants in road runoff adhere, there is a consequent beneficial improvement in the quality of the water that passes through the filter en route to the pipe at the bottom of the trench. During maintenance, the filter medium may be excavated and cleaned on site before being returned directly to the drain: hence, the graded aggregate is deemed sustainable and such filter drains are deemed to be part of a sustainable drainage system (SuDS) (see also Section 9.8).

With the relatively recent decision in the UK to convert many hard shoulders on motorways into permanent running lanes (HA, 2013), it is now specified that the loose stones at the tops of combined drains should be at least 1 m away from the trafficked edge of the nearest running lane. Nonetheless, stone scatter from the top of the exposed trench filter of a combined drain onto a carriageway can be a hazard to vehicles and, to alleviate this safety concern, it is not uncommon to spray for geogrids to be used to reinforce the surface layer of the filter material.

9.7. Surface drainage

Water on the carriageway surface poses a danger to moving vehicles due to the longer stopping distances required under wet conditions, and to the risk of skidding when brakes are applied at locations such as sharp curves, controlled intersections, roundabouts, exit lanes from high-speed roads and so on. Partial or full aquaplaning can occur on a wet carriageway if the water is sufficiently deep to cause the moving tyres to lose contact with the road surface. Spray from other vehicles, especially commercial vehicles, can dramatically reduce forward visibility, while water on the road surface obscures road markings (including reflectorised markings). Pedestrians and cyclists are also more at risk due to reduced visibility, as well as being vulnerable to splashing from passing vehicles.

Legal liabilities can arise if water from an inadequate road drain affects third-party properties. If the road drainage facilities are badly designed, constructed or maintained, water from the carriageway may encroach into adjacent properties or pond on the carriageway and reduce traffic flow. Also, as discussed previously, a bitumen-covered road surfacing or shoulder is not totally impervious, so that ponded water may seep into and through the pavement and soften unbound foundation layers and the subgrade.

9.7.1 Depth of water on the carriageway

Due to a loss of tyre contact with the road surface, driving becomes particularly dangerous when the water depth in the wheel path is greater than about 2.5 mm. Thus, road drainage design aims to remove as much surface water as possible, as quickly as possible, from the carriageway without causing other safety problems in the process.

Research on rolled asphalt and brushed concrete surfaces has resulted in the development of a basic formula (Ross and Russam, 1968) that relates the water depth to the rainfall intensity, the drainage length and the flow-path slope

$$d = 0.46(l_f I)^{0.5} n^{0.2} \tag{9.3}$$

where d is the flow depth above the tops of the surface texture asperities (mm), I is the rainfall intensity (mm/h), n is the flow-path slope (expressed as a ratio) and l_f is the length of flow path (m) $(= W(n_3/n_1) = W[1 + (n_2/n_1)^2]^{0.5}$, where W is the width of carriageway (m), n_1 is the crossfall (%), n_2 is the longitudinal grade (%) and n_3 is the flow-path slope (%)).

As the rainfall intensity is dependent upon hydrology, Equation 9.3 indicates that good geometric and drainage design should ensure that the flow path of the draining water is kept as short as possible (i.e. as the depth of flow is proportional to the square root of the flow-path length). As the depth also varies with the fifth root of the flow-path slope, there is relatively little benefit to be gained, in theory, from increasing this slope: in fact, the major benefit to be gained from having a steep crossfall is a reduction in the amounts of water ponding in deformations in, typically, vehicle wheel paths in the road surface.

The major factor affecting the length and depth of flow is the carriageway width, W: thus, the use of a cambered surface with two-way crossfalls is a very effective way of decreasing the flow depth on wide one-way carriageways. As the maximum depth occurs at the end of the flow path, it is important that the runoff water not be allowed to pond at, and be removed as quickly as possible from, the edge of the carriageway. On superelevated curved sections of dual carriageways with narrow central reserves, care needs to be taken to ensure that runoff water from the high carriageway is collected by a drain before it can find its way across the reserve and onto the lower carriageway.

9.7.2   Types and uses of surface drains

Unlike the situation in very cold countries subject to rapid temperature changes – where special consideration must normally be given to the provision of extra drainage facilities to cope with runoff water from melting snow accumulations – the road drainage engineer in the UK is mostly concerned with coping with runoff from rain. Road carriageways are normally designed with lateral (crossfall) and longitudinal slopes, so that rainwater can run off the pavement surface and over the edges under gravity, for onward disposal via a suitable outfall. Surface drainage design in the UK is therefore basically concerned with selecting a design storm, estimating the likely runoff from that storm from the design catchment area, and deciding how to collect and safely remove the water to a suitable outfall, so that it can be discharged and disposed of safely and economically.

The following discussion is concerned with the control of runoff water from relatively small catchment areas, mainly road carriageways, verges and cuttings, so that the water can be diverted to suitable disposal points with the minimum detrimental effect upon road users, the road or the environment. The following are the main types of edge-of-carriageway drainage systems used for this purpose.

9.7.2.1   Over-the-edge drainage

Usage of direct over-the-edge informal drainage is most applicable to low-height embankment conditions in non-built-up areas with gentle slopes (to preclude topsoil instability) that are formed from stable granular materials. If used on high embankments built with moisture-susceptible soils (e.g. clayey or silty soils), over-the-edge drainage will generally be the cause of erosion and softening of side slopes and, eventually, embankment instability. Direct over-the-edge (informal) drainage should never be used at locations where it is likely to interfere with the movement of people or vehicles (e.g. on bridge structures or where footways or segregated cycleways abut carriageways).

The construction, in combination with over-the-edge formal drainage, of gently sloping, grassed, dished channels in verges is becoming increasingly common with rural roads (HA *et al.*, 2006a). These are regarded as SuDSs due to their potential to control storm-water runoff rates and to reduce the sedimentation and pollution load carried in surface runoff. However, as vegetation growth on non-hard verges can inhibit the free drainage of direct over-the-edge systems, care needs to be taken to ensure that verges that use this form of drainage are well maintained so that standing water does not accumulate at carriageway edges.

9.7.2.2   Kerb and gully drainage

A kerb alongside the edge of a carriageway generally serves three functions: (a) it provides some structural support during the construction of a road pavement; (b) it gives some limited protection to footpaths and verges against vehicle overrun; and (c) it forms a channel that directs surface water to collection gullies that feed longitudinal carrier pipes that are set within the verge or below the footpath, for subsequent discharge at suitable outfall points.

Kerb and gully drainage is most usually employed in conjunction with footways in urban areas, or on high embankments or lay-bys in rural areas. In these locales, the drainage channel, which is normally composed of a different material (e.g. concrete) from that used in the surface course of the pavement (e.g. a bituminous material), is set flush with the pavement surface, and generally follows the longitudinal profile of the road. However, when the longitudinal profile is flat, then artificial falls of at least 0.5% have to be incorporated into the pavement edge.

Typically, a gully is a precast concrete or plastic pot with a concrete surround that comprises a sump and a trapped outlet above the normal sump water level. The gully outlet is generally connected to a longitudinal carrier pipe positioned in the verge, or below the footpath, that removes water to a suitable outfall point. The sump is water filled, and retains sediment for subsequent removal; floating debris is prevented from being washed out by the water trap. The water forms a seal that, in urban areas, prevents odours emanating from the drain. As gully inlets are vulnerable to clogging, good maintenance requires that carriageways be swept clean regularly and that gullies are regularly inspected, cleared, and sediment removed.

Water enters a gully either vertically through a bar grating set in the channel in the line of the flowing water or laterally through an open side-entry inlet set into the line of the kerb, or through a combination of arrangements. A grating inlet can collect up to 95% of the runoff water, provided that the width of flow does not exceed about 1.5 times the grating width (Russam, 1969). Unless specifically modified, the normal side-entry inlet on its own is much less efficient, as the small hydraulic head acting at right angles to the direction of flow is insufficient to move much water over the side weir, especially if the channel is on a steep slope.

The width of water flow against the kerb face generally increases in the direction of the longitudinal gradient until it is interrupted by a gully inlet. The permissible width of flow tends to be site specific, and varies according to various factors (e.g. the road standard, the speed limit, and whether or not a footpath or cycleway is nearby). Guidance regarding UK practice in relation to gully spacings on kerbed roads is available in the literature (HA *et al.*, 2000). Generally, however, gullies are often installed at low points (i.e. sags) on roads where water naturally accumulates, and upstream of lay-bys, bus stops, taxi ranks, and the radii of corner kerbs, so that the water is intercepted prior to locations that are heavily used by pedestrians, cyclists and/or turning vehicles.

Combined kerb and drainage block systems are precast (0.4–0.5 m long) concrete units of a profile similar to that of conventional kerbs but with an internal void below. Supplied as either a single unit or as separate top and bottom sections, the voids form a continuous, closed, internal 'pipe' channel when laid. Water enters the conduit through a preformed hole in each unit that is set at the pavement surface level, and subsequently outflows to a carrier pipe via an access box located above a gully. These systems are most effectively used where kerbs are needed at locations of little longitudinal gradient (e.g. at roundabouts). They may also be used in rock cuttings instead of (more expensive) carrier drain installations.

9.7.3   Surface water channels

Surface water channels are now the preferred edge-of-the-carriageway drainage solution for rural roads (including trunk roads and motorways) in the UK. They are usually of slip-formed concrete construction, with a triangular, trapezoidal or rectangular cross-section, and are installed at the edge of the hard strip or hard shoulder so as to be flush with the road surface. These channels are normally limited to side slopes of 1 : 5 on the carriageway side and 1 : 4.5 on the remote side: for vehicle safety reasons, all rectangular channels and others with side slopes steeper than 1 : 4 or deeper than 150 mm, are only used behind safety barriers.

Surface water channels must be provided with discharge outlets for the water when the channel flows reach capacity, and if suitable surface carriers (e.g. swales, or grassed channels or ditches) are not available at these outlet points, then below-ground carrier pipes will need to be provided to carry away the discharged water. The general philosophy underlying the use of surface water channel systems is that surface water should be kept on the surface, clear of the carriageway, for as much as possible of its journey to its ultimate outfall. The preference is to use natural drainage (see also Section 9.8) to transport surface runoff, and to minimise the use of manufactured products. Being natural, these carrier systems require a greater maintenance input; however, they can also result in reduced flood risk, as well as a reduction in pollution, as the slower velocities in natural drainage systems normally have less energy to convey sediments washed from the carriageway. The spacing between surface channel outlets can vary considerably, depending on the location and longitudinal gradient of the road section being drained.

Discharge from surface channels into intermediate outlets to carrier pipes (at, say, 100 m intervals) is usually unavoidable in cuttings that are longer than a few hundred metres. Surface channels may not be appropriate for use on long, flat stretches of road, as, to minimise the channel size, they can require frequent outlets at relatively close spacings to enable the collected water to be transferred into parallel ditches or carrier pipes. With dual-carriageway roads, it may be economical to design surface channels so that the outlet spacings in the verges are coincident with cross-carriageway discharges from the central reserve, thereby avoiding the need for a parallel pipe run in the central reserve.

Significant benefits associated with the use of surface water channels include ease of maintenance and the fact that long lengths devoid of interruption can be built quickly and (if the design is based on a uniform channel profile) fairly inexpensively. The construction of long channel lengths may, however, be hindered due to the need for their discontinuation at piers, abutments, slip roads, junctions, lay-bys, central reserve crossover points or emergency crossing points (HA et al., 2006a).

In practice, the hydraulic capacity of a surface water channel is normally based on a one in 1 year storm when the flow is contained within the channel cross-section in a verge, and one in 5 years when contained in a channel in the central reserve (HA et al., 1998). In the case of verges, the flooding of hard shoulders and hardstrips for widths of up to 1.5 and 1 m, respectively, is allowed adjacent to outfalls (where the depth of flow in the channel is greatest) under a one in 5 year storm. If longitudinal sags occur in cuttings, and a combined drain of any type is not used, it may be desirable to design outfall drainage for a design storm return period of, say, 10 years. With central reserves, the level of the back of channel can be set below the carriageway, so that any flooding occurs within the width of the reserve and does not encroach onto the offside carriageway lane.

Ideally, the rainfall intensities used to calculate the design storms should include an allowance for possible climate changes. Current climate change scenarios for the UK predict that summers will become hotter and drier, and winters will become wetter and extreme precipitation will become more frequent. Where, however, the rainfall data available are inadequate for this purpose, a sensitivity test may be carried out by increasing the intensities of the design storms by, say, 20% (HA *et al.*, 2006a).

The combined channel and pipe drain (HA *et al.*, 2005) is a variation of the conventional surface water channel that uses slip-forming techniques to construct a surface water drain with an internal unlined pipe that is formed within the base of the channel. As before, the function of the surface channel in this combined system is to collect and carry the runoff from the road surface; however, when the design runoff from a length of road reaches the flow capacity of the channel, the surface water must be intercepted at a suitable location by an intermediate outlet formed from gully gratings or slots set into the invert of the surface channel that allows the water to discharge into a shallow benched chamber constructed on the line of the internal pipe. At the point where the design rate of carriageway runoff reaches the capacity of the combined system, the flows from the pipe and the most downstream channel section are discharged into a terminal outfall chamber from which the water is conveyed to a watercourse or, if necessary, to a separate carrier pipe.

A combined channel and pipe drain system provides additional flow capacity for a channel of the same width: it may also reduce the number of outlets required, and eliminate the need for a separate carrier pipe. The larger the diameter of pipe used, the longer the length of road that can be drained without the need for an outfall. Where lateral space is limited, a combined system also allows a narrower or shallower channel to be formed (e.g. adjacent to vertical concrete barriers in the central reserve).

9.7.4   Some maintenance considerations: manholes and catchpits

Manholes or catchpits are constructed on linear carrier drains at sites where it is anticipated that there will be a need for cleansing, and where the chambers can be safely accessed by inspection and maintenance personnel. It is good practice to provide such access points at junctions of pipelines and at changes in the pipe gradient and/or direction and at changes in the pipe diameter. Conventionally, the maximum interval between manhole or catchpit chambers has been 90 m, to allow for the maintenance rodding of blocked pipes; however, the advent of pressurised water jetting has made longer intervals more practical, albeit, in some instances, not necessarily more desirable. The decision regarding the spacing distances between the chambers is generally taken nowadays on the basis of it being a balance of construction versus maintenance costs.

Manholes have a solid channel (invert) through the chamber between the entering and exiting pipe ends, and on both sides of the channel the chamber is filled with mass concrete to form a slightly sloping platform, called benching, that provides a dry working area just above the channel. Manholes are principally used on carrier drains, so as to maintain their flow characteristics through the chambers.

Catchpits are generally used with filter drains where a higher sediment inflow may be expected. The inclusion of a catchpit in the drainage run provides a means for the removal of detritus, debris and silt, which contributes toward improving the quality of the water as it exits the chamber. A catchpit chamber has no channel invert between the entering and exiting pipe ends, but has a base that is set some 300–450 mm below the invert level of the outgoing pipe. As the water flow in the pipe drain enters the chamber, its velocity drops due to the sudden expansion in flow width, and this allows much of the carried sediment to settle onto the catchpit base, from which it can be subsequently removed.

Manholes and catchpits should be kept out of the carriageway wherever possible, as the access covers are vulnerable to vehicular damage and can pose a risk to road users, especially motorcyclists.

9.8. Sustainable drainage systems

Nowadays, there is a greater knowledge regarding the polluting content of road runoff, and this has resulted in a growing awareness that the runoff water can have an adverse effect on the receiving environment, including on groundwater. Consequently, the legislative duty-of-care responsibilities of governmental agencies mean that it is now mandatory for road drainage engineers in the UK and other EU countries to ensure that routine road runoff complies with pollution legislation and does not lead to the deterioration of any receiving waters and/or impact detrimentally upon ecology and biodiversity over time.

Pollution effects can broadly be divided into two main groups (HA *et al.*, 2009b), namely those caused by metals that directly or indirectly affect water quality by chemically impairing biological functions, and those caused by sedimentation that smother feeding and breeding grounds and physically alter the habitat. Routine runoff from road surfaces typically carries with it degraded stone particles and mud and fine dust containing contaminants such as heavy metals derived from brake, tyre, engine and vehicle wear, incomplete fuel combustion, small oil/fuel leakages from vehicles, as well as pollutants from carriageway de-icing agents (e.g. salts and grit) and from adjacent buildings and the atmosphere. Chronic pollution is the term used to describe the non-lethal effects (such as reductions in feeding, growth rates and reproduction) that are imposed on various plants, insects, birds, and mammals and on the habitats in which they live, that are associated with routine road-surface runoff. The term acute pollution is most commonly ascribed to the toxic impact, including death, that the accidental spillage of readily-dissolved contaminants can have over a short period of time (e.g. hours/days) upon various organisms, when discharged in high concentrations into the water environment; however, the deposition of high loads of suspended solids (i.e. sedimentation) can also be the cause of acute pollution in certain circumstances (e.g. by smothering the breeding grounds of trout and salmon).

Research has shown that the scale of the pollution impacts on receiving waters from routine runoff appears to be broadly correlated with annual average daily traffic (AADT) flows on roads carrying above 11 000 AADT. What is not clear is whether there is an AADT flow below which potential pollution impacts are insignificant.

A practical tool for assessing both the short-term and long-term risks of routine surface runoff on receiving water ecology, and whether pollution mitigation measures are required in specific circumstances, has been developed for use on non-urban trunk roads and motorways in the UK (HA *et al.*, 2009b). This is known as HAWRAT (Highways Agency water risk assessment tool) and it enables the risks to each receiving watercourse to be assessed for each individual discharge, thereby allowing early identification of potential polluting sites and the more effective use of mitigation resources.

As a consequence of concerns about pollution from carriageway runoff, recent years have seen considerable attention being paid to the design and use of mitigating drainage systems that combine a sequence of management practices and control structures to drain surface water in a more sustainable fashion. Well-designed, constructed and maintained SuDS are more sustainable than conventional drainage methods because they work with the environment to mitigate many of the adverse effects of runoff water by (a) reducing runoff rates and the risk of downstream flooding; (b) encouraging natural groundwater recharge at locations that minimise the impacts on aquifers and stream baseflows; (c) reducing pollutant concentrations in runoff, thereby protecting the quality of the receiving water body; (d) acting as a buffer for accidental vehicle spillages by preventing the direct discharge of high concentrations of contaminants into outfall waters; (e) contributing to the enhanced amenity and aesthetic value of areas through which roads are located; and (f) providing habitats for wildlife in built-up areas and opportunities for biodiversity enhancement.

The SuDS Manual (Woods-Ballard *et al.*, 2007) provides extensive guidance on the principles underlying the selection of SuDSs, and their design and installation. Passive SuDS components (i.e. ones that do not require operators to activate them) that are commonly used in sustainable road drainage designs include

(a) Detention basins. A detention basin is a vegetated surface storage depression that is normally dry except following storm events. It is designed to store water temporarily, to attenuate runoff flows and allow the deposition of sediment and the infiltration of water into the ground. In some circumstances, the dry 'pond' may also be designed to function as a recreational facility.

(b) Bio-retention areas. These are shallow landscaped depressions, often planned as landscaping features, that are typically under-drained and rely on engineered soils and enhanced vegetation and filtration to (very effectively) remove pollution and reduce runoff downstream.

(c) Ponds. Ponds are permanently wet depressions that are designed for the temporary attenuation of road surface runoff. They cause pollutant removal via sedimentation, and provide the opportunity for biological uptake mechanisms vested in emergent and submerged aquatic vegetation along their shorelines to reduce nutrient concentrations.

(d) Reed beds. These are areas of grass-like marsh plants that are primarily adjacent to fresh water. Man-made reed beds are used to accumulate suspended particles and the heavy metals attached to them.

(e) Sediment sumps. A sediment sump is a lined or unlined pit that maintains a permanent water pool to promote the settling and storage of sediments carried into it by runoff water; it may also include a baffle to prevent oil and floating debris from entering an outlet pipe. Sediment sumps do little to improve water quality, as the dissolved pollutants or the colloidal contaminants and fine silts that carry pollutants tend to be stirred up and flow through at times of heavy rainfalls. They are most effective at reducing drainage flow velocities (e.g. on kerbed streets – see Section 9.7.2.2) and allowing coarse particles to settle out. All-weather access for maintenance vehicles should be provided to the sump, to allow regular maintenance activities to take place at all times. Sediment sumps may be contained within a SuDS system, or stand-alone from it.

(f) Swales. A swale is a linear vegetated drainage channel that is normally designed to retain and convey runoff surface water; however, at appropriate locations, it may also be designed to permit water infiltration into the ground below. The low-flow velocities of water in swales – these are necessary to prevent erosion of the base and side slopes – aided by vegetation, allow much of the water-carried particulates to settle, thereby providing for both effective sediment and pollutant removal. Well-designed and located swales can often replace carrier drainage pipes along the sides of roads.

(g) Wetlands. A wetland is a permanently wet pond and marsh area in which the water is sufficiently shallow to enable the growth of bottom-rooted plants. The soil stratum beneath a wetland must be sufficiently impermeable to maintain wet conditions; if the underlying soil is permeable, the addition of an impermeable material (e.g. puddle clay) will prevent the wetland from drying out and contaminated water from mixing with the water table below. Wetlands detain flows for an extended period, thereby allowing sediments to settle and contaminants to be removed by facilitating their adhesion to vegetation and aerobic decomposition.

(h) Soakaways. A soakaway is a man-made subsurface structure into which surface water is conveyed (see also Section 9.8.2). It is designed to promote the infiltration of runoff water into surrounding unsaturated ground.

A measure of the effectiveness of some passive SuDS mitigation components can be gained by noting the following reported percentage (in brackets) reductions in the risk of spillage causing a pollution incident to receiving waters (HA *et al.*, 2009b): pond and wetland (−50%); filter drain, grassed ditch/swale, infiltration basin or sediment trap

(−40%); and vegetated ditch (−30%). The percentage reductions associated with some active systems (i.e. which require an operator for their activation) are similarly reported as penstock/valve (−60%) and notched weir (−40%).

9.8.1 Coping with sediment

Sedimentation is the main removal mechanism employed in SuDSs to extract pollutants from runoff water. Most runoff pollution is attached to the finer sediment particles carried in the water, and so removal of sediment results in a significant reduction in pollutant loads. With SuDSs, sedimentation is achieved by reducing flow velocities to a level at which the sediment particles fall out of suspension. Care, however, has to be taken when designing a SuDS to minimise the risk of re-suspension when heavy rainfalls occur.

The surrounding land use is a significant factor influencing the amount of sediment that enters a road drainage system, and there are significant differences in the concentrations of sediment carried by surface water into edge-of-carriageway drains in rural versus urban locales (i.e. some 50 mg/l from rural roads compared with about 115 mg/l from urban roads (HA *et al.*, 2009c)). Also, the average particle size carried from rural surrounds tends to be larger than in urban locales (i.e. 0.9 and 0.5 mm, respectively). Of the sediment carried into drains on urban roads, up to 45% tends to be deposited in gully pots. Notwithstanding these differences, the emphasis placed on existing land usages when designing a drainage system for, say, a semi-rural road, should be treated with care, as it is not unlikely that current land usages can change significantly over the design life of a new road.

While poor drainage is the enemy of good pavement construction, sediment is the enemy of good surface drainage. Thus, the removal of sediment from the drainage system is a routine maintenance activity that needs to be programmed, and being able to assess the volume of sediment that may enter a drain at particular locations assists this programming activity.

Hydraulics research has shown that the discharge capacity of a pipeline may be reduced by up to 4% due to the increased energy losses resulting from the movement of sediment along the pipe invert. If sediment is deposited in a pipe, the impact is severe due to a combination of the reduced cross-sectional area and the increased invert roughness. For shallow deposits covering a small proportion of the pipe diameter, the reduction in the discharge capacity is mainly due to the increased bed roughness.

Ideally, carrier pipes in road drainage systems should be designed to be self-cleansing so that no sediment is deposited in the pipe network. In practice, however, the approach adopted in the UK is based on the assumption that a certain amount of deposition in pipes is not too detrimental, and that it is acceptable to design for maintenance on the basis that a bed deposit of up to 2% of the carrier pipe diameter is tolerable. A design procedure (with worked examples) applicable to drainage systems in the UK is available in the literature (HA *et al.*, 2009c): this calculates the volumes of sediment that can enter the drainage system, taking into account such variables as the geographical location (which affects rainfall characteristics), the land use adjacent to the road, the road type (e.g. whether a dual carriageway or three-lane motorway) and the road profile (i.e. whether the pavement is on embankment or in cutting).

When designing a road drainage system, the designer should always consider the requirements for future maintenance of the system. As discussed previously, sedimentation catchpits and gullies require the regular removal of accumulated sediment, and need to be capable of being accessed by vacuum tankers. Ponds and similar SuDS facilities that are installed for flow detention or other purposes eventually become filled with sediment and vegetation, and usually also require vehicular access for material removal purposes. From a maintenance aspect, the drainage systems needs to be safe and accessible, and comply with appropriate governmental regulations, especially when and where measures to activate a system to prevent, say, a polluting spillage from entering a watercourse need to be accessed quickly in the dark.

9.8.2   Infiltration outfall systems

By infiltration is meant the soaking of water into the ground following its transfer from a different part of the environment. Where there is no risk of contamination of protected water, the process of infiltration can be used to recharge underlying groundwater and feed the baseflows of local watercourses, thereby reducing (or sometimes eliminating) the volume of drained runoff that ultimately has to be dealt with at the final outfall. The rate at which ground infiltration occurs is dependent upon the soil type, the antecedent conditions and the time.

Groundwater is a particularly important natural resource, and can be so vulnerable to pollution that, even if the pollution source is removed, it can be especially difficult to clean up. Thus, in England, for example, the Environment Agency has defined groundwater source protection zones (SPZs) around potable abstraction sources such as springs, wells and boreholes, to protect them from pollution.

In general, the road drainage design ethos is to locate outfalls so that they can dispose of surface runoff via sustainable systems at or as close as practical to its source (i.e. where the rain falls). There is a preferred hierarchy for the disposal of surface water: first, into the ground via infiltration (e.g. through soakaways); second, into natural (fluvial) water courses (e.g. a stream or river); and third, to surface water sewers. Road drainage systems should never be allowed to outfall to lakes, ponds or canals, or be connected to a foul sewer.

A soakaway aids pollution control through processes of sedimentation, filtration, biodegradation and volatilisation. A soakaway that is designed to handle surface runoff tends to be either a circular excavation holding pre-formed polyethylene or pre-cast concrete rings (typically, 1–2.5 m in diameter) that are hollow or filled with single-sized stone or plastic high-void media (Figure 9.4), or infiltration trenches (i.e. linear soakaways) with high internal surface areas. When properly located and designed, these excavations will facilitate groundwater recharge and reduce the volume of water that needs to be disposed of downstream, by using below-ground spaces to store surface water runoff and allow for its infiltration into the ground through their bases and sides. Combined surface and subsurface drains, fin and filter drains, unlined ditches, grassed

Figure 9.4 Soakaway details, including a pre-treatment device (Woods-Ballard *et al.*, 2007)

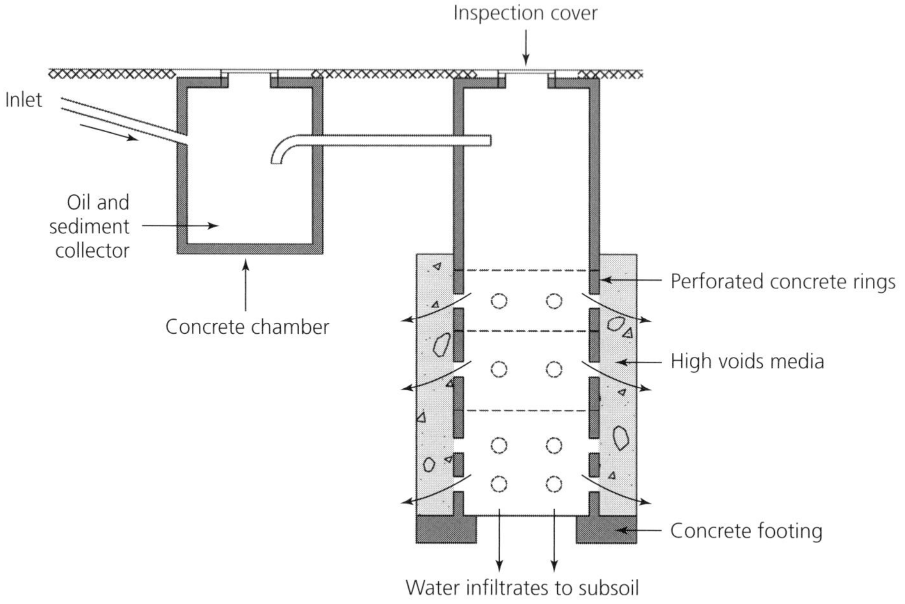

surface channels (including swales), detention basins and dry sedimentation ponds, and infiltration basins may also act as forms of de facto soakaways when in operation.

The time taken for the collected water to infiltrate the surrounding soil through the base and sides of a soakaway depends on its size and shape, and on the infiltration character-istics of the soil. Generally, soil surrounds that are regarded as good infiltration media are gravel, sand, loamy sand, sandy loam, loam, silt loam, chalk and sandy clay loam, while poor infiltration media include silty clay loam, clay, till and rock.

Individual designs of underground soakaways cannot be prescribed because of the site-specific factors that have to be taken into account (i.e. each situation needs to be considered separately). For example, questions that need to be answered include (HA *et al.*, 2006b)

▨ Is the substrate of the proposed site an aquifer?
▨ Is it within the boundaries of a drinking-water SPZ that requires additional protection to cope with untreated surface water?
▨ What is the thickness of the unsaturated zone (i.e. the depth to groundwater)?
▨ What is the type and thickness of the soil?
▨ What is the microclimate of the region like?
▨ Will the installation of a soakaway lead to the mobilisation of pollutants in existing contaminated land?
▨ Are the site dimensions adequate to meet the needs of the soakaway without bringing groundwater into contact with nearby road or building foundations?

To operate successfully, a soakaway must be sited in unsaturated permeable ground and be of sufficient depth and lateral extent to cope with the rapid dispersal of flow under heavy rain conditions. Its lowest infiltration surface should be not less than 1 m above the maximum anticipated, seasonal, groundwater level (Woods-Ballard *et al.*, 2007). The successful long-term performance of a soakaway depends on maintaining its initial storage volume, and any material that is likely to clog the pores of the surrounding drainage material, or seal the interface between the storage and the adjacent soil, should ideally be intercepted upstream before its discharge into the soakaway: for example, Figure 9.4 shows the characteristics of a lined soakaway and one type of a pre-treatment device for intercepting oil and sediment. While the particulates that enter a soakaway can be prevented from migrating outward into granular surround materials with the aid of geotextiles, the retained sediment needs to be removed as part of a regular maintenance schedule.

Where slow infiltration rates are used to optimise in-ground soakaway processes, the use of complementary balancing ponds or other flow attenuation mechanisms may be needed to cope with the runoff generated under peak flow conditions by the road drainage. Active control systems (e.g. notched weirs or penstocks located upstream from the soakaway) can also be used to prevent spillage pollution from entering the infiltration area.

9.9.   Design for maintenance: some additional comments

Road pavements are designed to have long lives and, as has been noted on a number of occasions in this chapter, to satisfactorily achieve their life expectancies they need regular maintenance. Thus, when deciding on the choice of an appropriate drainage system, the road drainage designer should place a high priority on including in the design the requirements for future maintenance. In fact, under the UK's *Construction Design and Maintenance Regulations 2015* it is the designer's responsibility to ensure that what is designed is not only buildable but can be safely maintained. In this context, the drainage systems need to be both safe and accessible, including when active control measures to prevent a polluting spillage from entering a watercourse need to be accessed at night.

It cannot be too strongly emphasised that, for example, manhole chambers, catchpits, gullies and soakaways require the regular removal of accumulated sediment, and they need to be located and designed so that they can be accessed by vacuum tankers. Many SuDS that are installed as flow detention or infiltration facilities will inevitably become filled with sediment and vegetation that require mechanical removal, so there needs to be a substantial vehicular access (including possibly ramps) into these facilities also. Ponds will contain deep water, so that a safe constructed access with handrails, especially around the headwall, will need to be provided.

REFERENCES

Balkham M, Fosbeary C, Kitchen A and Rickard C (2010) *C689: Culvert Design and Operation Guide*. Construction Industry Research and Information Association, London, UK.

BSI (British Standards Institution) (2007) BS EN 13242:2002 + A1: Aggregates for unbound and hydraulically-bound materials for use in civil engineering work and road construction. BSI, London, UK.

DfT (Department for Transport) (2014) *Transport Analysis Guidance (TAG) Unit A3: Environmental Impact Appraisal*. Department for Transport, London, UK. See https://www.gov.uk/transport-analysis-guidance-webtag (accessed 10/06/2015).

Faulkner D (1999) Rainfall frequency estimation. In *Flood Estimation Handbook*. Institute of Hydrology, Wallingford, UK, vol 2.

HA (Highways Agency) (2013) *Interim Advice Note 161/13: Managed Motorways – All Lane Running*. Stationery Office, London, UK.

HA, Transport Scotland, Welsh Government and Department for Regional Development Northern Ireland (1998) Section 2: drainage. Part 1: edge of pavement details. In *Design Manual for Roads and Bridges*, vol. 4. *Geotechnics and drainage*. Stationery Office, London, UK, HA 39/98.

HA, Transport Scotland, Welsh Government and Department for Regional Development Northern Ireland (2000) Section 2: drainage. Part 3: spacing of road gullies. In *Design Manual for Roads and Bridges*, vol. 4. *Geotechnics and drainage*. Stationery Office, London, UK, HA 102/00.

HA, Transport Scotland, Welsh Government and Department for Regional Development Northern Ireland (2004a) Section 2: drainage. Part 7: design of outfall and culvert details. In *Design Manual for Roads and Bridges*, vol. 4. *Geotechnics and Drainage*. Stationery Office, London, UK, HA 107/04.

HA, Transport Scotland, Welsh Government and Department for Regional Development Northern Ireland (2004b) Section 2: drainage. Part 1: drainage of runoff from natural catchments. In *Design Manual for Roads and Bridges*, vol. 4. *Geotechnics and drainage*. Stationery Office, London, UK, HA 106/04.

HA, Transport Scotland, Welsh Government and Department for Regional Development Northern Ireland (2005) Section 2: drainage. Part 6: combined channel and pipe system for surface water drainage. In *Design Manual for Roads and Bridges*, vol. 4. *Geotechnics and drainage*. Stationery Office, London, UK, HA 113/05.

HA, Transport Scotland, Welsh Government and Department for Regional Development Northern Ireland (2006a) Section 2: drainage. Part 3: surface and sub-surface drainage systems for highways. In *Design Manual for Roads and Bridges*, vol. 4. *Geotechnics and drainage*. Stationery Office, London, UK, HD 33/06.

HA, Transport Scotland, Welsh Government and Department for Regional Development Northern Ireland (2006b) Section 2: drainage. Part 8: design of soakaways. In *Design Manual for Roads and Bridges*, vol. 4. *Geotechnics and drainage*. Stationery Office, London, UK, HA 118/06.

HA, Transport Scotland, Welsh Government and Department for Regional Development Northern Ireland (2009a) Drainage and service ducts. In *Manual of Contract Documents for Highway Works*, vol. 1. *Specification for Highway Works*. Stationery Office, London, UK, Series 500.

HA, Transport Scotland, Welsh Government and Department for Regional Development Northern Ireland (2009b) Section 3: environmental assessment techniques. Part 10: road drainage and the water environment. In *Design Manual for Roads and Bridges*, vol. 11. *Environmental Assessment*. Stationery Office, London, UK, HD 45/09.

HA, Transport Scotland, Welsh Government and Department for Regional Development Northern Ireland (2009c) Section 2: drainage. Part 4: determination of pipe roughness and assessment of sediment deposition to aid pipeline design. In *Design Manual for Roads and Bridges*, vol. 4. *Geotechnics and Drainage*. Stationery Office, London, UK, HA 219/09.

Marshall DCW and Bayliss AC (1994) *Flood Estimation for Small Catchments*. Institute of Hydrology, Wallingford, UK, Report 124.

NAASRA (National Association of Australian State Road Authorities) (1983) *Guide to the Control of Moisture in Roads*. NAASRA, Sydney, Australia.

Ross NF and Russam K (1968) *The Depth of Water on Road Surfaces*. Road Research Laboratory, Crowthorne, UK, Report LR236.

Russam K (1969) *The Hydraulic Efficiency and Spacing of BS Road Gullies*. Road Research Laboratory, Crowthorne, UK, Report LR236.

Woods-Ballard B, Kellagher R, Martin P, Bray R and Shaffer P (2007) *The SuDS Manual*. Construction Industry Research and Information Association, London, UK.

Young CP and Prudhoe J (1973) *The Estimation of Flood Flows from Natural Catchments*. Transport and Road Research Laboratory, Crowthorne, UK, Report LR565.

Highways
ISBN 978-0-7277-5993-1

ICE Publishing: All rights reserved
http://dx.doi.org/10.1680/h5e.59931.305

Chapter 10
Introduction to pavement thickness design: some basic considerations

David Hughes Senior Lecturer in Geotechnical and Highway Engineering, Queen's University Belfast, UK

Coleman O'Flaherty Professor Emeritus, University of Tasmania, Australia

10.1. Evolution of the road pavement

Since the invention of the wheel in Mesopotamia (partly, today's Iraq) in about 5000 BC, and the subsequent development of the axle that joined two wheels and enabled heavy loads to be carried more easily, the 'holy grail' for travellers has been to move easily, quickly and safely on 'pavements' that can carry the imposed traffic loads in all-weather conditions.

Archaeological excavations show that diverse forms of road evolved in early times. Initially, these were natural trackways that became trade routes along which settlements developed. It is now well established that stone-paved streets were constructed in the Middle East about 4000 BC, brick pavings in India about 3000 BC and, in Europe, corduroy-log paths near Glastonbury in England *c.* 3300 BC, and stone roads in Crete *c.* 2000 BC. Starting in 312 BC, the Roman dictators used conscript labour to build a great military road system of some 78 000 km, based on 29 major roads radiating out from Rome; the function of these roads was to aid the imperial administration of the Roman Empire and enable its legions to quell rebellions after regions had been conquered. Many of these roads were built on embankments 1–2 m high, to give troops commanding views and make them less vulnerable to surprise attacks: in hindsight, this had the engineering by-product of keeping the pavement dry – which is probably why so many sections of Roman roads are still extant.

The lowest types of Roman road, *viae terrenae*, were made of levelled earth, those with gravel surfaces were termed *viae glareatae*, and the highest type, *viae munitae*, were paved with rectangular or polygonal stone blocks, and lime mortars were commonly used to fill the gaps between the surface slabs. In the case of the *viae munitae*, the roads were often flanked by longitudinal excavations, the soils from which were generally used to form the embankments upon which the pavements were placed; these erstwhile excavations not only protected the soldiers but (again, in hindsight) helped to protect the road construction against the natural elements. It is reported (Lay, 1993) that the Romans achieved desired compaction levels by rolling their pavements with large, heavy, cylindrical stones drawn by oxen or slaves and hand-guided by extended axle shafts.

It is not unlikely that the modern term pavement originated from the Latin word 'pavimentum', which means a rammed floor.

From Roman times until the 18th century, road-making simply involved heaping soil on existing paths and relying on traffic to compact it into hardened surfaces. Pierre Tresaguet, who was appointed Inspector General of Roads for France in 1775, appreciated the detrimental impact of subsoil moisture upon foundation stability and sought to mitigate the effect of water intrusion by digging side-ditches alongside roads and cambering the tops of soil foundations (i.e. subgrades). Tresaguet's design required large (150–180 mm deep) stones to be spread over the entire pavement width and covered with at least 25 mm of compacted stone that was then topped with walnut-sized stones to a depth of about 75 mm so that the convex-shaped pavement surface was as dense as possible.

Thomas Telford, who started his career as a stonemason and went on to found the Institution of Civil Engineers, perfected a method of pavement construction that refined Tresaguet's. Telford built his road pavements on even (not convex) formations, raised them above ground level to mitigate drainage problems, shaped the foundation stones so that they fitted more closely together, and made the pavements as dense as possible to minimise moisture penetration. When building the London-to-Holyhead road (started in 1815), Telford first laid a foundation layer of hand-packed stones, with each stone being placed with its broadest end downward. These stones varied in depth from 9 to 7 in at the pavement centre, and reduced to between 5 and 2 in at the haunches. The specification required that the top face of each foundation block be not more than 4 in wide, and that the interstices between adjacent stones be filled with fine chippings. The central 18 ft width of the convex-shaped foundation layer was covered with two layers of stones – 4 in and then 2 in thick, respectively, with individual stones sized so that each passed through a 2.5 in diameter ring. This central width formed the 'working' portion of the pavement, and its cambered layers were left to be compacted by traffic. The remaining 6 ft wide side portions of pavement comprised broken stone or gravel that was levelled to give a crossfall of not more than 1 in 60. A 1.5 in thick binding layer of gravel was then placed over the entire width of pavement, and its consolidation was again left to traffic. As the interstices of the foundation layer were big enough to admit rainwater percolating down from the surface, cross-drains were often provided beneath this layer (at ~91 m intervals) so that the water could drain into side ditches.

While Thomas Telford was a civil engineer who had a remarkable aptitude for road-making, his contemporary, John Louden McAdam, was a road-building specialist who took a great interest in all matters relating to roads, including their administration. Today, McAdam is best remembered for the method of pavement construction that he advocated and which, in modified form, now bears his name (macadam), and for the two construction principles that he espoused, namely (a) 'that it is the native soil which really supports the weight of the traffic; that while it is preserved in a dry state it will carry any weight without sinking' and (b) 'put broken stone upon a road, which shall unite by its own angles so as to form a solid hard surface'.

Under McAdam, the subgrade formation was shaped to the intended surface camber, using a crossfall that was about half that of Telford's: this provided lateral drainage to both the foundation and carriageway surfaces while ensuring a uniform thickness across the entire width of pavement. McAdam had little respect for Telford's costly (large) stone-paved foundations, and advocated two 4 in thick layers of 3 in broken stone in their stead. On top of the uppermost layer was placed another layer of angular fragments, each not greater in size than 1 in: these were compacted by hand-ramming them into interstices, and then the normal traffic was allowed to further consolidate them.

While both Telford and McAdam recognised the value of rolling in producing a smoother pavement surface, they considered that compaction by traffic was adequate for their pavements; that is, they were concerned that rolling would break down individual stones, destroy mechanical interlock, and result in shorter pavement life. In the 1830s, the French added rolling as an integral part of the McAdam process (which they adopted), but their horses had difficulty in pulling the heavy rollers required to produce the desired contact pressures: this problem was only resolved in the late 1850s with the use, initially in Bordeaux, of the self-propelled steam-roller.

While McAdam's pavements were often structurally inferior to Telford's the greatness of his construction approach was that, for efficiency and cheapness, it was a major improvement over the methods then employed by his contemporaries.

After the 1914–1918 Great War, the numbers of motor vehicles, especially lorries, grew dramatically, consequent upon the increased availability of trained drivers as they were demobilised. The demand for surfaced roads rose simultaneously, fuelled by the availability of cheap bitumen that was then a waste product of the burgeoning oil-refining industry. Nonetheless, prior to World War II, little attention was paid to pavement design, mainly because the roads were seen as being structurally satisfactory for the traffic volumes and loads of the era, and pavement 'design' often relied on the usage of a proved pavement thickness for particular classifications of road. Given that normal subgrade strengths in the UK are now known to vary over a range of at least 25 to 1, the classification approach meant that the same pavement thickness was as likely to be placed on a weak foundation as on a strong one.

The onset of World War II saw a greater focus on pavement design due to the urgent need to construct roads and airport runways for the heavy traffic loads demanded by the war effort. This resulted in the acceptance of empirical design methods based on the California bearing ratio (CBR) test (ASCE, 1950; Porter, 1938) and on the soil classification test (Steel, 1945) that related subgrade 'strength' to flexible pavement thickness and commercial vehicle volumes. The implicit assumption underlying these approaches was that, irrespective of the quality of the pavement materials used, the same pavement thicknesses distributed the applied wheel loads to a similar extent.

From the 1950s onward, world tensions and the accelerated worldwide growth in the numbers of private and commercial motor vehicles combined to create pressures on governments in developed countries for greater expenditures on road systems and on intensive road research, and on engineers for the development of more economic (and better researched) pavement designs that met the following performance criteria:

- The finished carriageway should have good skid resistance and provide the motorist with a comfortable and safe ride throughout its design life.
- The pavement should be able to carry its design traffic load throughout its design life without excessive deformation, and its component layers should not fail as a result of the stresses and strains imposed on them by heavy commercial vehicles and/or climatic conditions.
- A pavement foundation (including its sub-base and any capping layer that might be required to protect the subgrade) should have enough load-spreading capability for it to provide a satisfactory platform for construction vehicles while the road is being built.

The late 1950s and through to the early 1970s saw road engineers in many countries, especially in the USA and UK, carry out ground-breaking research involving observations of the performance under traffic of large full-scale pavement sections. For example, the then American Association of State Highway Officials (AASHO) road test (AASHO, 1962) involved experiments to determine how various axle loads contributed to the deterioration of highway pavements of different thicknesses. This particular road study involved the construction of six specially designed two-lane loops along what became the future alignment of Interstate 80. In the course of testing, each lane was subjected to 24 h controlled loading by specific vehicle types carrying known axle weights, and the pavement thicknesses and materials within each loop were varied so that the interaction of vehicle loads and the pavement structure could be investigated. The test results obtained were then used to develop a pavement design guide titled *AASHO Interim Guide for the Design of Rigid and Flexible Pavements*: this was first issued in 1961, with major updates issued in 1972 and 1993. The 1993 version (AASHTO, 1993), which is still in widespread use in the USA and in many other countries, is briefly described in Section 10.7.

The AASHO road test introduced many seminal concepts in pavement engineering – including the 'fourth power law' (see Section 10.3.1), which proposed that the damage caused to road pavements by commercial vehicles is related to the fourth power of their axle weights.

In the UK, the (then) Road Research Laboratory also devised a large number of full-scale test road experiments – one of these was on the A1 at Alconbury Hill (Croney and Loe, 1965) – that were different to the AASHO road test. In these, testing was carried out on road sections on major routes that were used by general traffic (rather than by dedicated vehicles operating on dedicated lanes with designated axle loads), and weighbridges were installed to determine the axle load distribution of the traffic. The UK work confirmed many relationships between layer thickness, pavement stiffness and subgrade strength, and related these to the ability of a pavement to carry traffic loading: for example, Figure 10.1 clearly shows how pavement deterioration (reported as 'surface deformation') is influenced by layer thickness and stiffness. In due course, this empirical research led to a major revision (Leigh and Croney, 1972; RRL, 1970) of the design procedure for new roads that had been in use in the UK since the early 1960s.

Figure 10.1 Deformation history of a flexible pavement constructed with bases of varying thickness at Alconbury Hill (Thompson *et al.*, 1972)

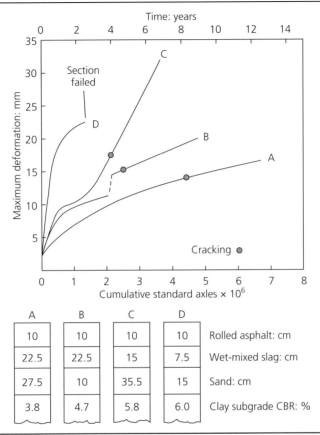

A	B	C	D	
10	10	10	10	Rolled asphalt: cm
22.5	22.5	15	7.5	Wet-mixed slag: cm
27.5	10	35.5	15	Sand: cm
3.8	4.7	5.8	6.0	Clay subgrade CBR: %

As traffic volumes and loadings increased and it became clearer that long-life pavements were required, the need for a more mathematical modelling approach to pavement design – which involved structural theory and a fundamental understanding of the behaviour of pavement materials under repeated stress – gained credibility. The publication of the Transport and Road Research Laboratory report LR1132 (Powell *et al.*, 1984) was a major step forward in the introduction of the analytical (or mechanistic) approach for the design of road pavements in the UK.

The philosophy underlying the analytical pavement design approach assumes that it is possible to use the fundamental engineering properties of pavement materials to calculate the final thicknesses of each pavement layer, taking into account the traffic loading, the environmental conditions and the expected performance and failure modes of the materials. With this approach, the stress and strain distributions are calculated under a standard wheel load, using either a multi-layer linear elastic model or a finite-element model. These stresses and strains are then compared with allowable values for the materials, and the design process is proceeded with on an iterative basis, by adjusting the thicknesses of the layers until allowable values for stress or strain for each of the pavement materials are not exceeded. The perceived advantage of this design approach is that it can be used to investigate a wide range of non-standard loadings, as well as different environmental conditions, materials and layer combinations, that would not be included in empirical design methods that rely on the observed performance of existing pavements.

In essence, the current analytical pavement design approach (see Sections 10.6 and 10.8) attempts to introduce more flexibility into the pavement design process by moving it away from empirical chart-based methods that are based on observations of the performance of trafficked pavements.

10.2. Component layers of modern road pavements

A road pavement is a structure comprising layers of selected materials that are superimposed on the parent material or subgrade soil. The term subgrade is applied to both the in situ (natural) soil exposed by excavation and to the soil that forms the top of an embankment.

The function of a pavement is to support the wheel loads applied to the carriageway and to distribute them to the subgrade. Pavement thickness design is concerned with determining the most economical combination of pavement layers that will ensure that the stresses and strains transmitted from the carriageway do not exceed the supportive capacity of each layer or the subgrade during the design life of the road. Major variables affecting the design of a pavement are therefore the traffic loads (i.e. vehicle volumes and compositions), the environmental conditions such as precipitation and temperature, the quality of the materials economically available for use in the component layers, the thickness of each layer and the subgrade strength.

As Figure 10.2 suggests, pavements can be divided into three main types: flexible, composite and rigid.

10.2.1 Flexible pavements

A flexible pavement comprises a number of unbound and/or bituminous-bound granular layers that are topped by a surface layer that, commonly (but not exclusively), is aggregate bound with bitumen. Ideally, this surface course, which forms the uniform carriageway surface upon which vehicles run, should (Nunn *et al.*, 1997) offer good skid resistance, allow for the rapid drainage of surface water, minimise traffic noise, resist cracking and rutting, withstand traffic turning and braking forces, protect the underlying road structure, require minimal maintenance, be capable of being recycled or overlaid and be durable and give value for money. No one material meets all of these requirements so, in practice, the selection of the material(s) to use in a surface course depends on the particular needs at a given site. Most modern-day surface courses in the developed world are bitumen bound, and only minor roads have surfacings composed of soil–aggregate materials.

On heavily trafficked roads, the surface course is commonly laid on top of a binder course; on minor roads this layer may be omitted. The binder layer is a structural

Figure 10.2 Basic elements of new flexible, flexible composite and rigid types of major road pavements (not to scale)

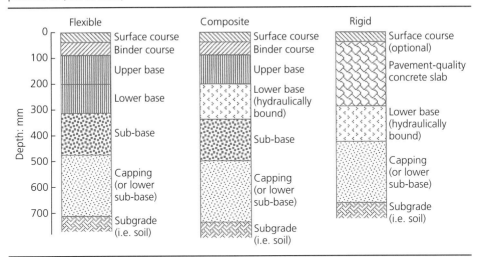

platform that regulates (i.e. makes even) the top of the underlying base, thereby ensuring that the surface course has a good riding quality; it also helps distribute the applied traffic loads to the base. If the surface course is impervious, the binder course may be composed of a more permeable material; if the surface course is permeable, the binder course is most usually impermeable.

The base layer (previously termed the *roadbase*) provides the platform for the binder course, as well as being the main structural layer in a flexible pavement. As the wheel-load-induced stresses decrease with depth, the main function of the upper and lower bases is to distribute the transmitted loads so that the strength capacities of the sub-base and subgrade are not exceeded. Bases in flexible pavements are normally designed to be very dense and stable, and to resist fatigue cracking and structural deformation.

A sub-base layer may or may not be present beneath the (lower) base in a flexible pavement; that is, if the subgrade is strong (e.g. if it comprises a well-graded granular soil) the sub-base may not be needed. While the material comprising the sub-base is of a lesser quality (and, thus, normally costs less) than the material used to form the base, as a structural layer the sub-base must always be inherently capable of resisting the stresses transmitted to it, as well as being stronger and stiffer than the subgrade on which it rests. The sub-base also acts as a working platform for construction vehicles as the pavement is being built; this is especially important when the subgrade comprises poor-quality soil, as the critical load-carrying period is then when the heavy axle loads from the construction equipment used in the placement and compaction of the base are transmitted to the sub-base during construction. While they may be relatively few in number, the magnitude of these loads can be great.

A well-graded dense sub-base may be designed (with or without a geotextile filter) to prevent the upward infiltration of fine-grained subgrade soil into the base above. This preventive function is again especially important during construction, when site traffic and compaction loadings are high. A dense sub-base can also be used to prevent moisture from migrating upward from the subgrade into a soil–aggregate base, or to protect a vulnerable subgrade from downward frost action.

An open-graded sub-base may be designed to act as a drainage layer to pass moisture that falls during and/or enters the pavement after construction. Removal of the water is best ensured by extending the sub-base through the shoulders into longitudinal drains located at their edges; these drains should not be allowed to clog up and should have periodic outlets that are well maintained. If the pavement has a non-porous bituminous surface course and a bituminous-bound base, the amount of water infiltrating after construction will be small, and the need for a drainage-oriented sub-base is lessened. If, however, both the surfacing and base are pervious, an open-graded sub-base may be required to protect the subgrade.

When the subgrade soil is weak, a capping layer may be created to provide a working platform for the construction equipment used to lay the sub-base. This is most commonly done by improving the top of the subgrade: for example, by admixing a layer of imported material that is stronger than the subgrade soil or by stabilising the upper reaches of the subgrade with, say, lime or cement.

The interface between the sub-base and the subgrade is termed the formation. Its cross-section is normally shaped to reflect the cross-slope(s) of the carriageway, so as to assist in the lateral drainage of water that might accumulate within the pavement.

In the UK, the pavement foundation is the collective term used to describe the sub-base, capping and subgrade.

10.2.2 Composite flexible pavements

If a 'flexible' pavement contains layers of hydraulically bound materials, the structure is often referred to as a 'composite' pavement (see Figure 10.1). The rationale underlying the construction of a composite pavement is that it economically combines some of the better qualities of both flexible and rigid pavements. In the UK, for example, a high-quality composite pavement in a major road would have its surface and upper base courses formed from bitumen-bound materials, and these might then be supported on a lower base and/or a sub-base of cement-bound material that rests on a lime-modified capping layer.

10.2.3 Rigid pavements

The main feature of a typical rigid pavement (see Figure 10.1) is a cement concrete slab with high flexural strength. If the concrete slab is designed to allow traffic to run directly on its surface (i.e. if the top of the slab is able to provide a smooth comfortable ride and has good skid resistance under all weather conditions), then the slab can be seen as analogous to the combined surface and base courses of a new flexible pavement.

The concrete slabs in rigid pavements are either jointed unreinforced (URC) or reinforced. Reinforced concrete slabs are usually described as being either jointed reinforced (JRC) or continuously reinforced. In the UK, the currently preferred rigid pavement construction is either a continuously reinforced concrete pavement (CRCP) – this is normally covered with a 30 mm thick (minimum) asphalt overlay – or a continuously reinforced concrete base (CRCB) with an asphalt overlay of 100 mm.

The sub-base in a rigid pavement is intended to provide a uniform, stable and permanent support for the concrete slab when subgrade damage is anticipated, for example, from one or more of the following: frost action, poor drainage, mud-pumping, swell and shrinkage, and construction traffic.

Practices regarding how to prevent frost action under concrete slabs vary from country to country. Generally, however, they involve ensuring that the water table is well below the formation and that an adequate sub-base thickness of non-frost-susceptible granular or stabilised material is provided between the concrete slab and the frost-susceptible subgrade within the zone of frost penetration. In the UK, this generally means ensuring that the total pavement thickness of non-frost-susceptible material below the carriageway is at least 450 mm in frost-susceptible locales.

Mud pumping can occur at slab joints, edges and cracks when there is (a) free water in the pavement, (b) a subgrade soil with a high clay content that is able to go into suspension and (c) traffic flows involving the frequent passage of heavy wheel loads. If any one of these three criteria is missing, mud pumping will not occur. Thus, the likelihood of pumping is obviated if the pavement is provided with an open-graded granular sub-base that can act as a drainage layer where there is danger of water entering the pavement and accumulating beneath the slab. Alternatively, pumping can be prevented if the sub-base is composed of a well-graded compacted material, or a cemented layer, that is essentially impervious to water.

Large areas of very expansive soils are quite common in many regions of the world. The excessive shrinking and swelling associated with these clayey soils during alternate dry and wet seasons often results in non-uniform subgrade support, consequent pavement distortion, and loss of carriageway smoothness and riding quality for asphaltic as well as concrete pavements. Minimising this problem can require the creation of a substantial capping layer thickness (e.g. by chemically stabilising the subgrade) and/or building a deep granular sub-base to form a pavement of sufficient thickness and mass to weigh down the subgrade and minimise its upward expansion.

Whatever other functions it may be required to perform, it is essential that a sub-base in a rigid pavement should always be able to perform as a working platform for the construction equipment that will lay the concrete slab, with minimum disruption by wet weather.

The subgrade (and any capping) beneath a concrete pavement must also be sufficiently stable to withstand the stresses caused by construction traffic while the sub-base and concrete slab are being put in place, and to provide the uniform support required by the pavement throughout its life. If the subgrade is strong, then, technically, a sub-base can be omitted from beneath the concrete slab; in practice, however, this rarely occurs in the UK.

10.3. Some factors affecting flexible pavement design

10.3.1 How flexible pavements deteriorate over time

An inadequately designed pavement does not fail suddenly. What happens (Figure 10.3) is that after the road is opened to traffic (a) the pavement may initially gain some strength until a stability phase is achieved; (b) a slow rate of structural deterioration then sets in and continues until (c) an 'investigatory' phase is reached, when the structural deterioration becomes less predictable and monitoring is indicated to determine if and when preventative remedial work needs to be carried out to prevent a failure phase from being entered into; and (d) if the repair work is not initiated in time, the pavement continues to get worse until a failure condition (defined by unacceptable levels of rutting, pot holes, general unevenness and roughness cracking, etc.) is reached from which it can only be 'rescued' by total reconstruction. This failure phase may be quite short or it can be long (i.e. the pavement can have several years of life left in it).

Whole-life costing studies have shown that it is more economical to intervene and repair a pavement during the investigatory phase than to reconstruct after failure at the end of a full design life. Thus, if the performance of a pavement is closely monitored so that the point in time is identified at which the deterioration begins to accelerate (i.e. the onset of the investigatory phase), and if maintenance in the form of, say, the provision of an overlay is then carried out to utilise the remaining existing strength of the pavement and preserve its structural integrity, the time to 'failure' can be very considerably extended.

Major determinants of the rate at which pavement deterioration takes place include the topography and the subgrade soil, the pavement materials and thicknesses, the drainage (surface and subsurface), the quality of construction and maintenance, the environment (rain, frost, solar radiation), the traffic (volume, axle loads and configuration) and the road condition (World Bank, 1988).

Figure 10.3 The pavement deterioration cycle

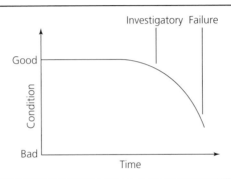

10.3.2   Traffic loads, pavement deterioration and the power law

Pavement design is basically concerned with protecting the subgrade and the various courses within the pavement structure from excessive stresses and strains imposed by the wheel loads of vehicles using the carriageway. The heaviest axles in a stream of commercial vehicles cause a disproportionate amount of damage to a flexible pavement – especially if they are loaded in an unbalanced way (e.g. see OECD, 1989). The load applied through any wheel assembly of a properly loaded commercial vehicle should be half of the load on the axle to which the assembly is attached. However, if a commercial vehicle has a heavy payload that is not correctly placed in the vehicle, the effect will be to significantly increase the axle and/or wheel load and, consequently, the pavement damage (or 'wear') factor.

By comparison, the structural damage caused by lighter vehicles (i.e. cars and light commercial vehicles) is negligible and can be neglected from the point of view of pavement thickness design.

All heavy vehicles do not cause equal distress to a pavement (because of variations in the wheel load, number and the location of axles, the type of suspension, etc.) and because the damage caused is specific to the properties of a pavement, and to operating and environmental conditions. Of the various forms of pavement distress, fatigue (which leads to cracking) and rutting (i.e. permanent deformation) are of considerable importance.

A major objective of the AASHO road test (AASHO, 1962) was to determine the relative damaging effects of different commercial vehicle axle loads. In this study, an axle load of 18 000 lb (8160 kg) was defined as a standard axle, and given a damaging factor of unity, and the number of repetitions of this axle load that caused the same amount of structural damage (i.e. wear) to different flexible pavement sections as was caused by many different axle loads were then determined using statistical analyses. The conclusion from the study was that the relative damaging effect of an axle load was approximately proportional to the nth power of the load, and that $n = 4$, irrespective of the type or thickness of the pavement. In other words, it was suggested that a 6 t axle load (i.e. about 75% of the standard axle load) only caused damage equivalent to less than 0.3 times that caused by a standard axle, whereas a 10 t load (i.e. about 25% more than a standard load) caused damage equivalent to 2.3 standard axles.

A subsequent re-analysis of the same data (Addis and Whitmash, 1981) confirmed the thrust of the above conclusion, but suggested that n lay in the range 3.2–5.6. Another study (Nunn et al., 1997) reported that n varied between 2 and 9, depending upon the degree and mode of pavement deterioration and its condition at the time that the comparison was made.

Irrespective of the exact value of n, the concept of a standard axle load is important, as it allows the range of commercial vehicle loads in normal traffic to be replaced by equivalent numbers of single axle loads, which are easier to handle in pavement design. Table 12.1 (see Chapter 12) lists the wear factors for commercial vehicles (i.e. those

Figure 10.4 Typical transverse deformation profiles on the same pavement section (composed of 100 mm rolled asphalt surfacing, 250 mm wet-mix slag base and 170 mm sand sub-base) at four different levels of traffic flow (Lister and Addis, 1977)

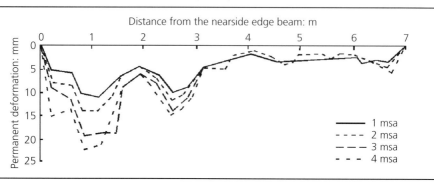

>3.5 t gross weight) in terms of the number of equivalent standard 80 kN axle load that are currently used in pavement design in the UK.

Figure 10.4 is an actual transverse deformation profile for one carriageway of a busy dual-carriageway road in the UK. It shows that the greatest deformation is along the wheel paths of the nearside (left) lane, which carries the greatest proportion of heavy commercial vehicles, and that the far-side traffic lane, which is mainly used by overtaking cars and light commercial vehicles, is little stressed. These measurements were taken after 1, 2, 3 and 4 million standard axles (msa) had passed the test site.

The data in Figure 10.4 also bring to mind the importance of (a) ensuring that the pavement base course is carried through the verge and (b) providing a shoulder with an impervious surfacing. Shear failure and the lateral displacement of the pavement and/or subgrade occurs more easily if the base is not extended beyond the carriageway edge, while the lack of a covered shoulder means that rainwater can find ingress to the subgrade and pavement from the verge.

10.3.3   Rutting and structural deformation

Rutting is caused by deformation in one or more pavement courses. Deformation that occurs within the uppermost courses of a pavement – termed surface rutting – affects the structural integrity of a pavement when it becomes large. Experience has shown that if the depth of rutting at the top of a bituminous-surfaced pavement is more than about 15 mm below the original carriageway level, cracking is likely to occur, and water can then enter the pavement and accelerate its deterioration. A rut of 10 to 20 mm or severe longitudinal cracking in the wheel path of a trunk road pavement is generally regarded as an indicator of a road section that has failed, and probably requires reconstruction to fix it (see also Table 10.1).

Deformation that arises in the subgrade due to the inadequate load-spreading abilities of bituminous-bound and granular courses in the pavement is termed structural deformation. This is commonly the main deformation component, and, if allowed to continue, will result in the creation of surface rutting and the break-up of the pavement.

Under traffic loading, the various courses in a bituminous-bound pavement are subject to repeated stressing, and the possibility of damage by fatigue cracking continually exists. When a wheel load passes over a flexible pavement, each course in the pavement responds in the same general way: an applied stress pulse is caused by the wheel mass while the resultant horizontal strain consists of resilient and permanent components. The permanent strain component, although tiny for a single-load application, is cumulative and becomes substantial after a great number of load applications. An excessive accumulation of these permanent strains from all layers can lead to fatigue cracking and pavement failure.

While fatigue damage to flexible (and rigid) pavements is most directly associated with large wheel loads and inadequate pavement thickness, other vehicle properties also have smaller but still significant influences on fatigue. Fatigue cracking in an inadequate bituminous-bound pavement layer is generally considered to originate at the bottom of the layer, with its onset controlled by the maximum tensile strain that is repeatedly generated by the passage of commercial vehicles. As the cracks propagate upward through the bituminous-bound courses to the carriageway, there is a progressive weakening of these structural layers, which, in turn, increases the level of stress transmitted to the lower layers and contributes to the development of structural deformation in the subgrade. As time progresses and the applied axle loads increase in number, the transmitted stresses increase, and the development of permanent structural deformation in the subgrade is accelerated, and this is reflected eventually in excessive surface rutting.

If, however, the traffic-induced strains in the subgrade are limited (i.e. too low to cause structural damage due to the quality and thickness of the main pavement layers), excessive rutting will not occur or, if it does, it may be due to non-structural deformation within the uppermost pavement layers.

Table 10.1 shows that pavements on weak subgrades (e.g. with a CBR of <5%) have rates of surface rutting that are greater than those on subgrades with a CBR of >5%, irrespective of the thickness of the bituminous-bound cover. (This table also shows the predicted

Table 10.1 Comparison of rates of surface rutting in flexible pavements on weak and strong subgrades (Nunn et al., 1997)

Roadbase type	Rate of rutting: mm/msa			Life to 10 mm rut: msa
	Mean	Standard deviation	Sample size	
HRA or DBM (subgrade CBR <5%)	0.58	0.18	20	17
HRA or DBM (subgrade CBR >5%)	0.36	0.12	21	28
Lean concrete	0.38	0.14	28	28

DBM, dense bitumen macadam; HRA, hot rolled asphalt

Figure 10.5 Measured changes with time of (a) the recovered bitumen penetration and (b) the elastic stiffness modulus for long-life pavements with dense bitumen macadam bases manufactured with a nominal 100 pen bitumen (Nunn *et al.*, 1997)

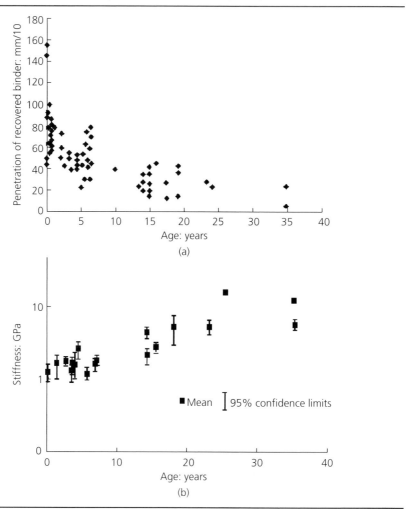

life to a 10 mm rut depth, assuming a linear relationship with traffic.) However, this does not necessarily mean that surface rutting automatically indicates pavement structural deficiency. The data in Figure 10.5 are crucial to understanding why, in well-built thick bituminous-bound flexible pavements that are sufficiently strong to resist structural damage in their early life, the conventionally accepted mechanisms of pavement deterioration (base fatigue and structural deformation) are far less prevalent than surface-initiated deterioration that is associated with excessive ageing of surface course(s).

As is illustrated in Figure 10.5(a), the penetration of the 100 pen binder used in the base reduced from about 70 pen, shortly after laying, to between 20 and 50 pen after some

15–20 years of 'curing', while Figure 10.5(b) shows the corresponding changes in the elastic stiffness modulus with time. These data indicate that time curing allowed the bitumen in the base to gradually harden, with the result that there were in-step increases in the base stiffness modulus: as the pavements became stiffer, their load-spreading abilities also improved steadily at the same time, so that there were consequent reductions in the traffic-induced stresses and strains in the base and subgrade and, therefore, fatigue and structural deformation did not become prevalent in the structures. Thus, as these bituminous-bound pavements are most vulnerable to structural damage during their early life, it can be expected that they will have a very long life if they are built initially to an above-threshold strength that is sufficient to withstand the traffic-imposed stresses and strains during this early period.

10.3.4   Course thickness and material quality

Figure 10.6 shows that, when the contact pressure is kept constant and the wheel load of a test vehicle is progressively increased, the measured stresses at the top and, especially, the bottom of a flexible pavement also increase. Thus, if the stress transmitted to the subgrade is not to be increased, the thickness of the pavement would appear to need to be made greater. However, it is not just the total thickness of the flexible pavement but also the quality and thickness of the materials used in each of its component courses that determine the stress distribution and resultant deformation.

The influence of the quality of material used in a pavement layer is reflected in Figure 10.7, which shows the dynamic stress measurements taken at pavement sections laid on a uniform clay subgrade, before a test road was opened to general traffic. The only differences between the sections related to the materials used in the pavement base. Note that the stresses transmitted through the rolled asphalt section were the lowest, lean concrete was next, while the soil–cement and crushed stone (wet-mix) sections were the least successful in spreading the wheel loads. When the road was opened to heavy traffic, it was found that the stress under the rolled asphalt base did not increase with time whereas that under the same-thickness lean concrete doubled during the first year of traffic and increased again during the second year. The deterioration of the load-

Figure 10.6 Effect of changing the applied wheel load on stresses measured at (a) the pavement surface and (b) the formation (Whiffin and Lister, 1962)

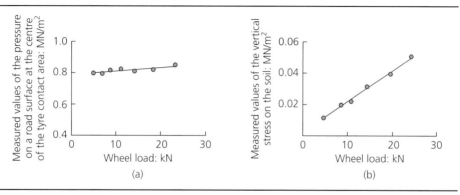

Figure 10.7 Effect of the pavement base material upon subgrade stress (Thompson *et al.*, 1972)

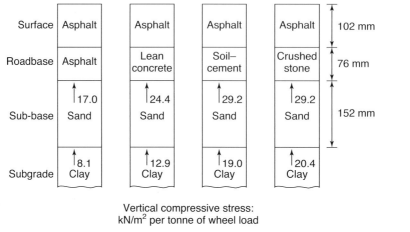

Vertical compressive stress:
kN/m^2 per tonne of wheel load

spreading properties of the lean concrete base was considered to be due to excessive cracking of the layer arising from tensile stresses.

While not graphically shown here, many early studies (e.g. Cooper and Pell, 1974; OECD, 1989) have demonstrated that the deformation history of a pavement of fixed material composition is influenced by the thickness of each layer as well as by its total thickness. In a given pavement, a slight increase in course thickness can result, for any given traffic load, in an appreciably lower probability of pavement distress (and lower subsequent maintenance costs) while, conversely, a slight decrease in course thickness will result in a higher probability of distress (and higher subsequent maintenance costs).

10.3.5   Tyre pressure and vehicle speed

The tyre inflation pressure and the tyre type have an impact on the stress applied to the pavement surface, and therefore on the stresses imposed on the pavement materials. Given that the contact area beneath the tyre of a wheel is approximately elliptical in shape, research has shown that under normal combinations of wheel load and inflation pressure (a) the average contact pressure is less than the inflation pressure, (b) at a constant inflation pressure the contact pressure increases with load and (c) at constant load an increase in the inflation pressure causes a contact pressure increase.

High inflation pressures, especially over-inflated, wide-base single tyres, greatly increase the fatigue damage caused to flexible pavements (Gillespie *et al.*, 1993), whereas they exert only a moderate influence upon rigid pavement fatigue. The effects of high contact pressures are most pronounced in the upper layers of a flexible pavement, and have relatively small differential effects at greater depths; that is, for a given wheel load the inflation pressure has little effect upon the total depth of pavement required above a subgrade, but it can influence the quality of material used in, for example, the surface course.

Generally, there is a decrease in pavement deflection with increasing vehicle speed; this is most obvious in pavements carrying vehicles at very low speeds. High speeds result in a reduction in the time that a moving wheel load 'rests' on a given part of a pavement, and this reduced exposure can reduce fatigue and rutting of the viscoelastic materials in a flexible pavement. The speed effect – which is more noticeable when the pavement base comprises bituminous-bound materials, in comparison with cement-bound ones – suggests that, in concept, greater thicknesses and/or higher-quality materials should be considered for pavements in urban areas and on uphill gradients. Overall, however, the effects are relatively minor, and flexible pavement fatigue remains fairly constant with speed in most cases, albeit rutting decreases as speed increases.

10.3.6 Temperature conditions

Ambient temperature conditions should be taken into account when designing a flexible pavement, especially when high axle loads and high air temperatures are involved. In general, the performance of bituminous-bound pavements deteriorates with rising temperature, and the resistance to permanent deformation/rutting drops rapidly. Generally, therefore, harder bituminous binders and lower binder contents tend to be used in pavements in hot climates; conversely, softer binders and higher contents are used in colder climates.

The characterisation of pavement layers, and the subgrade, for mechanistic analysis purposes is complicated by the fact that the ability of most flexible pavement materials to sustain and distribute load is greatly influenced by temperature conditions: for example, the seminal data in Table 10.2 show how the modulus of a bituminous material is highly dependent on temperature and the loading duration. Temperature-induced modulus variations also affect the ability of a bituminous material to resist cracking when subjected to repetitions of tensile stress or strain. The structural capacities of underlying unbound granular layers and of the subgrade are also affected by temperature-induced variations in the load-spreading capability of bituminous layers.

10.3.7 Moisture control

A certain amount of water must be present in subgrades, and in pavement courses containing soil, in order to lubricate the soil particles and ensure that the design densities are achieved when compaction is applied. Additional, unplanned, water should always be prevented from entering a pavement because of its deleterious impacts upon,

Table 10.2 Modulus values determined at different temperatures and loading frequencies for a bituminous surfacing used in the AASHTO road test (Finn *et al.*, 1977)

Temperature: °F (°C)	Dynamic modulus (psi × 10^5) for loading frequencies of		
	1 Hz	4 Hz	16 Hz
40 (4.4)	9.98	12.3	14.0
70 (21.1)	2.32	3.53	5.30
100 (37.8)	0.62	0.89	1.29

especially, unbound courses and cohesive subgrades and the consequent reduction in the structural conditions upon which the design of the pavement is based. It is essential, therefore, that appropriate drainage measures always be put in place to minimise the likelihood of surface water being the cause of surface damage to a pavement or verge, or finding its way into the pavement foundation.

The equilibrium moisture content of the soil, which is commonly the moisture amount used in the UK when determining the strength of a subgrade, should not be exceeded while the formation is being established and a sub-base and capping is being constructed, or during the service life of the pavement.

During construction, any aggregate to be used in the pavement should ideally be put in place before rain can enter and soften the foundation. If excess moisture is already in the foundation, a subsurface escape route should be provided for this water, and this drainage should normally be kept separate from the surface drainage. When the water table is high and the subgrade is moisture sensitive (e.g. if the plasticity index is >25), the installation of a granular, preferably aggregate, drainage blanket should be beneficial.

10.4.   Some factors affecting the design of rigid pavements

The major factor influencing fatigue cracking in concrete slabs and in lean concrete bases and sub-bases is the maximum tensile stress developed at the bottom of the cemented layer. Also, cracking of concrete slabs is encouraged by the enhanced stresses generated at the corners and edges of the slab, and by temperature stresses.

In practice, four types of cracking can be described in relation to concrete pavements: (a) hair cracks – usual slab features that are most clearly seen when concrete is drying out after rain; (b) fine cracks that are up to 0.5 mm wide at the slab surface; (c) medium cracks that are >0.5 mm wide and usually accompanied by a loss of aggregate interlock across the fracture; and (d) wide cracks that exceed 1.5 mm and are associated with a complete loss of aggregate interlock. Cracks allow water to enter sub-bases and subgrades, and lead to deformation and loss of structural support and, in cold climates, to an accelerated rate of crack formation associated with ice formation. Debris enters cracks and causes stress concentrations when the pavement flexes: this encourages spalling in the vicinity of those cracks, and this, in turn, affects riding quality. If a crack opens to the extent that the ability to transfer load is lost, then 'faulting' will occur; that is, the slab on one side of the crack will be displaced relative to the other, and a 'step' will be caused in the carriageway.

URC and JRC slabs are provided with transverse and longitudinal joints to control any cracking resulting from the contraction–expansion stresses induced by temperature and moisture content changes. A separation membrane is normally placed between the bottom of the concrete slab and the top of the sub-base in URC and JRC pavements. With these pavements, the membrane is usually an impermeable polythene sheeting whose functions are to (a) reduce the friction between the slab and the sub-base and allow free movement of the slab to occur when caused by temperature and/or moisture changes in the concrete, (b) prevent the underlying sub-base material from being admixed upward into the bottom of the freshly poured concrete, and/or (c) prevent the downward loss of moisture and fine material from the concrete mix into a porous sub-base (or subgrade, if there is no sub-base).

Individual bays of an URC pavement are usually assumed to have failed if any one of the following is present: (a) a medium or wide crack crossing the bay longitudinally or transversely; (b) a medium longitudinal and medium transverse crack intersecting, with both starting from an edge and being longer than 200 mm; or (c) wide corner cracking of more than 200 mm radius that is centred on the corner. UK experience, on average, is that up to 30% of URC bays can be expected to fail by the end of the design life of a pavement.

In the UK also, the end of the serviceable life of a JRC pavement is judged to be nigh when the rate of cracking begins to increase rapidly. Individual bays of JRC pavements are deemed to have failed when the length of wide cracking per bay exceeds one lane width. On average, up to 50% of JRC bays can be expected to have failed by the time that the design traffic load is achieved, albeit individual slabs are replaced in the interim.

A CRCP with a 30 mm (minimum) asphalt overlay and a composite pavement composed of a CRCB and a binder-plus-surface course asphalt overlay of 100 mm (minimum) are the only types of concrete pavement normally considered suitable by the Highways Agency for use where large or significant differential movement or settlement is expected, because they can withstand large strains while remaining substantially intact (HA *et al.*, 2006).

With CRCP and CRCB slabs, the need is to ensure a high level of friction between the slab and the sub-base and to encourage (i.e. control) the formation of a desirable crack pattern. While these slabs are normally provided with longitudinal joints, they have no transverse joints (except for construction joints). Because of the lack of transverse joints, the slabs develop a fine transverse shrinkage crack pattern soon after the concrete is laid; typically, this crack spacing is about 3 or 4 m, but it tends to increase as the concrete strength increases – and, hence, the amount of longitudinal crack control steel must be increased with strength, so as to maintain the appropriate crack spacing. The continuous longitudinal bars in the slabs hold the cracks tightly closed, thereby minimising corrosion of the steel and ensuring load transfer by aggregate interlock. Transverse bars, which are usually incorporated into the support arrangement for the steel, are also provided for ease and consistency of construction and to prevent longitudinal cracking and local deterioration. As the long central lengths of continuously reinforced slabs are fully restrained, only the ends have to be anchored to cater for thermal movements. A bituminous spray may, however, be applied to the top of the sub-base prior to concreting, to prevent water movement from the fresh concrete mix.

Assuming they are properly designed, CRCP and CRCR rigid pavements should not normally be expected to need structural strengthening before the end of their design lives. Nonetheless structural strengthening of both pavement types is usually required when (a) most of the cracks are wide, the reinforcement is showing signs of corrosion, and the sub-base and subgrade is affected by water penetration, or (b) settlement has resulted in a profile that seriously affects surface water drainage.

10.4.1 Temperature conditions

The thermal gradient in a concrete slab is important because of the thermal stresses that can be created in the slab. The need to insulate a cement-bound layer from thermal effects, and prevent subsequent reflection cracking, can influence the thickness of the overlying bituminous surfacing in a composite rigid pavement.

If a concrete slab in a rigid pavement is uniformly maintained at a constant temperature, it will rest flat on its supporting layer. If, however, the slab is at rest during a hot day and then the temperature drops significantly in the evening, the initial reaction of the top of the slab will be to try to contract. As the thermal conductivity of the concrete is fairly low, the bottom will stay at its initial temperature, and the slab corners and edges will try to curl upward. If the temperature conditions are reversed, the tendency of the slab will be to warp downward. In either case, temperature stresses will be induced that can result in the slab cracking, usually near its centre. Hence, warping stresses need to be taken into account in the slab design, especially if large diurnal temperature variations are anticipated.

As the temperature of a rigid pavement increases (or decreases) over the course of, say, a year, each end of the slab will try to move away from (or toward) its centre. If the slab is restrained, the friction generated between its underside and its supporting foundation will generate compressive or tensile stresses in the slab, depending upon whether it is increasing or decreasing in temperature; hence (and excepting continuously reinforced slabs), a plastic separation layer is normally put between the slab and its supporting layer to reduce this friction. The greater the long-term temperature difference between when the slab is laid and that which it experiences during the various seasons of a given year, the more important it is that this temperature-friction stress be taken into account in the design of URC and JRC slabs.

10.4.2 Joints in concrete pavements

Joints are deliberate planes of weakness in slabs that are created to control the stresses that result from (a) expansion and/or contraction volume changes associated with variations in temperature and, to a lesser extent, moisture content, and (b) warping induced by temperature and moisture content differentials between the top and bottom of a concrete slab. Figure 10.8 illustrates the main types of joints used in concrete pavements, namely expansion, contraction, warping and construction joints.

Expansion joints (Figure 10.8(a)) are transverse joints that provide spaces into which the thermal expansion of adjacent slabs can occur when the concrete temperature rises above that at which the slabs were laid. If construction takes place in hot weather, expansion joints may not be necessary; however, their use is assumed in the UK when specifying the modulus for URC and JRC slabs that are built in winter. Expansion joints are vertical, are kept open with firm compressible fillers and are sealed at the top.

Figure 10.8 Typical joints used in concrete pavements (not to scale): (a) expansion, (b) contraction, (c) warping and (d) construction joints

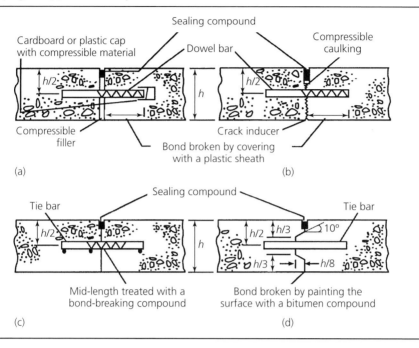

(a)   (b)   (c)   (d)

Smooth steel 'dowel' bars provide for load transfer across expansion joints, and keep adjacent surfaces at the same level during slab movement. The dowels are located at the mid-slab depth and parallel to the centreline of the road; they are placed across the joint openings so that one-half of the length of each is fixed in one slab, while the other half is provided with (usually) a protective plastic sleeve, to break the bond with the concrete so it can slide within the adjacent slab. The sliding end of each dowel normally terminates in a tight-fitting waterproof cap containing an expansion space equal to the width of the joint gap. There is ample evidence to indicate that distress at expansion joints is mainly associated with dowels that do not function properly.

Contraction joints (Figure 10.8(b)) are the most-used transverse joint in jointed concrete pavements. As concrete is weak in tension, the function of a contraction joint is to enable a slab to shorten under controlled conditions when its temperature falls below that at which it was laid, and to expand again by up to the same amount when the temperature rises. As with expansion joints, contraction joints are only required at a right angle to the axis of the pavement, and it is not normally necessary to allow for transverse contraction and expansion.

A common type of contraction joint involves placing a crack inducer at the bottom of a slab and a surface groove at the top, to narrow the thickness of the slab and deliberately create a vertical plane of weakness. As the slab thickness at that point is reduced by some

25–35%, controlled cracking results from the tensile stresses focused there. If the contraction joints are closely spaced (say, less than 4.5 m apart) the crack opening may be small enough for the interlocking aggregate particles at the faces to provide for load transference without the need for load-transfer bars. More commonly, however, contraction joints are provided with dowel bars similar to those used with expansion joints, except that a receiving cap is not provided; that is, each bar is covered with the plastic sheath for two-thirds of its length so that, when a slab contracts, the free end of the dowel leaves a gap into which it can return when the slab again expands to its original length.

Warping joints (Figure 10.8(c)) – also known as hinge joints – are simply breaks in the continuity of the concrete in which any widening is restricted by tie bars, but which allow a small amount of angular movement to occur between adjacent slabs.

Transverse warping joints are used only in URC pavements, to control cracks that result from the development of excessively high, longitudinal, warping stresses (e.g. in long, narrow slabs). As the longitudinal movements in these plain slabs are usually fairly small, it is not uncommon to allow up to three consecutive contraction joints to be replaced by warping joints with tie bars.

In reinforced slabs, the warping stresses – those associated with the development of temperature gradients in the slabs – are controlled by the reinforcing, and load transfer is maintained by aggregate interlock.

Longitudinal warping joints control the irregular cracking that occurs as a result of thermal warping and loading stresses within wide concrete pavements. This cracking, which should never be allowed to form in the wheel tracks of vehicles, is often induced so as to coincide with the intended lane lines on the carriageway.

The tie bars used in warping joints are intended to keep the adjacent slab surfaces at the same level, and to hold the joints tightly closed so that load transference is obtained via face-to-face aggregate interlock. Unlike the smooth dowels used with contraction joints, tie bars are firmly anchored at either end. They are normally located at the mid-depth of the slab; however, because they do not have to allow for movement, great accuracy in their placement is not essential.

Construction joints are those other than expansion, contraction or warping joints that are formed when construction work is unexpectedly interrupted (e.g. by machine breakdown or bad weather) at points where joints are not normally required by the design. Good construction planning ensures that normal end-of-day joints coincide with predetermined expansion or, preferably, contraction joint positions (i.e. the structural integrity of the pavement is better maintained if they are at contraction joints). When transverse construction joints coincide with contraction joints in non-dowelled pavements, the joints should be keyed, whereas they are dowelled in dowelled pavements. Transverse construction joints that are located between contraction joints should be keyed, and tie bars provided (Figure 10.8(d)).

When the full width of a pavement is not laid in one concreting operation, a formal longitudinal construction joint has to be established between the two abutting concrete slabs. These slabs are tied with tie bars.

Irrespective of the type of joint, the following should apply to all joints, to enable them to fulfil their functions:

■ Long-life sealing materials should be applied at the time of construction to make the joints waterproof and able to withstand repeated contraction–expansion of the concrete so that foreign materials (e.g. grit) is unable to enter and hinder the free movement of adjacent slabs in hot weather.
■ Joints should not cause uncomfortable riding conditions that are structurally undesirable and/or detract from the riding quality of the road, such as those due to excessive relative deflections of adjacent slabs or to repeated impacts by vehicles driven over numerous transverse ridges of sealant that project above the carriageway.
■ Joints should not be the cause of unexpected, non-designed, structural weaknesses in a pavement: for example, transverse joints on either side of a longitudinal joint should not be staggered from each other, as transverse cracking will be induced in line with the staggered joints, and no joint should unintentionally be constructed at an angle of less than 90° to an adjacent joint or edge of a slab.
■ Joints should interfere as little as possible with the continuous placing of concrete during construction.

Filler boards provide the gaps for expansion joints at the time of construction, and support for the joint sealing compound used over the design life of the pavement. The filler board material should be capable of being compressed without extrusion, be sufficiently elastic to recover its original thickness when the compressive force is released, and not affect/be affected by the covering sealant. Holes have to be accurately bored or punched out of the filler boards in order to provide a sliding fit for sheathed dowels.

Joint seals (also called sealants) should ideally (a) be impermeable, (b) be able to deform to accommodate the rate and amount of movement occurring at the joint, (c) be able to recover their properties and shape after cyclical deformations, (d) be able to bond to the faces of joints and neither fail in adhesion nor peel at areas of stress concentration, (e) not rupture internally (i.e. not fail in cohesion), (f) be able to resist flow or unacceptable softening at high service temperatures, (g) not harden or become unacceptably brittle at low service temperatures and (h) not be adversely affected by ageing, weathering or other service factors, for a reasonable service life under the range of temperatures and other environmental conditions that can be expected. As no sealant material is able to meet all these requirements, successful joints are those that get regular sealant maintenance (Hodgkinson, 1982).

Transverse joint spacing is a reflection of the capacity of a slab to distribute strain, rather than allow damaging strain concentrations. The joint spacings in plain (URC) slabs normally depend upon concrete strength, the coefficient of thermal expansion of the aggregate, the climatic conditions during construction, and the in-service environmental regime. Use of limestone aggregate, which has a lower coefficient of thermal expansion than other types of aggregate, results in less slab expansion–contraction and, consequently, a greater joint spacing can be used with concrete slabs containing this type of aggregate.

The effectiveness of reinforcement as a distributor of strain increases with the amount of steel used. Hence, greater joint spacings can also be used with larger amounts of reinforcing steel in JRC slabs.

10.4.3 Reinforcement in concrete pavements

The main function of the reinforcing steel in a concrete slab in a rigid pavement is not to contribute to the flexural strength of the concrete but to control the amount and scale of cracking. While concrete is strong in compression, it is weak in tension; hence, repeated stressing from commercial vehicle and temperature loading will eventually lead to crack initiation. By resisting the forces that pull the cracks apart, the steel reinforcement ensures that the interlocking faces of the cracks remain in close contact, thereby maintaining the structural integrity of the slab. The tight contact also makes it less easy for water on the carriageway to find its way through the cracks into the sub-base and subgrade, while the entry of foreign material and (in cold climes) freezing water, both of which promote crack widening, is also minimised.

Typically, a JRC slab in a road pavement is rarely wider than 4.65 m, and its length exceeds its width. With this configuration, the reinforcement is normally only required to control transverse cracking in the slab, as longitudinal cracking seldom occurs. Reinforcement is therefore placed in the longitudinal direction, and transverse steel is normally only provided for ease of construction (e.g. to give rigidity to mesh fabrics or to support and space deformed bars), except when there is a risk of differential settlement or where it is required to act as tie bars at joints in JRC pavements. If the 4.65 m width is exceeded (e.g. when constructing a three-lane carriageway in two equal widths), extra transverse steel may be needed to control longitudinal cracking.

As the reinforcing steel is not intended to resist induced flexural stresses, its exact location within a jointed reinforced slab is not critical, provided that it is reasonably close to the upper surface of the slab (because of its crack-control function), and it is well bonded and protected from corrosion induced by salt penetrating through cracks. As it is convenient for construction, UK practice is to place the steel 60–70 mm below the surface in single-layer slabs; with two-layer construction (e.g. when using two different aggregates and/or an air-entrained upper layer), the steel is laid on top of the lower concrete slab.

With both CRCP and CRCR slabs the steel is normally placed at the mid-depth of the slab to minimise the risk of corrosion from any water that may percolate down through the cracks.

10.5. Design life and service life

In practice, a new pavement is designed to last for a selected number of years (i.e. for its design life). At the end of this period it is assumed that the pavement will have deteriorated to its terminal or failure condition, and, in concept, it will then have to be rehabilitated or reconstructed to restore its structural integrity and serviceability.

In countries with mature road networks, new road construction typically accounts for around 50% of the road budget, and much of the remainder is spent on maintenance and rehabilitation of existing roads (OECD, 2005). The road construction methods and materials in use in many countries contribute to this distribution, as they lead to recurrent maintenance needs that can only be met at a relatively high cost. To pressured road administrators operating within limited budgets, the maintenance work that is likely to be incurred in future years often seems preferable to increasing the capital expenditure now on proposed new roads. As a consequence, it is not uncommon for the initial construction costs of some road pavements to be surpassed by the costs of its design life maintenance.

Apart from the direct costs of maintenance funded from road budgets, pavement maintenance also imposes significant costs on road users. On highly trafficked roads, in particular, maintenance causes traffic congestion and disruption to normal traffic flows, and, despite the measures taken to minimise them, the costs to users of busy roads are high and increasing. Hence, there are growing pressures for long-life pavements at higher construction costs that require minimal maintenance and therefore avoid many future costs to road administrations and road users.

In order to optimise road budgets, whole-life costing methods are now increasingly used to determine how, where and when to best spend budget funding on road construction and maintenance. Within this framework, the shift to full maintenance contracting has helped to reduce costs, and the adoption of long-term contracts has helped to establish an environment in which the development of more durable pavement types is stimulated.

The current practice in the UK is to use a standard design life of 40 years for trunk roads (including motorways) where the design traffic is heavy in relation to the capacity of the layout, and in all cases where whole-life value is taken into account (HA *et al.*, 2006). Twenty-year designs may be used for less heavily trafficked roads or for major maintenance where other site constraints apply. In addition to major maintenance, it can also be expected that surface treatments may be needed at about 10 and 20 year intervals, dependent upon the nature of the traffic; when a surface treatment is required will also depend on the need by a site for skidding resistance.

The service life of a pavement is of indefinite duration. In concept, it may equal the initial design life or, more likely, it may comprise a succession of design lives, for each of which the pavement is provided with the structural capacity needed to satisfactorily carry a specified traffic loading. For example, at the end of the second 20 years of the design life of a flexible pavement, a new 'investigatory' condition may be reached, at which time a further major maintenance operation may be carried out to extend its design life for another 20 years, and so on.

10.6. Some remarks on the approaches to flexible pavement design

As noted previously, there are a number of approaches to flexible pavement design. The design process varies considerably from country to country, and is still somewhat fluid and evolving. Some methods rely on assessments of subgrade strength that are based on CBR or subgrade stiffness measurements and on empirical relationships between the pavement material thickness and the traffic loading. Other methods use an analytical approach, and rely more heavily on fundamental theoretical relationships of material behaviour and performance. In practice, most methods are now evolving to use elements of both empirical and analytical design.

Today, these evolving methods normally use some form of penetration or bearing test to determine the subgrade strength, and then rely on empirically derived test road results that are supplemented by structural analysis to rationalise and extend the data, to determine the thickness design for different pavement layer materials. Two methods that are of particular interest are those used in the UK (HA, 2009; HA *et al.*, 2006) and in the USA (AASHTO, 1993, 2008), because of the influence that they have had/are having on flexible pavement design methodology in many countries.

With the method that is now used in the UK for major road pavement design, the designer must first select a 'foundation class' that relies on an assessment of the stiffness of the subgrade and on the thickness and moduli of one or more granular or hydraulically bound sub-base layers. The upper pavement is then designed, based on relationships (presented in the form of design charts) that require knowledge of the design traffic load, preferred base course materials and the selected foundation class. This method is discussed in detail in Chapter 12, and is therefore not further described in this chapter.

The American Association of State Highway and Transportation Officials design method (AASHTO, 1993) is an empirically derived procedure that is primarily based on results from the AASHO road test (AASHO, 1962). It has the key advantage of simplicity and transparency, and is briefly described here as it has introduced new concepts to, and influenced pavement design in, many countries outside the USA.

The general analytical approach to pavement design is also discussed here, to put it in perspective relative to other design methods.

10.7. The AASHTO approach to flexible pavement design

The AASHTO 1993 method (AASHTO, 1993), although now officially superseded in the USA by the AASHTO *Mechanistic–Empirical Pavement Design Guide* (MEPDG) method (AASHTO, 2008), is still in common use in the USA and around the world (Thom, 2014). It is essentially an empirical method that relies heavily on a measure of pavement strength that is expressed as a structural number (SN) and represents the pavement's structural ability to carry a given traffic loading for an assumed subgrade resilient modulus. This design method consists of selecting a combination of layers and thicknesses so that the sum of the product of thickness, structural coefficient and moisture coefficient for each layer adds up to the required design SN.

The objectives underlying the controlled road test investigations by the AASHO (i.e. by the predecessors of AASHTO – see Section 10.1) was to determine relationships between the three main elements of pavement design: loading, structural capacity and performance. Major outputs included design equations for flexible (and concrete) pavements as well as load and structural equivalence factors.

Prior to the actual AASHO road test, the important concept (Carey and Irick, 1962) of a present serviceability index (PSI) was developed as a means of measuring, in numerical terms, the performance and condition of a pavement at a given time. Panels of experienced drivers were used to subjectively assess the performance of sections of flexible (and concrete) pavements on the public road network that were in various states of serviceability, with each assessed section being rated on a scale of 0–5, with 0 and 5 representing very poor and very good ratings, respectively. Using multiple linear regression analysis, the subjective performance ratings were then combined with objective measurements of the pavement condition on the same sections, to develop serviceability equations for both types of pavement. The equation developed for flexible pavements was

$$\text{PSI} = 5.03 - 1.91 \log(1 + \text{SV}) - 0.01(C + P)0.5 - 1.38\text{RD}^2 \tag{10.1}$$

where SV is the average longitudinal slope variance (i.e. longitudinal roughness) in the two wheel tracks; C is the cracked area ($\text{ft}^2/1000 \text{ ft}^2$ of pavement); P is the patched area ($\text{ft}^2/1000 \text{ ft}^2$ of pavement); and RD is the average rut depth (in) for both wheel tracks measured under a 4 ft straight edge.

At the time that the pavement serviceability concept was subsequently applied to the flexible pavements in the AASHO road test, the average initial PSI was 4.2. The test track pavements were then subjected to continued traffic flows, and sections were considered 'terminal' and taken out of service when the PSI reduced to 1.5: this represented an average PSI loss of 2.7 under traffic. PSI values of 2.5 and 2.0 are now commonly specified as terminal conditions for high-volume and low-volume roads, respectively, with this design method.

Thickness design equations were developed from the AASHO road test data that related the structural number of a pavement (i.e. an abstract thickness index reflecting the relative strength contribution of all layers in the pavement) to the number of axle load applications, the PSI loss, the resilient modulus of the subgrade, and the drainage characteristics of unbound base and sub-base layers. The initial design equations were subsequently modified, and the following is now used in the design of flexible pavements:

$$\log W_{18} = Z_R S_0 + 9.36 \log(\text{SN} + 1) - 0.20$$
$$+ \log[\#\text{PSI}/(4.2 - 1.5)]/[0.4 + (1094/\text{SN} + 1)5.19]$$
$$+ 2.32 \log M_R - 8.07 \tag{10.2}$$

where W_{18} is the design equivalent standard axles loads; Z_R is the overall standard deviation; S_0 is the combined standard errors of the traffic and performance predictions; #PSI is the design serviceability loss (i.e. the difference between the initial and terminal PSI values); M_R is the effective subgrade resilient modulus (psi); and SN is a structural number ($= a_1D_1 + a_2D_2m_2 + a_3D_3m_3$, where D_1, D_2 and D_3 are the thickness (in) of the surfacing, base and sub-base layers, respectively, a_1, a_2 and a_3 are the corresponding layer structural coefficients, and m_2 and m_3 are the drainage coefficients for the unbound base and sub-base layers).

A range of structural coefficients was subsequently developed for use with differing materials in the various pavement layers. Drainage coefficients recommended for the unbound layers are based on the quality of the drainage (rated from 'excellent' to 'very poor') and the percentage of time that the pavement is exposed to moisture contents approaching saturation (ranging from <1% to >25%). Values assigned to Z_R and S_0 vary with the desired level of reliability: for example, they range from 0.40 to 0.50 for Z_R (with a low value being selected when traffic predictions are reliable), while S_0 values vary according to the functional classification of the road (typically 80–99.9 and 85–99.9 for rural and urban motorways, respectively, and 50–80 for both rural and urban local roads). Soil support is characterised in terms of the resilient modulus, M_R, and relationships are provided that allow the layer structural coefficients, a_1, a_2 and a_3, to be determined from modulus test data. The M_R value is representative of the monthly variations in modulus over a 1 year period, and the effect that these have on damage accumulation; hence, the term 'effective' subgrade modulus.

Various combinations of layer thickness are used to derive the overall structural number, SN, that is determined from Equation 10.2, and a procedure is described in the AASHTO guide for selecting the layer thicknesses that will provide a structurally balanced pavement (AASHTO, 1993). The steps involved in this procedure are as follows: (a) determine the M_R value of the subgrade; (b) select the #PSI; (c) estimate the equivalent standard axle loads for the design life; (d) select appropriate Z_R and S_0 values; (e) determine the SN required for the pavement; and (f) select the pavement material type and vary the thickness of individual layers until an SN value is calculated that is equal to or greater than the required SN.

The basic principle inherent in the AASHTO 1993 design process is that the quality of the material in each course, and the thickness of each course, should be sufficient to prevent overstressing (and early failure) of the underlying layer.

The AASHTO 1993 method of pavement design, albeit that it is still used in practice, has now been technically superseded in the USA by the MEPDG method (AASHTO, 2008). This new design approach is applicable to the evaluation of all types of pavement structures, whether they be new, reconstructed, or rehabilitated flexible, rigid or semi-rigid pavements.

The MEPDG approach is a very complex software-oriented design method that is based on mechanistic–empirical principles and uses project-specific inputs such as traffic loadings, climatic data and pavement performance and material distress prediction models, to aid the designer in optimising the desired design pavement cross-section. In so doing, the design method outputs predictions of many forms of pavement distress (e.g. rutting, cracking, surface ravelling, loss of evenness and faulting at joints) to estimate damage accumulation over the design life of each trial pavement cross-section that is specified for examination. All performance-indicator prediction models used in the MEPDG method were calibrated to observed field performance from a representative sample of pavement test sites located throughout North America: because of the consistency in the calibration results derived over time and the diversity of the test sections considered, these models are described as being 'globally calibrated'.

Most commonly, pavement designers in the USA use the *AASHTO Guide for Design of Pavement Structures* (AASHTO, 1993) to determine the initial trial design cross-section for evaluation when using the more sophisticated MEPDG.

10.8. The analytical approach to flexible pavement design

Presented here is a general overview of the analytical design process as generally applied to flexible pavements.

The approach used in the analytical (or mechanistic) pavement design process is to treat the pavement as one would any civil engineering structure, that is, the engineer assesses the stress and strain distributions in each of the pavement layers under the design traffic loading, and compares these stresses and strains against some limit state or ultimate failure values of stress or strain, to determine the adequacies of the various pavement materials and thicknesses used. In order to do this, it is necessary for the pavement designer to make many assumptions and simplifications about the loading and the properties of the materials used in the pavement.

Commercial vehicle traffic (or axle) loadings vary significantly, so a standard axle load is generally used in this 'structural analysis' of the pavement. In the UK, it is generally assumed that the standard axle loading is 80 kN, and that half of this loading is transferred to the pavement via a single standard wheel load that is assumed to be in contact with the carriageway over an area represented by a circular patch (radius = 0.151 m) with uniform stress (Nunn, 2004). As noted previously, this is a simplification of the actual loading on the pavement; that is, the tyre contact area on the pavement is actually non-circular and the tyre loading can be via a single wheel or a dual wheel or a 'super' single tyre with a higher contact pressure. In addition, the loading is not static and is not applied vertically: in practice, the pavement loads are applied via rolling wheels that exert a dynamic stress distribution to the pavement materials and, most significantly, cause a rotation of principal stresses in the pavement materials that impact on the way that these materials respond and subsequently deform or fail.

The locations of the critical stresses and strains developed in the pavement are shown in Figure 10.9 for both a composite and a flexible pavement. The analytical design process is predicated on the assumptions that the magnitude of the

■ horizontal tensile strain at the bottom of a bitumen-bound base is inversely related to the fatigue life of the pavement

Figure 10.9 Locations of critical stresses and strains for composite and flexible pavements

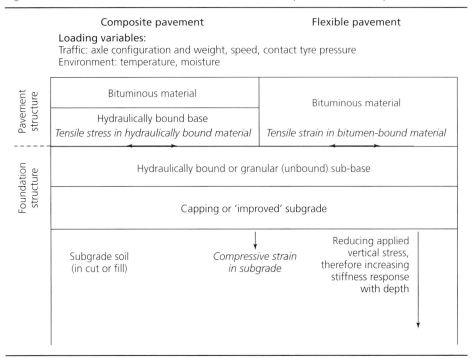

- horizontal tensile stress at the bottom of a hydraulically bound base is inversely related to the design life of the material
- vertical compressive strain at the top of the subgrade controls the deformation in the pavement.

These critical stresses and strains are calculated under a standard wheel load using multi-layer linear elastic analysis, assuming values of thickness, stiffness and Poisson's ratio for the various pavement layers. The analytical design process then relates these critical stresses and strains to performance models for the various pavement materials to estimate the number of repetitions of standard axles that are likely to cause failure in the pavement. This analytical design process is summarised in the flow chart in Figure 10.10.

10.8.1 The application of structural analysis

The first analytical model developed to solve for stresses in a semi-infinite single-layer homogeneous elastic half-space was developed by the French elastician Boussinesq (1885). However, it was not until some 60 years later that the first solution for a multi-layered elastic system was subsequently developed by Burmister (1943, 1945), who pioneered a solution for the deformation of specific two and three-layered systems. A useful, simplified, 'method of equivalent thickness' that extended the use of these equations to a multi-layered pavement structure was developed initially by Oedmark (1949) and subsequently elaborated by Ullidtz (1987). Ullidtz's method assumes that all layers in a multi-layer model can be transformed into a one-layered system with an

Figure 10.10 The basic analytical design process

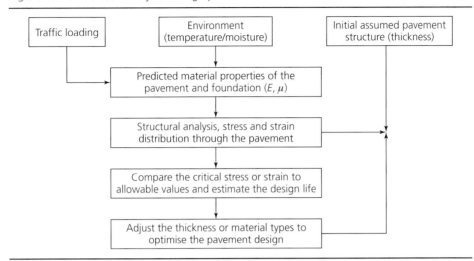

equivalent stiffness based on the stiffness ratios of the various layers, and that the stresses and strains in this 'equivalent layer' can then be estimated using Boussinesq's equations.

Today, computer programs are available (e.g. BISAR, CIRCLY, CHEVRON, ELSYM5 and WESLEA) that utilise multi-layer linear elastic analysis to calculate pavement responses (i.e. stress and strain distributions) under traffic loads. Such multi-layer linear elastic analyses assume that

- pavement layers are composed of homogeneous linear elastic materials
- the pavement extends infinitely horizontally and has infinite thickness or depth (commonly known as a semi-infinite elastic half-space)
- a load of uniform pressure is applied over a circular area
- the bonds between layers are perfect.

Although multi-layer linear elastic analysis provides a rigorous solution given the above assumptions, in practice many of these assumptions are not always defensible, so that outputs have to be treated with caution. As noted previously, traffic loading is dynamic and the tyre stress is applied via a rolling wheel and not as a simple normal stress – which means that the principal stresses in the pavement are actually rotating under the actions of traffic. The pavement layers do not extend infinitely in either the horizontal or vertical directions. Further, bituminous materials are viscoelastic at elevated temperatures while subgrade soils and granular materials exhibit non-linear, stress-dependent behaviour. Therefore, the 'linear elastic' characteristics chosen for each layer are essentially equivalent best estimates of the layer stiffness at the time of design, when used with analytical design methods.

Although not in standard use, finite-element analysis (FEA) methodology now allows the road pavement research community to introduce many more levels of complexity.

Constitutive models for materials can be non-linear, and the geometry of the pavement and dynamic loading can be more realistically defined. FEA also allows for slippage between layers if required, and viscoelasticity can be introduced. However, this is the domain of the research community at present, as the stress–strain distribution calculated using FEA will often give differing values to those calculated using multi-layer linear elastic analysis based on Boussinesq solutions.

10.8.2   Pavement material characteristics and deterioration models

The analytical design process requires the stresses and strains in the pavement materials to be calculated, using linear elastic theory, under a standard wheel load. In order to do this, it is assumed that the pavement materials behave in a linear elastic way and have properties of stiffness, E, and Poisson's ratio, μ. This is a gross simplification, as no materials used in flexible pavement construction are truly linear elastic, and, as they age (e.g. harden, crack, and vary in moisture content and density), their characteristics change. Bituminous materials exhibit viscoelastic characteristics, and their 'elastic' response is highly dependent on temperature and strain rate. The stiffness of unbound materials is non-linear elastic and highly dependent on confining stress. Subgrade soils are also non-linear elastic (i.e. their 'elastic' response is dependent on the rate of strain and on the magnitude of the applied deviatoric stress).

In practice, the designer must assume elastic stiffness and Poisson's ratio values for the materials to input into the linear elastic numerical model used to calculate critical stresses and strains, and then compare these critical stresses and strains with allowable values for the pavement materials using material deterioration models.

The most important mechanical characteristics of bituminous materials that are required for the numerical model in the analytical design process are the stiffness, E, and Poisson' ratio, μ. Other performance properties required are the fatigue life (i.e. resistance to cracking through repetitive traffic loading), the deformation characteristics (i.e. resistance to rutting under the action of traffic) and durability.

Bituminous materials are viscoelastic, and their stiffness is a function of temperature and the rate of loading: at low temperatures and short loading times they have relatively high stiffness, and behave elastically, whereas at high temperatures and longer loading times they have relatively lower stiffness, and tend to exhibit some viscous behaviour. In addition, bituminous materials initially age harden with time, and their stiffness can increase until they crack, after which the stiffness of the whole bituminous layer starts to reduce. In the analytical design process, therefore, it is necessary to assign an appropriate stiffness value (e.g. the predominant or weighted average stiffness) over the design life of the pavement. In the UK a single design temperature of 20°C and a traffic speed of 80 km/h (for loading times of 5 Hz) tend to be used to estimate the stiffness of the bituminous material, whereas more complex analytical methods (e.g. AASHTO, 2008) have requirements that vary with geographical area, day, night and seasonal temperature data, and traffic speed.

The analytical pavement design process therefore requires material performance models that relate the critical stresses and strains within the pavement structure, as estimated using the numerical model, to an allowable number of load (or standard axle) applications over the design life of the pavement. In the UK, the following performance models (Powell et al., 1984) have been developed for use in flexible pavement design for typical bituminous materials:

dense bitumen macadam: $\quad \log N_t = -9.38 - 4.16 \log_{10} \varepsilon_t$ $\qquad$ (10.3)

hot rolled asphalt: $\quad \log N_t = -9.38 - 4.16 \log_{10} \varepsilon_t$ $\qquad$ (10.4)

where N_t is the number of standard axles that causes fatigue cracking in the bituminous layer and ε_t is the tensile strain at the base of the bituminous layer.

In the USA, the following fatigue model (Asphalt Institute, 1991) has been defined for use with bituminous materials in flexible pavements:

$$N = 0.00432 C (1/\varepsilon_t)^{3.291} (1/E)^{0.854} \qquad (10.5)$$

where N is the number of standard axles that causes fatigue cracking in the bituminous layer; C is a constant that is a function of the bituminous mix binder and void content; ε_t is the critical tensile strain in the bituminous layer; and E is the stiffness of the bituminous material.

The critical mechanical characteristics of hydraulically bound materials that are required for the linear elastic numerical model are also the stiffness, E, and Poisson's ratio, μ. Other performance characteristics required are the flexural strength, thermal characteristics and durability. The following example of a performance model (Nunn, 2004) has been developed to control the fatigue cracking of hydraulically bound materials in composite flexible pavements in the UK:

$$\log N = 1.23[(\sigma_f/\sigma)K_{Hyd}K_{Safety} + 0.1626]^2 + 0.2675 \qquad (10.6)$$

where N is the number of load applications necessary to cause fatigue cracking; σ is the tensile stress at the bottom of the hydraulic base due to loading (MPa); σ_f is the flexural strength of the concrete material (MPa); K_{Hyd} is the material calibration factor, which includes temperature, curing and cracking effects; and K_{Safety} is the pavement design risk factor.

The mechanical characteristics required by the linear elastic numerical model for unbound (i.e. granular) materials are again the stiffness, E, and Poisson's ratio, μ. However, unbound materials exhibit markedly non-linear elastic behaviour, and the measured stiffness is primarily a function of the level of confining stress applied, as described in a general form in Equation 10.7 (Hicks and Monismith, 1971)

$$E = k_1 \theta^{k2} \qquad (10.7)$$

where E is the stiffness (or resilient) modulus; K_1 and K_2 are material constants commonly derived for experimental data; and θ is the confining stress applied to the unbound material. Other physical factors affecting the stiffness of granular materials are the density, grading, particle shape and texture.

Figure 10.11 Illustration of the effect of deviator stress on the resilient modulus of a cohesive subgrade soil (Seed *et al.*, 1962)

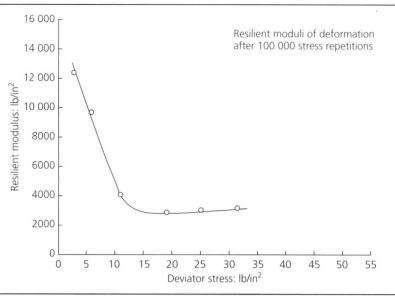

For the purposes of analytical pavement design, it is generally sufficient to assume a constant modulus for an unbound sub-base (i.e. stiffness values of between 100 and 150 MPa and a Poisson's ratio of between 0.3 and 0.35 are commonly used).

Subgrade soils also behave in a markedly non-linear way, and their measured stiffness is most significantly affected by the magnitude of the applied deviator stress, as well as by the density and moisture content of the material being evaluated. Thus, for example, Figure 10.11 shows how the measured modulus of a cohesive subgrade increases markedly with decreasing deviator stress: similarly, in relation to the behaviour of a subgrade soil below a pavement, the relative stiffness of the subgrade increases with depth as the vertical (i.e. deviator) stress under a wheel load decreases with depth.

The subgrade modulus can be determined from measurements made in situ or from laboratory testing. There are also empirical relationships for stiffness that are based on CBR test results, and these may be used (with caution) to assign an initial stiffness to the subgrade in the analytical pavement design process. (Note: the CBR test is essentially a strength test, and stiffness, particularly at small strains, does not necessarily correlate well with strength.) Two such relationships are given in Equations 10.8 (AASHTO, 1993) and 10.9 (Powell *et al.*, 1984)

$$E = 10 \times \text{CBR} \tag{10.8}$$

$$E = 17.6 \times \text{CBR}^{0.64} \tag{10.9}$$

where E is the stiffness modulus (MPa) and the CBR is a percentage.

Poisson's ratio for subgrade soils ranges from 0.35 to 0.45, and guidance on which value to use with particular soils is available in pavement design guides.

10.8.3   Some final comments on the analytical design approach

The analytical design approach uses linear elastic analysis to evaluate stresses and strains in a pavement structure for various pavement materials, layer combinations, traffic loading and environmental conditions, to ensure that materials in a pavement are not overstressed over the design life of a pavement. While this appears to be a logical approach, and is in line with practice in structural engineering design, its implementation is not a trivial undertaking, and the approach should be used with caution – and by experienced pavement designers.

Analytical pavement design methods are predicated on many assumptions that can be questioned. The accurate prediction of the stress and strain distribution over the life of a pavement is very difficult, and depends on the values of stiffness and of Poisson's ratio that are chosen for the materials used in the pavement (i.e. for materials that, in practice, do not conform to simple linear elastic behaviour). The performance models for pavement materials are also difficult to verify, particularly for new materials.

These limitations are clearly recognised in the Australian pavement design guide (Austroads, 2012), which states that 'Care must be taken to ensure that the sophistication of the analysis method is compatible with the quality of the input data. If not, then so many assumptions must be made to fill the gaps that the results of the analysis can be misleading, if not worthless.'

Notwithstanding the difficulties, however, the analytical approach does represent a significant step forward in flexible pavement design, and, at this time, can be very usefully employed in research and as a tool to illustrate the comparative performance of pavements using materials of differing mechanical properties.

10.9.   Rigid pavement design considerations

Worldwide, most countries now have standard approaches to rigid pavement design that are based on accumulated engineering experience gained by the study and appraisal of existing concrete pavements, controlled and experimental road tests under normal traffic, accelerated tests on existing or specially built pavement sections, laboratory experiments, and theoretical and rational analyses. The current standard approach used in the design of rigid pavements for new trunk road design in the UK is discussed in Chapter 12, while basic theoretical/semi-theoretical methods that can be used elsewhere (e.g. in industrial estate and off-road pavements) are covered in Chapter 11.

The following is an introductory discussion of some factors that are taken into account when designing a rigid pavement.

10.9.1   Wheel load stresses

A concrete slab will fail under load when the applied load or bending moment is so great that the developed flexural stress exceeds the modulus of rupture. When a vehicle travels

over a concrete pavement, the stresses caused vary with the position of the wheels at a given time, so that, in concept, the stresses in each slab should be analysed for when the wheel loads are at all points on the slab, allowing the most severe stresses to be evaluated and used for design purposes.

Commonly used methods of analysis are based on that derived by Westergaard (1926), in which the author assumed that

- a concrete slab acts as a homogeneous, isotropic, elastic solid in equilibrium
- subgrade reactions are vertical only, and proportional to the deflections of the slab
- the subgrade reaction per unit of area at any given point is equal to the deflection at that point multiplied by a constant k, termed the 'modulus of subgrade reaction' and assumed to be constant at each point, independent of the deflection, and the same at all points within the area of consideration
- the slab thickness is uniform
- a single wheel load at the interior or at the corner of the slab is distributed uniformly over a circular area of contact and, for the corner loading, that the circumference of this circular area is tangential to the edge of the slab
- a single wheel load at the slab edge is distributed uniformly over a semi-circular contact area, with the diameter of the semi-circle being at the edge of the slab.

Westergaard initially examined three critical conditions of single wheel loading – at the corners, edges and interior of the slab – and developed the following three equations:

$$\sigma_{\mathrm{c}} = (3P/h^2)[1 - (2^{1/2}a/l)^{3/5}] \tag{10.10}$$

$$\sigma_{\mathrm{i}} = (0.31625P/h^2)[4 \log(l/b) + 1.0693] \tag{10.11}$$

$$\sigma_{\mathrm{e}} = (0.57185P/h^2)[4 \log(l/b) + 0.3593] \tag{10.12}$$

where σ_{c} is the maximum tensile stress (lbf/in^2) at the top of the slab, in a direction parallel to the bisector of the corner angle, arising from a load applied at the unprotected corner; σ_{i} is the maximum tensile stress (lbf/in^2) at the bottom of the slab directly under the load, when the load is applied at a point in the interior of the slab at a considerable distance from the edges; σ_{e} is the maximum tensile stress (lbf/in^2) at the bottom of the slab directly under the load at the edge, and in a direction parallel to the edge; P is the point load (lbf/in^2); h is the thickness of the slab (in); μ is Poisson's ratio for concrete ($= 0.15$); E is the modulus of elasticity of the concrete (lbf/in^2); k is the modulus of subgrade reaction (lbf/in^2/in); a is the radius of the load contact area (in), noting that the area is circular for corner and interior loads and semi-circular for edge loads; b is the radius of equivalent pressure distribution at the bottom of the slab (in) ($= (1.2a^2 + h^2)^{1/2} - 0.675h$); and l is the radius of relative stiffness (in) ($= [Eh^3/12k(1 - \mu^2)]^{1/4}$), which relates the modulus of subgrade reaction to the flexural stiffness of the slab.

The most critical situation illustrated by the above equations (which were subsequently modified by Westergaard and by others) relates to the corner loading where, due to local depressions of the underlying material (e.g. from mud pumping) or to warping of the slab, the slab corners can become unsupported and, in extreme circumstances, behave as cantilevers and break. Edge loading produces stresses that are slightly less than those caused by corner loading, while a load at the interior of the slab, away from edges and corners, generates the least stress.

In the case of a wheel load applied at the slab edge, at a considerable distance from any corner, the edge deflects downward immediately under the load and upward at a distance. The critical tensile stress is on the underside of the slab directly below the centre of the (semi-)circle, and the tensile stresses at the upper surface of the edge at a distance are much smaller than the tensile stress at the bottom of the slab below the centre.

The least critical situation considered by Westergaard relates to when the load is applied at the interior of the slab. Here, the critical stress is the tensile stress at the bottom of the slab under the centre of the circle, except when the circle radius is so small that some of the vertical stresses near the top become more important; this exception is not a problem, however, in the case of a wheel load applied through a rubber tyre.

The modulus of subgrade reaction, k, is a measure of the stiffness of the supporting material. While the values of k are important, they vary widely, depending upon the soil density, moisture condition and type: for example, k values for plastic clays and for gravels can range from 14 to 27 MPa/m for the former, and to over 81 MPa/m for the latter. However, it should also be noted that a fairly large change in the value of k has a relatively small influence upon the calculated stress in a concrete slab.

10.9.2 Temperature-warping stresses

Westergaard also considered the warping effects brought about by temperature variations in a pavement slab (Westergaard, 1928). Assuming that the temperature gradient from top to bottom of a slab was a straight line, the author developed equations for three different cases, the simplest of which (for an infinitely large slab) was

$$\sigma_0 = E\varepsilon t/2(1 - \mu) \qquad (10.13)$$

where σ_0 is the developed tensile stress (lb/in^2); E is the modulus of elasticity of concrete (lb/in^2), ε is the coefficient of linear thermal expansion of concrete per °F, μ is Poisson's ratio for concrete, and t is the temperature difference between the top and bottom of the slab (°F). Since then, of course, it has been shown that the temperature gradient is closer to a curved line, and this results in calculated stress values that are much lower than those originally derived by Westergaard.

While a temperature gradient of 5°C can occur between the top and bottom of a 150 mm slab in the UK, the stresses induced by temperature warping are not as detrimental as might be expected. The reasons for this are that (a) at the corners of a slab, where the load stresses are actually the greatest, the warping stresses are negligible, as the tendency of a slab to curl at these locations is resisted by only a very small amount of concrete; (b) while significant warping stresses can be developed at the interior of the slab and along its edges that, under certain circumstances, are additive to load stresses, slabs normally have a uniform thickness that is able to cope with the corner loading needs, so that the margins of strength present in the interior and edges are usually enough to offset the warping stresses at these locations; and (c) long-term studies have indicated that the temperature at the bottom of a slab exceeds that at the top more often than the reverse, so that the warping stresses in the interior and at the edges are more frequently subtractive rather than additive.

10.9.3   Temperature–friction stresses

Stresses are also generated in concrete pavements due to long-term changes in the temperature of a slab that cause each end of the slab to try to move away from or toward its centre as the pavement temperature increases or decreases. If cooling takes place uniformly, a crack may occur about the centre of the slab. If the slab expansion is excessive and adequate expansion widths are not provided between adjacent slabs, blow-ups can result in these slabs being jack-knifed into the air.

Assuming that adequate widths of joint are provided, the stresses due to such long-term temperature changes could be considered negligible if there was no friction between the slab and its supporting layer. In fact, however, much friction can be developed between them; that is, as the slab tries to expand, it is restrained by friction, and compressive stresses are produced at its underside, and, as it tries to contract, the friction restraint causes tensile stresses in the bottom of the slab.

The stresses resulting from friction restraint are only important when the slabs are quite long, say >30 m, or if they are laid (unusually in the UK) at temperatures greater than, say, 32°C. They are only critical when conditions allow them to be applied at the time that the combined loading and warping stresses from other sources are at their maximum. As the maximum tensile stress due to frictional restraint only occurs when a slab is contracting, and as the warping stresses from temperature gradients are not at their maximum at this time, the net result is that, in practice, these restraint stresses are often neglected when calculating the maximum tensile stresses in a conventional concrete slab for thickness design purposes.

10.9.4   Moisture-induced stresses

Differences in moisture content between the top and bottom of a slab also cause warping stresses. This is due to the ability of concrete to shrink when its moisture content is decreased, and to swell when it is increased.

Generally, it can be assumed that the moisture effects will oppose those of temperature: for example, in summer a slab will normally shorten rather than lengthen, and, as it dries out from the top, will warp upward rather than downward. Overall, it is likely that the effects of moisture change on slab stresses are more important in hot regions of the world with pronounced wet and dry seasons than in the UK's more equitable climate.

REFERENCES

AASHO (American Association of State Highway Officials) (1962) *The AASHO Road Test*. Highway Research Board, Washington, DC, USA, Special Report 61A-E.

AASHTO (American Association of State Highway and Transportation Officials) (1993) *AASHTO Guide for Design of Pavement Structures*. AASHTO, Washington, DC, USA.

AASHTO (2008) *A Mechanistic–Empirical Design Guide (MEPDG): A Manual of Practice*, interim edition. AASHTO, Washington, DC, USA.

Addis RR and Whitmash RA (1981) *Relative Damaging Power of Wheel Loads in Mixed Traffic*. Transport and Road Research Laboratory, Crowthorne, UK, Report LR979.

ASCE (American Society of Civil Engineers) (1950) Symposium on development of the CBR flexible pavement design method for airfields. *Transactions of the American Society of Civil Engineers* **115(1)**: 453–589.

Asphalt Institute (1991) *Thickness Design: Asphalt Pavements for Highways and Streets – Manual Series No. 1*. Asphalt Institute, Lexington, KY, USA.

Austroads (2012) *Guide to Pavement Technology, Part 2: Pavement Structural Design*. Austroads, Sydney, Australia, AGPT02-12.

Boussinesq J (1885) *Application des Potentiels a l'etude de l'equilibre et du Movement des Solides Elastiques*, Gauthier-Villard, Paris, France.

Burmister DM (1943) The theory of stresses and displacements in layered systems and applications to the design of airport runways. *Proceedings of the Highway Research Board* **23**: 126–144.

Burmister DM (1945) The general theory of stresses and displacement in layered systems. *Journal of Applied Physics* **16(2)**: 89–94.

Carey WN and Irick PE (1962) Performance of flexible pavements in the AASHO road test. *Proceedings of the 1st International Conference on the Structural Design of Asphalt Pavements*. University of Michigan, Ann Arbor, MI, USA.

Cooper KE and Pell PS (1974) *The Effect of Mix Variables on the Fatigue Strength of Bituminous Materials*. Transport and Road Research Laboratory, Crowthorne, UK, Report LR633.

Croney D and Loe JA (1965) Full-scale pavement design experiment on the A-1 at Alconbury Hill, Huntingdonshire. *Proceedings of the Institution of Civil Engineers* **30**: 225–270.

Finn F, Saraf C, Kulkarni R *et al.* (1977) The use of distress prediction subsystems for the design of pavement structures. *Proceedings of the 4th International Conference on the Structural Design of Asphalt Pavements*. University of Michigan, Ann Arbor, MI, USA.

Gillespie TD, Karamihas SM, Sayers MW *et al.* (1993) *Effects of Heavy-vehicle Characteristics on Pavement Response and Performance*. National Academy Press, Washington, DC, USA, NCHRP Report 353.

HA (Highways Agency) (2009) *Interim Advice Note 73/06, Revision 1 (2009): Design Guidance for Road Pavement Foundations (Draft HD25)*. Stationery Office, London, UK.

HA, Transport Scotland, Welsh Government and Department for Regional Development Northern Ireland (2006) Section 2: pavement design and construction. Part 3: pavement design. In *Design Manual for Roads and Bridges*, vol. 7. *Pavement Design and Maintenance*. Stationery Office, London, UK, HD 26/06.

Hicks RB and Monismith CL (1971) Factors influencing the resilient response of granular materials. *Highway Research Record* **345**: 15–31.

Hodgkinson JR (1982) *Joint Sealants for Concrete Road Pavements*. Cement and Concrete Association of Australia, Sydney, Australia, Technical Note 48.

Lay MG (1993) *Ways of the World*. Primavere Press, Sydney, Australia.

Leigh JV and Croney D (1972) The current design procedure for flexible pavements in Britain, *Proceedings of the Third International Conference on the Structural Design of Asphalt Pavements*. University of Michigan, Ann Arbor, MI, USA, pp. 1039–1048.

Lister NW and Addis RR (1977) Field observations of rutting and their practical implications. *Transportation Research Record* **640**: 28–34.

Nunn ME (2004) *Development of a More Versatile Approach to Flexible and Flexible Composite Pavement Design*. Transport Research Laboratory, Crowthorne, UK, Report 615.

Nunn ME, Brown A, Weston D and Nicholls JC (1997) *Design of Long-life Flexible Pavements for Heavy Traffic*. Transport Research Laboratory, Crowthorne, UK, TRL Report 250.

OECD (Organisation for Economic Co-operation and Development) (1989) *Heavy Trucks, Climate and Pavement Damage*. OECD, Paris, France.

OECD (2005) *Economic Evaluation of Long-life Pavements: Phase 1*. OECD, Paris, France.

Oedmark N (1949) *Undersokning av Elasticitetegskaperna hos Olika Jordater Samt Teori for Berakning av Belagningar Eligt Elasticitesteorin*. Statens Vaginstitute, Stockholm, Sweden.

Porter OJ (1938) The preparation of subgrades. *Proceedings of the Highway Research Board* **18(2)**: 324–392.

Powell WD, Potter JF, Mayhew HC and Nunn ME (1984) *The Structural Design of Bituminous Roads*. Transport and Road Research Laboratory, Crowthorne, UK, Report LR1132.

RRL (Road Research Laboratory) (1970) *A Guide to the Structural Design of Pavements for New Roads*. Stationery Office, London, UK.

Seed HB, Chan CK and Lee CE (1962) Resilience characteristics of subgrade soils and their relation to fatigue failures in asphalt pavements. *Proceedings of the 1st International Conference on the Structural Design of Asphalt Pavements*. University of Michigan, Ann Arbor, MI, USA.

Steel DJ (1945) Discussion to classification of highway subgrade materials. *Proceedings of the Highway Research Board* **25**: 388–392.

Thom N (2014) *Principles of Pavement Engineering*, 2nd edn. ICE, London, UK.

Thompson PD, Croney D and Currer EWH (1972) The Alconbury Hill experiment and its relation to flexible pavement design. *Proceedings of the 3rd International Conference on the Structural Design of Asphalt Pavements*. University of Michigan, Ann Arbor, MI, USA, pp. 920–937.

Ullidtz P (1987) *Pavement Design*. Elsevier, Amsterdam, the Netherlands.

Westergaard HM (1926) Stress in concrete pavements computed by theoretical analysis. *Public Roads* **7(2)**: 25–35.

Westergaard HM (1928) Analysis of stresses in concrete pavements caused by variations in temperature. *Public Roads* **7(3)**: 54–60.

Whiffin AC and Lister NW (1962) The application of elastic theory to flexible pavements. *Proceedings of the 1st International Conference on Asphalt Pavements*. University of Michigan, Ann Arbor, MI, USA, pp. 499–521.

World Bank (1988) *Road Deterioration in Developing Countries – Causes and Remedies*. World Bank, Washington, DC, USA.

Highways
ISBN 978-0-7277-5993-1

ICE Publishing: All rights reserved
http://dx.doi.org/10.1680/h5e.59931.345

Chapter 11
Analysis of stresses in rigid concrete slabs and an introduction to concrete block paving

John Knapton Engineering Consultant, UK

This chapter deals primarily with the development of stresses in rigid pavements, but includes a section on concrete block paving, as this is a form of concrete pavement that behaves in a different way from rigid concrete.

Stresses in rigid concrete slabs result from applied load and from restraint to slab movement induced by moisture loss and temperature changes. This chapter explains how those stresses are calculated and how those calculated stresses are compared with concrete strength in order to proportion slab thickness. It demonstrates the use of Highways England's (formerly the Highways Agency) transfer function equation (see Figure 2.3 of HA *et al.*, 2006a) that relates the allowable stress to the number of repetitions of a wheel patch load on a fatigue basis. In other words, this chapter presents a mechanistic design method for rigid highway pavement slabs. In particular, pure tensile stresses and tensile flexural stresses are considered, since it is these stresses that can lead to concrete cracking and to subsequent deterioration of the pavement. Compressive stresses developed in the concrete slabs are usually so low that they can be regarded as negligible.

11.1.   Foundations
A rigid concrete road pavement comprises a concrete slab that is normally placed on a sub-base foundation that may be constructed directly on the subgrade. The strength of the subgrade is usually specified in terms of its California bearing ratio (CBR) value.

Sometimes the foundation comprises a sub-base that is constructed over an improved subgrade; that is, one that has a platform/capping layer – sometimes termed a 'lower sub-base' – atop a low-strength subgrade. Capping material typically comprises a locally available low-cost material with a CBR of 15% or more. The material forming the capping layer often comprises secondary materials or crushed rock of insufficient strength or stability to function in the sub-base: its primary functions are to strengthen the subgrade, to provide a working platform for construction equipment, and to reduce the thickness of (more expensive) sub-base material that would otherwise be required. When a capping or platform layer is specified, for mechanistic design purposes it is assumed that the subgrade has been strengthened to a degree whereby it has an equivalent CBR of 5%.

The subgrade is the naturally occurring ground or, in the case of an embankment, it comprises imported fill to the formation level. Homogeneity of the subgrade strength is particularly important, and avoiding hard and soft spots is a priority in subgrade preparation. Ideally, any subgrade should comprise suitable material of such particle size distribution (PSD) – referred to as 'grading' prior to the European Committee for Standardization (CEN) initiatives – that it can be well compacted. If the subgrade material is too soft or contains oversize particles or excessive fines that are difficult to compact, they can give rise to settlement and early failure of the pavement. On very good quality subgrades (e.g. well-compacted firm sandy gravels), a sub-base may be omitted – albeit, as noted above, one may actually be constructed for 'buildability' reasons (i.e. to provide a protected working surface for construction equipment usage when construct-ing the base).

The inclusion of a sub-base in the foundation is normally essential for rigid pavement construction. This layer consists of at least an inert, well-graded granular material; however, a cement-bound granular material or another hydraulically bound mixture is normally specified for sub-bases in the foundations of major concrete roads now built in the UK. Both the sub-base and the upper part of the subgrade can also be improved by cement stabilisation. While in situ cement stabilisation can be an economic means of improving (i.e. capping) a poor subgrade, in the case of sub-base stabilisation the cement should be plant mixed to ensure a uniform stable material.

For patch loads applied to the surface of a slab from wheel loading, the sub-base assists in reducing the vertical stress transmitted to the subgrade. However, where a distributed load is present, the pavement slab will achieve very little load spreading, so that the bearing capacity of the underlying subgrade may limit the maximum load that can be applied to the pavement.

A sub-base material is usually specified according to its CBR, and CBR values of 20–40% are commonly used in the design specifications for concrete pavements, even though materials used in sub-bases may achieve strengths in excess of these values. Other criteria such as frost resistance, particle size distribution, and a requirement to avoid the inclusion of plastic fines are often included in pavement specifications for sub-bases.

11.1.1 Modulus of subgrade reaction

When assessing the value of the stresses induced into a concrete slab when loaded by a wheel-patch load, the influence of the sub-base and subgrade is often treated as that of an elastic medium whose strength is defined by its modulus of subgrade reaction, K. The K values define the amount of deflexion of the foundation under a loaded pavement slab.

CBR tests and/or plate-bearing tests (see Chapter 4) can be used to establish K values. In many instances, however, the subgrades are variable, and the measured values obtained from in situ tests can often show scatter. When this occurs, it is common to use the lowest measured value in design once any obvious outliers have been removed from the data set.

Table 11.1 Assumed modulus of subgrade reaction K for typical UK soils. (Note: this is a very coarse assessment, and Table 11.5 may provide more accurate data)

Soil type	Typical soil description	Subgrade classification	Assumed K: N/mm^3
Coarse-grained soils	Gravels, sands, clayey or silty gravels/sands	Good	0.054
Fine-grained soils	Gravely or sandy silts/clays	Poor	0.027
	Clays, silts	Very poor	0.013

It is normal for soils to have the same or increased strength with depth, which means that CBR or plate-bearing tests carried out at the surface are usually valid. Nonetheless, from time to time, a layer of weak material may be present at a depth of a some metres below the formation (e.g. a layer of buried peat), and where this occurs it can have a deleterious effect on the subsequent performance of a pavement. CBR and plate-bearing tests introduce a shallow stress bulb, and thus their measurements may not reflect the influence of a deeper layer of subgrade soil that is vulnerable to stress beneath a loaded slab. In order to deal with this possibility, it is important that a full site investigation be carried out prior to any rigid pavement design, including, as appropriate, standard penetration tests (BSI, 1990) to assess the strength of the ground to a depth of, say, 12 m.

If no plate-bearing or CBR test results are available, the K values shown in Table 11.1 provide a general guide to the moduli measured on typical UK soils. Note that the units of K are newtons per cubic millimetre (N/mm^3): this can be considered as the vertical pressure (N/mm^2) required to produce a deflection of 1 mm.

It has been suggested (Chandler and Neal, 1988) that the sub-base can be taken into account by enhancing K, as in Table 11.2.

Rigid concrete pavement slabs normally comprise pavement quality concrete (PQC): this is a concrete with a cement content of 300–330 kg/m^3, and a relatively low water/cement

Table 11.2 Enhanced K values (N/mm^3) when a sub-base is used on top of the subgrade

K for the subgrade alone	Enhanced K when used in conjunction with							
	Cement-bound sub-base of thickness				Granular sub-base of thickness			
	150 mm	200 mm	250 mm	300 mm	100 mm	150 mm	200 mm	250 mm
0.013	0.018	0.022	0.026	0.030	0.035	0.050	0.070	0.090
0.020	0.026	0.030	0.034	0.038	0.060	0.080	0.105	–
0.027	0.034	0.038	0.044	0.049	0.075	0.110	–	–
0.040	0.049	0.055	0.061	0.066	0.100	–	–	–
0.054	0.061	0.066	0.073	0.082	–	–	–	–
0.060	0.066	0.072	0.081	0.090	–	–	–	–

Table 11.3 Modified thickness of a PQC slab when used with a lean-concrete sub-base

Calculated thickness of slab: mm	Modified thickness of PQC slab required (mm) when used in conjunction with a lean-concrete sub-base of thickness		
	100 mm	130 mm	150 mm
250	190	180	–
275	215	200	–
300	235	225	210

ratio (e.g. 0.45) so as to achieve a low-slump mixture. The reason for using PQC is to avoid excess shrinkage, while at the same time achieving a mixture that will have sufficient flexural strength.

When a lean concrete sub-base is specified as part of the foundation beneath a PQC slab, the K value of the subgrade material is used to calculate the required thickness of the concrete slab, and this calculated thickness is then apportioned between the PQC slab thickness (i.e. the higher-strength concrete) and the lean-concrete sub-base thickness. This relationship is shown in Table 11.3 for a C32/40 concrete slab and a C16/20 lean-concrete sub-base. (Note that the prefix C is commonly used to denote, first, the characteristic 28 day compressive strength of the concrete measured on a cylinder of length equal to twice its diameter and, second, the 28 day compressive strength measured on a 150 mm cube.)

Although PQC is specified by its compressive strength, it is the flexural strength of the material that is important in the pavement design process. Although there are formulae defining the relationship between compressive strength and flexural strength, these are approximate and depend upon many properties of the concrete and its aggregates. Table 11.4 includes flexural strength values for some commonly used plain PQC mixtures.

11.1.2 Plate-bearing and CBR testing

K is normally established by plate-bearing tests with a plate diameter of 750 mm. The plate-bearing test procedure involves loading the ground through a steel disk that is usually

Table 11.4 Concrete design flexural strengths at 7 days

Concrete grade and dosage	Flexural strength: N/mm^2
Plain or polypropylene-reinforced C25/30 concrete	2.0
20 kg/m^3 steel fibre-reinforced C25/30 concrete	2.4
30 kg/m^3 steel fibre-reinforced C35/30 concrete	3.2
40 kg/m^3 steel fibre-reinforced C25/30 concrete	3.8
Plain or polypropylene-reinforced C32/40 concrete	2.4
20 kg/m^3 steel fibre-reinforced C32/40 concrete	2.8
30 kg/m^3 steel fibre-reinforced C32/40 concrete	3.8
40 kg/m^3 steel fibre-reinforced C32/40 concrete	4.5

Table 11.5 K and CBR values for a number of common subgrade and sub-base materials

Soil	CBR: %	K: N/mm^3
Humus soil or peat	<2	0.005–0.015
Recent embankment	2	0.01–0.02
Fine or slightly compacted sand	3	0.015–0.03
Well-compacted sand	10–25	0.05–0.10
Very well-compacted sand	25–50	0.10–0.15
Loam or clay (moist)	3–15	0.03–0.06
Loam or clay (dry)	30–40	0.08–0.10
Clay with sand	30–40	0.08–0.10
Crushed stone with sand	25–50	0.10–0.15
Coarse crushed stone	80–100	0.20–0.25
Well-compacted crushed stone	80–100	0.20–0.30

mounted on the back of a vehicle, and recording the load and the corresponding deflexion. K (expressed in units of N/mm^3, MN/m^3 or kg/cm^3) is obtained by dividing the pressure exerted on the plate by the resulting vertical deflexion. A modification is needed if a different plate diameter is used: for example, for a 300 mm diameter plate, K is obtained by dividing the measured result by 2.3, and for a 160 mm diameter plate it is divided by 3.8.

Alternatively, CBR testing can be carried out, from which K values can be derived. Table 11.5 shows the relationship between CBR and K for a number of common soil types found in the UK.

11.2. Fibre-reinforced PVQ

The concept of reinforcing construction mortars and concretes with various types of fibres is not new (e.g. Exodus 5 says that the pharaoh ordered his foremen to cease supplying reinforcing straw to Israelite brickmakers). The Romans are also known to have used hair fibres in structural mortars.

In recent years, pavement construction methods involving the addition of polypropylene or steel fibres to concrete mixtures have become common worldwide. The inclusion of fibres in concrete has been driven by the need to meet quicker construction programmes and the consequent mechanisation of concrete placing and compaction, particularly of highway pavement slabs. For example, with the use of slip-form pavers or laser-guided screeding machines, steel or polypropylene fibres are often specified instead of conventional steel fabric because of the inconvenience associated with positioning individual mats of steel fabric immediately in front of the paving machine as the placing of the concrete progresses. Slip-form pavers and laser-guided screeding machines have difficulty in constructing conventional steel fabric/bar-reinforced concrete pavements efficiently because the steel bars impede progress of the work.

Tests by a manufacturer of polypropylene fibres reveal the changes (Table 11.6) in strength characteristics associated with the use of these fibres in concrete. Compressive

Table 11.6 Test results comparing the strength of polypropylene fibre-dosed concrete and conventional plain concrete

Measurement	Strength of polypropylene fibre-reinforced concrete: N/mm^2	Strength of unreinforced concrete: N/mm^2
Compressive strength (equivalent cube method)		
1 day	16.5	16.0
3 day	28.5	24.5
7 day	34.0	35.0
28 day	43.5	39.5
Cube compressive strength		
1 day	16.0	14.5
3 day	28.0	27.5
7 day	34.0	36.0
28 day	48.5	44.5
Flexural strength		
1 day	2.3	2.1
3 day	4.0	3.7
7 day	4.2	4.8
28 day	4.6	6.2

strength tests conducted in accordance with BS 1881 (BSI, 1983) have indicated that the polypropylene fibres, when used at the normally recommended dosage rate of 0.9 kg/m^3, slightly increase the early strength gain of concrete. The fibres have no significant effect on the 28 day compressive strength of concrete cubes, nor do they have any substantial effect on the flexural strength of concrete.

Steel fibres are now commonly used in place of steel fabric reinforcement. The stresses occurring in a pavement slab are complex, and depending on the type of load, flexural tensile stresses can occur at the top and at the bottom of the slab. There are, in addition, stresses that are difficult to quantify, arising from a number of causes such as vehicles executing sharp turns, moisture loss, thermal effects and impact loads. The addition of steel wire fibres to a concrete slab results in a homogeneously reinforced slab achieving a considerable increase in flexural strength and enhanced resistance to shock and fatigue.

A manufacturer of anchored steel fibres commissioned the Netherlands Organisation for Applied Scientific Research (TNO) to undertake flexural strength tests using fibres embedded in C25/30 concrete. The TNO tests resulted in values of flexural strength of up to 4.5 N/mm^2, depending on dosage, type and size of fibre. Partly from this study and partly from work undertaken at the Cement and Concrete Association in the UK, the flexural strength values shown in Table 11.4 have been developed.

11.3.   Thermal and moisture-related stresses in concrete slabs

The basic premise underlying most concrete pavement design methods is that stresses developed as a result of a concrete slab changing its temperature or moisture content are contained by the provision of stress-relieving joints, while stresses developed by traffic and other applied loads are controlled by proportioning the thickness of the slab and its underlying supporting courses. The exception is in continuously reinforced concrete pavements, where temperature and moisture-loss stresses are contained by the composite action of the reinforcement and the concrete.

Whether temperature or moisture loss stresses are predominant depends upon many factors that are difficult to calculate. A simple rule of thumb states that temperature stresses are more important in the case of 'external' concrete that is used in roads, ports and airports, whereas moisture-loss stresses are more important for 'internal' concrete (e.g. as used in ground-bearing industrial floors). Moisture-related stresses are potentially greater than temperature-related ones by an order of magnitude. However, it is often the case that a highway pavement retains much of its moisture throughout its life; also, moisture-related stresses develop slowly, so that creep (i.e. the propensity for semi-permanently stressed concrete to change its shape so as to reduce the stress) can lead to a progressive lowering of stress. Temperature stresses, on the other hand, are often at their most severe immediately following construction, as the setting concrete cools: furthermore, temperature-related stresses are usually diurnal, so that creep has little mitigating effect. For this reason, temperature-related effects are normally the ones of most concern in most concrete road pavement projects.

Although moisture-loss effects are usually less important than temperature-related effects in the case of external concrete, they need careful consideration in the case of highway pavements constructed in a dry climate. Both temperature and moisture can cause a slab to shrink uniformly, to curl upwards at its perimeter or to curl downwards at its perimeter. The way in which these three conditions impart stress into the slab is now considered.

11.3.1   Uniform shrinkage

As a result of uniform temperature fall or moisture loss, a concrete slab will shrink uniformly about its centre on plan while, theoretically, the centre remains stationary. At a distance from the centre, the slab will attempt to displace horizontally, and this displacement will increase uniformly towards the edge of the slab. Frictional restraint between the underside of the concrete slab and the surface of the sub-base will inhibit or prevent this movement, and so tensile stress is generated within the slab. This stress in the concrete, resulting from the frictional restraint to shrinkage, can be calculated. The force required to overcome friction is given by $F_f = w\mu$, where F_f is the friction force, w is the mass of the concrete (calculated by assuming a density of 24 kN/m^3) and μ is the coefficient of friction. As the weight of concrete generating frictional restraint increases with the distance from the slab edge, the stress gradually increases to a maximum at the slab centre, and there is zero stress at the edge of the slab or at the joints. Assuming the values for the coefficient of friction between concrete and sub-base and polythene are 0.65 and 0.15, respectively, the theoretical stresses that result from uniform shrinkage friction are

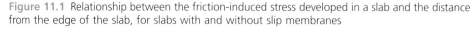

Figure 11.1 Relationship between the friction-induced stress developed in a slab and the distance from the edge of the slab, for slabs with and without slip membranes

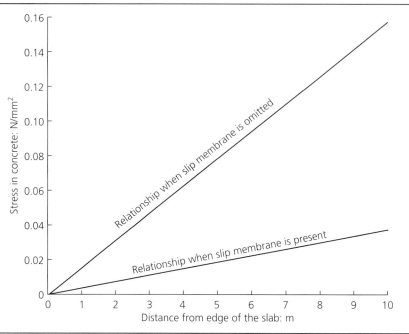

shown in Figure 11.1. Note that even without a slip membrane, the stresses are low, attaining a value of less than 0.05 N/mm² in a 6 m long slab.

Figure 11.1 illustrates why the provision of a slip membrane is not a crucial issue from a stress perspective. Indeed, in an unreinforced concrete pavement with closely spaced joints (e.g. at 5 m spacings), the provision of a slip membrane may be detrimental in that it may concentrate movement at one joint – known as a dominant joint – which can then become a maintenance problem, while other dormant joints never operate. (In such an instance, it may be preferable to provide a layer of polyethylene, to prevent concrete water loss into the sub-base.)

11.3.2   Slab perimeter curling downwards (hogging)

As a result of the underside of the concrete slab cooling or drying faster than the top, non-uniform shrinkage develops throughout the slab, with the lower concrete shrinking more than the upper. The result of this non-uniform shrinkage is a tendency for downward curling (i.e. hogging) of the slab. Assuming that the hogging slab can be represented by a simply supported beam of length L with a uniformly distributed load equal to the concrete weight, then the maximum moment $M = wL^2/8$, where L is the length of the slab (i.e. the distance between joints) and w is the dead weight of the concrete, calculated on the basis of 24 kN/m³. (In the extreme case, L would be the distance between joints allowing rotational freedom.) The stress is calculated from the classic bending equation

Figure 11.2 Relationship between joint spacing and the stress developed as a result of restraint to hogging

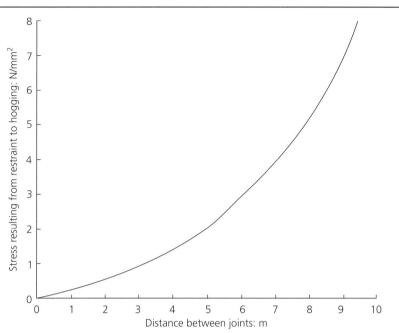

Figure 11.2 shows a graph with "Stress resulting from restraint to hogging: N/mm²" on the vertical axis (scale 0 to 8) and "Distance between joints: m" on the horizontal axis (scale 0 to 10).

$\alpha/y = M/I$, where α is the stress, y is the depth to the neutral axis (usually half of the thickness of the slab), M is the bending moment and I is the second moment of area.

Figure 11.2, which shows stresses that would develop in the hogging situation for a 200 mm thick slab, demonstrates that such behaviour would crack each bay. In fact, the temperature fall or moisture loss is usually insufficient to cause the slab to separate from its sub-base, because this extreme condition rarely occurs. One of the factors that prevents this extreme condition is the fact that the slab does not curl to a degree whereby the simple support condition applies.

11.3.3   Slab perimeter curling upwards (curling)

The result of the upper side of the slab cooling or drying faster than the underside is that the slab attempts to curl upwards at its edges. This curling can be represented by a cantilever of length L equal to the curled length. Assuming that this cantilever carries a uniformly distributed load generated by the weight of the concrete (based upon an assumed density of 24 kN/m³), the bending moment at the point of contact of the slab with the ground is given by the expression $M = wL^2/2$. As in the case of hogging, the stress can then be calculated from the expression $\sigma/y = M/I$. Figure 11.3 shows curling stresses calculated for a 200 mm thick slab.

As with the hogging situation, this is an extreme example and testing has shown that it is often the case that approximately 1 m width of a slab curls to such a degree that the

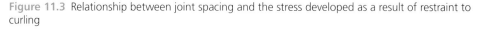

Figure 11.3 Relationship between joint spacing and the stress developed as a result of restraint to curling

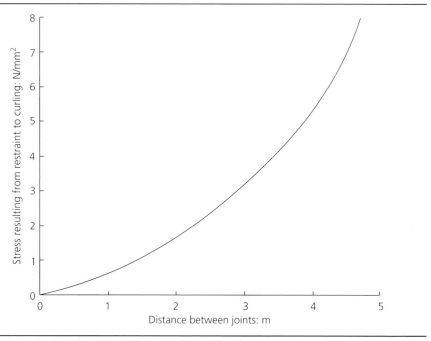

underside of the slab separates from the sub-base. Figure 11.3 confirms that the resulting maximum flexural stress at a distance of 1 m will be less than 1 N/mm^2.

11.3.4   Calculation of slab temperature changes

In order to determine temperature-related stresses through the depth of a slab, the temperature profile at the time that the concrete sets needs to be known. From that time forward, whenever the 'at-set' temperature profile is replicated, temperature stresses will disappear. The 'at-set' temperature profile has been investigated by several researchers, and they have concluded that there can be no standard profile that would be of value to the designer. The profile depends upon the type of concrete, the curing regime, the weather during concreting, and the time of day at which the concrete sets.

Figure 11.4 illustrates differing temperature profiles 'at set' and at subsequent times in different climatic conditions. This figure shows that concrete that sets during a warm mid-afternoon may have a locked-in profile as shown in Figure 11.4(b). However, if this slab were to be laid in a warm climate, then it might subsequently be subjected to a profile as in Figure 11.4(e) during the night: in such a case, the temperature of the slab surface will have fallen by, say, 20°C, and the temperature of the underside of the slab will have increased by 17°C, and this will cause the slab to attempt to curl upwards at its perimeter (i.e. curling). The self-weight of the slab together with the applied loading will attempt to keep the slab in contact with its sub-base, so that tensile stresses will

Figure 11.4 Concrete slab temperature profiles 'at set' and at subsequent times

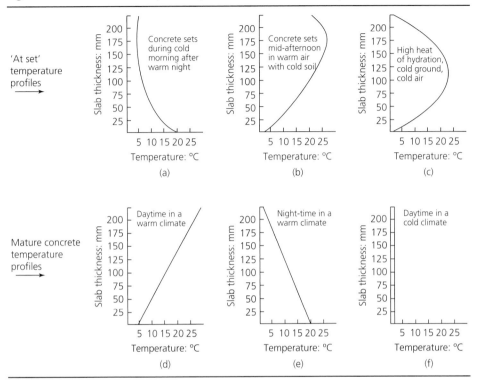

develop at and near the upper surface of the slab, and their values will depend upon the relative elastic properties of the slab and the underlying sub-base material.

The opposite effect would apply to slabs that developed their initial set during a cold morning following a relatively warm night (Figure 11.4(a)). If, however, this slab were to be subsequently subjected to the temperature profile shown in Figure 11.4(d), where the surface temperature rises by, say, 25°C and the underside temperature falls by 15°C, this would cause the slab to attempt to curl downwards at its perimeter (i.e. hogging).

The above two examples represent extremes that slabs might be expected to sustain in normal situations. In the first case, the upwards perimeter curling is caused by a temperature differential of 37°C, and in the second case the downwards perimeter curling temperature differential is 40°C. In order to gauge the magnitude of the stresses that might be generated, consider the extreme case in which the slab is fully restrained against hogging. If the concrete has a coefficient of thermal expansion of 0.000009/°C, the maximum tensile strain equals $40 \times 0.000009 = 0.00036$. Concrete with a Young's modulus of 20 000 N/mm^2 would develop a tensile stress of approximately 7 N/mm^2, which would be sufficient to crack the concrete. This value is never attained in the concrete, however, because full restraint is never achieved and because curing regimes

Table 11.7 Transverse joint spacings that can control temperature and moisture-related cracking in concrete pavements

(a) Conventionally reinforced concrete pavements

Weight of conventional steel fabric reinforcement: kg/m^2	Spacing of joints: m
2.61 (6 mm diameter wires at 100 mm c/c) – C283	16.5
3.41 (7 mm diameter wires at 100 mm c/c) – C385	21
4.34 (8 mm diameter wires at 100 mm c/c) – C503	27.5
5.55 (8 mm diameter wires at 80 mm c/c) – C636	35

(b) Plain and steel fibre-reinforced pavements

Type of concrete	Spacing of joints: m
Plain or polypropylene-reinforced C25/30 concrete	6
20 kg/m^3 hooked steel fibre-reinforced C25/30 concrete	7
30 kg/m^3 hooked steel fibre-reinforced C25/30 concrete	8
40 kg/m^3 hooked steel fibre-reinforced C25/30 concrete	10
Plain or polypropylene-reinforced C32/40 concrete	6
20 kg/m^3 hooked steel fibre-reinforced C32/40 concrete	8
30 kg/m^3 hooked steel fibre-reinforced C32/40 concrete	10
40 kg/m^3 hooked steel fibre-reinforced C32/40 concrete	12

c/c, centre to centre

ensure that the 'as set' profiles shown in Figure 11.4 are not attained in well-controlled projects. In practice, it is found that providing transverse warping joints (i.e. joints that permit rotation at spacings of 5–30 m (depending upon the level of reinforcement)) is sufficient to control temperature stresses. Table 11.7(a) and 11.7(b) shows joint spacings that have been found to be sufficient to control temperature and moisture-related cracking.

Difficulties can occur when different concretes are used in two-layer construction. The use of low-heat concrete below PQC can lead to excessive tensile stresses: for example, difficulties were experienced on the London Orbital Motorway (M25) as a result of transverse cracking.

Moisture-related shrinkage occurs when a concrete slab loses its free water through evaporation. Typically, PVQ will have a water/cement ratio of 0.45. Approximately two-thirds of this water is combined chemically with the cement during hydration, and the remainder acts as a lubricant that creates concrete workability. All of the lubricant water is free to evaporate, and when this happens the volume that it occupied is no longer present and shrinkage occurs. There may be 50 l/m^3 of water evaporating over a period of several months following construction: as this represents 5% of the volume of the concrete, significant shrinkage can occur. In fact, creep and precipitation reduce the effect of shrinkage to a controllable level, so that the joint spacings discussed earlier are usually sufficient to prevent distress.

11.4. Crack-control methods

Most concrete pavements include joints in order to reduce the effect of temperature and moisture-related stress. The joints are characterised by the type of movement that they permit. Full-movement joints permit expansion and contraction but not rotational movement. Contraction joints permit contraction but not expansion. Warping joints permit rotation and a little contraction but no expansion. Usually, nearly all of the joints in a concrete pavement are either contraction joints or warping joints.

A cost-effective way of constructing such joints is to cast the concrete continuously through the joint, and then to saw through part of the concrete before the stresses become severe. In some rigid pavements, all of the joints – except for the end-of-day construction joints – are formed in this way. The saw cut acts as a line of weakness that is incorporated into the slab at the position of the joint, so that the slab will crack at that point owing to an increase in the tensile stress in the remaining depth of slab. The induced groove has a depth of between one-quarter and one-third of that of the slab (Figure 11.5). As well as the joint being the most durable crack-induction form, a sawn joint is also very serviceable, as there is no difference in level between each side of the cut.

Joints should be cut when the concrete has gained sufficient strength to support the weight of the cutting equipment but before it has gained such strength that sawing might loosen or pull out aggregates and fibre reinforcement. A suggested sawing time-scale is between 24 and 48 h after the initial concrete set. For example, consider a pavement construction in which the concrete is placed between 8.00 a.m. and 5.00 p.m. on day 1: then, assuming that (a) all of the concrete is mixed 1 h before placing, (b) the first concrete is mixed at 7 a.m. and the last at 4 p.m., and (c) the concrete takes 6 h to reach initial set, the earliest initial set is at 1.00 p.m. and the latest initial set is at 10.00 p.m. If the concrete is to be sawn between 24 and 48 h after the initial set, the first cut can be performed 24 h after the latest initial set (to ensure that all of the concrete has gained

Figure 11.5 A typical induced joint formed by the saw-cut technique

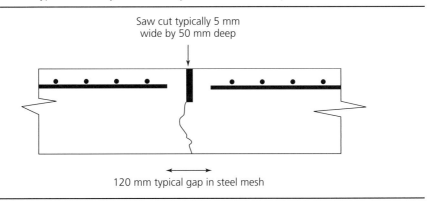

Saw cut typically 5 mm
wide by 50 mm deep

120 mm typical gap in steel mesh

sufficient strength), and all saw-cutting must be finished 48 h after the earliest initial set. Thus, in this example, saw-cutting can commence at 10.00 p.m. on day 2 but must be finished by 1.00 p.m. on day 3: this gives a window of 15 h of sawing time.

11.4.1   Spacing of joints

Table 11.7 shows transverse joint spacings that have been used successfully for many years with various types of concrete. Some engineers consider that the spacings can be greater than 12 m for steel fibre-reinforced concrete pavements. While 14 m joint spacings are probably acceptable for most road pavements, the additional movement that would occur at joints could result in loss of aggregate interlock and joint degeneration in cold climates (e.g. as can be experienced in the UK).

In the case of conventional steel fabric reinforcement, longitudinal joint spacing is usually dictated by mesh size, and a mesh length of 4.8 m is common. In the case of fibre-reinforced concrete, no such considerations apply, and longitudinal joints can be located at transverse joint spacings. This provides the designer with the opportunity to locate longitudinal joints at lightly loaded positions (e.g. at lane markings on road pavements).

In the case of plain PVQ , it is normal to include transverse induced joints at 6 m spacing when the concrete contains limestone aggregates, and at 5 m spacing in the case of other types of aggregates.

11.5.   Mechanistic rigid pavement design method

The design method presented here is based upon assessing the stresses developed in rigid concrete slabs and comparing such stresses with those known to be suitable in different categories of concrete. It uses Westergaard equations (Westergaard, 1947) in the case of patch loading, and has been formulated into a set of design charts to facilitate use. Examples of design from first principles and the use of the design charts are presented. The loading of highway pavements will be one of, or a combination of, the following: (a) uniformly distributed load, (b) patch loading from vehicle tyres and (c) horizontal loading. Providing conventional joint details are specified, the stresses induced by moisture-related shrinkage or by temperature effects will remain minimal, using this design method.

Pavement design comprises assessing the loading regime that the pavement is predicted to sustain, and selecting materials, thicknesses and a joint configuration that will sustain those loads while at the same time satisfying flatness, durability, abrasion, drainage and riding-quality requirements. An essential part of the design process is the assessment of the anticipated load regime. For example, horizontal loading may be introduced into the pavement at locations where vehicles brake and/or undertake turning manoeuvres. Areas subjected to these forces are often referred to as stress lengths of pavement (e.g. at the approaches to road junctions). At such locations, special care is needed when detailing joints that must be able to transmit tension between neighbouring slabs. Also, small slabs may need to be restrained; the design of the restraint can be undertaken conservatively by assuming that the pavement slab spreads the applied horizontal load through a projected 90° path.

The concrete pavement design procedure set out in this section involves the calculation of stresses resulting from the loading regime and ground conditions, and comparing those stresses with the strength of the concrete. Thus, the following factors need to be taken into account in this method of design: (a) the loading regime, (b) the strength of the concrete and (c) the strength of the existing ground and the effect of the sub-base.

11.5.1   Loading regime

Fatigue often leads to pavement distress, so, to cater for this, pavement design is simplified by introducing a load safety factor into the design method. A load safety factor of 2.0 is recommended for an infinite number of load repetitions, and may therefore be used conservatively in all cases. An alternative design approach is to use a transfer function equation that relates the permissible stress to the number of passes of a patch wheel load (see Figure 2.3 in HA *et al.* (2006a)).

The damage to road pavements is essentially caused by commercial vehicles, so, assuming that the numbers of the different categories of commercial vehicles expected to use a given road during its design life are known, it is standard practice to express the often-complex axle configurations of these vehicles in terms of the number of standard 8000 kg axles that would cause the same damage (albeit, nowadays, it is common for an axle load to exceed 8000 kg). In the UK, this conversion number is termed a wear factor, and wear factors recommended for use by the (now) Highways England (HA *et al.*, 2006b) are listed in Table 11.8.

However, in many regions of the world (including the USA and UK), the maximum allowable non-steering axle weight is approximately 11 000 kg, and the maximum steering axle load is 6500 kg. In the examples that follow, 11 000 kg (i.e. 110 000 N) standard axle loads are therefore assumed for the design process, and the consequent wear factors shown in Table 11.9 are expressed in terms of 11 000 kg axle loads.

Table 11.8 Wear factors, based on a standard 8000 kg axle load, recommended by the (now) Highways England for use on pavement maintenance and new pavement designs

Commercial vehicle type	Maintenance wear factor, W_M	New pavement design wear factor, W_N
Buses and coaches	2.6	3.9
2-axle rigid	0.4	0.6
3-axle rigid	2.3	3.4
4-axle rigid	3.0	4.6
3 and 4-axle articulated	1.7	2.5
5-axle articulated	2.9	4.4
6-axle articulated	3.7	5.6

Table 11.9 Commercial vehicle wear factors, based on an 11 000 kg axle load, used in the design examples in this chapter

Commercial vehicle type	11 000 kg axle wear factor for pavement maintenance	11 000 kg axle wear factor for new pavement design
Buses and coaches	0.78	1.17
2-axle rigid	0.12	0.18
3-axle rigid	0.69	1.02
4-axle rigid	0.90	1.38
3 and 4-axle articulated	0.51	0.75
5-axle articulated	0.87	1.32
6-axle articulated	1.11	1.68

The position of the load relative to the slab edge is critical, and in the case of patch loads from vehicle wheels, three alternative cases need to be considered, namely internal loading (i.e. >0.5 m from the slab edge), edge loading and corner loading. With internal or edge loading, the maximum stress occurs beneath the heaviest load at the underside of the slab. Corner loading creates tensile stress at the upper surface of the slab a distance away from the corner. This distance can be calculated from

$$d = 2[(2^{0.5})rl]^{0.5} \qquad (11.1)$$

where r is the radius of loaded area (mm), l is the radius of relative stiffness (mm), and d is the distance from the slab corner to the position of maximum tensile stress (mm).

11.5.2 Strength of concrete

Highway pavement slabs are frequently constructed from C25/30 or C32/40 PVQ with a minimum cement content of 300 kg/m^3 and with a slump of 25 mm or less (often referred to as zero-slump concrete). Design is based upon comparing material flexural strength with calculated flexural stresses, and specification is by characteristic compressive strength. Table 11.4 shows flexural strength values for a range of commonly used concretes.

11.5.3 Strength of the existing subgrade and effect of the sub-base

The design method requires a value for the modulus of subgrade reaction, K, that defines the deformability of the material beneath the pavement. The following four values of K are used in the design procedure:

- $K = 0.013$ N/mm^3 – very poor ground
- $K = 0.027$ N/mm^3 – poor ground
- $K = 0.054$ N/mm^3 – good ground
- $K = 0.082$ N/mm^3 – very good ground (no sub-base needed).

The beneficial effect of a granular sub-base is taken into account by increasing K according to the thickness and strength of the sub-base, as shown in Table 11.3.

11.5.4   Design methods: flexural stress in the concrete slab

The magnitude of the flexural stress developed in a pavement slab depends upon (a) the properties of the subgrade, (b) the loading regime (i.e. whether the applied loads are uniformly distributed loads or patch loads), (c) the thickness of the pavement slab and (d) the strength of the sub-base.

11.5.4.1 Designing for uniformly distributed loading

While most highway pavements are required to withstand only patch loads, some may need to be designed to also withstand distributed loading (e.g. if industrial roads are to be used for storage).

A common loading system comprises alternate unloaded and loaded storage areas. The maximum negative bending moment (hogging), M_{hog}, occurs within the centre of the unloaded areas

$$M_{hog} = -q/2\lambda^2 (B\lambda a' - B\lambda b') \tag{11.2}$$

where a' is the half-width of the unloaded area a ($a' = a/2$ (mm)), b' is the width of the loaded area b plus $a/2$ (mm), q is the uniformly distributed load ($=$ characteristic load $\times$ load factor (N/mm^2)) and λ is the radius of relative stiffness ($= (3K/Eh)^{1/4}$, where K is the modulus of subgrade reaction (N/mm^3), E is the concrete modulus (N/mm^2) ($= 10\,000$ N/mm^2 for a sustained load) and h is the concrete slab thickness (mm)).

The maximum positive bending moment (sagging), M_{sag}, occurs beneath the centre of the loaded area

$$M_{sag} = q/2\lambda^2 B_{(1/2)\lambda b} \tag{11.3}$$

where B_x is an exponential function ($= e^{-x} \sin x$).

The two moment equations can be simplified into a single conservative equation for any combination of unloaded and loaded width

$$M_{max} = -0.168q/\lambda^2 \tag{11.4}$$

and the corresponding maximum flexural stress (N/mm^2) is given by

$$\sigma_{max} = 6M_{max}/h^2 = 1008q/(\lambda^2 h^2) \tag{11.5}$$

which cannot exceed the relevant value from Table 11.4.

To ease calculation, Table 11.10 shows values of $\lambda^2 h^2$ for common combinations of K and slab thickness.

11.5.4.2 Uniformly distributed load design example

Consider a 150 mm thick pavement slab carrying a distributed load of 50 kN/m^2 between unloaded areas on very poor ground.

Table 11.10 Values of $\lambda^2 h^2$ for combinations of slab thickness and modulus of subgrade reaction K

K: N/mm³	$\lambda^2 h^2$ for slab thickness of				
	150 mm	175 mm	200 mm	225 mm	250 mm
0.082	0.061	0.066	0.070	0.074	0.078
0.054	0.049	0.053	0.057	0.060	0.063
0.027	0.035	0.038	0.040	0.043	0.045
0.013	0.024	0.026	0.028	0.029	0.031

In Equation 11.5, $q = 0.05 \times 2$ (i.e. the load safety factor) $= 0.10 \text{ N/mm}^2$. Extrapolate in Table 11.5 for very poor ground, to get $K = 0.013 \text{ N/mm}^3$. From Table 11.10, use the $\lambda^2 h^2$ value of 0.024 for $K = 0.013 \text{ N/mm}$ and a 150 mm thick slab. Then, using Equation 11.5,

$$\sigma_{max} = (1008 \times 0.10)/0.024 = 4.2 \text{ N/mm}^2$$

From Table 11.4, this flexural stress can be withstood by a C32/40 concrete incorporating 40 kg/m² of 60 mm long, 1 mm diameter anchored steel fibre that can be used for any combination of aisle and stacking zone width.

11.5.4.3 Designing for patch loading, using the Westergaard equations

The maximum flexural stress occurs at the bottom of the slab under the heaviest wheel load. The maximum stress under the wheel can be calculated using the following equations:

(a) point load at mid-slab (i.e. >0.5 m from the slab edge),

$$\sigma_{max} = \frac{0.275(1 + \mu)P\log(0.36Eh^3/Kb^4)}{h^2} \tag{11.6}$$

(b) point load at the edge of the slab,

$$\sigma_{max} = \frac{0.529(1 + 0.54\mu)P\log(0.20Eh^3/Kb^4)}{h^2} \tag{11.7}$$

(c) point load at a slab corner,

$$\sigma_{max} = \frac{3P\{1 - 1.41[12(1 - \mu^2)]^{1/4}\}Kb^4/Eh^3}{h^2} \tag{11.8}$$

where σ_{max} is the flexural stress (N/mm²), P is the point load (N) (i.e. a characteristic wheel load × load factor), μ is Poisson's ratio (usually 0.15), h is the slab thickness (mm), E is the elastic modulus (usually 20 000 N/mm²), K is the modulus of subgrade reaction (N/mm^3) and b is the radius of the tyre contact zone (mm) ($= (W/\pi p)^{1/2}$, where W is the patch load (N) and p is the contact stress between the wheel and the pavement (N/mm^2)).

Twin wheels that are bolted side by side are assumed to be one wheel transmitting half of the axle load to the pavement. In certain cases, wheel loads at one end of an axle magnify the stress beneath wheels at the opposite end of the axle, at a distance S away (where S is measured between load patch centres). To calculate the stress magnification, the characteristic length (i.e. the radius of relative stiffness, l) has to be found from

$$l = [Eh^3/12(1 - \mu^2)K]^{1/4} \tag{11.9}$$

Once Equation 11.9 has been evaluated, the ratio S/l can be determined, so that Figure 11.6 can be used to find M_t/P, where M_t is the tangential moment. The stress under the heaviest wheel has to be increased to account for the other wheel, and this is determined by adding an appropriate flexural stress calculated from

$$\sigma_{add} = (M_t/P)(6/h^2)P_2 \tag{11.10}$$

where P is the greater point load and P_2 is the other point load.

Figure 11.6 Relationship used to calculate the effect of two patch loads in close proximity

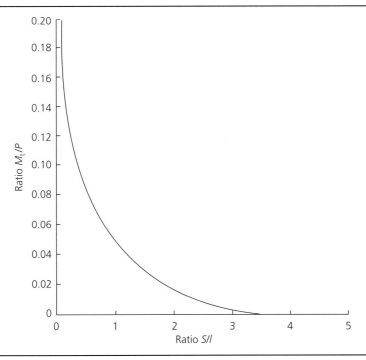

Then, sum the stresses and verify that the flexural strength has not been exceeded (see Table 11.4) for the prescribed concrete mix.

11.5.4.4 Road vehicle (patch loading) design example, using a fatigue (safety) factor of 2

Consider a highway vehicle with a rear axle load of 11 000 kg, and assume a slab thickness of 200 mm on good ground (i.e. $K = 0.054$ N/mm^3).

The load safety factor is 2.0, and the design axle load is 220 000 N (i.e. 110 000 N $\times$ 2.0 (the load safety factor)), so the design wheel patch load is 110 000 N.

Assume $p = 0.7$ N/mm^2, a (commonly used) contact stress between the wheel and pavement, then the radius of contact is

$$b = (W/\pi p)^{1/2} = (110\,000/\pi \times 0.7)^{1/2} = 224 \text{ mm}$$

Substituting known values into Equation 11.7

$$\sigma_{max} = \frac{0.529(1 + 0.54 \times 0.15)110\,000 \times \log[(0.2 \times 20\,000 \times 200^3)/0.054 \times 200^4]}{200^2}$$

Therefore, $\sigma_{max} = 3.71$ N/mm^2 beneath one wheel.

Assuming that the wheel at the other end of the axle is 2.7 m away (i.e. $S = 2700$ mm), then, using Equation 11.9, the radius of relative stiffness is

$$l = [20\,000 \times 200^3/12(1 - 0.15^2)0.054]^{1/4} = 709 \text{ mm}$$

and

$$S/l = 2700/709 = 3.8$$

Then, from Figure 11.6, $M_t/P = 0.001$, so that, using Equation 11.10,

$$\sigma_{add} = (M_t/P)(6/h^2)P_2 = 0.001 \times (6/200^2) \times 110\,000 = 0.02 \text{ N/mm}^2$$

Thus,

$$\text{total flexural stress} = 3.71 + 0.02 = 3.73 \text{ N/mm}^2$$

Therefore, a C25/30 PQC incorporating 40 kg/m^2 of 60 mm long, 1 mm diameter anchored steel fibres is adequate for this design.

11.5.4.5 Road vehicle design example using the HA transfer function fatigue equation

The Highways Agency's (now Highways England) jointed rigid pavement design method (HA et al., 2006a) employs a transfer function equation that relates the ratio of the flexural stress/flexural strength to the number of load repetitions that the slab can sustain. Transfer function equations, which are used commonly by road pavement designers in mechanistic design, are based upon observation of the long-term performance of pavements and are therefore empirical.

The HA transfer function equation for both PQC and cement-bound granular materials is

$$\log N = 17.61 - 17.61(R) \tag{11.11}$$

where R is the ratio of flexural stress to flexural strength and N is the number of stress repetitions. Applying this transfer function equation to the previous design example leads to the following design procedure.

The design wheel patch load is 55 000 N, and p is the contact stress between wheel and pavement ($= 0.7 \text{ N/mm}^2$). Thus, the radius of the contact area

$$b = (W/\pi p)^{1/2} = (55\,000/\pi \times 0.7)^{1/2} = 158 \text{ mm}$$

Substituting known values into Equation 11.7

$$\sigma_{\text{max}} = \frac{0.529(1 + 0.54 \times 0.15)55\,000 \times \log(0.2 \times 20\,000 \times 200^3)/0.054 \times 158^4}{200^2}$$

$$= 2.34 \text{ N/mm}^2 \text{ beneath one wheel}$$

Using Equation 11.9, the radius of relative stiffness is

$$l = [20\,000 \times 200^3/12(1 - 0.15^2)0.054]^{1/4} = 709 \text{ mm}$$

Assuming that the wheel at the other end of the axle is 2.7 m away (i.e. $S = 2700 \text{ mm}$), then

$$S/l = 2700/709 = 3.8$$

From Figure 11.6,

$$M_t/P = 0.001$$

Therefore, using Equation 11.10,

$$\sigma_{\text{add}} = (M_t/P)(6/h^2)P_2 = (0.005)(6/200^2) \times 110\,000 = 0.02 \text{ N/mm}^2$$

Therefore,

$$\text{total stress} = 2.34 + 0.02 = 2.36 \text{ N/mm}^2$$

Assume that the road is required to sustain 30 million repetitions of a wheel load of 55 000 N. Then, using the transfer function (Equation 11.11)

$$7.48 = 17.61 - 17.61(R)$$

and $R = 0.58$.

Therefore, the concrete must achieve a 28 day flexural strength of

$$2.36/0.58 = 4.07 \, \text{N/mm}^2$$

From Table 11.4, this can be achieved by using a C32/40 PQ concrete with 40 kg/m³ steel fibres.

11.5.4.6 Designing for patch loading using a design chart derived from the Westergaard equations

The Westergaard design procedure for patch loading is as follows.

Step 1: assess the existing conditions. Determine the modulus of subgrade reaction K (from Table 11.1 or 11.5), and the actual point load (APL).

Step 2: verify the distance between the point loads and determine σ_{add}. If the distance between loads is greater than 3 m, the APL can be used directly (depending on the radius of the contact zone – see step 6) to calculate the thickness of the slab using the relevant design chart. If the distance between loads is less than 3 m, the radius of relative stiffness, l, can be calculated from Equation 11.9 or determined from Table 11.11. Calculate σ_{add} using Equation 11.10.

Table 11.11 Values of the radius of relative stiffness, l, for different slab thicknesses and K values

Slab thickness: mm	Radius of relative stiffness, l: mm			
	$K = 0.013 \, \text{N/mm}^3$	$K = 0.027 \, \text{N/mm}^3$	$K = 0.054 \, \text{N/mm}^3$	$K = 0.082 \, \text{N/mm}^3$
150	816	679	571	515
175	916	763	641	578
200	1012	843	709	639
225	1106	921	774	698
250	1196	997	838	755
275	1285	1071	900	811
300	1372	1143	961	865

Note: Elastic modulus, $E = 20\,000 \, \text{N/mm}^2$ and Poisson's Ratio, $m = 0.15$

Step 3: select a proposed concrete mix – hence σ_{flex}. The mix can be determined from Table 11.4. In the example that follows, plain C40 concrete has been selected.

Step 4: calculate σ_{max}. When two point loads are acting in close proximity (i.e. less than 3 m apart), the greater point load, P, produces a flexural strength σ_{max} directly beneath its point of application, and the smaller point load, P_2, produces additional stress σ_{add} beneath the larger load. Then,

$$\sigma_{max} = \sigma_{flex} - \sigma_{add}$$

Step 5: calculate the equivalent single point load (ESPL) that, when acting alone, would generate the same flexural stress as the actual loading configuration. Thus,

$$ESPL = APL(\sigma_{flex}/\sigma_{max}) \qquad (11.12)$$

where APL is the actual point load.

Step 6: prior to using the (selected) design chart shown in Figure 11.7, modify the APL to account for the contact area as well as wheel proximity, to obtain the ESPL. The design chart applies directly when point loads have a radius of contact of between 150 and 250 mm; however, if the radius of contact is outside this range, multiply the patch load by a factor from Table 11.12 prior to using the design chart.

Step 7: use the appropriate design chart for the mix selected in step 3 to determine the slab thickness, and return to step 3 if an alternative concrete mix is required.

11.5.4.7 Design example of multiple patch loading, using Figure 11.7

Consider a situation where two patch loads are applied to a concrete pavement so that a 60 kN patch load is located 1 m away from a 50 kN patch load. The 60 kN point load has a contact zone radius of 100 mm, and the 50 kN point load has a 300 mm radius. The existing ground conditions are poor.

Step 1. From Table 11.1, use a modulus of subgrade reaction K of 0.027 N/mm^3 for poor ground conditions, and, initially, assume a slab thickness of 200 mm.

For a slab thickness of 200 mm and $K = 0.027$ N/mm^3, Table 11.11 gives a radius of relative stiffness $l = 843$ mm.

Step 2. The patch load distance separation $S = 1000$ mm. Therefore,

$$S/l = 1000/843 = 1.186$$

From Figure 11.6, $M_t/P = 0.05$: therefore, from Equation 11.10,

$$\sigma_{add} = (0.05)(6/200^2) \times 50\,000 = 0.4 \text{ N/mm}^2$$

Figure 11.7 Concrete slab design chart for C40 concrete (derived by the author using Equation 11.7)

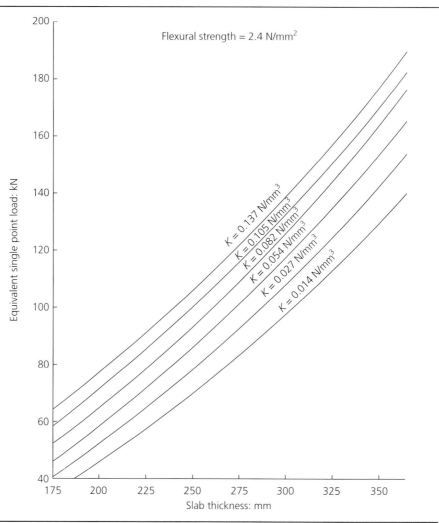

Step 3. Try plain C32/40 concrete with a flexural strength $\sigma_{\text{flex}} = 2.4\ \text{N/mm}^2$ (see Table 11.4).

Step 4.

$$\sigma_{\text{max}} = \sigma_{\text{flex}} - \sigma_{\text{ad}} = 2.4 - 0.4 = 2.0\ \text{N/mm}^2$$

This is the maximum flexural stress that the 60 kN load can be allowed to develop.

Step 5. From Equation 11.13, calculate the ESPL

$$\text{ESPL} = \text{APL}(\sigma_{\text{flex}}/\sigma_{\text{max}}) = 60 \times (2.4/2.0) = 72\ \text{kN}$$

Table 11.12 Point load multiplication factors for loads with a radius of contact outside the range 150–250 mm

Radius of contact: mm	Point load multiplication factor			
	$K = 0.013$ N/mm^3	$K = 0.027$ N/mm^3	$K = 0.054$ N/mm^3	$K = 0.082$ N/mm^3
50	1.5	1.6	1.7	1.7
100	1.2	1.2	1.3	1.3
150	1.0	1.0	1.0	1.0
200	1.0	1.0	1.0	1.0
250	1.0	1.0	1.0	1.0
300	0.9	0.9	0.9	0.9

Step 6. From Table 11.12, the modified factor to be applied to the ESPL is 1.2 for $l = 1$ m and $K = 0.027$ N/mm^3. Therefore,

$$\text{ESPL} = 72 \times 1.2 = 86.4 \text{ kN}$$

Step 7. From Figure 11.7, the thickness of slab required to carry a static load of 86.4 kN is 200 mm.

A 200 mm thick C40 concrete slab with 0.9 kg/m^3 of polypropylene fibre reinforcement is therefore adequate for this design.

11.6. Concrete block paving

Since the mid-1950s, concrete block paving has become a significant surfacing material for roads, and it is estimated that, worldwide, over 750 million square metres are now installed annually. Concrete block paving was introduced by the city of Rotterdam shortly after World War II as an expedient substitute for the brick pavers that had been previously used throughout the city centre – a post-war shortage of coal led to a shortage of brick pavers, and the bricks that could be manufactured at that time were needed to rebuild the city's buildings. Road engineers then noted that the new concrete pavers had superior engineering properties, particularly in relation to dimensional stability and skidding resistance. This led to the development of a significant Dutch market in which – because the pavers had been introduced as a replacement for paving bricks – the use of rectangular pavers predominated.

German building block manufacturers turned to producing concrete paving blocks following a collapse in demand for their traditional building product in 1962. This eventually led to Germany becoming the world's largest consumer of pavers, with over 100 million square metres now being installed annually: in Germany, also, pavers of non-rectangular shapes became popular, and different cities now have their preferred shapes. The worldwide adoption of pavers, which began in the 1970s, was the result of German manufacturers of paver-making equipment marketing their products internationally at a time when there was a general perception that roads and streets should have more interesting appearances. The subsequent introduction of pedestrian-only schemes and traffic-calming measures into town and neighbourhood centres in Europe further stimulated this usage.

The technology of concrete block paving has evolved since the early 1970s, and the reader is referred to the literature – especially to the proceedings of a series of international conferences and workshops (ICCBP, 1980–2015) – for details of ongoing technological developments.

11.6.1   Components of a concrete block pavement

The components of a typical concrete block pavement are shown in Figure 11.8. Note, however, that not all of the component layers illustrated in this figure need be present in every pavement.

Typically, concrete block road pavements comprise pavers that are bedded in a 'laying' course of sand on top of a lean concrete or asphalt concrete base (which is still termed 'roadbase' by block pavers), and the foundation usually comprises a crushed rock or cement-bound sub-base (which, if appropriate, may be atop a capping layer) that is designed as for a flexible pavement. In the main, this type of pavement has proved successful for all types of road pavements except in the case of heavily trafficked urban streets where channelised commercial vehicle traffic has caused unacceptable rutting and surface irregularity. (During the 1980s, for example, the suitability of block pavers as a surfacing material for roads trafficked mainly by buses in city centres was queried as result of this concern, and this led to a search for improved specifications for paver bedding sands.) The author has developed enhanced specifications for the bedding sand and roadbase that are included in the next section.

Figure 11.8 Typical cross-section of a concrete block pavement

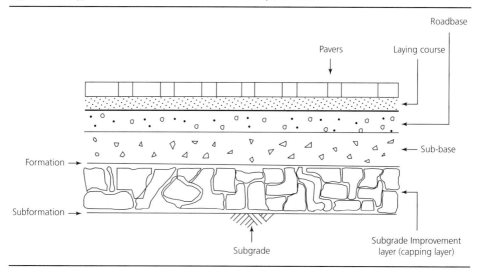

11.6.2 Basis of design

In the UK, concrete block pavements are designed and laid in accordance with criteria specified in British standard publications (BSI, 2001a, 2001b, 2005, 2009). The design of new pavements is based upon the flexible pavement design procedure developed by the Transport and Road Research Laboratory for asphalt roads (Powell *et al.*, 1984) but with block pavers used as the surfacing in place of conventional asphalt.

The design process normally proceeds according to the flow charts shown in Figure 11.9. In the case of a new road, the first steps require (a) estimating the numbers of commercial vehicles expected to use the pavement during its design life, expressed in millions of standard axles (msa) (see also Chapter 12), and (b) assessing the strength of the subgrade.

11.6.2.1 Subgrade capping and sub-base courses

The design CBR of the subgrade is most commonly obtained by direct laboratory test measurements. Alternatively, in the case of a cohesive subgrade soil, the CBR can be estimated from measurements of the plasticity index or of the equilibrium suction index (Powell *et al.*, 1984). In locations where it is possible that the subgrade may become saturated during all or part of the design life of the pavement, the CBR test that employs the soaking procedure should be used.

Figure 11.9 Design procedure for a new concrete block pavement. (a) Table 3 of BS 7533-1:2001 (BSI, 2001a): the main pavement design table. (b) Figure 2b of BS 7533:2001: the pavement foundation thickness when only a crushed rock sub-base is installed as the foundation. (c) Figure 2a of BS 7533-1:2001: foundation layer thicknesses for a foundation composed of a crushed rock sub-base installed over a capping layer. All thicknesses are in mm

Start

Determine cumulative standards axles (msa) for design

Design msa	0.5–1.5				>1.5–4				>4–8		>8–12	
CBM3 roadbase thickness (in mm) or	130	130	130	130	160	150	145	130	195	180	245	230
Dense bitumen macadam roadbase 100 pen (in mm)	130	130	130	130	160	150	145	130	170	155	185	170
Laying course (in mm)	30	30	30	30	30	30	30	30	30	30	30	30
Paver thickness (in mm)	50[a]	60	65	80	50[b]	60	65	80	65	80	65[b]	80

End

[a] For clay pavers type PB and concrete pavers only
[b] For clay pavers type PB only
[c] There is no long-term evidence concerning the performance in service of 65 mm pavers beyond 4 msa, so this design information has been extrapolated

(a) BS 7533-1: Table 3. Roadbase, laying course and paver thickness design chart

Figure 11.9 Continued

Start

Determine design subgrade CBR and number of standard axles using sub-base as an access road

Foundation design option 2b

Subgrade CBR design value	2% or less[a,b]	>2%– <3%[a]	≥3%– <4%[a]	≥4%– <5%	≥5%– <10%	≥10%– <15%	≥15%– <30%	30% or greater
Untrafficked	Figure 2a	170	150	150	150	150	150	0
Up to 4 dwellings or 2000 m² commercial or standard axles (sa)	Figure 2a	250	190	160	150	150	150	0
Up to 20 dwellings or 200 sa	Figure 2a	310	240	210	180	150	150	0
Up to 50 dwellings or 5000 m² commercial or 500 sa	Figure 2a	350	270	230	200	160	150	0
Up to 80 dwellings or 8000 m² commercial or 1000 sa	Figure 2a	400	310	270	225	180	150	0
Large development or 5000 sa	Figure 2a	450	350	310	270	240	225	0

Go to Figure 3 in BS 733-1 for structural design

[a] Below 4% a separating membrane is required
[b] Heavy grade support mesh or fabric may be necessary to enable construction to proceed

(b) BS 7533-1: Figure 2b. Sub-base-only option

If a capping layer is to be used in conjunction with a sub-base, the thickness of each is selected according to the subgrade CBR value, as indicated in Figure 11.9(c). For a subgrade with a CBR of <5%, the designer is allowed a choice of capping thickness: this permits engineering judgement to be used according to when, where and how the CBR was measured, local knowledge regarding the performance of the subgrade soil, the extent to which the capping layer and sub-base are likely to be trafficked before the remaining layers of the pavement are laid, the proximity of subgrade drainage, and the time of year for construction. The capping layer material should provide a CBR of >15% before laying the specified thickness of sub-base, when designed according to the values given in Figure 11.9(c).

If the sub-base is not to be used as an access road, then its minimum thickness should be either 150 mm if a capping layer is provided, or 225 mm if there is no capping layer. If a capping layer is provided and the sub-base is to be used as an access road during

Figure 11.9 Continued

```
Start
```

Determine design subgrade CBR and number of
standard axles using sub-base as an access road

Foundation design option 2a								
Subgrade CBR design value	2% or less[a,b]	>2%– <3%[a]	≥3%– <4%[a]	≥4%– <5%	≥5%– <10%	≥10%– <15%	≥15%– <30%	30% or greater
Untrafficked	150/150	150/150	Sub-base alone – see Figure 2b					0/0
Up to 4 dwellings or 2000 m² commercial or standard axles (sa)	150/210	150/180	150/150	Sub-base alone – see Figure 2b				0/0
Up to 20 dwellings or 200 sa	150/370	150/250	150/170	150/160	150/150	150/150	150/0	0/0
Up to 50 dwellings or 5000 m² commercial or 500 sa	150/470	150/340	150/250	150/220	150/200	150/150	150/0	0/0
Up to 80 dwellings or 8000 m² commercial or 1000 sa	150/600	150/450	150/350	150/300	150/250	150/180	150/0	0/0
Large development or 5000 sa	200/600	200/450	150/450	150/350	150/300	150/250	150/150	0/0

Go to Figure 3 in BS 7533-1 for structural design

NOTE Sub-base and capping are shown thus: 150/350
[a] Below 4% a separating membrane is required
[b] Heavy grade support mesh or fabric may be necessary to enable construction to proceed

(c) BS 7533-1: Figure 2a. Capping and sub-base option

construction, then the sub-base thicknesses should be as in Figure 11.9(c), according to the amount of traffic that is expected to travel directly over the sub-base during construction: this traffic flow can be determined as the number of equivalent standard axles (esa) trafficking the sub-base or, as indicated in Figure 11.9(c), by either the number of dwellings or the equivalent size in square metres of the industrial or commercial property being constructed. If a site cannot be categorised in any one of these ways, it should be assumed that the sub-base will serve a large development trafficked daily by 5000 esa. (Note that, unless detailed planning data are available, engineering judgement may be needed to qualify the assessment of the amount of construction traffic to which the sub-base is likely to be subjected.)

If the roadbase of the pavement is to be an unbound material, a type 2 sub-base should not be used if the design traffic load is expected to exceed 2 msa: instead, the sub-base should be composed of a bound material. The CBR of the constructed sub-base should be at least 30% (HA *et al.*, 2014). The detailed preparation of the sub-base should be in accordance with the recommendations in BSI (2005).

When rainfall is expected, the sub-base should be covered as quickly as possible to prevent its saturation and to protect any capping layer and the subgrade. Prior to the construction of the roadbase, remedial work may be required on the surface of the sub-base if it has been used as an interim construction access road. If excessive trafficking of the sub-base has caused rutting of the capped/non-capped subgrade, or contamination of the sub-base by subgrade material, the complete removal of the sub-base may be necessary, to enable remedial works to be carried out on the subgrade.

11.6.2.2   Roadbases, laying courses, and pavers

Figure 11.9(a) shows, for various traffic loads over the design life of the paving, the recommended roadbase thicknesses for cement-bound material and dense bitumen macadam. Note that in cases where the design traffic load is <0.5 msa, a roadbase need not be provided. A minimum roadbase thickness of 130 mm is specified in Figure 11.9(a) for practical construction reasons.

The bedding sand used in the laying course of a concrete block pavement should be in accordance with the recommendations in BSI (2005). The nominal laying course thickness is either 50 mm for unbound roadbases or, as shown in Figure 11.9(a), 30 mm for bound roadbases. In the case of pavements likely to be subjected to channelled traffic, particularly buses, the bedding sand is required to be a naturally occurring uncrushed silica-based material, and the fraction passing the 75 μm sieve must be less than 0.3% by mass.

Most commonly, trafficked concrete pavers are 60, 65 or 80 mm thick. While there is considerable debate regarding the preferred size and shape of paver, the author has found 100 mm × 200 mm × 80 mm rectangular blocks to be suitable for all trafficked situations, and recommends that pavers of other sizes and shapes be used with care.

11.6.3   Permeable block pavements

Since the mid-2000s, the use of permeable concrete block pavements (BSI, 2009) has grown in popularity. These pavements are designed to deliberately allow the infiltration of water through wider joints between individual pavers, and this water flows through bedding grit into a storage medium directly below the pavers. The way in which this water is then dealt with depends upon site-specific circumstances. In some pavements, the water is allowed to drain through each of the pavement courses and to infiltrate the underlying subgrade; in others, the water is held within the pavement, from whence it is discharged in a controlled manner to downstream drainage, or harvested and used as grey water.

Advice on the design of permeable concrete block pavements is also available from suppliers of the pavers. Because of the environmental implications, permeable pavement guidance tends to be specific to particular countries and regions.

REFERENCES

BSI (British Standards Institution) (1983) BS 1881-116:1983: Testing concrete. Method for determination of compressive strength of concrete cubes. BSI, London, UK.

BSI (1990) BS 1377-9:1990: Methods for test for soils for civil engineering purposes. In-situ tests. BSI, London, UK.

BSI (2001a) BS 7533-1:2001: Pavements constructed with clay, natural stone or concrete pavers. Guide for the structural design of heavy duty pavements constructed of clay pavers or precast concrete paving blocks. BSI, London, UK.

BSI (2001b) BS 7533-2:2001: Pavements constructed with clay, natural stone or concrete pavers. Guide for the structural design of lightly trafficked pavements constructed of clay pavers or precast concrete paving blocks. BSI, London, UK.

BSI (2005) BS 7533-3:2005 + A1:2009: Pavements constructed with clay, natural stone or concrete pavers. Code of practice for laying precast concrete paving blocks and clay pavers for flexible pavements. BSI, London, UK.

BSI (2009) BS 7533-13:2009: Pavements constructed with clay, natural stone or concrete pavers. Guide for the design of permeable pavements constructed with concrete paving blocks and flags, natural stone slabs and setts and clay pavers. BSI, London, UK.

Chandler JWE and Neal FR (1988) *The Design of Ground Supported Concrete Industrial Floor Slabs: BCA Interim Technical Note 11.* British Cement Association, Crowthorne, UK.

HA (Highways Agency), Transport Scotland, Welsh Government and Department for Regional Development Northern Ireland (2006a) Section 2: pavement design and construction. Part 3. Pavement design. In *Design Manual for Roads and Bridges*, vol. 7. *Pavement Design and Maintenance.* Stationery Office, London, UK, HD 26/06.

HA, Transport Scotland, Welsh Government and Department for Regional Development Northern Ireland (2006b) Section 2: pavement design and construction. Part 1: traffic assessment. In *Design Manual for Roads and Bridges*, vol. 7. *Pavement Design and Maintenance.* Stationery Office, London, UK, HD 24/06.

HA, Transport Scotland, Welsh Government and Department for Regional Development Northern Ireland (2014) *Manual of Contract Documents*, vol. 1. *Specification for Highway Works.* Stationery Office, London, UK.

ICCBP (1980–2015) *Proceedings of the 1st–11th International Conferences on Concrete Block Paving.*

Powell WD, Potter JF, Mayhew HC and Nunn ME (1984) *The Structural Design of Bituminous Roads.* Transport and Road Research Laboratory, Crowthorne, UK, Report LR1132.

Westergaard HM (1947) New formula for stress in concrete pavements of airfields. *Proceedings of the American Society of Civil Engineers* **73(5)**: 687–701.

Highways
ISBN 978-0-7277-5993-1

Chapter 12
Current UK thickness design recommendations for new bituminous and concrete pavements

Nick Thom   Lecturer, Nottingham Transportation Engineering Centre, UK

12.1.   Introduction

Highways England (formerly the Highways Agency (HA)) is the dominant voice in road pavement design in the UK, and this chapter will therefore concentrate primarily on its specifications, standards and advice notes. As well as being used in England, these HA documents are still the primary source of structural design advice in Scotland, Wales and Northern Ireland, and in all the various local highway authorities, albeit some have their own distinctive guidance, notably on surface course material type.

The current documents have developed incrementally over the course of more than 50 years, and the design thicknesses are therefore based on a large body of evidence of past performance – although naturally the continuing changes to heavy goods vehicle volumes, types and tyres, as well as pavement materials, mean that past experience is not a foolproof guide to future performance. Since the 1980s, experience has therefore been supplemented by analytical calculations, effectively extrapolating for a changing future, calculations that have been reviewed more than once as performance experience has accumulated. The basis of the approach is therefore undeniably sound.

Nevertheless, problems have occurred. Probably the most notable was the adoption in the 1990s of a conventional dense bitumen macadam – now known as asphalt concrete – with a 15 pen grade binder as a roadbase material for use in the UK; this was a variation on a material with a proved track record in France. Unfortunately, the decision to reduce the binder content from the level used in France to the standard dense bitumen macadam roadbase binder content used in the UK led to numerous early failures due to insufficient resistance to moisture attack. Although the specification was altered as soon as the issue became apparent, consequent problems are still occurring today.

Thus, lessons have been learned, and current design documents should be seen as presenting reasonably conservative state-of-the-art pavement designs. They are conservative because they are primarily intended for use in the most important and heavily trafficked roads, where disruptive maintenance must be kept to a minimum. The intention is that the designs represent 85% reliability (i.e. there is a 15% chance of an early failure), accepting that variability in materials, thickness, compaction and so on inevitably lead to a degree of uncertainty in performance.

12.2. Derivation of current designs

The current HA design standards stem from work done at the Transport Research Laboratory (Nunn, 2004). This research report (TRL615) paved the way for the adoption of a much more versatile approach than that previously used, which had been highly restrictive and was seen as a barrier to innovation. Previously, only a single standard pavement foundation design had been permitted whereas TRL615 proposed four classes, on the basis that the higher the class of foundation the thinner the upper pavement could be. Previously, the benefits of hydraulically bound foundation materials (e.g. lean concretes) had not been properly included: TRL615 developed a system of design factors that allowed for the realistic computation of pavement life and, therefore, realistic pavement design. Previously, only a restricted range of asphalt options had been allowed: TRL615, by introducing a factor related to the flexibility of each material, opened the door to the introduction of alternatives.

The actual basis of the designs in TRL615 is analytical. The major step forward that it embodied consisted of the way it combined analytical design with a thoroughly pragmatic set of adjustments, based on experience, that gave the Highways Agency the confidence that it demanded – as well as flexibility for designers.

12.2.1 Key design documents

The bulk of this chapter is based on the following (now) Highways England's HA publications:

■ *Specification for Highway Works*, comprising volume 1 of the *Manual of Contract Documents for Highways Works* (HA *et al.*, 2014)
■ *HD 24, 'Traffic assessment'* (HA *et al.*, 2006a), which is found in volume 7 of the *Design Manual for Roads and Bridges* (HA *et al.*, 2006a)
■ *IAN 73, Design Guidance for Road Pavement Foundations*, an interim advice note that may eventually become HD 25 in volume 7 of the *Design Manual for Roads and Bridges* (HA, 2009)
■ *HD 26, 'Pavement design'*, which is also found in volume 7 of the *Design Manual for Roads and Bridges* (HA *et al.*, 2006b).

The first of these documents (*Specification for Highway Works*) describes the permissible range of materials for each layer in the pavement; the roles of the latter three documents are obvious from their titles. In most respects the *Specification for Highway Works* is independent of the pavement design process, that is, the designer calls up a certain material, and IAN 73 and HD 26 assume that the material has the properties it is supposed to have; in other words, the *Specification for Highway Works* is there to make sure that there is a way of controlling material production so that the desired properties really are achieved.

The design philosophy underlying the HA approach is simple, and expressed in the flow chart in Figure 12.1.

Figure 12.1 The HA pavement design philosophy applied in the UK

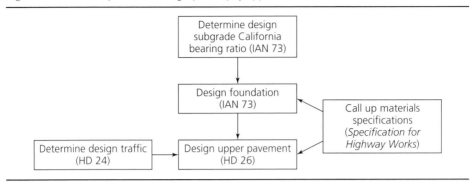

12.3. Estimating the design traffic load

Pavement designs tend not to be highly sensitive to traffic estimation – which is fortunate since accurate estimation is impossible. Figure 12.2, which shows the governmental estimates of future national traffic growth as determined in 2013, illustrates the current range of uncertainty when estimating traffic growth over the design life of a road.

As with most pavement design methods, the HA approach is to rationalise the huge variety of actual vehicles on the roads by assigning to each a damaging power (known in HD 24 as a wear factor) in terms of numbers of equivalent standard (80 kN) axles. Table 12.1 replicates the information in HD 24. The numbers are derived from data collected from a number of weigh-in-motion (WIM) sites at various locations across the motorway and trunk road network.

Figure 12.2 UK traffic growth forecasts (DfT, 2013)

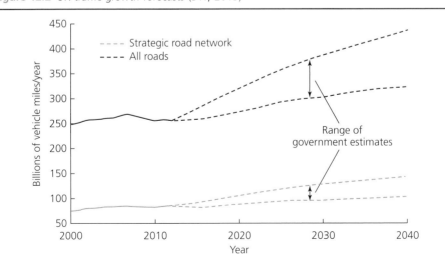

Table 12.1 Vehicle wear factors currently used in the design of pavements for existing and new roads (including road widenings)

Vehicle category	Vehicle type	Wear factor (= number of equivalent standard axles)	
		Maintenance	New road
Public service vehicles (PSV)	Buses and coaches	2.6	3.9
Other goods vehicles category 1 (OGV1)	2-axle rigid	0.4	0.6
	3-axle rigid	2.3	3.4
OGV1 + PSV average		0.6	1.0
Other goods vehicles category 2 (OGV2)	3-axle articulated	1.7	2.5
	4-axle rigid	3.0	4.6
	4-axle articulated	1.7	2.5
	5-axle articulated	2.9	4.4
	6-axle articulated	3.7	5.6
OGV2 average		3.0	4.4

It will be noted that Table 12.1 contains two sets of wear factors. Those shown for maintenance design are the real average factors determined from WIM evidence. Those shown for new road design are artificially high, and, in effect, this is where a safety factor has been applied to take account of uncertainty in the prediction of future traffic flows.

From here, the designer has two routes to the estimation of traffic:

■ Base the estimate on actual traffic counts of the numbers of commercial vehicles per day (cvd); that is, of the numbers of PSV and OGV but excluding vehicles below 3.5 t. This information would normally be available in the case of a rehabilitation/maintenance design for an existing road, but it may also be possible to generate it where the traffic on a new road link can sensibly be estimated from counts on adjacent links.
■ If detailed counts are not available (e.g. for new road schemes), generate an estimate of the average annual daily flow (AADF) in cvd. In the absence of count data, this estimate has to be based on local network analysis, the principles of which are given in the *Traffic Appraisal Manual* (HA *et al.*, 1997). The estimated AADF then has to be subdivided into PSV + OGV1 and OGV2, based on whatever traffic studies have been carried out. For new road schemes, HD 24 stipulates that the percentage of OGV2 used in design cannot be less than a certain percentage of the AADF (in cvd), as derived from a chart that can be closely approximated using

$$\text{Min}\%_{\text{OGV2}} = 42.5 \times [\log_{10}(\text{AADF}) - 2] \tag{12.1}$$

If the second approach is taken, a chart in HD 24 can be used to directly obtain the 40 year design traffic load in millions of standard axles (msa): this chart has standard (national average) traffic growth assumptions built into it. The following equations can be used to give close approximations to the results obtained from the chart:

$$\text{design traffic (msa)} = [1.8 + 2.1 \times (\%_{\text{OGV2}}/25)] \times 10^{0.96 \times [\log_{10}(\text{AADF}) - 2]} \tag{12.2}$$

or

$$\text{design traffic (msa)} = 10^{\{0.12 + 1.58 \times [\log_{10}(\text{AADF}) - 2]^{0.76}\}} \tag{12.3}$$

whichever the greater.

However, where more information is available, or where a design life other than 40 years is to be used, the design traffic has to be summed year by year for each category or class of vehicle. This may either be a simple division between PSV + OGV1 and OGV2 or it can involve summing across individual vehicle types, depending on the details available. It is therefore necessary to know (a) the AADF for each category or type of vehicle, (b) the appropriate wear factor from Table 12.1 and (c) a growth factor that takes account of anticipated future changes in traffic.

HD 24 gives growth factors for both PSV + OGV1 and OGV2 as a function of design life in graphical form. Again, the following equations are offered as close approximations:

$$\text{growth factor for PSV} + \text{OGV1} = 1 + (\text{design life}/180)^{1.1} \tag{12.4}$$

$$\text{growth factor for OGV2} = 1 + (\text{design life}/54.5)^{1.3} \tag{12.5}$$

The design traffic load in each category or type is therefore

$$\text{design life (years)} \times 365 \times \text{AADF} \times \text{wear factor} \times \text{growth factor} \tag{12.6}$$

This is then summed for both categories or across all vehicle types, to give the total design traffic load (in msa).

For multi-lane carriageways, the final step is then to read off the percentage in the most heavily trafficked lane (lane 1) as a function of the total commercial vehicle flow in one direction. While again this is given in graphical form in HD 24, the equations below give acceptable approximations:

for $\leq 10\,000$ cvd: $\%$ in lane $1 = 97 - 2.5 \times [\log_{10}(\text{AADF}) - 2]^{3.35}$ (12.7)

for $\geq 10\,000$ cvd: $\%$ in lane $1 = 50 + 215\,000/\text{AADF}$ (12.8)

For new road design, it is the traffic load in the heaviest loaded lane that controls the overall pavement design thickness (including that of the shoulder). The heaviest loaded lane for carriageways with two or three lanes in one direction is normally the nearside lane (i.e. lane 1); for carriageways with four or more lanes, it is not always lane 1 (due to the influence of lane drops/gains). For maintenance design purposes, it is sometimes necessary to estimate the traffic load in the other lanes separately.

12.3.1 Example (using equations)

Consider the case of a new dual two-lane road with an expected AADF of 5000 cvd in each direction, with a 30 year design life.

Step 1. Using Equation 12.1, obtain the percentage OGV2 for design (assuming no other information)

$$\text{Min}\%_{\text{OGV2}} = 42.5 \times [\log_{10}(5000) - 2] = 72.2\%$$

Step 2. Determine the growth factors, using Equations 12.4 and 12.5

$$\text{growth factor for PSV} + \text{OGV1} = 1 + (30/180)^{1.1} = 1.14$$
$$\text{growth factor for OGV2} = 1 + (30/54.5)^{1.3} = 1.46$$

Step 3. Using Equation 12.6 and Table 12.1, multiply up to obtain the total traffic load

PSV + OGV1: $30 \times 365 \times (0.278 \times 5000) \times 1.0 \times 1.14 = 17$ msa

OGV1: $30 \times 365 \times (0.722 \times 5000) \times 4.4 \times 1.46 = 254$ msa

The total traffic load is therefore

$$17 + 254 = 271 \text{ msa}$$

Step 4. Using Equation 12.7, determine the percentage in lane 1

$$\% \text{ in lane } 1 = 97 - 2.5 \times [\log_{10}(5000) - 2]^{3.35} = 82.2\%$$

Therefore, combining steps 3 and 4

$$\text{lane 1 design traffic load} = 0.822 \times 271 = 223 \text{ msa}$$

A design traffic load at this level represents a very major road. By comparison, major links within cities in the UK might be expected to have design traffic values in the 20–80 msa range. Other connector roads might be at 5–20 msa, while less than 5 msa represents a minor road that would commonly be designed for a life of less than 30 years.

12.4. Materials used in road pavements

Figure 10.2 (see Chapter 10) illustrates the different layers that make up a typical heavy duty pavement in the UK. The current HA design philosophy, which is based on the 'versatile' approach espoused in TRL615, is to open up pavement construction to as wide a choice of materials as possible. The following sections summarise the material options currently available to the designer.

12.4.1 Asphalt materials

The surface course can be composed of any one of a large number of HA-approved thin surfacing systems. These are not described in any of the design standards, nor in the *Specification for Highway Works*, but are covered by the highway authorities product approval scheme (HAPAS). The intention of this scheme is that the surface course market will be very versatile, and it seeks to achieve this by simply ensuring that all materials used have adequate skid resistance and durability. It might be noted that this approach has attracted criticism, and is subject to ongoing investigation since there have been a number of cases of unanticipated short life, possibly associated with unexpectedly high air voids content.

Here, it should be noted that many other highway authorities allow their own materials to be used in surface courses, notably hot rolled asphalt. In Scotland, use is now made of stone mastic asphalt to German specifications, which provides a significantly more durable material compared with the HAPAS-approved systems, while avoiding some of the problems (notably spray) associated with the use of hot rolled asphalt. These variations reflect the fact that surface course design is a relatively fast-developing field. Nevertheless, future changes in surface course specification are unlikely to impact on the pavement thickness design approach covered in this chapter.

The following materials are included in HD 26 for use in binder courses and base courses:

- DBM125 – this dense bitumen macadam would be classified as an asphalt concrete in current EN standards (the '125' stands for a nominal binder penetration of 125 dmm).
- DBM50 – this is also an asphalt concrete with a 50 dmm binder penetration.
- HDM50 – this heavy duty macadam is a very similar material to DBM50, but has a marginally different aggregate gradation and binder content.
- HRA50 – this hot rolled asphalt is a traditional gap-graded UK material that uses a 50 dmm penetration binder.
- EME2 – *enrobé à module élevé* is basically an asphalt concrete of French design that uses a nominal 15 dmm penetration binder and a slightly increased binder content.

In practice, the two most common materials used in the binder and base courses of heavy duty pavements in the UK are DBM50/HDM50 – these cause no differences in design thickness requirements – and EME2. Other authorities make use of DBM125 designs, and some may still use hot rolled asphalt.

12.4.2 Cement and hydraulically bound materials

In HD 26 it is generally assumed that pavement quality concrete designs will comprise continuously reinforced concrete with either 0.4% or 0.6% longitudinal reinforcing steel, depending on how much overlying asphalt is to be provided. The strength is specified in terms of its 28 day mean flexural strength. A jointed concrete pavement is no longer given as a standard design case, and is only used on UK roads in special cases, for example, to match up with an existing jointed construction.

In the EN standards there is a wide range of allowable materials that can be used in hydraulically bound bases, sub-bases and cappings. In the HA designs, hydraulically bound materials (HBMs) for use in pavement bases can be either cement, slag or fly ash-bound, and strength grades can range from C8/10 (i.e. minimum 28 day compressive strengths of 8 or 10 N/mm² for cylinder or cube specimens, respectively) to C16/20. When used in sub-bases or cappings, the range is effectively infinite, since, as explained in Section 12.5, the contractor has an option to use a performance-based specification that tests for the in situ condition rather than for material composition. This approach also allows in situ mixed materials to be used.

12.4.3   Unbound materials

The principal unbound material specified (HA *et al.*, 2014) for use in the sub-bases of major road pavements is termed 'type 1 unbound mixture' (see the gradation in Table 12.2). Type 1 material is normally made from crushed rock of high quality that has its durability specified in terms of a Los Angeles abrasion value (see Chapter 6), and the size fraction passing the 0.425 mm sieve must be non-plastic. Type 2 aggregate, which is a lower quality sand and gravel material for which the fraction passing the 0.425 mm sieve can have a plasticity index of up to 6, is acceptable in designs for low (<5 msa) traffic loads.

In practice, a performance-based specification option for sub-bases is also allowed, and this opens the door to any number of contractor-proposed alternatives.

In the case of cappings, materials are specified that range from very coarse to very fine that are acceptable under many circumstances but, as with sub-bases, these can be extended where a performance-related foundation specification is used.

Table 12.2 Summary grading requirements for type 1 and 2 aggregates

Sieve size: mm	Overall grading range: % by mass passing	
	Type 1	Type 2
63	100	100
31.5	75–99	75–99
16	43–81	50–90
8	23–66	30–75
4	12–53	15–60
2	6–42	NA
1	3–32	0–35
0.063	0–9	0–9

NA, not applicable

12.5. Foundation design

12.5.1 Foundation classes

The philosophy that has been adopted in UK design since the 1980s is that the foundation design and upper pavement design are entirely separate activities. The justification for this is that one of the principal roles of the foundation is to facilitate construction of the overlying pavement courses, no matter what the make-up of those layers may be and no matter what traffic service life the pavement is to be designed for. In fact, until the early 1990s this philosophy meant that only two grades of material were normally used in the foundations of flexible and flexible composite pavements, namely a 'type 1 crushed rock sub-base' and a 'capping', and their thicknesses depended solely on the quality of the subgrade as expressed by its California bearing ratio (CBR) value. In effect, this meant that there was just a single standard 'class' of foundation for flexible pavements. In the case of rigid pavements, the only difference was that the crushed rock sub-base had to be replaced by a cement-bound material.

This was a very simple and well-understood system – but it was anything but versatile. There was absolutely no scope for a contractor to propose any alternative materials (e.g. a recycled industrial by-product or an in situ stabilised material), since there was no mechanism to allow any thickness design other than that for a standard foundation using standard materials. Thus, following a number of research projects, it was decided that the system should be broadened out. As a result, instead of a single standard class of foundation, there are now four classes designated in IAN 73, each of which is associated with a particular long-term minimum value of equivalent foundation modulus, as follows:

- class 1– 50 MPa (minimum) equivalent foundation modulus that is intended to represent a capping-only foundation
- class 2 – 100 MPa (minimum) equivalent foundation modulus that is equivalent to the old standard foundation using a type 1 crushed rock sub-base or a sub-base plus capping design
- class 3 – 200 MPa (minimum) equivalent foundation modulus representing foundations using a low-strength HBM sub-base
- class 4 – 400 MPa (minimum) equivalent foundation modulus representing foundations using a high-strength HBM sub-base.

Clearly, the fact that there are now four classes of foundation means that the upper pavement design is dependent on the class of foundation used, which means that for the designer there will be a trade-off between money spent on the foundation and money saved in the upper pavement – or vice versa.

For pavements relating to IAN 73 designs there is the further restriction that class 1 foundations may only be used on roads taking traffic loads up to 20 msa, because of the likelihood of damage during construction.

12.5.2 Performance foundations

Having selected a foundation class, the designer's next task is to determine which materials and thicknesses should be used to produce it – and here it is fair to say that the method of so doing is genuinely versatile. The term performance foundation, which is related to the performance-related specification of materials, basically means that the contractor has a free choice of material source – subject to appropriate checks on durability and on the potential of the material to cause environmental damage – but the quality of the proposed foundation has to be checked on site before and during construction. This check is carried out by measuring the surface modulus, initially on a demonstration trial foundation, using a lightweight deflectometer (LWD) – a simple drop-weight test that measures the ground surface deflection under an applied load pulse. On major road projects, the LWD must be calibrated against a falling-weight deflectometer (FWD) – see Chapter 16. Figure 12.3 describes the LWD equipment. Details of the test and its specified frequency and timing are given in IAN 73, which includes draft clauses for Series 800 of the *Specification for Highway Works*.

Further surface modulus performance testing must be carried out during the main works stage of the foundation within the 24 h before the foundation is covered by the construction of the upper pavement. Table 12.3, which is reproduced from IAN 73, lists the unadjusted mean and minimum surface modulus values for each foundation class and different categories of materials that must be achieved at the top of the built foundation at that time. In general, a test spacing of 20 m along each lane is specified, and the modulus limits in IAN 73 relate both to the absolute minimum value and the mean of any five consecutive tests: for example, 50 and 80 MPa, respectively, for a class 2 foundation with unbound materials.

In theory, this is an excellent system and should provide a guaranteed level of performance from the foundation throughout the service life of the pavement. In practice, however, it is more complicated than this: for example, (a) the LWD modulus obtained depends on the moisture state of the layer, which in turn depends on the weather; (b) the

Figure 12.3 The LWD test

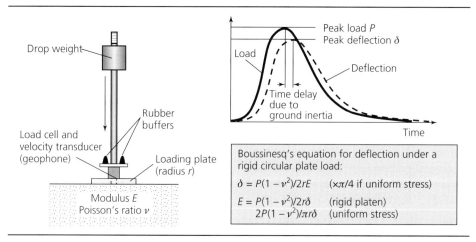

Drop weight

Load cell and velocity transducer (geophone)

Rubber buffers

Loading plate (radius r)

Modulus E
Poisson's ratio ν

Load

Peak load P
Peak deflection δ

Deflection

Time delay due to ground inertia

Time

Boussinesq's equation for deflection under a rigid circular plate load:

$\delta = P(1 - \nu^2)/2rE$ ($\times \pi/4$ if uniform stress)

$E = P(1 - \nu^2)/2r\delta$ (rigid platen)
     $2P(1 - \nu^2)/\pi r\delta$ (uniform stress)

Table 12.3 Top-of-foundation surface modulus requirements from LWD testing

	Class 1	Class 2	Class 3	Class 4
Long-term equivalent foundation modulus: MPa	50	100	200	400
Mean LWD-determined surface modulus: MPa				
Unbound materials	40	80	NA	NA
Fast-setting materials	50	100	300	600
Slow-setting materials	40	80	150	300
Minimum LWD-determined surface modulus: MPa				
Unbound materials	25	50	NA	NA
Fast-setting materials	25	50	150	300
Slow-setting materials	25	50	75	150

NA, not applicable

nominal equivalent foundation modulus is intended to represent the situation once the upper pavement has been constructed, whereas the LWD test induces quite different stresses and, in granular materials particularly, this leads to a quite different apparent modulus; and, (c) as a further complication, cement-bound foundation layers are commonly rather stiff in their early life (i.e. while they are still largely intact), and their long-term state is expected to be that of a layer that has been partially cracked by trafficking, whereas a foundation layer that is bound using a slower-setting hydraulic binder may have a lower early life modulus than that applying in service. All of these complications have to be taken into account in formulating the requirements for the LWD modulus – and this is far from straightforward.

Note that the unbound and slow-setting materials are expected to measure less under the LWD than their eventual stiffness in the finished pavement. In the case of unbound materials, this is due to the stress-dependent nature of all granular materials; in the case of slow-setting mixtures, this is because further strength/stiffness gain is expected after the date of test. For fast-setting mixtures, on the other hand, which are defined as those including Portland cement, a higher value is expected in the short term since they are expected to crack during their in-service lives.

Note also that the requirement in Table 12.3 is specified in two forms. The mean is specified as the average of five consecutive LWD tests, and ensures a satisfactory average performance, while the minimum requirement places a limit on foundation variability. The intention of these limits is to ensure that the foundation is of adequate stiffness to allow the compaction of the overlying materials, and that it has a reasonable chance of providing appropriate long-term support to the upper pavement. Naturally, this cannot be absolutely guaranteed simply by passing an LWD test during construction; hence the additional durability requirements for materials. Compaction also has to be checked. But the most direct check of all is given by the requirement that a 60 m long trial area is produced for each different type of foundation, and that this area is tested thoroughly and also trafficked by 1000 equivalent standard axles.

From the point of view of Highways England, this performance system gives a good measure of assurance that the foundation promised is actually delivered in practice. Inevitably, however, problems have arisen with the implementation of this approach, the most significant of which are (a) the costs for all the additional testing, trafficking trials and so on, combined with the added risk to the contractor, make this performance foundation approach financially unattractive for all but the largest projects; (b) it is very difficult to achieve the class 2 required mean of 80 MPa from LWD tests on a type 1 sub-base; and (c) it is extremely easy to achieve the requirements for class 3 and 4 foundations using HBMs. These three points represent serious problems for different reasons: the first means that the intended opening of doors to new materials and innovative designs is stifled; the second suggests that the class 2 requirements are too onerous and that perfectly valid designs are being excluded; and the third suggests that the performance-related specification does not always provide a meaningful check on in-service performance for HBM layers. There is therefore a likelihood of future change.

12.5.3 Designing performance foundations

When designing performance foundations, it is necessary to predict the LWD surface modulus in advance, since it is squarely the responsibility of the contractor to ensure that the modulus requirement is met. IAN 73 gives guidance for this in the form of charts and design equations: Equations 12.9 to 12.17 are these equations. In the equations, the following definitions apply: H_{cap} is the capping layer thickness (mm), H_{sb} is the sub-base layer thickness (mm), E_{cap} is the capping layer stiffness (MPa), E_{sb} is the sub-base layer stiffness (MPa) and CBR is the California bearing ratio of the subgrade (%). The design equations are as follows.

Foundation class 1 (capping only). For a subgrade CBR of $\geq 2.5\%$ and $\leq 5\%$, the capping thickness (mm) is the greater of the thicknesses given by Equations 12.9 and 12.10

$$H_{cap} = 1.845 \times 10^3 \times E_{cap}^{-0.25}[1 - 0.395E_{cap}^{-0.025} \ln(\text{CBR})] \tag{12.9}$$

$$H_{cap} = 2.0 \times 10^2 \times E_{cap}[\ln(\text{CBR}) - 1.538] - 10.918 \times 10^3[\ln(\text{CBR}) - 1.541] \tag{12.10}$$

For a subgrade CBR of $>5\%$ and $\leq 15\%$

$$H_{cap} = 1.016 \times 10^3 \times E_{cap}^{-0.214}[1 - 0.23E_{cap}^{-0.026} \ln(\text{CBR})] \tag{12.11}$$

The minimum permissible value of H_{cap} is 150 mm, and the relationships in Equations 12.9 to 12.11 are valid for sub-base material with a layer stiffness modulus of between 50 and 100 MPa.

Foundation class 2 (sub-base only). For a subgrade CBR of $\geq 2.5\%$ and $\leq 15\%$

$$H_{sb} = 2.85 \times 10^3 \times E_{sb}^{-0.341}[1 - 0.389E_{sb}^{-0.021} \ln(\text{CBR})] \tag{12.12}$$

(Note: an error in IAN 73 has been corrected in this equation.)

For a subgrade CBR of $>5\%$ and $\leq 30\%$

$$H_{sb} = 9.25 \times 10^2 \times E_{sb}^{-0.202} - 69 \ln(CBR) \tag{12.13}$$

The minimum permissible value of H_{sb} is 150 mm, and the relationships in Equations 12.12 and 12.13 are valid for sub-base material with a layer stiffness modulus of between 150 and 250 MPa.

Foundation class 2 (sub-base on capping). For a subgrade CBR of $\geq 2.5\%$ and $\leq 15\%$

$$H_{sb} = 8.27 \times 10^4 \times [0.4123 \ln(E_{cap}) - 1]E_{sb}^{-[0.2075 + 0.1933 \ln(E_{cap})]}$$
$$- 21.39 E_{cap}^{1.745} E_{sb}^{[0.271 - 0.335 \ln(E_{cap})]} \ln(CBR) \tag{12.14}$$

$$H_{cap} = 3.01 \times 10^2 - 56 \ln(CBR) \tag{12.15}$$

where the same restrictions on layer thickness and stiffness modulus range apply as for the sub-base only.

Foundation class 3 (sub-base only). For a subgrade CBR of $\geq 2.5\%$ and $\leq 30\%$

$$H_{sb} = 8.44 \times 10^3 \times E_{sb}^{-0.48}[1 - 0.261 E_{sb}^{-0.008} \ln(CBR)] \tag{12.16}$$

where the minimum permissible value of H_{sb} is 175 mm, and the relationship is valid for sub-base material with a layer stiffness modulus of between 500 and 1000 MPa.

Foundation class 4 (sub-base only). For a subgrade CBR of $\geq 2.5\%$ and $\leq 30\%$

$$H_{sb} = 1.53 \times 10^4 \times E_{sb}^{-0.4833}[1 - 0.234 E_{sb}^{-0.025} \ln(CBR)] \tag{12.17}$$

where the minimum permissible value of H_{sb} is 200 mm, and the relationship is valid for sub-base material with a layer stiffness modulus of between 1000 and 3500 MPa.

The thickness designs given by Equations 12.9 to 12.17 should not be seen as more than providing guidance. The contractor's designer has to carry out a proper evaluation of the subgrade in order to assign a CBR, and then has to assess the likely quality of any sub-base or capping material that is to be used in terms of a modulus value. This represents a significant shift of responsibility, although it also opens up significant new opportunities for the use of previously non-approved but more cost-effective materials. It means, for example, that contractors can propose in situ recycling of an existing subgrade – but it also means that someone has to evaluate whether there is enough time between stabilis-ation and upper pavement construction for the material to reach the modulus it needs to pass the LWD test, not just the moving average but also the absolute minimum require-ment. Furthermore, the contractor has to ensure that the layer in question is protected adequately from the elements during construction, avoiding both excess water and frost damage. In short, while the financial savings may be enticing, the risk has to be thought about seriously.

12.5.3.1   Tools to aid in performance foundation design

The above design equations have the subgrade CBR as an input parameter, which means that a value for the CBR has to be first determined by the designer before the equations can be used. Measurement of the CBR itself is therefore a valid option, either directly or from interpretation of dynamic cone penetrometer (DCP) data. Yet, in reality, it is the stiffness modulus of the subgrade that is needed in order to generate a prediction of the foundation performance under the LWD, and the assumption underlying development of the design equations is the following relationship:

$$\text{modulus (MPa)} = 17.6 \times \text{CBR}^{0.64} \tag{12.18}$$

The problem is that the uncertainty inherent in this relationship is at least a factor of two (Brown *et al.*, 1987) – which means that a correct and realistic measure of the CBR can lead to a very optimistic estimate for subgrade modulus with certain soil types (notably heavy clays), which can then lead to an optimistic estimate of the LWD-determined foundation modulus.

Direct LWD testing of the subgrade may give an improved level of foundation design confidence; alternatively, for reconstruction or widening projects, an FWD test on the existing road (see Chapter 16) would also provide excellent data. The difficulty, however, is that these moduli would almost certainly need to be adjusted to give the value expected beneath the foundation layers alone. This would be a higher modulus than that given by direct LWD tests on the subgrade but lower than that derived from FWD testing on an existing road. Then, whichever method is used to determine a realistic modulus, the above equation has to be used to convert it back to a CBR for use in the foundation design equations.

The only other type of test described by IAN 73 in relation to foundation layers is a form of confined compression test known as the 'springbox' (Edwards *et al.*, 2005), which is able to measure the modulus of unbound or lightly bound materials. In this test, and others of the same type that have come onto the market (Semmelink and de Beer, 1995; Thom *et al.*, 2012), the specimen is confined by spring-loaded walls, giving stress conditions simulative of those in situ, and moduli measured in this way have been shown to be realistic (Thom *et al.*, 2012).

12.5.3.2   Example performance foundation design

Using the equations, design a class 2 foundation over a subgrade with an average LWD-measured modulus of 45 MPa and a DCP-derived CBR of 3%.

Step 1: select a design CBR. This could be 3% if the DCP-derived measure is accepted. Alternatively, the LWD modulus of 45 MPa converts to a CBR of about 4.3%, and one could argue that this should be increased slightly to allow for the difference in stress conditions between direct subgrade testing and whole-foundation testing. A value of 4.5% would therefore appear to be a sensible design value based on the modulus.

Step 2: select the foundation materials. In this example, it is assumed that conventional capping and type 1 crushed rock are available, for which the assumed values of modulus are 150 and 75 MPa, respectively. However, let it be assumed that an alternative by-product material with a modulus of 350 MPa (as measured in a confined compression test) is also available.

Step 3: calculate the design options. Using Equation 12.12 for a type 1 sub-base only

$$H_{sb} = 2.85 \times 10^3 \times 150^{-0.341}[1 - 0.389 \times 150^{-0.021} \ln(4.5)] = 244 \text{ mm}$$

Using Equations 12.14 and 12.15 for a type 1 sub-base and conventional capping

$$H_{sb} = 8.27 \times 10^4 \times [0.4123 \ln(75) - 1]150^{-[0.2075 + 0.1933\ln(75)]} - 21.39 \times 75^{1.745}$$
$$\times 150^{[0.271 - 0.335\ln(75)]} \ln(4.5) = 182 \text{ mm}$$

$$H_{cap} = 3.01 \times 10^2 - 56 \ln(4.5) = 217 \text{ mm}$$

Using Equation 12.12, check the alternative sub-base. While Equation 12.12 is technically only usable to a modulus of 250 MPa, check it out with 350 MPa also. Thus,

for $E_{sb} = 250$ MPa: $H_{sb} = 208$ mm

for $E_{sb} = 350$ MPa: $H_{sb} = 187$ mm

IAN 73 advises that a thickness should only be quoted after rounding up to the nearest 10 mm, so the options would appear to be either (a) 250 mm of a type 1 sub-base, (b) 190 mm of the alternative sub-base or (c) 190 mm of a type 1 sub-base over 220 mm of capping. In this case, the alternative sub-base may well be the most cost-effective.

12.5.4 Restricted foundation designs

The preceding sections have described the rather complicated but extremely versatile approach that is termed 'performance design'. When originally conceived, it was intended that most contracts let in the UK would use this foundation design approach. However, the testing requirements, particularly those for the construction of trial demonstration areas, combined with the additional contractor risk, have effectively prevented this from happening. At the same time, an alternative foundation design was also proposed that is much simpler to use. Termed 'restricted design', this alternative approach was originally intended for small projects only, but has since become widely adopted for large projects as well.

The restricted design process is extremely simple. Charts are included in IAN 73 for class 1, 2 and 3 foundations that use a restricted range of standard materials, and design thicknesses are derived from these charts once the strength of the subgrade (as measured by its CBR) is known. Because of the greater likelihood of damage during construction, class 1 designs are limited to pavements carrying traffic loads not exceeding 20 msa. Class 2 sub-base designs may make use of granular materials (types 1 and 2), cement-bound granular materials (CBGMs) and soil cement; however, cementitious materials cannot be used if they achieve compressive strengths of less than C3/4 after 28 days, and use of type 2 sub-base is limited to traffic loads not greater than 5 msa. Class 3 designs are restricted to CBGMs achieving the compressive strength class C8/10. The cement used in class 2 and 3 cement-bound designs is currently limited to CEM1 as the primary binder: this recognises the greater uncertainty in the use of other hydraulic binders. While a designed capping may be incorporated in class 2 foundations, the use of cappings is encouraged with all designs because of the practical benefits they provide in terms of working and compaction platforms, and, possibly, as a drainage layer below cement-bound materials.

The foundation thicknesses derived from the restricted foundation design charts are typically about 100 mm greater than the equivalent class 2 and 3 performance designs and 120 mm greater than the equivalent class 1 designs. Thus, in the example given in Section 12.5.3.2, the 250 mm thickness for a type 1 sub-base would increase to 340 mm with a restricted design. These additional thicknesses are effectively a method of obtaining assurance without the need to test.

Although IAN 73 has been in place since 2006 (updated in 2009 (HA, 2009)), it is still an interim document, and it would not be surprising to see revisions in the future that opened up the more versatile approach to a wider range of projects.

12.6. Upper pavement designs

Design of the upper part of the pavement is definitely less complicated than that of the foundation, and so is also a little less versatile. It is simply a matter of inputting the design traffic and the chosen foundation class into one of two charts (Figures 12.4 and 12.5), and reading off the thicknesses of the appropriate upper pavement material.

12.6.1 Flexible pavements

As shown in Figure 12.4 in relation to the design of fully flexible pavements, the designer simply has to enter the chart with the value of the design traffic load (50 msa in the example shown), come down to the selected foundation class (class 4 in the example), then move horizontally to the right to the appropriate curve for the chosen base material (DBM50 or HDM50 in this example), and then down to read off the total asphalt thickness, in this case 260 mm. Note that this thickness is for all the asphalt layers including the surface course and, where present, the binder course. The assumptions behind the chart are that all the asphalt layers are bonded so that they act as a single monolithic layer, that the surface course will be an approved material, and that the binder course will comprise a similar material to the base but with a smaller aggregate size.

Note that the minimum allowable total asphalt material thickness for flexible pavements with asphalt bases is 200 mm and that, once the traffic load reaches 80 msa, the design thickness attains a limiting value (i.e. it remains the same). Designs over 80 msa are termed 'long-life' designs. The concept is that the base and binder course will remain intact for a long but indeterminate period of time, and that the only foreseeable maintenance will be the periodic replacement or renewal of the surface course. It is a concept

Figure 12.4 Flexible and flexible composite upper pavement designs

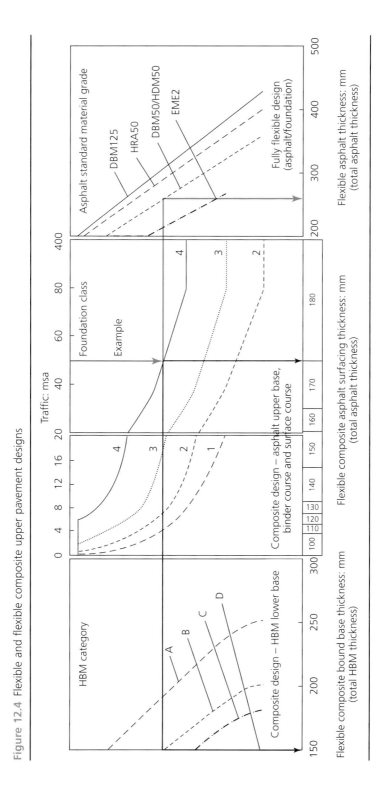

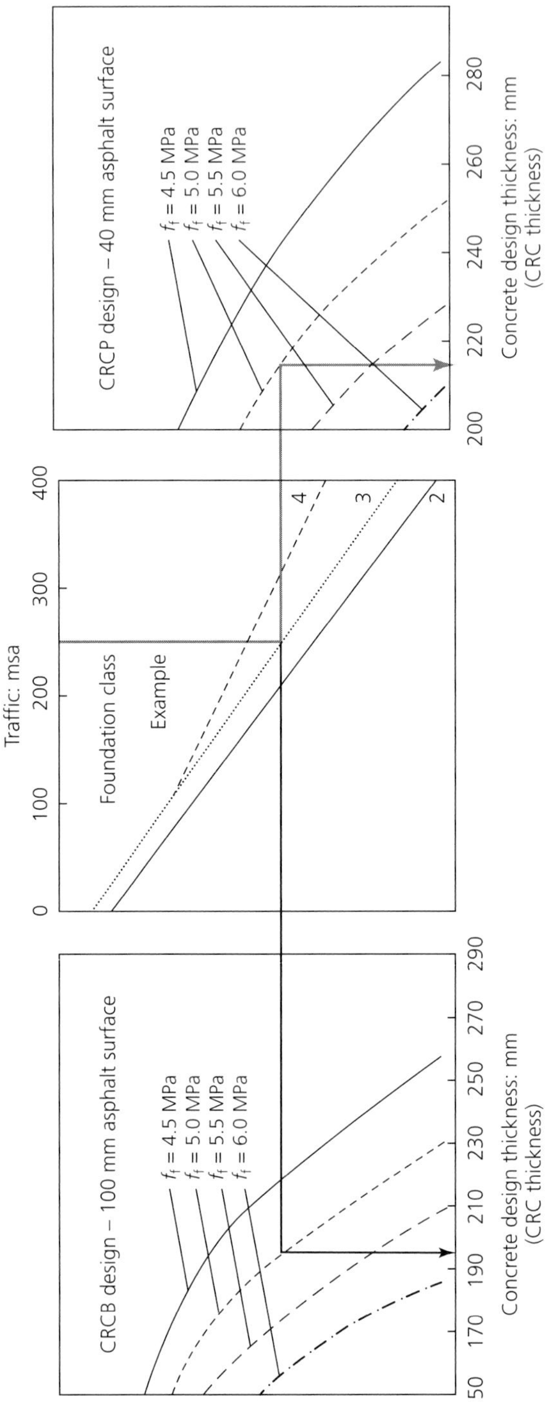

Figure 12.5 Rigid and rigid composite upper pavement designs

that has now been adopted by other countries (termed 'perpetual pavements' in the USA, for instance), and has been shown to have genuine validity (Nunn *et al.*, 1997) although not absolute certainty.

12.6.2 Flexible composite pavements

Figure 12.4 is also used for the thickness design of flexible composite pavements, except that the left-hand path is taken in this instance. Thus, for the example traffic load of 50 msa and the selected foundation (class 4), the path goes across to the relevant curve representing the chosen HBM (code B) proposed for use in the base, and then down to give the thickness of the base (i.e 150 mm). The meaning of the code letters A to D used in Figure 12.4 is given in Table 12.4. For full specification details, see the *Specification for Highway Works* (HA *et al.*, 2014), but the key point is that A represents the weakest permitted material and D the strongest.

The total thickness of asphalt above the HBM base is obtained by continuing the vertical line in Figure 12.4 straight down from the traffic load value: this is 180 mm in the example shown. It is not explicit in the standard exactly what this asphalt should consist of, the thickness being largely controlled by the need to limit reflective cracking from cracks in the HBM material. It is advised here that a DBM50 or HDM50 base should be used on heavily trafficked roads, but that a DBM125 base may be satisfactory on less

Table 12.4 HBMs permitted in the bases of flexible composite pavement designs

HBM category	Crushed rock coarse aggregate (with coefficient of thermal expansion $<10 \times 10^{-6}/°C$)	Gravel coarse aggregate (with coefficient of thermal expansion $\geq 10 \times 10^{-6}/°C$)
A	–	CBGM B – C8/10 (or T3) SBM B1 – C9/12 (or T3) FABM1 – C9/12 (or T3)
B	CBGM B – C8/10 (or T3) SBM B1 – C9/12 (or T3) FABM1 – C9/12 (or T3)	CBGM B – C12/15 (or T4) SBM B1 – C12/16 (or T4) FABM1 – C12/16 (or T4)
C	CBGM B – C12/15 (or T4) SBM B1 – C12/16 (or T4) FABM1 – C12/16 (or T4)	CBGM B – C16/20 (or T5) SBM B1 – C15/20 (or T5) FABM1 – C15/20 (or T5)
D	CBGM B – C16/20 (or T5) SBM B1 – C15/20 (or T5) FABM1 – C15/20 (or T5)	–

CBGM, cement-bound granular material; FABM, fly ash-bound material; SBM, slag-bound material; Cx/y, minimum 28 day compressive strength of x or y N/mm² for cylinder or cube specimens, respectively; Tz, minimum 28 day tensile strength of z N/mm² heavily trafficked roads. The uppermost part will consist of an approved surface course and, usually, a binder course.

It should be noted that clause 818 of the *Specification for Highway Works* now requires a rapid-setting HBM base to be pre-cracked, generally at 3 m centres, during or immediately following construction. This is to ensure that no single large crack develops, thereby protecting the overlying asphalt surfacing against reflective cracking. Pre-cracking is essentially a cheap way of forming a joint, and it can be achieved in a number of ways. Most commonly, a slot will be formed to about one-third of the depth of the HBM base immediately after paving (e.g. using a blade that is welded beneath a vibrating plate compactor), and bitumen emulsion is poured into the slot, and the layer is then compacted. The bitumen emulsion forms a weakness, which means that under the action of hydration and thermal shrinkage the HBM will tend to crack at those locations.

12.6.3 Rigid and rigid composite pavements

HD 26 specifies that a rigid pavement design must be considered when the design traffic load is >30 msa, especially when the advantages of lower maintenance during the design life appear worthwhile. The preferred rigid pavement construction (see Figure 12.5) is either a standard continuously reinforced concrete pavement (CRCP) with a 30 mm (minimum) asphalt overlay (for noise reduction) or a composite pavement comprising a continuously reinforced concrete base (CRCB) with a binder-plus-surface course asphalt overlay of 100 mm (minimum). The use of these preferred pavements is especially recommended at locales where the ground is expected to experience significant differential movement or settlement, as they can withstand large strains while remaining intact.

A standard rigid pavement CRCP design assumes the use of continuously reinforced concrete with at least 0.6% of the cross-sectional area of the slab comprising T16 (16 mm diameter) deformed longitudinal steel bars for crack control, a tied hardshoulder or a 1 m hard strip (i.e. with the concrete slab extending 1 m beyond the edge of the heavily trafficked left-hand design lane), and specially designed ground anchorages at either end of a continuously reinforced concrete section. Furthermore, it is assumed that the concrete will be overlaid by 30 mm of asphalt surface course, although this does not affect the continuously reinforced concrete design thickness.

In Figure 12.5, the procedure is to come down from the design traffic load (250 msa in the example shown) to the chosen foundation class (class 3 in the example) and across to the right until the appropriate flexural strength curve is encountered – in this example, it is 5.0 MPa. The design concrete thickness is then read off – in this example, it is 220 mm (after rounding up). If there is no shoulder or 1 m hard strip, the design thickness has to be increased by 30 mm: the purpose of the hard strip is to ensure that the untied edge is away from the wheel paths of vehicles, with a consequent reduction in the stresses imposed at the slab edge.

Note the restricted choice of foundation class for rigid pavements. In fact, a class 2 foundation can only be used with specific permission, and it has to include a bound sub-base.

A rigid composite CRCB pavement comprises a 100 mm of asphalt (surface course + binder course) overlying a continuously reinforced concrete slab base containing at least 0.4% T12 (12 mm diameter) reinforcement. The example design shown in Figure 12.5 gives a 200 mm concrete slab of 5.0 MPa flexural strength. If a 1 m hard strip or a shoulder is not constructed, the thickness again has to be increased by 30 mm.

12.6.4 Example designs

To illustrate the breadth of different options provided by the current pavement design procedures, Figure 12.6 shows two series of designs, one for a traffic load of 20 msa and the other for 100 msa, both over a 4% CBR subgrade and assuming that 'performance designs' are used for the foundation.

The range of designs shown in Figure 12.6, which is only a small sample from the full spectrum of possibilities, illustrates the way that designers can tailor their selection with regard to the overall pavement thickness as well as to the availability, cost and inherent risk of different materials. For example, the 100 msa concrete option is the thinnest pavement, at around 400 mm total, but continuously reinforced concrete may be seen as high risk, in that failure could be very costly indeed for the contractor. In contrast, a design

Figure 12.6 Examples of comparative pavement designs

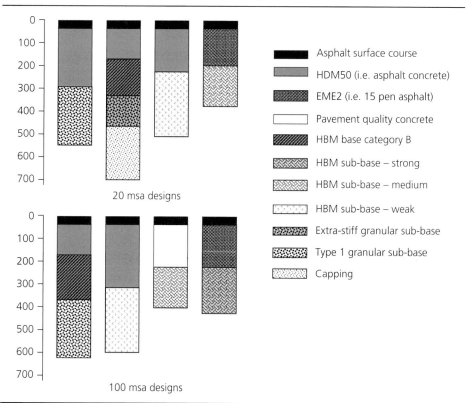

with DBM50/HDM50 asphalt over a type 1 sub-base would be around 600 mm thick, with implications for the construction programme and so on, but is probably much less risky for the contractor.

12.7. The HA design approach: a critique

As stated at the beginning of this chapter, the HA designs are based on a considerable body of performance evidence, and are therefore reasonably trustworthy in the UK environment and with traffic loads similar to those in the UK. Indeed, when developing any extension to these designs (e.g. using geogrid reinforcement or a so-called stress-absorbing membrane interlayer), the first check is to see that predictions without these additional features approximately match the standard assumptions. Nevertheless, these designs are simply state-of-the-art new pavement designs that neither preclude future developments nor give anything other than approximate guidance when designing for strengthening or partial reconstruction. These are serious limitations.

Consider the case of a partial reconstruction of one lane of a multi-lane carriageway, where it is necessary to make a judgement regarding the condition of the remaining pavement materials. In a simple case, all remaining materials can be considered as a foundation, and the equivalent foundation modulus can be estimated from FWD data obtained from the original road. From these data, a foundation class is assigned, and the upper pavement design charts can then be used. The procedure is iterative, since the carriageway surface level is fixed by the adjacent lane, and the depth to which materials have to be removed may have to be re-evaluated, but the procedure is clearly a reasonable one.

More difficult is the case where a significant thickness of asphalt material has to be left in place, potentially in a damaged state. Again, use can be made of FWD test data or core information, and the left-in-place material can be assigned an equivalence (e.g. to a DBM125 base). The underlying foundation also has to be assigned a design class. Designs can then be derived using the DBM125 base and whatever other base material is proposed, and an interpolation must then be made between the two design thicknesses according to the relative thicknesses of the remaining and new base materials.

Another potential extension to the HA design method is where an alternative un-approved material is proposed (e.g. a polymer-modified asphalt material). This is a real problem, and it would be fair to state that the relative inflexibility of the design documents has inhibited the use of such materials: this compares unfavourably with what is allowed in several other countries. Logically, the requirement would be to add an extra design curve – one that gave a lower thickness – to the right-hand part of Figure 12.4, but there is currently no mechanism to do this. Although the basis of the charts in Figure 12.4 is analytical, the way that this has been pragmatically applied makes it very difficult to extend to other materials without extensive field data.

Another limitation associated with the HA design approach is that it is restricted to hot-mix asphalt, and there is no scope for the use of cold-mix materials that are well suited to

Figure 12.7 TRL611 designs for cold-mix asphalt base superimposed on HA designs

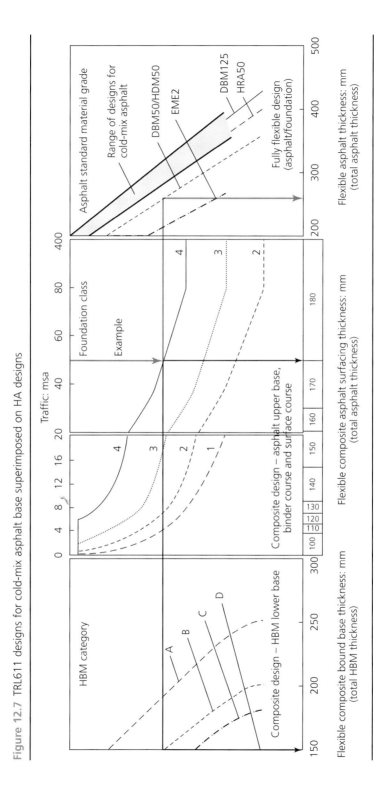

the use of recycled or secondary aggregates. The Transport Research Laboratory (TRL) has produced advice regarding the use of cold-mix recycled materials (Merrill *et al.*, 2004) that effectively allows for the introduction of additional design curves to Figure 12.4 (see Figure 12.7), and these are widely used in practice in the design of non-HA roads. This TRL research publication (TRL611) has identified three grades of cold bituminous material with characteristic stiffness moduli ranging from 1900 to 3100 MPa, which assume a thickness of hot-mix asphalt surface course of between 50 and 100 mm, depending on the road category. The designs are claimed to be suitable provided that the traffic load does not exceed 30 msa. However, this advice has no official status, and this limits the potential use of secondary and recycled materials in major road pavement construction.

Notwithstanding the above criticisms, the current design documents represent an excellent starting point for pavement design. They are generally slightly conservative, the foundation design element is probably too complicated, and there are limitations in the range of upper pavement alternatives that can be accommodated. There is therefore plenty of scope for improvement and for enabling the use of other materials. The problem that the current Highways England is faced with is that, by virtue of its authority, it has to ensure a reasonable level of performance assurance, and this is not easy to maintain while allowing the introduction of new materials. The memory of DBM15 and the problems it led to is still too fresh. It will probably always be the case therefore that Highways England remains a step or two behind the latest developments – but for a national body this is not unreasonable. The challenge for all government agencies is to maintain the right degree of support for innovation, including the monitoring of performance, so that developments that do indeed prove successful have a direct route into the standards: this means that they must continue to accept added risk where an innovative proposal is put forward that genuinely has the potential for success.

REFERENCES

Brown SF, Loach SC and O'Reilly MP (1987) *Repeated Loading of Fine Grained Soil*. Transport and Road Research Laboratory, Crowthorne, UK, Report CR72.

DfT (Department for Transport) (2013) *Road Transport Forecasts 2013*. DfT, London, UK.

Edwards JP, Thom NH, Fleming PR and Williams J (2005) Accelerated laboratory based mechanistic testing of unbound materials within the newly developed NAT Springbox. *Transportation Research Record* **1913**: 32–40.

HA (Highways Agency) (2009) *Interim Advice Note 73/06, Revision 1 (2009): Design Guidance for Road Pavement Foundations (Draft HD25)*. Stationery Office, London, UK.

HA, Transport Scotland, Welsh Government and Department for Regional Development Northern Ireland (1997) Section 1: traffic appraisal manual. Part 1: the application of traffic appraisal to trunk road schemes. In *Design Manual for Roads and Bridges*, vol. 12. *Traffic Appraisal of Road Schemes*. Stationery Office, London, UK.

HA, Transport Scotland, Welsh Government and Department for Regional Development Northern Ireland (2006a) Section 2: pavement design and construction. Part 1: traffic assessment. In *Design Manual for Roads and Bridges*, vol. 7. *Pavement Design and Maintenance*. Stationery Office, London, UK, HD 24/06.

HA, Transport Scotland, Welsh Government and Department for Regional Development Northern Ireland (2006b) Section 2: pavement design and construction. Part 3: pavement design. In *Design Manual for Roads and Bridges*, vol. 7. *Pavement Design and Maintenance*. Stationery Office, London, UK, HD 26/06.

HA, Transport Scotland, Welsh Government and Department for Regional Development Northern Ireland (2014) *Manual of Contract Documents for Highway Works*, vol. 1. *Specification for Highway Works*. Stationery Office, London, UK.

Merril D, Nunn ME and Carswell I (2004) *A Guide to the Use and Specification of Cold Recycled Materials for the Maintenance of Road Pavements*. Transport Research Laboratory, Crowthorne, UK, Report TRL611.

Nunn ME (2004) *Development of a More Versatile Approach to Flexible and Flexible Composite Pavement Design*. Transport Research Laboratory, Crowthorne, UK, Report TRL615.

Nunn ME, Brown A, Weston D and Nicholls JC (1997) *Design of Long-life Flexible Pavements for Heavy Traffic*. Transport Research Laboratory, Crowthorne, UK, Report TRL250.

Semmelink CJ and de Beer M (1995) Rapid determination of elastic and shear properties of road-building materials with the K-Mould. In *Unbound Aggregates in Roads* (Dawson AR and Jones RH (eds)). University of Nottingham, Nottingham, UK, pp. 151–161.

Thom NH, Cooper C, Grafton P *et al.* (2012) A new test for base material characterisation. *Proceedings of the International Symposium on Heavy Duty Asphalt Pavements and Bridge Deck Pavements*. International Society for Asphalt Pavements, Nanjing, China, paper 12-09.

Highways
ISBN 978-0-7277-5993-1

http://dx.doi.org/10.1680/h5e.59931.403

Chapter 13
Design and construction of asphalt bases and surfacings

Gordon Airey Director, Nottingham Transportation Engineering Centre, UK

The term 'asphalt' is generally taken to refer to a bituminous material comprising a mixture of aggregate bound with bitumen that is almost exclusively used in road construction. In addition to the graded aggregate and bitumen, the mixture also contains a small proportion of air. Thus, bituminous materials are three-phase materials, and their properties depend upon the properties of the individual phases as well as the mixture proportions. The two solid phases are quite different in nature – the aggregate is stiff and hard while the bitumen is flexible and soft and is susceptible to temperature change. Therefore, the proportion of bitumen in the asphalt mixture has a great influence on the mixture properties and is crucial in determining the performance of the material.

Bitumen can be supplied in a number of forms, either to facilitate the mixing and laying process or to provide a particular performance, while aggregates may come from a wide range of rock types or from artificial sources such as slag. The grading of the aggregate is important, and ranges from continuous grading for mixture types known as asphalt concretes, through to gap grading for mixtures such as hot rolled asphalts or stone mastic asphalts. The very fine component of the aggregate (passing a 63 μm sieve) is called filler: while the graded aggregate normally contains some filler material, it is usually necessary to add extra amounts of filler in the form of limestone dust, pulverised fuel ash, hydrated lime or even ordinary Portland cement to the asphalt mix.

Most asphalts used in major roads in the UK are hot-mixed and hot-laid bituminous materials; these are produced and mixed at temperatures ranging from 120 to 190°C, depending on the bitumen used. In recent years, however, the use of lower-temperature asphalts (i.e. materials that are produced and mixed at temperatures ranging from ambient to 100–140°C) has come to the fore. Thus, while much of the discussion in this chapter is primarily concerned with hot-mix asphalts, the use of low-temperature asphalts is also briefly discussed.

13.1.   Design objective
The overall objective underlying the design of bituminous materials for both pavement bases and surfacings is to determine, within the limits of the project specifications, a cost-effective blend of bitumen and graded aggregate that yields a mixture having the following properties:

- sufficient bitumen to ensure a durable pavement
- sufficient mixture stability to satisfy the demands of traffic without incurring distortion or displacement
- sufficient voids in the total compacted mixture to allow for a slight amount of bitumen expansion due to temperature increases, without the occurrence of flushing, bleeding and loss of stability
- a maximum void content to limit the permeability of harmful air and moisture into the mixture and the pavement layers below
- sufficient workability to permit the efficient placement of the mixture without segregation and without sacrificing stability and performance.

For surfacings, there are a number of additional requirements

- sufficient aggregate texture and hardness to provide good skid resistance during bad weather conditions
- an acceptable level of tyre/road noise
- an acceptable ride quality.

Ultimate pavement performance is related to durability, imperviousness, strength, stability, stiffness, flexibility, fatigue resistance and workability. The goal of bituminous mixture design is to select a unique design bitumen content that will achieve an appropriate balance between all of the desired properties. The selection of the aggregate gradation and the grade and amount of bitumen can be accomplished by two general methods, namely a 'recipe' approach (see Chapter 8) or an engineering design approach.

13.2.   Terminology associated with bituminous mixture design

It is very important that the engineer engaged in the design of a bituminous material be very clear on the terminology that is used and its precise meaning. The following are the main terms and their meanings.

13.2.1   Specific gravity of aggregate

Mineral aggregate is porous and can absorb water and bitumen to a variable degree. Furthermore, the ratio of water-to-bitumen absorption varies with each aggregate. The three methods of measuring the aggregate specific gravity (i.e. bulk, apparent and effective) take these variations into consideration. The differences between the specific gravities arise from different definitions of the aggregate volume.

The *bulk specific gravity* (G_{sb}) is the ratio of the mass in air of a unit volume of a permeable material (including both permeable and impermeable voids normal to the material) at a stated temperature to the mass in air of equal density of an equal volume of gas-free distilled water at a stated temperature.

The *apparent specific gravity* (G_{sa}) is the ratio of the mass in air of a unit volume of an impermeable material at a stated temperature to the mass in air of equal density of an equal volume of gas-free distilled water at a stated temperature.

The *effective specific gravity* (G_{se}) is the ratio of the mass in air of a unit volume of a permeable material (excluding voids permeable to bitumen) at a stated temperature to the mass in air of equal density of an equal volume of gas-free distilled water at a stated temperature.

When the total aggregate consists of separate fractions of coarse aggregate, fine aggregate and mineral filler, all having different specific gravities, the bulk specific gravity for the total aggregate is calculated from

$$G_{sb} = \frac{P_1 + P_2 + \cdots + P_n}{P_1/G_1 + P_2/G_2 + \cdots + P_n/G_n} \tag{13.1}$$

where G_{sb} is the bulk specific gravity of the total aggregate, P_1, $P_2, \ldots,$ P_n are the individual percentages by mass of the aggregates and G_1, $G_2, \ldots,$ G_n are the individual bulk specific gravities of aggregates. The bulk specific gravity of a mineral filler is difficult to determine accurately: however, if the apparent specific gravity of the filler is substituted, the error is usually negligible.

When based on the maximum specific gravity of a paving mixture, G_{mm} (as measured using ASTM D2041 (ASTM, 2011)), the effective specific gravity of the aggregate, G_{se}, includes all void spaces in the aggregate particles except those that absorb bitumen. G_{se} is determined from

$$G_{se} = \frac{P_{mm} - P_b}{P_{mm}/G_{mm} - P_b/G_b} \tag{13.2}$$

where G_{se} is the effective specific gravity of the aggregate, G_{mm} is the maximum specific gravity (ASTM D2041) of the paving mixture (no air voids), P_{mm} is the percentage by mass of the total loose mixture ($=100$), P_b is the bitumen content measured using ASTM D2041 (expressed as a percentage by total mass of the mixture) and G_b is the specific gravity of the bitumen.

It is important to note that the volume of bitumen absorbed by an aggregate is almost invariably less than the volume of water absorbed. Hence, the value of G_{se} will normally lie between its bulk and apparent specific gravities.

13.2.2 Maximum specific gravity of mixtures with different bitumen contents

When designing a paving mixture containing a given aggregate, the maximum specific gravity, G_{mm}, at each bitumen content has to be determined in order to calculate the percentage of air voids for each bitumen content. While the maximum specific gravity can be determined for each bitumen content using the ASTM D2041 test, the test precision is at its best when the mixture is close to the design bitumen content. Because of its sensitivity, the determination of the maximum specific gravity measurement should, preferably, be duplicated or triplicated.

After calculating the effective specific gravity of the aggregate from each measured maximum specific gravity and averaging the G_{se} results, the maximum specific gravity for any other bitumen content can be obtained from Equation 13.3. For practical purposes, G_{se} is constant, because the bitumen absorption does not vary much as the bitumen content changes, thus

$$G_{mm} = \frac{P_{mm}}{P_s/G_{se} + P_b/G_b}$$ (13.3)

where G_{mm} is the maximum specific gravity of the paving mixture (no air voids), P_{mm} is the percentage by mass of the total loose mixture ($=100$), P_s is the aggregate content (percentage by total mass of the mixture), P_b is the bitumen content (percentage by total mass of the mixture), G_{se} is the effective specific gravity of the aggregate and G_b is the specific gravity of the bitumen.

13.2.3 Bitumen absorption

Bitumen absorption is expressed as a percentage by mass of the aggregate rather than as a percentage by total mass of the mixture. Thus, the bitumen absorption, P_{ba}, is determined from

$$P_{ba} = 100 \times \frac{G_{se} - G_{sb}}{G_{sb} G_{se}} G_b$$ (13.4)

where P_{ba} is the absorbed bitumen (percentage by mass of the aggregate), G_{se} is the effective specific gravity of the aggregate, G_{sb} is the bulk specific gravity of the aggregate and G_b is the specific gravity of the bitumen.

13.2.4 Effective bitumen content of a paving mixture

The effective bitumen content, P_{be}, of a pavement material is the total bitumen content minus the quantity of bitumen lost by absorption into the aggregate particles. As this is the portion of the total bitumen content that remains as a coating on the outside of the aggregate particles, it is the bitumen content that governs the performance of a bituminous mixture. This bitumen content is determined from

$$P_{be} = P_b - \frac{P_{ba}}{100} P_s$$ (13.5)

where P_{be} is the effective bitumen content (percentage by total mass of the mixture), P_b is the bitumen content (percentage by total mass of the mixture), P_{ba} is the absorbed bitumen (percentage by mass of the aggregate) and P_s is the aggregate content (percentage by total mass of the mixture).

13.2.5 Percentage air voids in the mineral aggregate in a compacted mixture

The VMA (voids in the mineral aggregate) is defined as the intergranular void space between the aggregate particles in a compacted bituminous mixture that includes the air voids and the effective bitumen content, expressed as a percentage of the total volume (Figure 13.1). The VMA is calculated on the basis of the bulk specific gravity of the aggregate, G_{sb}, and is expressed as a percentage of the bulk volume of the compacted paving mixture. Therefore, the VMA can be calculated by subtracting the volume of the

Figure 13.1 Representation of the volumes in a compacted bituminous material

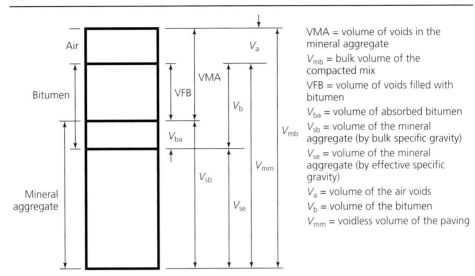

VMA = volume of voids in the mineral aggregate
V_{mb} = bulk volume of the compacted mix
VFB = volume of voids filled with bitumen
V_{ba} = volume of absorbed bitumen
V_{sb} = volume of the mineral aggregate (by bulk specific gravity)
V_{se} = volume of the mineral aggregate (by effective specific gravity)
V_a = volume of the air voids
V_b = volume of the bitumen
V_{mm} = voidless volume of the paving

aggregate determined by its bulk specific gravity from the bulk volume of the compacted paving mixture. If the mixture composition is determined as a percentage by mass of total mixture, then

$$VMA = 100 - \frac{G_{mb}P_s}{G_{sb}} \qquad (13.6)$$

where VMA is the voids in the mineral aggregate (percentage of bulk volume), G_{sb} is the bulk specific gravity of the total aggregate, G_{mb} is the bulk specific gravity of the compacted mixture (measured by AASHTO T 166, ASTM D1188 or ASTM D2726) and P_s is the aggregate content, percentage by total mass of the mixture.

If a mix composition is calculated as a percentage by mass of the aggregate

$$VMA = 100 - \frac{G_{mb}}{G_{sb}} \frac{100}{100 + P_b} \times 100 \qquad (13.7)$$

where P_b is the bitumen content (percentage by mass of the aggregate).

13.2.6 Percentage air voids in the compacted mixture

The air voids, V_a, are the total volume of the small pockets of air between the coated aggregate particles throughout a compacted bituminous mixture, expressed as a percentage of the bulk volume of the compacted paving mixture (Figure 13.1). The volume percentage of air voids in a compacted mixture is determined from

$$V_a = 100 \times \frac{G_{mm} - G_{mb}}{G_{mm}} \qquad (13.8)$$

where V_a is the air voids in the compacted mixture (percentage of the total volume), G_{mm} is the maximum specific gravity of the paving mixture (as determined earlier or as measured directly by ASTM D2041) and G_{mb} is the bulk specific gravity of the compacted mixture.

13.2.7 Percentage voids filled with bitumen in the compacted mixture

The VFB (voids filled with bitumen) is the percentage of the intergranular void space between the aggregate particles (VMA) that is filled with bitumen (Figure 13.1). The VFB, which does not include the absorbed bitumen, is determined from

$$\text{VFB} = \frac{100 \times (\text{VMA} - V_a)}{\text{VMA}} \tag{13.9}$$

where VFB is the voids filled with bitumen (percentage of the VMA), V_a is the air voids in the compacted mixture (percentage of the total volume) and VMA is the voids in the mineral aggregate (percentage of the bulk volume).

13.3. The recipe approach and its limitations

As noted previously, there are two general approaches to the creation of bituminous materials, namely the recipe approach and the engineering design approach. As implied by its name, the recipe approach to selecting the types and proportions of the materials comprising a bituminous mixture is a 'cookbook' procedure (i.e. it is based wholly on experience, not on engineering principles), and the mixtures are combined according to proportions that are prescribed in appropriate standards and/or specifications. Provided that the ingredients meet their specifications, the resultant hot-mix, hot-laid asphalts will achieve the required performance in most situations when used in pavement bases and surfacings.

However, there are limitations to the use of recipe mixtures that are analogous to the limitations on the empirical structural design of roads, namely

- The conditions to which the in-service mixture is subjected (traffic, climate, etc.) may not be the same as those that existed when the specifications were developed.
- Non-specified materials cannot be used (e.g. if a locally available sand does not meet the grading requirements of a specification, it cannot be used even if it produces a satisfactory mixture).
- Notwithstanding the trend in the highway industry towards the inclusion of innovative materials such as polymer-modified bitumens, reclaimed asphalt and fibre-modified mixtures in bituminous paving mixtures, specification by recipe may not be suitable for such materials.
- No procedure is available to assess the causes of layer failure.
- No procedure is available to optimise the mixture proportions.

The last criterion is very important so far as bitumen is concerned, as it is the most expensive ingredient and has a strong bearing on mixture performance, especially of dense materials.

Table 13.1 Designation of bituminous mixtures under BS EN 13108

Mixture type	Size	Pavement layer	Bitumen grade e.g. 40/60
AC	D	base/bin/surf	binder (xx/yy)
HRA	Grading designation	base/bin/surf	binder (xx/yy)
SMA	D	base/bin/surf	binder (xx/yy)
PA	D	surf	binder (xx/yy)

The move to European standardisation has meant that bituminous recipe mixtures are now classified into seven material specifications as described in the BS EN 13108 series (BSI, 2006a, BSI, 2006b, BSI, 2006c, 2006d) according to the grading of the aggregate: that is, mixture type, upper sieve size (i.e. the maximum nominal aggregate size), the intended use of the material and the binder used in the mixture. Table 13.1 summarises the designations of the four main bituminous mixtures used in the UK: AC, asphalt concrete; HRA is hot rolled asphalt; PA is porous asphalt; SMA, stone mastic asphalt; note also that 'binder' denotes the full bitumen grade designation, D is the aggregate size, 'surf' is the surface course, 'bin' is the binder course and 'base' is the base course; an example designation would be SMA 14 base 40/60.

13.3.1   Asphalt concretes

Asphalt concretes, which are characterised by relatively low binder contents and continuously graded aggregates, rely on the packing and interlock of the aggregate particles for their strength and stiffness. The bitumen coats the aggregate, and acts as a lubricant when hot and as an adhesive and waterproofing agent when cool. Because of their lower binder contents, asphalt concretes are cheaper than hot rolled asphalt and stone mastic asphalt materials. In general, asphalt concretes have higher void contents than either hot rolled asphalts or stone mastic asphalts, and are therefore more permeable and less durable.

Asphalt concrete, which is specified in BS EN 13108-1 (BSI, 2006a), is used in pavement surfacings and bases. A summary of asphalt concrete options used in the UK is listed in Chapter 8 (Table 8.1).

13.3.2   Hot rolled asphalts

Hot rolled asphalts (HRA) are dense materials that are characterised by their high bitumen and high fines/filler contents, and are used in surfacings and bases as specified in BS EN 13108-4 (BSI, 2006b). They derive their strength and stiffness from a dense stiff mortar of bitumen, fines and filler. The coarse aggregate content is relatively low so that the overall particle size distribution is gap graded. A summary of hot rolled asphalt material options used in the UK is listed in Table 8.3, and the grading specifications for the preferred mixes are given in Table 8.4.

13.3.3   Porous asphalts

Porous asphalt (PA) is a bituminous material that has a large volume (at least 20%) of interconnected air voids so that water can drain through the material and run off within the thickness of the layer: this criterion requires the underlying layer to be impermeable. Porous asphalt is used exclusively for surfacings, and can be laid in more than one layer. The very high content of interconnected voids allows not only the passage of water but also the movement of air, thereby exhibiting noise reduction characteristics. Other advantages of porous asphalt are that it minimises spray in wet weather, reduces surface noise, improves skidding resistance, and offers lower rolling resistance than dense mixtures.

Porous asphalts are specified in BS EN 13108-7 (BSI, 2006d). The aggregate grading consists predominantly of coarse aggregate, with about 75% retained on the 2 mm sieve. The fine aggregate fractions are added to enhance the cohesion and stiffness of the mixture, but in sufficiently small quantities that they do not interfere with the interlock of the coarse particles, and will leave enough voids to provide a pervious structure. Because of its porous nature, the porous asphalts are vulnerable to ageing through oxidation of the bitumen: to counteract this, the bitumen content must be sufficient to provide a thick coating on the coarse aggregate particles.

13.3.4   Stone mastic asphalts

Stone mastic asphalt (SMA), which is used in pavement surfacing and binder courses in the UK, is specified in BS EN 13108-5 (BSI, 2006c). A summary of stone mastic asphalt options is listed in Table 8.5, and target gradations for England, Wales and Scotland are listed in Table 8.6.

Stone mastic asphalt has a coarse aggregate skeleton, but the voids are filled with a fines/filler/bitumen mortar. While stone mastic asphalt somewhat resembles hot rolled asphalt, particularly the high stone content mixtures, it may best be considered as having a coarse aggregate structure similar to porous asphalt but with the voids filled. Stone mastic asphalt differs from hot rolled asphalt in that the quantity of mortar is just sufficient to fill the voids in the coarse aggregate structure. It provides high stiffness due to the interlock of the coarse aggregate particles, and good durability because of its low void content.

13.4.   The engineering design approach

The drawbacks associated with the recipe approach led to the development of procedures for the design of asphalt mixtures at a time that coincided with the development of analytical procedures for the structural design of roads. The engineering design approach requires knowledge of certain properties of the ingredient materials, and it follows, therefore, that asphalts are now produced with particular characteristics.

The objectives underlying the engineering design of an asphalt material (see Section 13.1) can be achieved by ensuring that the mixture contains sufficient bitumen to adequately coat all of the aggregate particles and provide good workability so that, when the mixture is compacted, it possesses adequate stiffness, deformation resistance and air voids.

Many empirical and semi-empirical design procedures have attempted to evaluate various properties of bituminous mixtures and then sought to base the bitumen content determination on these evaluations. Some of the more widely known of these procedures are the Marshall (AI, 2014), Hveem (AI, 2014), SHRP Superpave (AI, 2014), LCPC (LCPC, 1997), CROW (Hopman *et al.*, 1992), Texas (Mahboub and Little, 1990), EXXON (Eckmann, 1989) and University of Nottingham methods (Gibb and Brown, 1994; Read and Collop, 1997); there are many others (Franken, 1998).

Regardless of the method of mixture design and testing procedure adopted, there is consensus that it is important to analyse the volumetric properties of the bituminous mixture in its compacted state. The objective of the volumetric analysis is to aid the designer in determining, among other factors, the efficiency of bitumen utilisation in the mix, the degree and efficiency of compaction, and the quantity of air voids present in the mix. These and other volumetric parameters have been shown to have a direct influence on the stiffness, stability and durability of a bituminous mixture. The steps generally involved in analysing a compacted bituminous mixture are as follows:

(a) measure the bulk specific gravities of the coarse aggregate (AASHTO T 85 or ASTM C127) and of the fine aggregate (AASHTO T 84 or ASTM C128)
(b) measure the specific gravity of the bitumen (AASHTO T 228 or ASTM D 70 (23)) and of the mineral filler (AASHTO T 100 or ASTM D854)
(c) calculate the bulk specific gravity of the aggregate combination in the paving mixture
(d) measure the maximum specific gravity of the loose paving mixture (ASTM D2041, BS DD 228)
(e) measure the bulk specific gravity of the compacted paving mixture (ASTM D1188 or ASTM D2726)
(f) calculate the effective specific gravity of the aggregate
(g) calculate the maximum specific gravity of the mix at other bitumen contents
(h) calculate the bitumen absorption of the aggregate
(i) calculate the effective bitumen content of the paving mixture
(j) calculate the percentage voids in the mineral aggregate in the compacted bituminous mixture
(k) calculate the percentage air voids in the compacted paving mixture
(l) calculate the percentage air voids filled with bitumen.

In the above list of steps, the abbreviations AASHTO and ASTM refer to the test procedures sponsored by the American Association of State Highway and Transportation Officials and the American Society for Testing and Materials, respectively.

13.4.1 Marshall method of mix design

The Marshall method of mixture design, which is currently the most widely accepted mix design procedure, is intended both for the laboratory design and the field control of hot-mix, dense-graded asphalt materials. Originally developed by Bruce Marshall of the Mississippi State Highway Department, the US Army Corps of Engineers subsequently refined and added certain features to Marshall's approach, and it was then eventually formalised in the ASTM D1559 and AASHTO T 245 standards.

13.4.1.1 Outline of the method

The Marshall method uses standard cylindrical test specimens (64 mm high by 102 mm diameter) that are prepared using a specified procedure for heating, mixing and compacting the bituminous mixture. The two principal features of the Marshall method of mixture design are a density voids analysis and a stability–flow test of the compacted test specimens. The stability of a test specimen is the maximum load resistance (in newtons) that the standard test specimen will develop at 60°C when tested as outlined below. The flow value is the total movement or displacement (in units of 0.25 mm) occurring in the specimen between when there is no load and when the point of maximum load is achieved during the stability test.

When determining the design bitumen content for a particular blend or gradation of aggregates, a series of test specimens is prepared for a range of different bitumen contents so that the test data curves show well-defined relationships. Tests are normally planned on the basis of 0.5 percentage increments of bitumen content, with at least two bitumen contents above the expected design value and at least two below. The expected design bitumen content is estimated using any or all of the following sources: experience, a computational formula, or performing the centrifuge kerosene equivalency and oil soak tests in the Hveem procedure (AI, 2014).

A rule-of-thumb method of arriving at a starting point for testing is to use the filler-to-bitumen ratio guideline (normal range: 0.6–1.2). The anticipated design bitumen content (as a percentage by total mass of the mixture) is then estimated as approximately equivalent to the percentage of the aggregate in the final gradation passing the 75 µm sieve.

Equation 13.10 is one example of a method of estimating the design content using a computational formula

$$P = 0.035a + 0.045b + Kc + F \tag{13.10}$$

where P is the approximate mixture bitumen content (percentage by weight of the total mixture); a is the percentage (expressed as a whole number) of the mineral aggregate retained on the 2.36 mm sieve; b is the percentage of the mineral aggregate passing the 2.36 mm and retained on the 75 µm sieves; c is the percentage of the mineral aggregate passing the 75 µm sieve; $K = 0.15$ for when 11–15% passes the 75 µm sieve, 0.18 for when 6–10% passes the 75 µm sieve and 0.20 for when 5% passes the 75 µm sieve; and $F = 0$–2.0%, based on the absorption of light or heavy aggregates (in the absence of other data, $F = 0.7$ is suggested).

13.4.1.2 Preparing test specimens

The aggregates to be used for testing are first dried to a constant weight at 105–110°C, and then separated, by dry sieving, into the desired size fractions. The amount of each size fraction required to produce a batch that will give a 63.5 ± 1.27 mm high compacted specimen is then weighed in a separate pan for each test specimen; this is normally about 1.2 kg of dry aggregate. At least three specimens are prepared for each combination of aggregate and bitumen. The batch pans are then placed in an oven until the dry aggregates achieve the requisite mixing temperature, after which a mixing bowl is charged with the heated aggregates, the required quantity of bitumen is added, and mixing is carried out until all the aggregate particles in each batch are fully coated.

Depending upon the design traffic category (light, medium or heavy) that the compacted bituminous mixture is being designed to withstand, 35, 50 or 75 blows are applied, respectively, with the Marshall compaction hammer to each end of the specimen. The design traffic categories are: 'light', traffic conditions giving an equivalent axle load (EAL) of $<10^4$; 'medium', design EAL of 10^4–10^6; and 'heavy', design EAL $>10^6$. After compaction, the specimens are allowed to cool in air at room temperature until no deformation results when removing each from its mould.

13.4.1.3 Marshall test procedure

With the Marshall test method, each compacted test specimen is subjected to the following tests and analysis in the order listed: (a) the bulk specific gravity test, (b) the stability and flow test and (c) density and voids analysis. The bulk specific gravity test may be performed as soon as the freshly compacted specimens have cooled to room temperature, after which the stability and flow tests are carried out.

Prior to stability and flow testing, the specimens are immersed in a water bath at $60 \pm 1°C$ for 30–40 min. The Marshall testing machine, which is a compression-testing device, is designed to apply loads to specimens through cylindrical segment testing heads at a constant rate of vertical strain of 51 mm/min (Figure 13.2). Loading is applied until specimen failure occurs: the point of failure is defined by the maximum load reading obtained. The force (in newtons) that produces failure of the test specimen is recorded as its Marshall stability value, and the deformation of the specimen at the point of failure (in units of 0.25 mm) is recorded as its flow value.

Following testing, the stability and flow values for all specimens with a given bitumen content are averaged. A 'best fit' curve is then plotted for each of the following relationships: stability versus bitumen content, flow versus bitumen content, unit weight of the

Figure 13.2 Marshall stability test

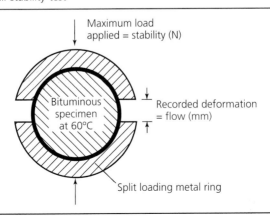

Figure 13.3 Test property curves for hot-mix design data using the Marshall procedure

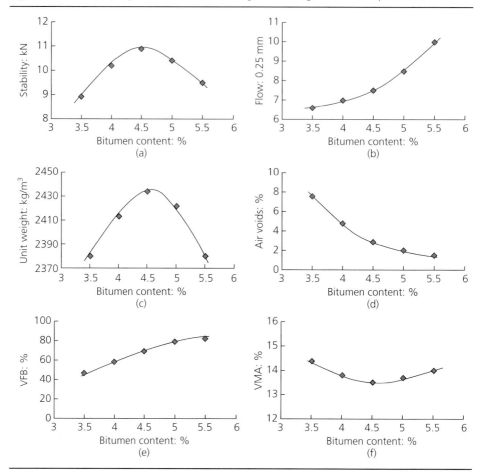

total mix versus bitumen content, percentage air voids (V_a) versus bitumen content, percentage voids filled with bitumen (VFB) versus bitumen content, and percentage voids in the mineral aggregate (VMA) versus bitumen content. These graphs (Figure 13.3) are used to determine the design bitumen content of the mix.

Note that the test property curves follow a reasonably consistent pattern for (in this case) dense-graded bituminous mixtures, but variations can and will occur. Trends to be noted are that

- The stability value increases with increasing bitumen content up to a maximum, after which the stability decreases.
- The flow value consistently increases with increasing bitumen content.
- The unit weight relationship follows a similar trend to the stability curve, except that the maximum unit weight normally (but not always) occurs at a slightly higher bitumen content than the maximum stability.

- The V_a steadily decreases with increasing bitumen content.
- The VFB steadily increases with increasing bitumen content, due to the VMA being filled with bitumen.
- The VMA generally decreases to a minimum value then increases with increasing bitumen content.

In evaluating mixtures it may also be helpful to consider the Marshall quotient, Q_m. This is derived from the stability and flow:

$$Q_m = \text{stability/flow} \tag{13.11}$$

Note that Q_m bears some resemblance to a modulus (the ratio of stress to strain), and may be taken as a measure of the mixture stiffness.

13.4.1.4 Determining the preliminary design bitumen content

The design bitumen content of the bituminous paving mixture is selected after considering all of the data discussed previously. As an initial starting point, the Asphalt Institute recommends choosing the bitumen content at the median of the percentage air voids criteria in Table 13.2 (i.e. 4%); in the example given in Figure 13.3(d), this equates to a bitumen content of 4.2%. All of the calculated and measured mixture properties at this bitumen content (stability = 10.5 kN, flow = 7 × 0.25 mm, VFB = 62% and VMA = 13.7%) are then compared with the relevant design criteria in Tables 13.2 and 13.3. If any criteria are not met, then some adjustment or compromise is required, or the mixture is redesigned.

In relation to Table 13.3 note that the nominal maximum particle size is one size larger than the first sieve to retain more than 10% of the aggregate: for example, if 11% by

Table 13.2 Marshall asphalt mixture design criteria

Marshall method: mix criteria	Light traffic: surface and base		Medium traffic: surface and base		Heavy traffic: surface and base	
	Minimum	Maximum	Minimum	Maximum	Minimum	Maximum
Compaction: number of blows to each end of specimen	35		50		75	
Stability: N	3336		5338		806	
Flow: 0.25 mm	8	18	8	16	8	14
Air voids: %	3	5	3	5	3	5
VMA: %	See Table 13.3					
VFB: %	70	80	65	78	65	75

Table 13.3 Minimum values for the VMA

Nominal maximum particle size: mm	Design air voids: %		
	3.0	4.0	5.0
1.18	21.5	22.5	23.5
2.36	19.0	20.0	21.0
4.75	16.0	17.0	18.0
9.5	14.0	15.0	16.0
12.5	13.0	14.0	15.0
19.0	12.0	13.0	14.0
25.0	11.0	12.0	13.0
37.5	10.0	11.0	12.0

mass of the aggregate is retained on the 12.5 mm sieve, the nominal maximum particle size for the mix is 19 mm.

13.4.1.5 Selecting the final mixture

The bitumen content finally selected from the mix design procedure is desirably the most economical one that will satisfactorily meet all of the established design criteria. However, this bituminous mixture should not be selected to optimise one particular property: for example, mixtures with abnormally high values of stability are often less desirable because pavements with such materials tend to be less durable, and may crack prematurely, under heavy volumes of traffic. This situation is especially critical if the courses beneath the bituminous layers of the pavement are weak and permit moderate to relatively high deflections under the traffic.

The design bitumen content should be a compromise selected to balance all of the mixture properties. Normally, the mix design criteria will produce a narrow range of acceptable bitumen contents that meet all of the guidelines, and the bitumen content selection is then adjusted within this narrow range (Figure 13.4) to achieve a mixture property that will satisfy the needs of the pavement being designed. Different properties are more critical for different circumstances, depending on the traffic loading and volume, the pavement structure, the climate, the construction equipment and other factors. Thus, the balancing process that is carried out prior to establishing the final design bitumen content is not the same for every pavement and for every mix design.

In many cases, the most difficult mix design property to achieve is the minimum amount of voids in the mineral aggregate. The goal is to furnish enough space for the bitumen so it can provide adequate adhesion to bind the aggregate particles, but does not bleed when temperatures rise and the bitumen expands. Normally, the curve exhibits a flattened U shape, decreasing to a minimum value and then increasing with increasing bitumen content, as shown in Figure 13.5(a).

Figure 13.4 Example of the narrow range of acceptable bitumen contents arising from the Marshall test

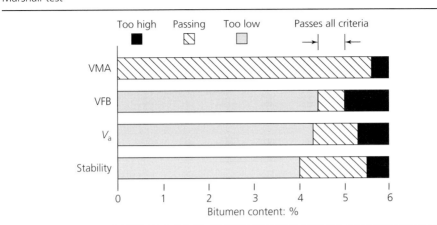

It might be anticipated that the VMA would remain constant with varying bitumen content, and that the air voids would simply be displaced by bitumen. In reality, the total volume changes across the range of bitumen contents. With the increase in bitumen, the mixture actually becomes more workable and compacts more easily (meaning more mass is compressed into less volume) and, up to a point, the bulk density of the mixture increases and the VMA decreases.

At some point as the bitumen content increases (the bottom of the U-shaped curve), the VMA begins to increase because more-dense material (aggregate) is displaced and pushed apart by the less-dense material (bitumen). Bitumen contents on the 'wet' or right-hand increasing side of this VMA curve should be avoided, even if the minimum air void and VMA criteria are met. Design bitumen contents in this wet range have a tendency to bleed and/or exhibit plastic flow when placed in the field; that is, additional compaction from traffic leads to inadequate room for bitumen expansion, loss of aggregate-to-aggregate contact and, eventually, 'rutting' and 'shoving' in high-traffic areas. Thus, the design bitumen content should be chosen slightly to the left of the low point of the VMA curve in Figure 13.5(a), provided none of the other criteria are violated.

The design bitumen content should never be selected at the extremes of the acceptable range, even though the minimum criteria are met. On the left-hand side, the mixture would be too dry, prone to segregation and, probably, too high in air voids while, on the right-hand side, the mixture would be expected to rut.

In some mixtures, the bottom of the U-shaped VMA curve is very flat, meaning that the compacted mixture is not very sensitive to the bitumen content in the range being evaluated. In the normal range of bitumen contents, compactibility is more influenced by the aggregate properties. However, at some point, the quantity of bitumen will become critical to the behaviour of the mixture, and the effect of the bitumen will dominate as the VMA increases drastically.

Figure 13.5 Relationship between the VMA and the specification limit

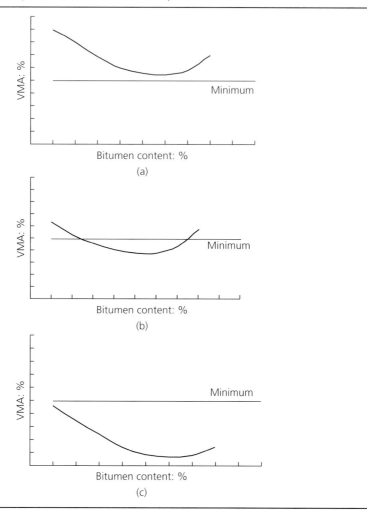

If the bottom of the VMA curve falls below the minimum criteria level required for the nominal maximum aggregate size of the mix (Figure 13.5(b)), it is an indication that changes to the mix formula are necessary and that the aggregate grading should be modified to provide extra VMA.

If the minimum VMA criteria are completely violated over the entire bitumen content range (i.e. if the curve is completely below minimum (Figure 13.5(c))), a significant redesign and/or change in materials sources is normally warranted.

It needs to be emphasised that the desired range of air voids (3–5%) is the level desired after several years of traffic, and that this range does not vary with traffic (see Table 13.2). This air void range will normally be achieved if the mixture is designed at the correct compaction effort (which is meant to be reflective of the traffic load) and the percentage air voids immediately after construction is at about 8%, as some consolidation can be expected under traffic.

Research has shown that asphalt concrete mixtures that ultimately consolidate to less than 3% air voids are likely to rut and shove if placed in heavy traffic locations. Several factors can contribute to this: for example, an accidental increase in the bitumen content at the mixing facility or an increased amount of ultra-fine particles passing the 75 μm sieve (which acts as a bitumen extender). Problems can also occur if the final air void content is above 5% or if the pavement is constructed with over 8% air voids initially: brittleness, premature cracking, ravelling and stripping are all possible under these conditions.

Although the VFB, VMA and V_a are all interrelated, and only two of the values are needed to calculate the third, including the VFB criteria helps prevent a mix design with a marginally acceptable VMA; that is, the main effect of the VFB criteria is to limit maximum levels of the VMA, and, subsequently, maximum levels of the bitumen content. The VFB also restricts the allowable air void content for mixtures that are near the minimum VMA criteria.

Mixtures designed for lower-traffic volumes will not pass the VFB criteria with a relatively high percentage of air voids (5%) even though the air void criteria range is met: this helps ensure that less-durable mixtures are avoided in light-traffic situations. Mixtures designed for heavy traffic will not pass the VFB criteria with a relatively low percentage air voids content (less than 3.5%), even though that air voids content is within the acceptable range: because a low air void content can be very critical in terms of permanent deformation, the VFB criteria help avoid those mixtures that would be susceptible to rutting under heavy traffic.

13.4.2 Alternative bituminous mixture testing/design methods
13.4.2.1 Gyratory compaction
As traffic conditions change and road user expectations rise, the use of alternative mixture design procedures, test methods and systems may become more appropriate. As is illustrated in Figure 13.6, the main characteristic of the gyratory compaction equipment is that it facilitates the application of an axial static pressure at the same time as it subjects a specimen to a dynamic shear 'kneading motion' while compacting a bituminous mixture: this compares with the impact compaction used in the Marshall design method. The bituminous mixture is thus subjected to shearing forces similar to those encountered under the action of a roller during field compaction (D'Angelo et al., 1995).

The energy applied by a gyratory compactor should be close to the energy applied in the field, and, consequently, compaction details such as angle/speed of gyration and axial pressure vary from one country to another. A good example is the US Superpave gyratory compactor, which produces 600 kPa compaction pressure for 150 mm diameter specimens, and rotates at 30 rpm with the mould positioned at a compaction angle of 1.25°. The required number of gyrations using the Superpave compactor is determined

Figure 13.6 Cross-section of a gyratory compactor

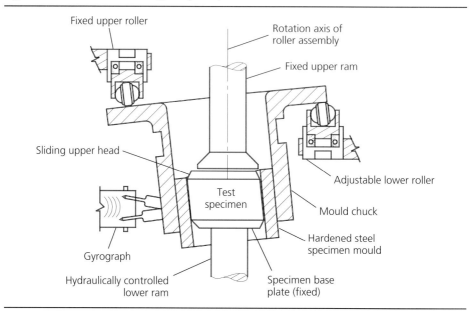

from the traffic level expected on the pavement, expressed as the number of equivalent standard axle loads (ESAL), and the design 7-day maximum air temperature for the site (Table 13.4).

In Table 13.4, N_d (i.e. N design) is the number of gyrations required to produce a density in the mixture that is equivalent to the expected density in the field after the indicated amount of traffic. In the mix design process, a bitumen content is selected that will provide 4% air voids when the mixture is compacted to N_d gyrations. N_i (i.e. N initial) is a measure of the mixture compactibility. Mixtures that compact too quickly are considered to be too 'tender' during construction, and may be unstable under traffic. A mixture that has 4% air voids at N_d should have at least 11% air voids at N_i; mixtures that fail this requirement are often finer mixtures that tend to have a large amount of natural sand.

Table 13.4 An example of the number of initial (N_i), design (N_d) and maximum (N_m) gyrations required for two selected traffic levels and high air temperature environments

Traffic equivalent standard axle load (ESAL)	Design 7-day maximum air temperature (°C)					
	<39			39–41		
	N_i	N_d	N_m	N_i	N_d	N_m
<3 × 10⁶	7	86	134	8	95	150
<1 × 10⁷	8	96	152	8	106	169

N_m (i.e. N maximum) is the number of gyrations required to produce a density in the laboratory that should never be exceeded in the field, whereas N_d provides an estimate of the ultimate field density. N_m provides a compacted density with a safety factor that ensures that the mixture does not densify too much and have an in-place voids content that is so low that it leads to rutting. The air voids at N_m are required to be at least 2%, as mixtures that have fewer air voids at N_m are believed to be more susceptible to rutting than those that exceed 2% air voids.

The number of gyrations for N_d was established on the basis of laboratory and field testing that compared the in-place density with the laboratory density for various numbers of gyrations. Once N_d was established for each traffic level and air temperature, the numbers of gyrations for N_i and N_m were determined from

$$N_i = N_d \times 0.45 \tag{13.12}$$

$$N_m = N_d \times 1.10 \tag{13.13}$$

To summarise, for optimum aggregate gradations and bitumen contents, the Superpave gyratory compaction requirements are as follows:

- at N_i the paving mix must attain 89% of the theoretical maximum specific gravity G_{mm} or less
- at N_d the mixture must attain 96% of G_{mm}
- at N_m the mixture must attain <98% of G_{mm} or an air voids content >2%.

By monitoring the mixture densification curve during gyratory compaction, a first estimate of the optimum bitumen content can be obtained by selecting a bitumen content that will produce a 4% air voids content at N_d.

13.4.2.2 Bituminous material ageing

To adequately characterise the ageing of a bituminous material, samples that are being evaluated in the laboratory must be aged to simulate the in situ properties of the mixture. Since the rate of ageing of the bitumen in a mixture is affected by such mixture properties as in-place air voids, the entire mixture must be aged.

The Superpave short-term ageing procedure for bituminous mixtures involves exposure of the loose mix, immediately after mixing and prior to compaction, for 4 h at 135°C in a forced draft oven. The process is intended to simulate the ageing of the mixture during production and placement in the field. It also allows the bitumen to be absorbed into the coarse aggregate, to simulate what happens during construction.

With the Superpave long-term ageing procedure, compacted specimens that have undergone short-term ageing (in loose form) are subsequently exposed to a temperature of 85°C in a forced draft oven. The time for which the compacted specimens are kept in the oven is varied according to the duration of the working life of the pavement being simulated. The recommended time is 2 days: this is considered to correspond to roughly 10 years of pavement service. After ageing the samples in this way, they are tested in a condition that is considered similar to that in the roadway, and, thus, should provide a better measure of the expected performance.

13.4.2.3 Mixture workability

The problem of measuring the workability of bituminous mixtures is not new: for example, measuring the torque required to mix mineral aggregates with bitumen has been proposed (Marvillet amd Bougault, 1979), as have parameters obtained from the triaxial test of bituminous mixtures.

Another method used to assess the workability of gap-graded mixes involves monitoring the specimen volume reduction during gyratory compaction so that the percentage air voids in the compacted mix can be calculated at any given number of revolutions. The rate of reduction of the air voids with increasing compaction effort is then used as an indicator of the workability of the mixture.

13.4.2.4 Measuring permeability

Permeability can be defined as the volume of gaseous or liquid fluid of unit viscosity that passes through, in unit time, a unit cross-section of a porous medium under the influence of a unit pressure gradient. Defined in this way, the permeability of a porous medium is independent of the absolute pressure or velocities within the flow system, or of the nature of the fluid, and is characteristic only of the structure of the medium. However, it has physical meaning only if the flow is of a viscous rather than a turbulent character.

Permeability is an important parameter of a bituminous mixture because it controls the extent to which both air and water can migrate into the material. (Note: how to measure the significance of exposure to air (ageing) is discussed in Section 13.4.2.2.) The intrusion of water may also bring about deterioration by causing the bitumen to strip from the aggregate particles (i.e. adhesive failure) or by causing weakening (damage) of the bitumen and of the bitumen/filler/fines mastic (i.e. cohesive failure).

In dimensions, permeability corresponds to an area, carrying the dimensions of length squared. The problems generally encountered in engineering deal with the flow of water in cases where the unit weight and the viscosity of water vary within fairly narrow limits. Thus, it has become customary to use a factor called the 'coefficient of permeability' when dealing with this problem; this is related to the intrinsic permeability by the expression

$$k = K \frac{\gamma}{\eta} \qquad (13.14)$$

where K is the intrinsic permeability (cm^2), k is the coefficient of permeability (cm/s), γ is the unit weight of the permeating fluid (g/cm^3) and η is the viscosity of the fluid (g s/cm^2).

The measurement of permeability is, in essence, a simple task, achieved by applying a fluid under pressure to one side of a specimen of a bituminous mixture and measuring the resulting flow of fluid at the opposite side. Both air and water have been used as the

Table 13.5 Classification of voids in terms of permeability of asphalt mixtures

Permeability, k: cm/s	Permeable condition	Voids	Mixture
$<10^{-4}$	Impervious	Impermeable	Dense
10^{-4}–10^{-2}	Poor drainage	Semi-effective	Stone mastic asphalt
$>10^{-2}$	Good drainage	Effective	Porous asphalt

Chen et al. (2004)

permeating fluid. Table 13.5 presents typical ranges of the coefficient of permeability for three common types of asphalt mixture (Chen et al., 2004).

If the pore system of a bituminous sample is fully saturated with water, and if sufficient pressure is applied, water may flow, and then

$$V = \frac{Ka\Delta p}{\eta L} \qquad (13.15)$$

where V is the flow rate (m^3/s), K is the intrinsic permeability (m^2), a is the cross-sectional area of the specimen (m^2), Δp is the fluid pressure head across the specimen (N/m^2), η is the viscosity of the fluid (Ns/m^2) and L is the length of the specimen (m). (Note that the intrinsic permeability (K) has a unit of area, and is dependent on the properties of the porous media only, as the viscosity of the fluid is included in the equation; this is in contrast to the coefficient of permeability, which has a unit of velocity, and is dependent on both the properties of the fluid and of the porous media.)

Air permeability measuring techniques of dense bituminous materials are mostly based on concepts that were originally developed to measure the permeability of mortars and concretes using differential pressure techniques. Air permeability testing is non-destructive, allows the determination of permeability in a very short period (Zoorob et al., 1999) and permits the specimens to be further tested for measurements of other parameters (e.g. stability or creep stiffness). The air permeability test is also very effective in detecting variations in the binder content, and, therefore, can be used to control the bituminous mixture composition. Small changes in the air voids content cause very large changes in the permeability.

An air permeability apparatus used with bituminous mixtures needs a specimen holder that will confine and seal the curved sides of the cylindrical compacted specimen, while maintaining the flat ends free. Pressurised gas (usually air) is then introduced at one flat end of the specimen, and, because of the side confinement, gas flow is then only possible through the interconnected air voids within the specimen and not around the sides. Other requirements are an accurate pressure gauge, a stable gas supply and an accurate flow meter at the downstream side. Knowing the viscosity of the gas at the test temperature allows pressure and flow rate readings to be entered into Equation 13.15, to determine the permeability of the bituminous specimen.

Figure 13.7 Apparatus for measuring the vertical water permeability of bituminous specimens

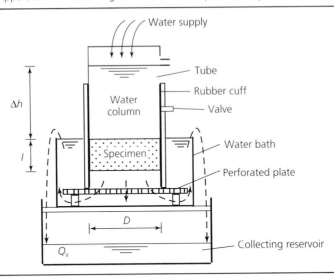

In dense mixtures, air voids have to be kept to a minimum while leaving sufficient voids for the bitumen to expand and to accommodate void reduction due to over-compaction from traffic. On the other hand, a porous asphalt mix must incorporate sufficient voids to maintain a permeable structure for adequate drainage; that is, have adequate water permeability, but not so much as to impair its strength and stability. Using the apparatus in Figure 13.7, the coefficient of vertical permeability k_v (m/s) can be calculated from

$$k_v = \frac{4Q_v l}{\Delta h \pi D^2} \tag{13.16}$$

where Q_v is the vertical flow rate through the specimen (m³/s), l is the thickness of the specimen (m), Δh is the height of the water column (m) and D is the diameter of the specimen (m).

Note that the water permeability values for porous asphalt gradations used in the UK are typically expected to lie between 0.5×10^{-3} and 3.5×10^{-3} m/s.

The permeability of a bituminous mixture depends on a large number of factors. Of particular importance are the quantity of voids, the distribution of void sizes and the continuity of the voids. Figure 13.8 shows how the permeability varies with the total voids in the mixture for a range of typical asphalt concrete, hot rolled asphalt, stone mastic asphalt and porous asphalt mixtures. It can be seen that there are significant differences in the air void size distribution and connectivity, and therefore the permeability, in mixtures with the same percentage of air voids. Figure 13.8 also shows that the relationship between permeability and air voids is exponential.

The voids are also affected by the nature of the aggregate; that is, the shape, texture and grading of the particles govern the packing and, hence, the void content at a

Figure 13.8 Relationships between the coefficient of permeability, *k*, and air voids for a range of asphalt mixtures used in the UK (Caro *et al.*, 2008)

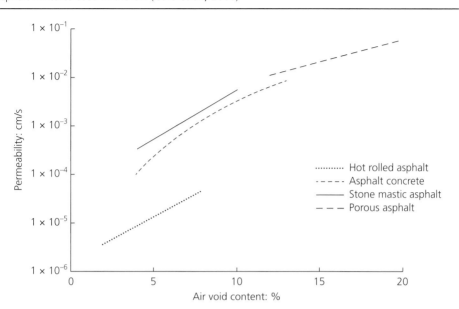

particular bitumen content. The amount of compaction effort employed is also important.

13.4.2.5 Indirect tensile stiffness modulus

A very important performance property of a bituminous base is its stiffness modulus, which is a measure of the load-spreading ability of the layer. The modulus controls the levels of the traffic-induced tensile strains at the underside of the lowest bituminous bound layer: these are responsible for fatigue cracking, as well as the stresses and strains induced in the subgrade that can lead to plastic deformations. The non-destructive indirect tensile test illustrated in Figure 13.9 is probably the most popular means of measuring this property. With this test, as detailed in BS EN 12697-23 (BSI, 2012–2015), the stiffness modulus, S_m (MPa), is given by

$$S_m = \frac{L(v + 0.27)}{Dt} \tag{13.17}$$

where *L* is the peak value of an applied vertical load (N), *D* is the mean amplitude of the horizontal deformation obtained from two or more applications of the load pulse (mm), *t* is the mean thickness of the test specimen (mm) and *v* is Poisson's ratio (i.e. 0.35).

During testing (which is carried out at 20°C), care is taken to ensure that the magnitude of the load pulses applied do not cause the specimen to deform outside the elastic recoverable range, as this may result in irrecoverable plastic deformations or micro-cracks. To achieve this, the rise time (which is the time taken for the applied load to

Figure 13.9 Indirect tensile stiffness modulus test configuration (LVDT, linear variable differential transformer)

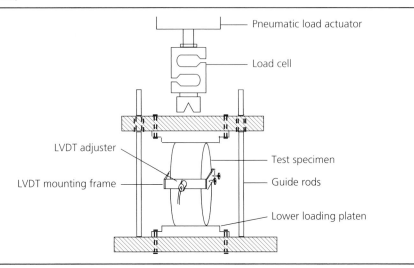

increase from zero (i.e. from when the load pulse commences) to its maximum value) is set at 124 ± 4 ms, and the load pulse (which is the time from the start of a load application until the start of the next load application) is set at 3.0 ± 0.05 s. The target load factor area is 0.6 – see the shaded area in Figure 13.10. The peak load value is adjusted to achieve a peak transient horizontal deformation of 0.005% of the specimen diameter.

13.4.2.6 Permanent deformation

Shear deformations resulting from high shear stresses in the upper portion of a bituminous pavement are the primary cause of rutting in flexible pavements. Repeated applications of these stresses under conditions of comparatively low mixture stiffness (e.g. stiffness at long loading times and high temperatures), as well as the balance between the viscous (non-recoverable) and elastic (recoverable) components of the mixture's deformation, lead to the accumulation of permanent deformations at the pavement surface.

Figure 13.10 Form of the load pulse showing the rise time and the peak load

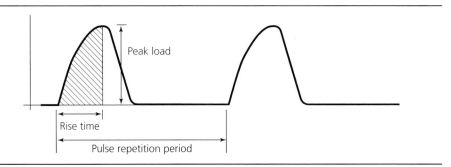

Bituminous mixtures (e.g. asphalt concretes), which use a continuously graded aggregate, rely mainly on aggregate interlock for their resistance to deformation: thus, aggregate grading and particle shape are major factors governing the deformation of these materials. The characteristics of the fine aggregate are particularly important in gap-graded materials such as hot rolled asphalt and stone mastic asphalt, which rely on a dense bitumen and fines mortar for their strength. Note that sand particles can vary considerably from spherical glassy grains in dune sands to angular and relatively rough grains from some pits, and that mixtures made with a range of sands, at the same bitumen content, have been shown to give deformations that varied by a factor of 4 from the best to the worst sands (Knight *et al.*, 1979).

Two tests that are commonly used to determine the permanent deformation properties of bituminous mixtures are the creep test (usually under repeated loading) and the wheel-tracking test.

The creep test is carried out either in the static or dynamic mode of loading. Each test typically lasts for 2 h, namely 1 h loading plus 1 h recovery, and gives results that allow the characterisation of the mixtures in terms of their long-term deformation behaviour as described in BS EN 12697-25 (BSI, 2012–2015). Tests are normally conducted at 40 or 60°C, and a typical value of applied creep stress is 0.1 MPa. During the test, axial deformation is continuously monitored as a function of time. Thus, knowing the initial height of the specimen, the axial strain, ε, and, hence, the stiffness modulus, S_{mix}, can be calculated at any loading time from

$$S_{mix} = \frac{\text{applied stress}}{\text{axial strain}} \qquad (13.18)$$

Better correspondence, in terms of mixture ranking, with respect to creep deformation or strain rate, has been shown to exist between the repeated load axial creep test and the wheel-tracking test (BS EN 12697-22 (BSI, 2012–2015)) than the ranking produced by the static creep test. For the repeated load test, which is known as the repeated load axial test (RLAT) in the UK (BSI, 1996), the applied stresses are as before, but the loading regime varies slightly (i.e. the pulse width = 1 s and the pulse period = 2 s), and the test is ended after 3600 pulses so as to give an accumulated loading time of 1 h.

Although simple to carry out, the repeated load creep test is extremely convenient and allows the relative performance of different bituminous mixtures to be easily determined. However, the test is often criticised as being too severe, as it does not employ a confining stress: in situ materials are clearly confined, and the effect of the confining stress on the vertical strain may be important. Thus, unless testing is conducted purely for the sake of ranking the performance of samples with identical gradations, specimen confinement (in a similar manner to that used in the triaxial test apparatus in soil mechanics) during creep loading is recommended. This is especially important for porous asphalt mixes that have been shown to fail in the unconfined creep mode of testing and yet, when properly designed, can perform as well as dense graded mixes in the wheel-tracking test. The actual amount of confinement required varies, but it is hypothesised to be dependent upon both the angle of internal friction of the aggregates and the mixture cohesion, and

Figure 13.11 Comparison of permanent strain and the number of test cycles for asphalt concrete (AC) and hot rolled asphalt (HRA) mixtures at different test temperatures

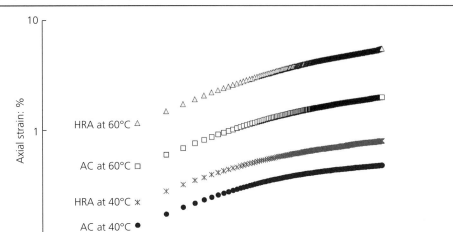

these, in turn, are influenced by the aggregate gradation and angularity and the stiffness of the binder.

Figure 13.11 shows permanent strain against the number of test cycles in a repeated load axial test. It can be seen that permanent strain increases with temperature: this is due to a reduction in the bitumen viscosity, the consequent reduction in the bitumen stiffness, and the accumulation of repeated, non-recoverable viscous deformations. The figure also indicates the effect of the aggregate grading. At low temperatures, the permanent strain in continuously graded asphalt concretes and gap-graded hot rolled asphalts will be similar; that is, the high degree of aggregate particle interlock in the asphalt concrete and the high-viscosity bitumen in the hot rolled asphalt provide a similar resistance to deformation. However, at higher temperatures, the hot rolled asphalt deforms more due to the reduced bitumen viscosity, which is not compensated by the aggregate interlock effect. In the asphalt concrete, however, although the bitumen will also be less stiff and viscous, the aggregate grading continues to provide a compensating resistance to deformation.

13.4.2.7 Resistance to disintegration

The Cantabro test of abrasion loss (BS EN 12697-17 (BSI, 2012–2015)) is used to quantitatively assess the resistance to disintegration of porous asphalt specimens in the laboratory. By varying the bitumen content for a given porous asphalt gradation, the minimum amount of bitumen required to ensure resistance against particle losses resulting from traffic can be estimated.

The procedure consists of subjecting a specimen to impact and abrasion in the Los Angeles abrasion test drum (omitting the steel balls) at a controlled temperature. The aim of the test is to determine the Cantabro loss (i.e. the percentage weight loss in relation to the initial weight of the sample) after 300 drum revolutions. Typically, 30% and 25% Cantabro abrasion losses are the limits for mixture acceptance at 18 and 25°C, respectively.

13.4.2.8 Wheel-tracking test

The wheel-tracking test (BS EN 12697-22 (BSI, 2012–2015)) seeks to measure the resistance of a bituminous material to plastic deformation. It can be considered a severe, simulative test, and is the preferred method for ranking surfacing mixtures that are expected to perform under high temperatures and/or at heavily stressed locations with channelled traffic and very heavy axle loads. Wheel tracking illustrates conclusively, as opposed to conventional creep testing, the advantages of using polymer-modified bitumens in asphalt surfacings.

Figure 13.12 is a diagrammatic representation of a laboratory-scale wheel-tracking test. The test involves subjecting a 50 mm thick slab of material – this can be either prepared in the laboratory or a 200 mm diameter core taken from a bituminous surfacing layer – to a rolling standard wheel that traverses the specimen at a constant temperature of either 45 or 60°C and an applied wheel load of 520 N. The performance of the bituminous mixture is assessed by measuring the resultant rut depth after a given number of passes or the rate of tracking (mm/h).

Equation 13.19, which was developed from test road data, relates the results from the wheel-tracking test (carried out at 45°C) to the number of commercial vehicles travelling

Figure 13.12 Wheel-tracking test

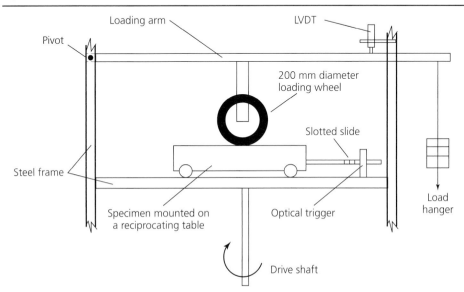

at 60 mph that is required to form a 10 mm rut in a sand-based hot rolled asphalt surface course at the end of a 20 year design life

$$\text{maximum WTR} = \frac{14\,000}{C_v + 100} \tag{13.19}$$

where WTR is the wheel-tracking (deformation) rate (mm/h) and C_v is the number of commercial vehicles per lane per day (cv/lane/day). This equation says that a road pavement carrying 7000 cv/lane/day should be covered with a surface course having a wheel-tracking rate of less than 1.97 mm/h at 45°C.

As the effect of loading time on mixture stiffness and, thus, deformation is significant, a correction factor needs to be applied to Equation 13.19 for road speeds other than 60 mph.

13.4.2.9 Fatigue testing

Under traffic loading the layers of a flexible pavement structure are subjected to continuous flexing. The magnitudes of the strains are dependent on the overall stiffness and nature of the pavement construction, but analysis confirmed by in situ measurements has indicated that tensile strains of the order of 30×10^{-6} to 200×10^{-6} occur under a standard wheel load. Under these conditions, the possibility of fatigue cracking exists: the phenomenon of fracture under repeated or fluctuating stress having a maximum value generally less than the tensile strength of the material.

Fatigue tests are carried out by applying a load to a specimen in the form of an alternating stress or strain, and determining the number of load applications required to induce failure of the specimen. As is suggested in Figure 13.13, a number of tests have been developed to measure the dynamic stiffness and fatigue characteristics of bituminous mixtures: these include bending tests using beams or cantilevers, compressive, compressive–tensile (push–pull), and indirect tensile tests (Read and Collop, 1997). With bending tests, the maximum stress occurs at a point on the surface of the specimen, and its calculation, using standard beam bending formulae, depends on the assumption of linear elasticity.

Fatigue tests may be conducted in two ways. They may be constant-stress tests, where each load application is to the same stress level regardless of the amount of strain developed. Alternatively, they may be constant-strain tests, where each load application is to the same strain level regardless of the amount of stress required. These two alternatives produce quite different results.

Figure 13.14(a) shows the general pattern of results from constant-stress tests: in this case, each line represents a different test temperature (i.e. a different stiffness), and it can be seen that mixtures with higher stiffness have longer lives. Figure 13.14(c) shows the general pattern of results from constant-strain tests with, again, each line representing a different temperature or stiffness: in this instance, however, the mixtures with higher stiffness have the shortest lives. This contrast may be explained in terms of the failure

Figure 13.13 Different configurations for measuring the fatigue life of bituminous specimens: (a) beam (three-point vibration) test; (b) trapezoidal cantilever (two-point vibration) test; (c) direct tension–compression test; (d) rotating bending test; (e) indirect tensile test

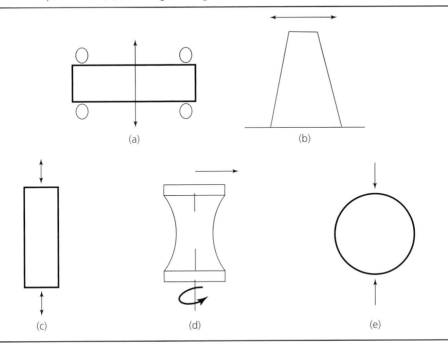

mechanism. Note that cracks are initiated at points of stress concentration, and propagate through the material until fracture occurs: thus, if the stress level is kept constant, the stress level at the tip of the crack continues to be high, and propagation is rapid. However, in a constant-strain test, the development of a crack causes a steady reduction in the applied stress level because the cracks contribute more and more to the strain as they propagate and, therefore, the stress at the crack tips reduces, and the rate of propagation is slow.

Thus, in concept, it is important to establish which test condition is most relevant to actual pavement behaviour. It has been shown (Monismith and Deacon, 1969) that strain control is appropriate to thin layers (e.g. bituminous surfacings), whereas stress control is appropriate to thicker structural layers (bituminous bases). This is because pavements are subject to a stress-controlled loading system, so that the main (and normally thick) structural layers are stress controlled. However, the thin surface layer must move with the lower structural layers, and so is effectively subject to strain control. Nevertheless, under low-temperature conditions giving high stiffness, crack propagation is relatively quick even under strain-controlled loading, so that the difference between the two loading conditions is small.

If the results of a controlled stress test are expressed in terms of an equivalent strain, then a log–log plot of strain versus the number of load cycles (Figure 13.14(b)) produces a

Figure 13.14 Fatigue lines representing the number of cycles to failure, N_f, under different test conditions (Brown, 1980): (a) controlled stress tests (log σ versus log N_f) at different stiffnesses S_1, S_2, S_3 and S_4; (b) controlled stress tests (log ε versus log N_f); (c) controlled strain tests (log ε versus log N_f)

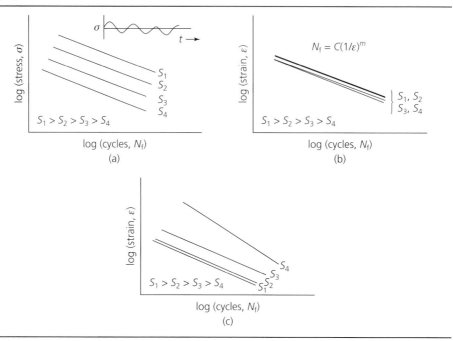

single linear relationship for all test conditions for a particular mixture (i.e. it shows that the relationship is independent of mixture stiffness). This suggests that strain is the principal criterion governing fatigue failure, and it has been demonstrated (Cooper and Pell, 1974) that flexure tests on a wide range of mixtures produce unique fatigue lines for each mixture. The general relationship defining the fatigue line is

$$N_f = C\left(\frac{1}{\varepsilon_t}\right)^m \tag{13.20}$$

where N_f is the number of load cycles to initiate a fatigue crack, ε_t is the maximum applied tensile strain, and C and m are constants depending on the composition and properties of the asphalt mixture.

Fatigue lines for a range of asphalt mixtures are shown in Figure 13.15. While a large number of variables associated with each asphalt mixture affect these fatigue lines, it has been shown (Cooper and Pell, 1974) that two are particularly important, namely (a) the volume of bitumen in the mixture, and (b) the viscosity of the bitumen as measured by the softening point. Thus, as the volume of bitumen increases to 15%, the fatigue life increases, and as the bitumen becomes more viscous, with the softening point increasing up to 60°C, the fatigue life also increases.

Figure 13.15 Fatigue lines under controlled stress loading conditions for typical asphalt mixtures

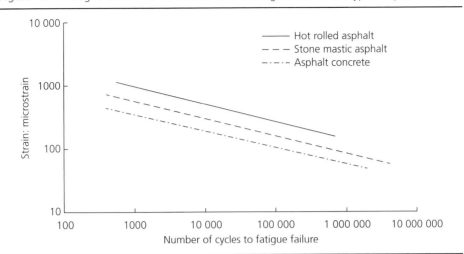

Other factors are also important in so far as they affect the above two main variables. The void content of the mixture has an effect on the volume of bitumen. The total void content is, in turn, affected by the particle shape and the grading of the aggregate, the compaction effort and the temperature. In other words, there is a link between workability, compaction effort and void content that is controlled by the bitumen content. However, while a higher bitumen content improves the fatigue life, it also reduces the stiffness, and that leads to increased strain.

Figure 13.16 illustrates the double influence of void content. Figure 13.16(a) shows the effect that increasing the bitumen content has on stiffness; that is, the stiffness is reduced, which increases the strain under constant stress conditions and causes a shift to the left along the fatigue line, thus reducing the fatigue life. Figure 13.16(b) shows the influence that the void content has on the fatigue line, so that for the same strain the fatigue life is reduced as the void content increases. The change in position of the fatigue line

Figure 13.16 Influence of voids on the fatigue life

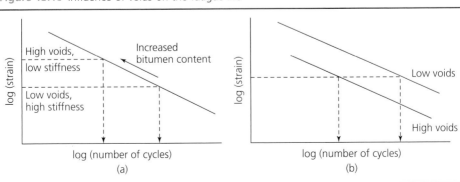

corresponds to a change in the material type, as was seen in Figure 13.15, whereas the shift along a fatigue line due to a stiffness change corresponds to a change in the degree of compaction. However, in practice, both effects occur if the change in the void content is associated with a change in the bitumen content.

13.4.2.10 Indirect tensile strength test

The indirect tensile strength test is performed by loading a cylindrical specimen with a single or repeated compressive load that acts parallel to and along the vertical diametral plane (see ASTM D4123 and EN 12697-23 (BSI, 2012–2015)). The loading configuration shown in, for example, Figure 13.13(e) develops a relatively uniform tensile stress perpendicular to the direction of the applied load and along the vertical diametral plane, which ultimately causes the specimen to fail by splitting along the vertical diameter. Based upon the maximum load carried by a specimen at failure, the indirect tensile strength is calculated from

$$ITS = \frac{2P_{max}}{\pi t d} \tag{13.21}$$

where ITS is the indirect tensile strength (N/mm^2), P_{max} is the maximum applied load (N), t is the average height (in this case, length) of the test specimen (mm) and d is the diameter of the specimen (mm).

The indirect tensile test provides two mixture properties that are useful in characterising bituminous mixtures. The first of these is tensile strength, which is often used to evaluate the water susceptibility of mixtures; that is, the tensile strength is measured before and after water conditioning of samples, to determine the retained tensile strength as a percentage of the original tensile strength. The second property derived from this test is tensile strain at failure, which is useful for predicting cracking potential; that is, mixtures that can tolerate high strains prior to failure are more likely to resist cracking.

13.4.2.11 Adhesion testing

Stripping (i.e. the detachment of bitumen from the aggregate) is associated with mixtures that are permeable to water. If stripping occurs, it can result in a loss of internal cohesion and, possibly, disintegration of the surfacing. The degree of adhesion is thus a function of the affinity between the aggregate and the bitumen within the bituminous material.

The mineralogical and physical nature of aggregate particles has an important bearing on adhesion, with the adhesive capacity being a function of the chemical composition, shape and structure, the residual valency, the surface energy, and the surface area presented to the bitumen. Generalisations about the effect of mineralogy are difficult because the effects of grain size, shape and texture are also important. However, in general, the more siliceous aggregates such as granites, rhyolites, quartzites, siliceous gravel and cherts tend to be more prone to moisture-related adhesive failures. Nonetheless, good results have been reported with these materials, and stripping failures have occurred with supposedly good rocks (e.g. limestones and basic igneous rocks).

The absorption of bitumen into an aggregate depends on several factors, including the total volume of permeable pore space and the size of the pore openings. The presence of a fine microstructure of pores, voids and micro-cracks can result in an enormous increase in the absorptive surface available to the bitumen. This depends on the petrographic characteristics of the aggregate as well as on its quality and state of weathering.

It is generally accepted that rougher surfaces exhibit a greater degree of bitumen adhesion and that the presence of a rough surface texture can mask the detrimental effects of mineralogy. A balance is, however, required between the attainment of good aggregate wettability, with smooth surfaces being more easily wetted, and a rougher surface that holds the binder more securely once wetting is achieved.

The important characteristics of the bitumen that affect its adhesion to aggregate are its viscosity, surface tension and its polarity. The viscosity and surface tension will govern the extent to which the bitumen is absorbed into the pores at the surface of the aggregate particles: both of these properties vary with temperature, and mixing of aggregate and bitumen is therefore done at high temperature to ensure that the bitumen coats the aggregate surface readily.

Various tests have also been developed to address the adhesive and stripping capabilities of bituminous materials. Immersion mechanical tests involve measurements of changes in a mechanical property of a compacted bituminous mixture after its immersion in water, and the ratio of the property after immersion to the initial property is an indirect measure of stripping (e.g. the ratio of the indirect tensile strengths of conditioned and dry specimens is one simple measure that is used). However, what is possibly the most popular measurement is the 'retained Marshall stability': this is the ratio (expressed as a percentage) of the Marshall stability of a bituminous specimen after wet conditioning to that of an identical specimen that is not subjected to wet conditioning.

Immersion trafficking tests recognise that traffic can play an important role in stripping. A test that simulates the effect of traffic is the immersion wheel-tracking test (see Section 13.4.2.8): in this case, however, the test specimens are kept immersed horizontally in a water bath at 40°C during testing.

13.5. Construction methods for hot-mix bituminous materials

The basic steps in the construction of asphalt bases and surfacings are as follows: (a) manufacture the bituminous mixture; (b) prepare the underlying or levelling course on site; (c) transport and place the asphalt mixture on site; (d) construct joints; and (e) compact and finish the laid material.

The process of manufacturing hot-mix asphalts involves three stages: (a) the aggregates must be proportioned to give the required grading; (b) the aggregates must be dried and heated; and (c) the correct amount of binder must be added to the aggregate and mixed to thoroughly coat the particles and produce a homogeneous material. These three stages of the production of bituminous mixtures are undertaken within an asphalt-mixing plant (AI, 1986, 1987, 1989).

13.5.1 Preparing the mixture at an asphalt-mixing plant

There are two general types of asphalt-mixing plants in common use: drum mix plants and batch plants.

A schematic diagram (Figure 13.17) of the components of a typical drum hot-mix plant shows that the flow of materials is basically from left to right, after the cold aggregates have been moved from the stock-pile area to the metal cold feed bins by, commonly, a front-end loader. The number of bins depends on the number of different aggregates to be blended in the bituminous mixture; in practice, most large facilities have a minimum of four bins. Each bin has slanted sides, and an attached vibrator (if gravity feed is inadequate) to ensure a constant downward flow of aggregate. Typically, an adjustable gate and a variable-speed feeder are located at the bottom of each cold bin to proportion the materials to meet the job-mix formula gradation. A gathering conveyor brings the job-mix material to a cold feed elevator that, in turn, moves it to the drum mixer. An automatic weighing system on this conveyor continuously weighs the amount of aggregate plus moisture going into the drum mixer. In the control room, a correction to the total weight is made for the moisture content of the aggregates, so that the proper amount of bitumen can be pumped into the drum mixer.

The drum mixer typically has a 'parallel flow' design whereby the aggregates move in the same direction as the exhaust gases; that is, the aggregate is fed into the drum at the burner end and it is then dried and heated as it moves down towards the discharge end. Heated bitumen, which is pumped from a bitumen storage tank, enters the drum at a point about one-third the drum length from the discharge end. When the bitumen is added into the drum, it is pumped into the bottom of the drum at about the same location that the mineral filler and/or fines are reintroduced. Adding the bitumen and dust in close proximity allows the bitumen to trap and coat most of the fines before they are picked up by the high-velocity exhaust gas stream. The aim of the dust collection system is to ensure that external (environmental) dust emission criteria are met.

In relation to the addition of the bitumen, it should be noted that if the viscosity of a bitumen is too high during mixing, the aggregate will not be properly coated, and, if the viscosity is too low, the bitumen will coat the aggregate easily but may subsequently

Figure 13.17 Basic drum hot-mix plant. (Courtesy of the Asphalt Institute)

1 Automatic weighing system   6 Asphalt pump
2 Cold feed conveyor   7 Dust collector
3 Cold feed bins   8 Hot-mix conveyor
4 Asphalt storage tank   9 Mix surge silo
5 Drum mixer   10 Control van

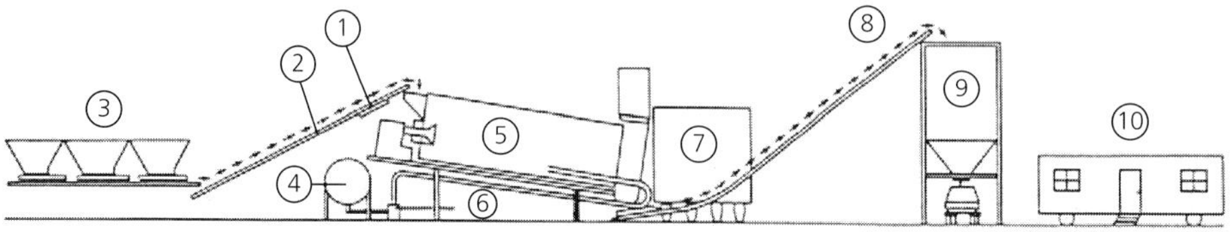

drain off during storage or transport to the pavement site. Thus, for satisfactory mixing, the temperature should be such that the viscosity of the bitumen is about 0.2 Pa. The mixing temperature can be estimated from a plot of the viscosity versus temperature relationship for the bitumen being used.

The bitumen (plus fines/dust) coats the aggregates as they move down the lower third of the drum, and the hot-mix material then exits the drum through a discharge chute into a conveyor system that carries it to a surge silo, from which it is discharged into trucks, via an automatic weighing system, for transport to the construction site.

All movements of material from the cold feed to the surge silo are monitored from the control room with the aid of sensors that monitor conveyor speeds, aggregate weights, temperature and other critical functions that affect the efficient operation of the mixing process.

The main components of a typical batch hot-mix plant are shown in Figure 13.18. A comparison of the first few components in the batch and drum facilities shows that the cold feed proportioning systems are generally similar for both types prior to the drying operation. However, the dryer in a batch facility is typically of a 'counter-flow' design whereby the aggregate flow in the drum is against the flow of the exhaust gases; that is, the drier, which is mounted at an angle to the horizontal, is essentially a large rotating cylinder that is equipped with a heating unit at the lower end. The heating unit is usually a low-pressure air-atomisation system using fuel oil, and the hot gases from the burner pass from the lower end up the cylinder and out at the upper end. The cold aggregate is fed into the upper end of the drier, picked up by steel angles or blades set on the inside face of the cylinder, and dropped through the burner flame and hot gases as it moves down the cylinder. The hot aggregate is then discharged from the lower end of the drier onto an open conveyor or an enclosed 'hot elevator' that transports it to screens and storage bins that are mounted at the top of the power plant.

In the preparation of these hot materials, as well as of those described earlier, the temperature of the aggregates may be raised to 160°C or more, to ensure that practically all of the moisture is removed.

The hot dust-laden exhaust gases from the dryer are passed through a dust collection system to remove dust particles so that external emission standards are met. The collected dust is returned to the hot material elevator or filler storage silo for subsequent reintroduction into the mixture as required. Mineral filler that is added to the mixture is not normally passed through the aggregate drier: instead, it is fed, via a separate device, directly into the mixing unit or into the aggregate batching unit. Separate feeder units may also be used to supply liquefier, other fluxes or hydrated lime to the mixture, should these be required.

After the aggregates exit the hot elevator at the top of the tower, they are discharged onto vibrating screens that separate them into a number of sizes, and these are then stored in hot material bins. The stored aggregates are later proportioned, by the control

Figure 13.18 Asphalt batch hot-mix plant. (Courtesy of the Asphalt Institute)

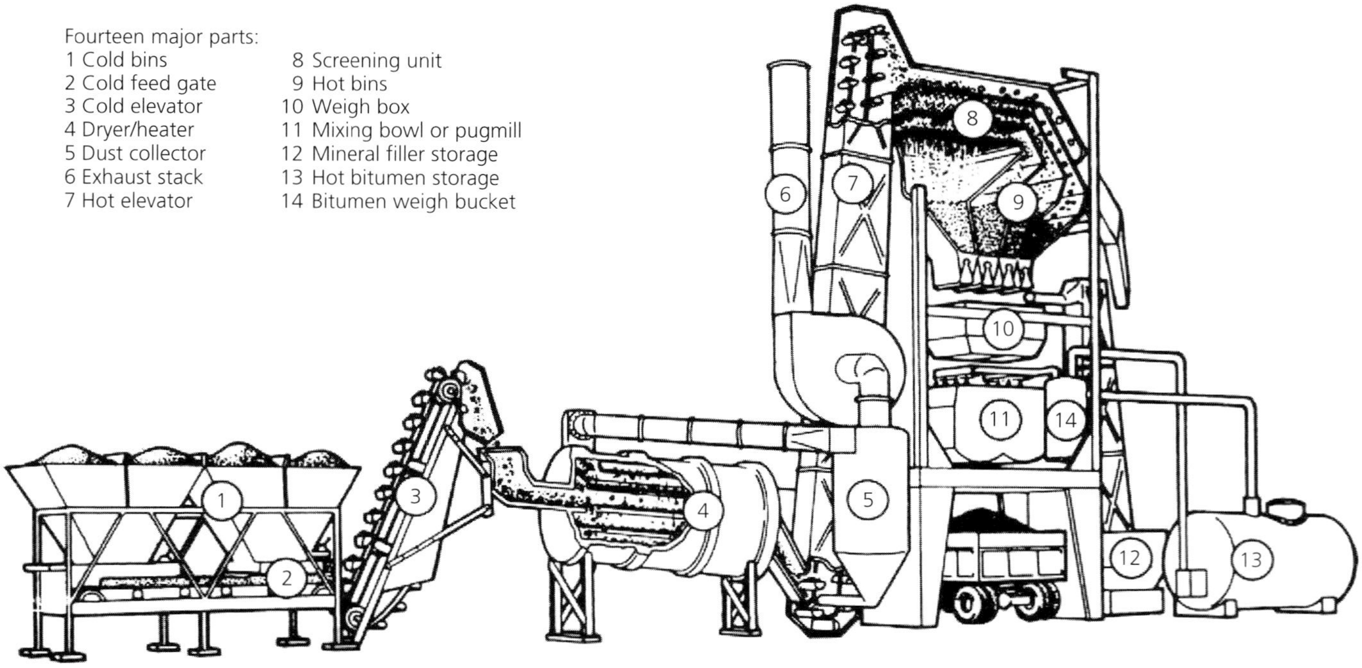

Fourteen major parts:
1 Cold bins
2 Cold feed gate
3 Cold elevator
4 Dryer/heater
5 Dust collector
6 Exhaust stack
7 Hot elevator
8 Screening unit
9 Hot bins
10 Weigh box
11 Mixing bowl or pugmill
12 Mineral filler storage
13 Hot bitumen storage
14 Bitumen weigh bucket

system, from the bins into a weigh box that is mounted on a set of scales. These dry aggregates are then discharged into a pugmill, and the hot bitumen, which has been weighed and stored in a weigh bucket, is sprayed into the pugmill after a few seconds of dry mixing. The pugmill is typically a twin-shaft, counter-rotating mixer that is designed to coat the aggregate quickly (typically, in about 45 s) with bitumen. After mixing, the bituminous mixture is ready to be discharged into a truck for transport to the pavement site.

Batch plants may be manually, semi-automatically or fully automatically operated. In manual batch plants, air cylinders or hydraulic cylinders, which are electrically actuated by the operator, control the bin gates, fines feeders, bitumen supply and spray valves, the weigh box discharge gate and the pugmill discharge gates. In semi-automatic batch plants, the various operations constituting each mixing cycle are under automatic control; that is, the quantities of bitumen and aggregate introduced into the mixes, the mixing times, the sequencing of the mixing functions, and the operation of the pugmill discharge gate. The fully automatic batch plant repeats the weighing and mixing cycle until the operator stops it, or until it stops itself because of material shortage or some other extraordinary event. Automatic plants also normally provide records of the amounts of the component materials in each batch.

Asphalt-mixing plants for the production of bituminous paving mixtures are also frequently described as being portable, semi-portable or stationary. Portable plants can be either relatively small units that are self-contained and wheel mounted, or larger mixing plants whose component units are easily moved from one place to another. Semi-portable plants are those in which the separate units are more difficult to dismantle, but can be taken down, transported on trailers, trucks or railway cars to a new location, and then reassembled, in a process that may require a few hours or several days, depending on the scale of the plant involved. Stationary plants are those that are permanently constructed in one location, and are not designed to be moved from place to place. The capacities of these plants range up to about 400 t/h of material, and the majority of hot-mix bituminous mixtures are prepared at temperatures ranging from 110 to 185°C.

13.5.2 Control of hot-mix uniformity

Control of the uniformity of hot asphalt mixtures is vital, as any appreciable variation in the aggregate gradation or the bitumen content will be reflected in a change in some other characteristic of the mixture. Sampling and testing are therefore among the most important functions in plant control.

Samples are typically obtained at various points in the asphalt-mixing plant to ensure that the processing is in order up to those points. A final extraction test of the mixture is normally carried out to confirm its uniformity, gradation and bitumen content. The extraction test measures the bitumen content, and provides enough aggregate for gradation testing. The extraction and gradation results should fall within the job mix tolerance specified; if they do not, corrective measures will need to be taken to bring the mixture within the tolerance limits.

13.5.3   Transporting hot-mix asphalt

After the final assembly, weighing, blending and mixing of the aggregates and the bitumen, the final bituminous material is discharged into hauling units for transport to the job site.

Dump trucks or trailers are normally used to transport the hot-mix material from the plant. These vehicles should have smooth metal beds that have previously been cleaned of all foreign material. The vehicle bed may be sprayed with a light coat of lime water, soap solution or some similar substance to prevent adherence of the mixture: fuel oils should not be used as they have a detrimental effect on the mix. The vehicle should be insulated against excessive heat loss in the mixture during hauling, and covered to protect the asphalt from inclement weather.

13.5.4   Placing hot-mix asphalt

Surface courses of asphalt are frequently placed on new or existing bases that require very little preparation other than a thorough sweeping and cleaning to remove loose dirt and other foreign materials. On absorbent surfaces, such as granular bases, a prime coat – typically a cutback bitumen – is normally applied prior to placing the new mixture. On existing paved surfaces, a tack coat (e.g. a light application of emulsified bitumen) is usually applied. In some cases, a bituminous 'regulating' (i.e. levelling) course may be laid, to correct irregularities in an existing surface. Placement of an asphalt should only be allowed when the underlying layer is dry and the weather conditions are favourable. The placement of hot-mix materials is normally suspended when the ambient air temperature is less than 4°C.

Asphalt bases and surfacings are placed and compacted in separate operations. In certain cases (e.g. in bituminous bases), thick layers of the same mixture may be placed in two or more layers.

Bituminous mixtures are normally put in place by an asphalt paver that spreads the mixture in a uniform layer of the desired thickness, and 'finishes' (i.e. shapes) it to the desired elevation and cross-section, so that it is ready for immediate compaction while still hot. The wheelbases of the pavers are sufficiently long to eliminate the need for forms, and to minimise the occurrence of irregularities in the underlying layer. The machines can process thicknesses of up to 250 mm over a width of up to 4.3 m at working speeds ranging from 3 to 21 m/min.

As is shown in Figure 13.19, the transport unit tips the asphalt mixture into a receiving hopper at the front of the paver, and it is then fed from the hopper towards the finishing section of the machine, where it is spread and agitated by screws that ensure the uniformity of the spread material over the full processing width. The loose 'fluffed-up' material is then struck off at the desired elevation and cross-section by one or more oscillating or vibratory screeds that employ a tamping mechanism to strike off and initially compact the material: these screeds are usually equipped with heating units to prevent the asphalt material from being picked up during the spreading and finishing operations. The pavers are fully adjustable, to ensure a uniform flow-through of material and the production of a smooth, even layer of the desired thickness and cross-section.

Figure 13.19 Mode of operation of an asphalt paver

Asphalt pavers are normally fitted with electronic screed control systems. These sensors operate on a reference profile, sense changes in the position of the floating screed element of the paver or of the reference profile, and then automatically apply corrections to the angle of the screed so that the surface being laid is continuously parallel to the reference profile. Usually, the reference profile controls the longitudinal profile of the surface at one side of the machine while a slope sensor controls the transverse slope (i.e. the cross-section).

When placing the bituminous mixture, special attention must be given to the construction of joints between old and new surfaces or between successive days' work. It is essential that a proper bond be secured at longitudinal and transverse joints between a newly placed mixture and an existing bituminous mat. The best longitudinal joint is obtained when the material at the edge being laid against is still warm enough for effective compaction; this means that two pavers working in echelon can produce a very satisfactory joint. Various other procedures are used when laying against a cold edge: for example, the cold edge may be cut back and painted with bituminous material, or an infrared joint heater may be attached to the front of the finisher to heat the cold material along the edge prior to placing the new layer.

13.5.5 Compacting the asphalt layer

When the spreading and finishing operations are completed, and while the mixture is still hot, rolling is begun. The temperature at which rolling is initiated is important as, if the viscosity of the bitumen is too low during compaction, the mixture will be excessively mobile and the result will be 'pushing' before the roller: on the other hand, if the viscosity is too high, it will significantly reduce the workability of the mix, and poor compaction will be achieved. It is now widely recognised that the optimal bitumen viscosity for compaction is between 2 and 20 Pa, and this means that the appropriate compaction temperature can be estimated from a plot of the viscosity versus temperature relationship for the bitumen being used.

Rolling may be carried out by steel-wheeled or pneumatic-tyred rollers or by a combination of the two. Conventional steel-wheeled rollers are, typically, of three types: three-wheel rollers, two-axle tandem rollers and three-axle tandem rollers. Vibratory steel-wheeled rollers, which provide a centrifugal force of up to 210 kN and up to 3000 vibrations per minute, are increasingly replacing conventional rollers to compact asphalt layers. Pneumatic-tyred rollers provide a closely knit surface by kneading aggregate particles together; the tyre contact pressures generally range from 276 to 620 kPa.

Most of the rolling is done in a longitudinal direction, beginning at the edges and gradually progressing towards the centre – except on super-elevated curves, where rolling begins on the low side and progresses toward the high side. Rolling procedures vary with the properties of the mixture, the thickness of layer and other factors. In modern practice, rolling is divided into three stages that follow closely upon each other: initial or 'break-down' rolling, intermediate rolling and finish rolling. The breakdown and intermediate phases primarily provide the desired asphalt density, while the final rolling gives the smoothness.

Specifications for the finished surface frequently stipulate that it should be smooth, even, and true to the desired grade and cross-section, with no vertical deviation being permitted that is more than 3 mm from a line established by a 3 m straight edge. While this seems to be a very high standard, these surfaces are readily attainable with modern construction equipment. Deviations of more than 6 mm from the specified thickness of a surface course are generally not allowed.

The densities to be obtained in the compacted layers are usually stipulated as a percentage of the theoretical maximum density or of the laboratory-derived density. Thus, use is often made of nuclear gauges to measure the density and to control the rolling process. In addition to checking the densities of the compacted materials on samples cored or sawn from the completed layers, many agencies also determine (for control purposes) the percentage of voids and of voids filled with bitumen, and thus extract samples of sufficient size to enable the gradation and bitumen content of the compacted mixture to be determined.

Rolling completes the construction of most asphalt pavements, and traffic is normally permitted on the surface as soon as the compacted mixture has cooled.

13.6. Use of low-temperature bituminous materials

In addition to conventional hot-mix asphalts, lower-temperature asphalts are increasingly being used in road pavements.

Low-temperature asphalt mixtures are defined as those that are intentionally manufactured and laid at temperatures less than the standard temperatures traditionally used for hot-mix bituminous bases and surfacings. Low-temperature asphalts are generally categorised into three types, depending on the extent of the temperature reduction (EAPA, 2010) which, when compared with the traditional mixtures, give the following:

- hot-mix asphalt – produced and mixed at temperatures roughly between 120 and 190°C, with the production temperatures depending on the bitumen used
- warm-mix asphalt – produced and mixed at temperatures roughly between 100 and 140°C
- half-warm-mix (aka semi-warm-mix) asphalt – produced with heated aggregate at a mixing temperature (of the mixture) between approximately 70 and 100°C.

Figure 13.20 Classification for hot, warm, half-warm and cold-mix asphalt mixtures

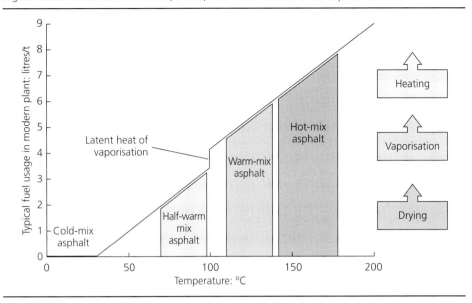

■ cold-mix asphalt – produced with unheated aggregate and bitumen emulsion or foamed bitumen.

The technologies behind these systems, which can be characterised schematically (Figure 13.20), vary quite markedly. There is no direct correspondence between the classifications and the technologies, but the most common technologies applied to the different classifications involve

■ incorporation of chemical additives that can be pre-blended with the bitumen or added at the time of mixing
   – warm – to modify the bitumen viscosity (e.g. waxes)
   – warm – to modify the frictional resistance to compaction
■ foamed bitumen
   – warm – by adding moisture-releasing additives (e.g. zeolites) within the mixture
   – warm or half-warm – using foam generation equipment
   – half-warm – by the addition of moist aggregates
   – half-warm or cold – other
■ bitumen emulsion
   – cold.

While most of the current specifications are still written around hot-mix asphalt, there is no reason why specifications for hot-mix asphalt cannot be modified to become applicable to the various categories of low-temperature asphalt.

Very little specification modification is required for warm-mix asphalt, but increasing modification is required as the mixture temperature departs (downward) from that for hot-mix asphalt. The properties of warm-mix asphalt should be judged against the same criteria as hot-mix asphalt. The difference is the reduced temperature at which the material is mixed, transported, laid and compacted. Drafts covering the various levels of categories of low-temperature asphalt that are modelled on current hot-mix asphalt standards, supporting national guidance documents and specifications, have already been produced (Nicholls *et al.*, 2013).

13.6.1 Current specifications
For bituminous road materials, the European standards are the relevant parts of BS EN 13108 (BSI, 2006a, 2006b, 2006c, 2006d), the national guidance document is PD 6691 (BSI, 2010a) and the national specification is usually the 900 series in the *Specification for Highway Works* (Highways Agency *et al.*, 2008) for roads; there are also national standards for airfields (Defence Estates, 2010). For the transport, laying and compaction of these asphalts, there is no unique European standard: rather, the European standard and national guidance levels are combined in BS 594987 (BSI, 2010b). The national specification level documents cover the materials and their transport, laying and compaction.

The BS EN 13108 series was originally drafted primarily for hot-mix asphalt mixtures, although this limitation is not explicitly acknowledged. However, in the related BS EN 12697 test methods (BSI, 2012–2015), the series does acknowledge the limitation in its title of 'Bituminous mixtures: Test methods for hot mix asphalt', although most of the test methods (as opposed to the specimen preparation procedures) are equally applicable to low-temperature asphalt mixtures.

It is expected that the BS EN 13108 series will be extended to explicitly cover warm-mix asphalt, rather than the current situation whereby it is not excluded. It is likely that cold-mix asphalt will require new parts – a part on cold emulsion-based mixtures has already been started. The position of half-warm mixtures is less clear. The specimen preparation procedures in the BS EN 12697 series are being reviewed to make them applicable to low-temperature asphalts although, again, some new tests may be required for cold-mix asphalt, and the title of the series will need to be changed.

At the national specification level, reference is made to low-temperature asphalt, but most clauses were drafted explicitly for hot-mix asphalt. However, with the increasing use of low-temperature asphalt – production of warm and half-warm mix asphalt in the USA grew from 16.8 million tons in 2009 to over 100 million in 2015 – a more comprehensive set of requirements is clearly needed if the use of these materials is to be encouraged in the UK and elsewhere.

13.6.2 Proposed specifications
While there are plans to update various documents to incorporate the various categories of low-temperature asphalt, a series of interim documents has been drafted in the UK to cover the European standard, national guidance and national specification levels for warm-mix, half-warm-mix and emulsion-based cold-mix asphalt (Nicholls *et al.*, 2013). Only a slight modification has been made in relation to warm-mix material

because the existing hot-mix documents can effectively be used. For warm-mix asphalt, the only requirement that will not be complied with is the minimum temperature of the mixture, as given in clause 5.2.10 of **BS EN 13108-1** (BSI, 2006a) for asphalt concrete.

More changes are required for half-warm-mix asphalts, with the draft approach specifying half-warm-mix asphalt in terms of component materials and proportions with additional requirements based on performance-related tests. The draft specification includes reference to BS 594987 (BSI, 2010b) for transporting, laying and compaction together with requirements to verify compliance. For half-warm-mix asphalt there is an extra requirement relating to the water content of the aggregate, and that the reheating of specimens should be avoided.

Half-warm mixtures will slowly eliminate water after their manufacture, during the delivery and application process. Therefore, it is important not to compact laboratory-prepared mixtures immediately because the excess water present may have a negative impact on some comparative tests. This phenomenon has been studied, and conditioning at 95°C for 2 h prior to compaction at 95°C has produced test samples with superior water sensitivity performance compared with samples compacted immediately.

Similar to half-warm mixes, the emulsion-based cold-mix approach specifies emulsion-based cold-mix asphalt in terms of component materials and proportions, with additional requirements based on performance-related tests. The main text has the same technical approach as is used in the various existing parts of BS EN 13108 (BSI, 2006a, 2006b, 2006c, 2006d), while the specification includes reference to BS 594987 (BSI, 2010b) for transporting, laying and compaction with compliance requirements to verify compliance.

The use of foamed-bitumen cold-mix asphalt is already covered by a well-established specification (Merrill *et al.*, 2004) that is supported by clauses 947 and 948 of the *Specification for Highway Works* (Highways Agency *et al.*, 2008).

REFERENCES

AI (Asphalt Institute) (1986) *Asphalt Plant Manual*. AI, College Park, MD, USA.
AI (1987) *Asphalt Paving Manual*. AI, College Park, MD, USA.
AI (1989) *Asphalt Handbook*. AI, Lexington, KY, USA.
AI (2014) *Mix Design Methods for Asphalt Concrete and Other Hot-mix Types*, 7th edn. AI, Lexington, KY, USA.
ASTM (2011) ASTM D2041: Standard test method for theoretical maximum specific gravity and density of bituminous paving mixtures. ASTM International, West Conshohocken, PA, USA.
Brown SF (1980) *An Introduction to the Analytical Design of Bituminous Pavements*. Department of Civil Engineering, University of Nottingham, Nottingham, UK.
BSI (British Standards Institution) (1996) BS DD 226: Method for determining resistance to permanent deformation of bituminous mixtures subject to unconfined dynamic loading. BSI, London, UK.
BSI (2006a) BS EN 13108-1:2006: Bituminous mixtures. Material specifications. Asphalt concrete. BSI, London, UK.

BSI (2006b) BS EN 13108-4:2006: Bituminous mixtures. Material specifications. Hot rolled asphalt. BSI, London, UK.

BSI (2006c) BS EN 13108-5:2006: Bituminous mixtures. Material specifications. Stone mastic asphalt. BSI, London, UK.

BSI (2006d) BS EN 13108-7: Bituminous mixtures. Material specifications. Porous asphalt. BSI, London, UK.

BSI (2010a) PD 6691:2010: Guidance on the use of BS EN 13108 bituminous mixtures. Material specifications. BSI, London, UK.

BSI (2010b) BS 594987:2010: Asphalt for roads and other paved areas. Specification for transport, laying, compaction and type-testing protocols. BSI, London, UK.

BSI (2012–2015) BS EN 12697-1 to BS EN 12697-47. Bituminous mixtures. Test methods for hot mix asphalt. BSI, London.

Caro S, Masad E, Bhasin A and Little DN (2008) Moisture susceptibility of asphalt mixtures, part 1: mechanism. *International Journal of Pavement Engineering* **9(2)**: 81–98.

Chen J, Lin K and Young S (2004) Effects of crack width and permeability on moisture-induced damage of pavements. *Journal of Materials in Civil Engineering* **16(3)**: 276–282.

Cooper KE and Pell PS (1974) *The Effect of Mix Variables on the Fatigue Strength of Bituminous Materials.* Transport and Road Research Laboratory, Crowthorne, UK, TRRL Laboratory Report 633.

D'Angelo JA, Paugh C, Harman T and Bukowski, J (1995) Comparison of the Superpave gyratory compactor to the Marshall for field quality control. *Proceedings of the Association of Asphalt Paving Technologists* **64**: 611–635.

Defence Estates (2010) *Asphalt for Airfields: SPEC 12, SPEC 13, SPEC 40 and SPEC 49.* Defence Infrastructure Organisation, Sutton Coldfield, UK.

EAPA (European Asphalt Pavement Association) (2010) *The Use of Warm Mix Asphalt: EAPA Position Paper.* EAPA, Brussels, Belgium.

Eckmann B (1989) EXXON research in pavement design – Moebius software: a case study reduction of creep through polymer modification. *Proceedings of the Association of Asphalt Paving Technologists* **58**: 337–350.

Franken L (ed.) (1998) *Bituminous Binders and Mixes.* Spon, London, UK, Rilem Report 17.

Gibb JM and Brown SF (1994) A repeated load compression test for assessing the resistance of bituminous mixes to permanent deformation. *Proceedings of the 1st European Symposium on the Performance and Durability of Bituminous Materials* (Cabrera JG and Dixon JR (eds)). Spon, London, pp. 199–209.

Highways Agency, Transport Scotland, Welsh Government and Department for Regional Development Northern Ireland (2008) Road pavements – bituminous-bound materials. In *Manual of Contract Documents for Highway Works*, vol. 1. *Specification for Highway Works.* Stationery Office, London, UK, Series 900.

Hopman PC, Valkering CP and Van der Heide JPJ (1992) Towards a performance related asphalt mix design procedure. *Proceedings of the Association of Asphalt Paving Technologists* **61**: 188–216.

Knight VA, Dowdeswell DA and Brien D (1979) Designing rolled asphalt wearing courses to resist deformation. In *Rolled Asphalt Road Surfacings* (Nicholls JC (ed.)). Institution of Civil Engineers, London, UK.

LCPC (Laboratoire Central des Ponts et Chaussées) (1997) *French Design Manual For Pavement Structures*. LCPC, Paris, France.

Mahboub K and Little DN (1990) An improved asphalt concrete mix design procedure. *Proceedings of the Association of Asphalt Paving Technologists* **59**: 138–175.

Marvillet J and Bougault P (1979) Workability of bituminous mixes: development of a workability meter. *Proceedings of the Association of Asphalt Paving Technologists* **48**: 91–110.

Merrill D, Nunn ME and Carswell I (2004) *A Guide to the Use and Specification of Cold Recycled Materials for the Maintenance of Road Pavements*. Transport Research Laboratory, Crowthorne, UK, TRL Report TRL611.

Monismith CL and Deacon JA (1969) Fatigue of asphalt paving mixtures. *Journal of Transport Engineering Division, ASCE* **95**: 154–161.

Nicholls JC, Bailey HK, Ghazireh N and Day DH (2013) *Specification for Low Temperature Asphalt Mixtures*. Transport Research Laboratory, Crowthorne, UK, TRL Published Project Report PPR666.

Read JM and Collop AC (1997) Practical fatigue characterisation of bituminous paving mixtures. *Proceedings of the Association of Asphalt Paving Technologists* **66**: 74–108.

Zoorob SE, Cabrera JG and Suparma LB (1999) A gas permeability method for controlling quality of dense bituminous composites. *Proceedings of the 3rd European Symposium on Performance and Durability of Bituminous Materials and Hydraulic Stabilised Composites* (Cabrera JG and Zoorob SE (eds)). AEDIFICATIO Publishers, Zurich, Switzerland, pp. 549–572.

Highways
ISBN 978-0-7277-5993-1

ICE Publishing: All rights reserved
http://dx.doi.org/10.1680/h5e.59931.449

Chapter 14
Constructing concrete pavements

Martin O'Connell   Project Manager, Airport Infrastructure Group, Amey Ltd, UK

Whatever the type of concrete pavement employed in the making of a major road or parking area, the keys to its successful construction are the use of good equipment and materials, quality site planning and management, easy communication with specialist subcontractors, and trusting coordination with the client authority. This last criterion is particularly relevant at sites subject to limited or short-rolling possessions (i.e. when completion of the work is a matter of considerable urgency).

Most concrete pavements constructed in the UK now use sophisticated construction processes involving electronically guided paving machines: hence the main emphasis in this chapter is on the construction of conventional cement concrete (BSI, 2013) rigid pavements using these types of equipment. In recent years, however, there has been significant growth in the use of some secondary cementitious materials in lieu of Portland cement in conventional concrete pavements, so these are briefly discussed here also. Also covered are non-conventional concrete pavements known as roller-compacted concrete and block paving, whose usage is now very common for particular pavement purposes.

14.1.   The conventional rigid pavement construction process

The construction of conventional cement concrete pavements in major infrastructure projects is now largely a mechanised process in which the use of sophisticated electronically guided paving machines is commonplace. It should never be forgotten, however, that the hand-laying of concrete may still be necessary at locations where a constrained site or the presence of obstacles (e.g. manholes and utility pits) complicate the paving operations, and the decision has to be made as to whether to employ mechanised or hand-laying techniques at these places.

There are two types of mechanised paving equipment used to lay concrete in rigid pavements: fixed-form and slip-form. Slip-form paving is used on major road and airport projects where long, continuous and uninterrupted stretches of concrete laying are most likely to occur. Fixed-form paving is used on smaller sections, often in urban areas, or on complicated trunk road interchanges. Both forms of paving require a high standard of subgrade and sub-base preparation to ensure uniform slab support, as well as high-quality finishing and curing to ensure good performance throughout the design life of the pavement. The steps in the construction process for a fixed form paver can be summarised as follows:

(a) prepare the foundation for the slab

(b) place the forms (if fixed form paving is to be used)

(c) install the joint assemblies (where applicable)

(d) batch the cement and aggregates, mix and transport them to the site

(e) lay and finish the concrete

(f) texture the running surface of the slab

(g) cure the concrete.

In the case of a slip-form paver, some of the above steps are combined.

14.1.1 Preparing the foundation and placing the sub-base

Irrespective of the type of paver employed, the preparation of the foundation for the concrete slab in a rigid pavement involves clearing and grading the site, installing any needed subgrade drainage, stabilising (if necessary), compacting and finishing the subgrade, and constructing (in most cases) a sub-base. Prior to this, any required deep services should be installed in the subgrade, and the backfill compacted so as to ensure continuity in slab support.

In some instances, the subgrade will be composed of a coarse-grained soil or uniform rock, and a sub-base may not be deemed essential. More often than not, however, a sub-base is required.

Sometimes, what are otherwise good subgrades may have 'soft spots' from which the poor soil has to be removed. These soft spots may also occur in areas where existing services have been installed and the desired quality of compaction of the back-fill material was not achieved. Such materials of inadequate bearing strength should be removed and replaced with a suitable fill material – either from good excavation or imported filling material – by a granular sub-base (e.g. type 1) material, or by lean concrete. After the soft spots have been replaced and compacted, additional plate-bearing or California bearing ratio tests should be carried out to ensure that the desired bearing strength is achieved. On occasion, in very soft subgrades, it may be necessary to further excavate below the formation and replace the weak material with (in UK practice) a class 6F2 selected granular fill prior to laying the sub-base with a type 1 material (HA, 2009).

If the subgrade as a whole is soft, then a capping layer may have to be installed over the entire area – after removing soil to an appropriate depth – to increase its stiffness modulus and strength, and prevent it from waving under construction traffic. (Guidance is available regarding the required thicknesses (HA, 2009; HA et al., 2014).) If the subgrade soil is cohesive and reactive, a lime or cement stabilisation treatment may be an economic correction option. If a capping layer is placed on the subgrade but it is not designed as a drainage blanket, drains may have to be installed below the bottom of the capping to ensure drainage of the foundation, if the capping is permeable and rain is likely during construction.

The sub-base, whether it is a compacted granular fill or a cement-bound material, will need to be laid to a regular and accurate level. This is achieved using a combination of a rotating laser level and a dumpy level, for enhanced accuracy, measured at regular intervals. Where a layer of wet lean concrete is laid over a granular sub-base, it will be necessary for side forms to be set in position, using steel anchor pins, to a line and level that is usually set by an engineer using a robotic total station (i.e. an electronic theodolite integrated with an electronic distance measure). The formed sub-base layer should be maintained in a clean condition, and its use by construction traffic strictly limited until the concrete is laid. Good control of the sub-base levels improves the uniformity and quality of performance of a concrete pavement, as any unevenness in the finished sub-base surface will be reflected in variations in the thickness of the poured concrete slab above it.

UK practice is for the sub-base above a capping layer in a rigid pavement to be 150 mm deep and for it to be composed of cement-bound granular material or wet lean concrete. Cement-bound materials (e.g. dry lean mixes) are normally plant mixed using paddle or pan-type mixers and laid by a paving machine similar to that used to place bituminous materials. As with unbound materials, the amount of water applied to a dry lean concrete mix is normally the optimum moisture content (OMC) required for satisfactory compaction by smooth-wheeled, rubber-tyred or vibrating rollers; typically, the OMC is in the range of 5–7% by mass, and is determined from trial mixes. The mix must not be too dry or the surface will remain loose after compaction: if it is too wet, the mix material will be picked up by the roller, and its surface will be irregular.

If the sub-base is to comprise wet lean concrete, this material may be mixed in drum mixers: in this case, however, the moisture content of the mix is typically 7–11%. If it is intended to use a slip-form paver to lay the concrete slab, it is normal to use this paver to lay the wet lean concrete also. The compaction and finishing of the sub-base is achieved with the aid of a vibratory truss screed that is winched over the surface of the lean concrete. Whatever the laying method employed, the maximum time that may elapse between the admixing of water and the compacting of the wet lean concrete should meet the criteria specified in Table 14.1.

Table 14.1 Maximum working times for the laying of concrete slabs in the UK (HA et al., 2014)

Temperature of the concrete at discharge from the delivery vehicle	Reinforced concrete slabs constructed in two layers without retarding admixtures		All other concrete slabs	
	Mixing the first layer to finishing the concrete	Between layers	Mixing the first layer to finishing the concrete	Between layers in two-layer work
<25°C	3 h	0.5 h	3 h	1.5 h
>25°C but <30°C	2 h	0.5 h	2 h	1 h
>30°C	Unacceptable for paving	–	Unacceptable for paving	–

14.1.2   Placing the forms for fixed-form paving trains

After the subgrade and sub-base have been compacted and graded, side forms must be put in place for the conventional fixed-form paver. Occasionally, the side forms are permanent edge strips, or kerbing, that are precast in advance of the paving. Typically, however, these side forms are wide-based steel sections that are temporarily fixed to the sub-base with steel pins driven through holes in their flanges: the forms are custom built, reuseable, clean and oiled, and 3 m long by the depth of the slab. With fixed-form paving trains (see Figure 14.1), the side forms are left in position during the preliminary curing of the concrete slab, and carefully removed after completion of construction – taking care to avoid damage to the finished concrete – for cleaning and reuse. Consequently, the contractor must always have enough side forms on site for about 3 days' work ahead of the paving train.

The side forms define the width of the slab and the level of its top surface, contain the concrete as it is being laid, and sometimes define the shape of joints specified for the pavement. The forms should be set to the correct line and, where necessary, raised up to the correct level using steel or plastic shims. The gap between the sub-base and the formwork should be packed with a sand–cement mixture that is strong enough to retain the concrete but weak enough to break out later: this packing should be completed 12 h before use, to allow the sand–cement to harden.

Side forms used with wheeled paving vehicles have flat-bottomed metal rails firmly fixed at a constant height below the top of the forms along which many of the paving machines operate in sequence during construction. Since the equipment has to be supported and guided by the rails, the side forms must be vertical and free from movement in any direction while spreading and compacting the concrete.

A separation membrane is placed on the surface of the foundation ahead of a fixed-form train (but not a slip-form one) prior to placing and fixing any joint crack inducers or assemblies. In the case of unreinforced concrete (URC) and jointed reinforced concrete (JRC) slabs, the membrane is commonly 125 μm thick plastic sheeting; however, too much membrane should not be laid in advance if there is a risk of damage or ponding of rainwater. In the case of either a continuously reinforced concrete pavement (CRCP) or a continuously reinforced concrete roadbase (CRCR) construction, the waterproof membrane is normally a bituminous spray. For site safety reasons, a bituminous spray may also be preferred as the separation membrane for use with URC and JRC slabs because of the likelihood of worker slips and falls occurring when the plastic sheeting gets wet.

14.1.3   Joint assemblies and reinforcement

After the forms have been put in place, the various joint assemblies may be placed in position, depending upon the method of construction employed. Before doing so, however, care should be taken to ensure that any transverse joints in the surface slabs and the sub-base are not coincident vertically but staggered at least 450 mm apart.

Irrespective of the method used to lay the concrete, expansion joint assemblies with dowel bars and filler boards are normally prefabricated and securely supported on

Figure 14.1 Typical arrangement of a fixed-form paving train used in concrete pavement construction

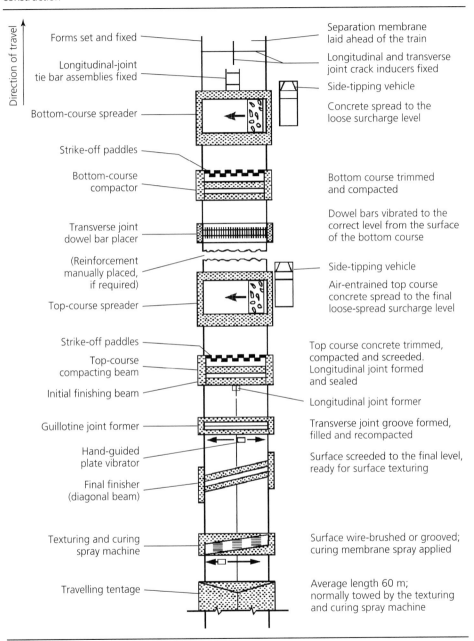

Direction of travel

Forms set and fixed ——— Separation membrane laid ahead of the train

Longitudinal-joint tie bar assemblies fixed ——— Longitudinal and transverse joint crack inducers fixed

——— Side-tipping vehicle

Bottom-course spreader ——— Concrete spread to the loose surcharge level

Strike-off paddles ———

Bottom-course compactor ——— Bottom course trimmed and compacted

Transverse joint dowel bar placer ——— Dowel bars vibrated to the correct level from the surface of the bottom course

(Reinforcement manually placed, if required) ———

——— Side-tipping vehicle

Top-course spreader ——— Air-entrained top course concrete spread to the final loose-spread surcharge level

Strike-off paddles ———

Top-course compacting beam ——— Top course concrete trimmed, compacted and screeded. Longitudinal joint formed and sealed

Initial finishing beam ———

——— Longitudinal joint former

Guillotine joint former ——— Transverse joint groove formed, filled and recompacted

Hand-guided plate vibrator ———

Final finisher (diagonal beam) ——— Surface screeded to the final level, ready for surface texturing

Texturing and curing spray machine ——— Surface wire-brushed or grooved; curing membrane spray applied

Travelling tentage ——— Average length 60 m; normally towed by the texturing and curing spray machine

453

cradles and fixed to the sub-base so that they will not shift when the concrete is being spread and compacted. To work effectively, the dowels in each assembly must be aligned parallel to both the surface of the slab and the centreline of the pavement. Any misalignment will result in the generation of very high stresses as adjacent slabs expand and contract. Filler boards, which are normally 25 mm thick and have holes that form a sliding fit for the sheathed dowel bars, are set vertically to ensure that adjacent slab ends are also vertical and that there is no likelihood of one riding up over the other.

The bottom crack inducers of contraction and longitudinal joints are securely fastened to the sub-base at the spacing specified in the design prior to any placement of the concrete, if this method of forming joints is employed. Typically, the crack inducer is a triangular or inverted Y-shaped fillet with a base width not less than its height; it is normally made of timber or a rigid synthetic material.

Dowels are commonly placed on prefabricated metal cradles and fastened to the sub-base ahead of the paving train. In the UK, as an alternative to using prefabricated assemblies for contraction joints, the dowels may be vibrated into position as the slab is being formed with the fixed-form process. The tolerances to which dowel bars are specified in the UK are quite stringent, especially for those on cradles (Bannon, 1991). Cradles supporting dowel bars should not extend across the joint line.

Tie bars in warping joints and wet-formed longitudinal joint assemblies are normally made up into rigid assemblies with adequate supports and securely attached to the sub-base in advance of the paving train. Alternatively, the tie bars at longitudinal joints may be vibrated into position using a method that ensures compaction of the concrete about the bars. In the case of construction joints, the tie bars are often cranked and held at right angles at the fixed-form face until the form is removed and the bars can be bent back. Alternatively, the tie bars may be fixed to the side forms or inserted into the side of the slab and re-compacted.

Heavy duty, thick slab pavements are sometimes constructed with key joints consisting of a male-to-female edge profile, so that, rather than using dowels to transfer loads from one concrete bay to the next, the transfers are achieved by the shape of the joint itself. A number of joint shapes have been trialled with varying success rates, but have loosely remained a variation on a sinusoidal wave-form or a recessed notch. The combination of a formed longitudinal key joint and aggregate interlock through a crack-induced contraction joint helps to achieve load transfer across a pavement without the use of dowels.

It is important to ensure that contraction joints are not spaced too far apart, as this may result in the creation of a joint gap that is too wide, and this, in turn, may hinder sufficient load transfer across the concrete bays. The bay size for concrete pavement construction is therefore critical to its performance, and is a function of its depth, aggregate type and temperature differentials associated with the seasons and the time of day. Induced thermal stresses in the concrete differ depending on the coefficient of thermal expansion of the aggregate used, and these can significantly affect the performance of the pavement. Thus, the bay sizes should be established at the design stage and correctly set out on site to ensure adequate pavement performance.

On space-constrained sites (e.g. at busy parking areas for commercial vehicles), increased pavement depth may have to be considered should the direction of traffic movement be perpendicular to the longitudinal joints: this is on account of longitudinal joints being less efficient at load transfer than contraction joints. Such a construction needs to be carefully considered at the design stage in conjunction with the client and contractor, as it is dependent on which direction the contractor intends laying the pavement slabs. While increasing the pavement depth may result in additional material costs, savings may be achieved by the ability of the contractor to lay longer stretches of pavement, thus speeding up the construction process.

After the joint assemblies have been put in place, any reinforcing steel used in single-course construction is securely fastened to supporting metal structures at the specified depth below the finished surface and distance from the edge of the slab, to ensure that the desired cover is achieved. The metal support cradles for the reinforcement must be stable and robust enough to carry workmen during construction; also, the cradles should not move when the concrete is being placed. The steel is most commonly either fabric (or mesh) reinforcement that is supplied to the site in large sheets or rolls for placement in front of the paving machinery, or bar reinforcement that is manually assembled on site.

The reinforcement, which is fixed higher than the tie bars and dowels, is terminated about 125 mm from the slab edges and from longitudinal joints with tie bars: it is also terminated about 300 mm from any transverse joint (excluding emergency construction joints). Lapping in any transverse reinforcement is normally about 300 mm, while longitudinal bars are overlapped by at least 450 mm. No longitudinal bars should be located within about 100 mm of a longitudinal joint. In all instances, the steel used should be clean, so as to provide a good bond with the concrete, and the spacing of the bars should not be less than twice the maximum size of the aggregate used in the concrete.

If two-layer construction is being employed, the JCR, CRCP or CRCR reinforcement, in the form of prefabricated sheets, is placed in or on top of the bottom course.

14.1.4 Preparing the concrete

On large contracts, well-organised fixed-form (and slip-form) trains are able to construct many kilometres of paving per day. For this to happen, concrete of a consistent quality must be provided at the site at a steady rate. Steady progress requires the careful and realistic matching of the productive capacity of the train with the output capability of the mixing plant – and the availability of enough suitable vehicles to transport the fresh concrete.

In the UK, wet mixing of the concrete most commonly takes place in a stationary batch-type mixer at a plant that is either on site or at a central plant located at a quarry or gravel pit that, ideally, is located close to the construction site. In the case of an existing central plant, the material is fed through the batching and mixing equipment, and the mixed concrete discharged into waiting vehicles for covered (to prevent evaporation of water or wetting by rain) transport to the construction site. The use of an existing off-site plant is preferable, where this is an economic and efficient option, as the relative permanency of the plant encourages the production of good-quality concrete. Existing plants also tend to have more elaborate storage facilities and equipment that can automatically control the flow of materials from the stockpiles.

A central plant should be as close as possible to the construction site, as long hauls require more vehicles to keep the paving train in continuous operation. If the run is through an urban area, the risk of vehicles bunching in traffic congestion increases, with the consequent risk of an irregular supply to the paving train. If the haul road is long and bumpy, agitation is essential to avoid segregation in the wet mixture while travelling to the site. If travelling takes too long, the concrete will begin to stiffen, and its consistency and workability may be detrimentally affected unless it is agitated during transportation. The workability of the concrete at its on-site placement point should be such as to enable it to be fully compacted and finished without undue flow. Low workability ensures that the inserted dowel bars are more easily kept in position, while high workability allows the use of smaller surcharges, as well as better texturing and finishing.

The compacting factor test (BSI, 2009) provides a useful measure of the on-site workability of grade C40 concrete when the aggregate is crushed rock or gravel. The conventional slump test is more commonly used with wet lean concrete of grade C20 or below.

When placing fresh concrete on a sub-base, or on the lower layer of two-course slabs, the engineering objective is to obtain an even depth of uniform density with the minimum of aggregate segregation. The spreading, compacting and finishing procedures employed will vary according to whether a fixed-form or a slip-form paving train is used.

14.1.5 Using fixed-form pavers to lay and finish the concrete

With a fixed-form paving train (Figure 14.1), the concrete is usually deposited to a uniform 'loose' density by a traversing box-hopper spreader. This spreader comprises a hopper of 3 or 4.5 m^3 capacity that is mounted on a self-propelled rail-mounted frame so that the box can move across the frame and discharge the concrete at a controlled rate as the frame moves forward. The underside of the hopper then strikes off the deposited concrete to the desired surcharge level (i.e. the design height by which the initial depth of spread concrete exceeds the ultimate thickness of the slab) as it returns with the frame remaining stationary. Good surface level control is obtained by ensuring that the spreader is able to spread the concrete evenly to the correct surcharge. It is bad practice to rely on regulating beams and the diagonal finisher to subsequently achieve the correct levels by major planing. If the slab is to be constructed in two layers, care needs to be taken to ensure that the spreading of the concrete in the top layer is carried out within the time limits given in Table 14.1.

In a conventional fixed-form train, the spreader is followed by a piece of equipment with rotary strike-off paddles and a transverse vibrating compaction beam. The function of the paddles – they are mounted on independently operated levelling screws at each side of the equipment that allow adjustments to be made for carriageway crossfalls – is to trim any minor irregularities in the surface of the surcharged concrete. The function of the compaction beam is to apply vibration to the surface of the concrete, with the amplitude and frequency used being adjusted to suit the characteristics of the wet mix, with the aim of obtaining a dense homogeneous slab of concrete that is free from voids, honeycombing and surface irregularities.

For air-entrained road pavements constructed in the UK, the compacted concrete should meet the requirements of class XF4 concrete (BSI, 2013). According to current specifications (HA *et al.*, 2014), the minimum air content for a 20 mm aggregate should be 3.5%, and for a 40 mm aggregate it should be 3%.

After compaction, transverse dowel bars and longitudinal-joint tie bars that have not been previously attached to the sub-base are vibrated into position. If the design calls for two-course construction with reinforcement, the steel is manually placed on the surface of the compacted bottom layer prior to the placement of the second course.

With the fixed-form paver, the initial regulation and finishing of the slab surface is carried out using a beam that oscillates transversely or obliquely to the longitudinal axis of the pavement. This beam is a simple oscillating box-section float that is either mounted on the carriage of the compactor machine or is carried on a trailing articulated framework: it is only used with full-depth or top-course paving. The beam provides a regulated and part-finished surface that may be adequate as a final finish on minor roads or in industrial estates, but requires further treatment for high-quality trunk road pavements.

Wet-formed longitudinal joint grooves are formed immediately after the initial oscillating beam has completed its work. The longitudinal joint is formed in one continuous operation using a hollow vertical 'knife' that travels submerged in the plastic concrete. The knife is attached to the underside of a flat plate on which is mounted a small vibrator unit: it inserts a pre-formed cellular permanent strip as it moves through the concrete and then re-compacts the concrete around the seal. A bottom crack inducer is provided at each wet-formed longitudinal joint position. Caution is advised in the specification of wet-formed joints: it can lead to localised over-working of the concrete and long-term durability issues.

If the transverse and longitudinal joints have not been wet-formed, a diamond saw blade is used to cut the concrete and allow a sharp-edged groove to be formed, thereby preventing the formation of uncontrolled cracking. The sawing is normally conducted after the initial set of the concrete: an experienced works manager should be able to determine the most appropriate time-frame to carry out the work so as to produce grooves without spalling – this normally ranges from 6 to 24 h after the concrete has been laid – however, it must be closely monitored if the weather is warm, dry and windy. Typically, a groove is about one-fifth of the slab depth and not less than 3 mm wide. Hot- or cold-poured bituminous sealants that can cope with normal expansion and contraction changes without allowing the entry of water or grit are then inserted into the sawed joints.

While sawed joints are more expensive to produce, when properly cut they are considered to give better results than wet-formed joints; however, it requires a skilled operator to ensure that grooves are cut perpendicular to the longitudinal joints. Also, some grooves require further widening prior to a joint sealant being applied, and an erratic initial groove can prove difficult to rectify. From a cost and planning perspective, the aggregate used in the concrete can also affect production rates (e.g. gravel aggregates such as flint are especially difficult to cut through). The sawing operation itself may damage the concrete, as aggregate pop-outs may occur along the cut line.

The final regulation and finishing of the concrete is usually carried out by a machine that has two oblique finishing beams mounted on an articulated mobile framework that oscillate in opposing directions. The leading beam vibrates, makes dense and smooths the surface, while the rear beam acts as a float finisher and removes any imperfections prior to texturing. Diagonal rather than transverse beams are preferred, as they minimise the area of finisher that is in contact with a joint former at any given time, thereby reducing the likelihood of joint damage. The shearing action of a diagonal screed also results in a more uniform surface finish.

14.1.6 Using slip-form pavers to lay and finish the concrete

With slip-form pavers, many of the operations of a fixed-form paving train are replaced by a single machine frame, and within this frame the main operations prior to the establishment of transverse joints are carried out, namely spreading, trimming, compacting and finishing. Waterproof membranes can also be fitted to the front of the equipment so that they unroll directly under the laid concrete as the paver moves forward. The slip-form paver is self-propelled and mounted on two or four caterpillar tracks that travel outside sliding side forms attached within the length of the machine. Functions such as the forming and finishing of transverse joint grooves, surface texturing and spraying of the curing compound are carried out by equipment that follows behind the paver.

In UK roads (and airports), concrete pavement operations are normally carried out using conforming-plate slip-form paving trains (see Figure 14.2(a)). With the conforming-plate paver (Figure 14.2(b)), the concrete is deposited into a hopper at the front of the moving paver, and a hydraulically adjusted plate that is attached to the hopper maintains a constant surcharge of concrete above the level of the conforming plate. As the paver moves forward, the slab edges are formed by the moving slip-forms, and the concrete between the forms is forced under the conforming plate. Immersed vibrators between the strike-off and conforming plates keep the concrete in a fluid state and help compact it as it is forced into the space between the sub-base and the conforming plate. In large pavers, the hopper may be fitted with a spreader to help distribute the concrete across its full width.

Current slip-form pavers now make use of what is known as 'stringless' technology, which eliminates the traditional guide-wire approach to laying concrete. The paving operation is almost entirely digitised, with the machine requiring only elevations, and north and east coordinates, to correctly lay the concrete. The digitisation also allows the

Figure 14.2 Typical arrangement of (a) a slip-form paving train, (b) a conforming-plate paver (section) and (c) an oscillating beam paver (plan) used in concrete pavement construction

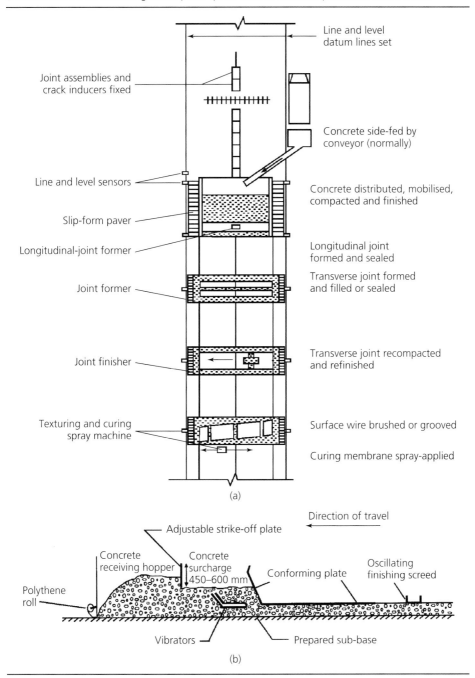

Line and level datum lines set

Joint assemblies and crack inducers fixed

Concrete side-fed by conveyor (normally)

Line and level sensors

Concrete distributed, mobilised, compacted and finished

Slip-form paver

Longitudinal-joint former

Longitudinal joint formed and sealed

Joint former

Transverse joint formed and filled or sealed

Joint finisher

Transverse joint recompacted and refinished

Texturing and curing spray machine

Surface wire brushed or grooved

Curing membrane spray-applied

(a)

Direction of travel

Adjustable strike-off plate

Concrete receiving hopper

Concrete surcharge 450–600 mm

Conforming plate

Oscillating finishing screed

Polythene roll

Vibrators

Prepared sub-base

(b)

Figure 14.2 Continued

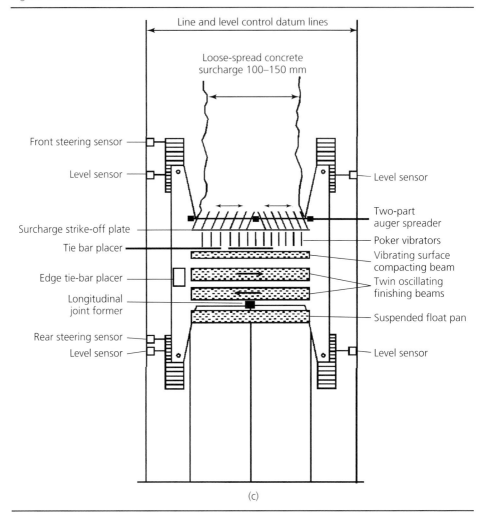

pavement designer's original files to be used to generate a paving model, so that little or no design interpretation is required on site. A software programme interfaces with the controller on the paver, allowing it to communicate with up to two robotic total stations that are set up appropriately along the paving lane: these command the paver to make travel adjustments when necessary, thereby ensuring that the paver follows the required line and level. A third total station is often used to record 'as-built' information, thereby allowing up-to-date paving records to be downloaded at the end of each working day.

Where contraction-joint dowel bars are required in the concrete pavement, the computer controls their insertion at predetermined positions using an in-the-pan dowel bar inserter. The reliability of this operation has steadily improved over time, with accurate placement and excellent compaction now readily achievable around the bars.

While equally good standards of pavement product have been obtained with both types of paver, the following are the main advantages of slip-form pavers as opposed to fixed-form pavers:

■ Slip-form paving is less labour intensive, involves fewer pieces of equipment and there is no need to provide for the (expensive) purchase, maintenance, fixing and stripping of large numbers of fixed forms.
■ Higher daily outputs of concrete slab can be achieved with slip-form pavers: for example, under summer weather conditions and with no joint assembly or plant hold-ups, a slip-form paver in the UK is typically able to lay an 11.2 m wide by 280 mm thick slab (with no expansion joints) at the rate of 65–75 m/h, as compared with 35–40 m/h with fixed-form paving.

The main disadvantages of slip-form paving are as follows:

■ The paving equipment is more complex than that used with a conventional fixed-form train and requires more skilled (and expensive) personnel for its proper operation and maintenance.
■ There is a danger of edge slump after the concrete leaves the paver.
■ A minor failure in the control system can cause the slip-form paving operations to come to a halt.
■ Greater stockpiles of cement and aggregate are required in advance to ensure a high output from the slip-form paver.
■ The paving contractor is more vulnerable to bad weather hold-ups.

14.1.7   Texturing the running surfaces

If a surface of a road pavement is to have good skid resistance, it must have adequate macro-texture and micro-texture. The macro-texture allows for the rapid drainage of the water trapped between vehicle tyres and the pavement surface, while the micro-texture penetrates the remaining film of water and maintains the tyre–carriageway contact.

The micro-texture at the surface of a concrete slab is ensured by using a fine aggregate in the concrete mix that is more resistant to abrasion than the matrix of the hardened cement paste. This ensures that the sand grains stand proud of the matrix as the softer paste abrades under traffic.

The macro-texture of the carriageway is obtained by wire brushing or grooving the concrete surface at right angles to the centreline. This must be carried out upon completion of the final regulation of the slab surface and before the application of the curing membrane. In the UK, the adequacy of the macro-texture depths obtained is normally checked using the volumetric patch technique (BSI, 2010).

Wire brushing is carried out manually or mechanically from a travelling bridge that moves at right angles to the centreline of the carriageway. Grooving is carried out mechanically using a profiled vibrating plate that traverses the width of the finished slab and forms grooves with random longitudinal spacing. From a construction aspect, grooving has the advantage that concrete of variable quality can be grooved, even when fairly stiff, as the speed and vibration of the plate can be varied to suit the consistency of the compacted concrete. Measurements and subjective assessments of the ride quality of concrete carriageways have shown that there is no significant difference between transversely grooved and brushed surfaces when both texturing processes are carried out correctly.

14.1.8   Curing the concrete

Curing aims to ensure that a satisfactory moisture content and temperature are maintained in the newly placed concrete slab so that hydration of the cement continues until the design strength is achieved. Moisture loss results in drying–shrinkage and the development of plastic cracks in the cement paste. Conditions conducive to plastic shrinkage include exposure to the direct rays of the sun, strong wind, low humidity, high air and concrete temperatures, and the use of concrete with low bleeding tendencies and/or a large surface area in relation to depth.

In practice, curing usually involves covering the surface of the concrete to keep it damp for at least 7 days. The covering material is, typically, either a resin-based compound, an aluminised curing compound, polythene sheeting, or a sprayed plastic film that can be peeled off before carriageway markings are applied. Care should be taken when specifying which product to use, as some compounds can leave a slippery surface, while others require light sand-blasting or high-pressure water jetting to remove them: the former can constitute a residual hazard for road users while the latter may artificially age the surface of the concrete if not carried out by a skilled operative.

The curing method that is now most used in the UK is to mechanically spray the surface and exposed edges of the concrete slabs with an aluminised compound immediately after the finishing and texturing processes. The spray contains finely dispersed flake aluminium particles that completely cover the surface with a metallic membrane that reflects the radiant heat of the sun: the membrane is eventually worn away by vehicle tyres when traffic is allowed to use the carriageway. Ideally, the aluminised compounds are completely removed before the application of road markings: if not, it may be necessary to return and apply additional coats to ensure their satisfactory application.

If heavy rain is anticipated, damage to the freshly textured concrete can be minimised by towing about 60 m of low-level travelling tentage behind the curing compound sprayer. The tentage also provides additional protection against drying by the sun and the wind.

14.2.   Using pozzolanic cements in road construction

Pulverised fuel ashes (PFAs) and ground granulated blastfurnace slag (GGBS) are secondary cementitious materials that are increasingly being considered as a partial substitute for conventional Portland cement in major concrete construction projects. The particular advantage of these pozzolanic materials is that they are available as 'waste' by-products from other industrial processes and, as such, they can be more economical to use on site than conventional Portland cement, thereby reducing the construction cost. An additional environmental benefit is that, being waste materials, their use in concrete pavement slabs with reduced amounts of Portland cement reduces the total carbon dioxide footprint of the construction project (because Portland cement generates high carbon dioxide emissions in the course of its production). While concretes produced using pozzolanic cements containing these agents produce strong, dense and durable slabs in the long term, they have the disadvantage that the slabs have reduced early-age strengths and require longer curing times to attain equivalent qualities, including resistance to frost action, in comparison with pavement slabs made with conventional Portland cement.

14.2.1   Using PFA cements

PFA, also known as fly ash, is the solid fine ash emitted in the flue gases from power station boilers that are fired with pulverised coal in the generation of electricity. The main ingredients of PFAs obtained from power stations are finely divided glassy spheres of silica and alumina: calcium oxide derived from limestone present in the original coal may also occur in some fly ashes. PFAs are pozzolans: that is, in the presence of lime and water a chemical reaction takes place that results in the formation of hydrous calcium aluminates and silicates that are similar to the reaction products of hydrated cement (ASTM, 2012). Thus, some fly ashes containing significant quantities of water-soluble lime and calcium sulphate can become involved in pozzolanic reactions, and will self-harden. When a pozzolanic cement containing PFA is mixed with water, the reaction is characterised by the hydration of the Portland cement powder and the release of free calcium hydroxide that reacts with the PFA particles to form cementitious products.

PFA used in pozzolanic cements in the UK must meet certain specified requirements (BSI, 2011, 2012). The use of a pozzolanic cement comprising PFA and Portland cement in a concrete pavement is dependent upon high early strength not being a priority: it is a slow-hardening, low-heat cementitious agent that permits the use of lower water-to-cement ratios for a desired workability that, in the longer term, produces a denser concrete of lower permeability and greater durability.

PFA concretes are generally less workable than conventional Portland cement concretes; however, when vibrated they will become mobile and move easily within the forms. While the addition of PFA also reduces the rate of bleeding within the concrete, excessive water loss should be prevented during construction, as this can lead to shrinkage cracking (Quality Ash Association, 2004). Curing products (e.g. an aluminium-based protective film such as Chemcure) that is sprayed onto the concrete can help to prevent early-age crack formation.

PFA concrete is most effectively used in road construction close to where the fuel ash is economically produced and/or where it is desired to impart a degree of resistance to chemical attack from sulphates and weak acids.

14.2.2   Using blastfurnace cements

GGBS is a by-product of the iron-smelting process, and, by itself, is a relatively inert material. However, its significant characteristic is that it is pozzolanic (i.e. it will react with lime in the presence of water, to form cementitious products). The hardening of blastfurnace cement, which is a mixture of conventional Portland cement and up to 95% GGBS (BSI, 2006, 2011) is characterised by two processes: (a) the cement clinker hydrates when water is added, and (b) free calcium is released that reacts with the ground slag as it hydrates.

In general, GGBS cement concrete hardens more slowly than ordinary cement concrete, and, hence, should not be used in locales likely to be subject to frost action before adequate slab strength is gained. Concrete made with GGBS cement is typically more fluid and generally easier to place in formwork and compact. The ability to pump the concrete is enhanced on account of its improved flow characteristics, and, generally, it retains its workability for an extended duration. Its usage is most justified in pavement construction when the cement is economically available and where high early slab strength is not a prerequisite.

In benign climatic areas in which road construction is taking place close to where blastfurnace slag is produced, it may be appropriate (for economic reasons) to simply add GGBS during the mixing process as a replacement for upward of 50–70% ordinary Portland cement in, for example, the sub-bases of road pavements.

14.3.   Other types of concrete pavements

In recent years there has been a rapid growth in the use of two other types of concrete pavement, namely roller-compacted concrete (RCC) pavement and concrete block pavement. These types of road pavement are constructed differently from conventional concrete pavements.

14.3.1   Using RCC

Internationally, RCC construction has been increasingly used (e.g. Petrie and Matthews, 1990) on diverse projects ranging from ports and haul roads to dams and airport taxiways. Although it has actually been in use since the US Army's Corps of Engineers developed its use as a paving material in the 1980s, it is only relatively recently that it has begun to enjoy widespread appeal in the UK (Donegan, 2013).

The principal benefits of RCC are inherent in its material properties. The material proportions of an RCC mix typically comprise 12–14% Portland cement by dry mass, 81–84% of well-graded aggregate and 4–6% water. If a compacted RCC slab is to be covered with an asphalt surface course, the maximum aggregate size may be limited to 20 mm, to avoid a harsh or 'bony' surface: if the RCC pavement is not to have a covering layer, the maximum aggregate size is generally 14 mm, to ensure a closed, textured, finish. RCC is stiffer than a zero-slump concrete, produces higher flexural and compressive strengths, requires no reinforcement, dowels or tie bars, and has no need for formwork.

A principal benefit of RCC is that it can be laid with a wide variety of paving plant. Ideally, placement is carried out using purpose-built equipment, similar to bituminous pavers, with tamping bars that carry out the initial compaction, and further compaction being achieved with steel-drum vibrating rollers; however, the formless edges must be compacted first to provide confinement for the material to be rolled in the interior of the pavement. The RCC surface texture can be improved by having a pneumatic-tyred roller follow closely behind the drum roller, to work some fines to the top and help to close any voids, cracks or tears in the surface. Should a paver not be available, or simply because it is more economical not to use one, it is not uncommon for an electronically guided dozer to spread the concrete mix: the dozer follows the design levels set by total stations and spreads the mix to the required thickness, after which it is followed by the appropriate rolling plant to achieve the desired degree of compaction.

RCC generally does not normally require special surface finishing, unless there is a special need brought about by the intended usage of the pavement. The as-built surface texture is generally sufficient for slow-moving vehicles or stationary objects such as those seen in container ports, albeit this does not preclude its use by more high-speed vehicles. (If a binder and/or surface course of asphalt is applied to regularise the friction characteristics and improve the ride quality, it is the author's opinion that future research developments could possibly see the extension of designed RCC pavements to higher-speed road pavements.)

RCC is particularly suitable for use on projects with short possession and tight completion periods. Early trafficking (after about 7 days) is possible, due to the mechanical stabilisation associated with its highly compact and dense cement-aggregate matrix.

The placement of RCC is a time-critical process, and a well-organised operation is crucial to ensure that the desired output is achieved. The concrete must be placed, levelled and compacted within 1 h of mixing, and a maximum lift thickness of about 250–270 mm should not be exceeded if good compaction is to be achieved. Good compaction is crucial to achieving the required design strength, and there is a rule of thumb to the effect that there is a 5–7% reduction in strength for every 1% loss of density. Any granular sub-base below the RCC slab must be neither too wet nor too dry before placing the mix, as this can result in significant and unwanted compaction difficulties. If the underlying material is too wet, moisture may be drawn into the rolled concrete, thereby increasing the water:cement ratio; if it is too dry, moisture from the RCC may be drawn into the granular sub-base, leaving behind an insufficient water content to properly hydrate the mixture.

Curing compounds are applied to RCC in a similar fashion to conventional concrete. Curing products that are suitable for pavement quality concrete are generally adequate, albeit the application rates may have to be raised (by up to 50%) on account of the more open surface texture of RCC. For composite pavements, where it is intended that a final asphalt surface course is to be laid over the RCC, a layer of bitumen emulsion will act as an acceptable curing membrane.

While the application of curing compounds remains standard practice for RCC pavements, the same cannot be said for the inclusion of air-entraining agents in the mix design. Air-entraining agents are not suited to RCC, and, hence, its preferred usage is in locales with relatively mild climates not subject to severe frost action.

In cases where the 1 h placement time limit may be exceeded because of, say, the likelihood of longer haul times from the batching plant to the construction site, consideration should be given in the specification to the use of a retarding agent in the concrete mix.

As noted previously, RCC pavements do not need tie bars or dowel bars to transfer the load across joints, and aggregate interlock is the primary contributor to load distribution. Sawn contraction joints, one-quarter to one-third the slab depth, need to be provided to ensure desired planes of weakness and to prevent random cracking in the surface of the pavement: thus, joint cutting with a diamond saw should be carried out as soon as the RCC can be cut without ravelling, typically within 24 h after mix placement. Spanish experience has shown that joint spacings of 2.5–3.5 m are effective in controlling the onset of cracking in RCC pavements, while the saw cutting of 4 m × 4 m bays in 200 mm deep slabs in large open areas has been shown to work well for UK conditions. Wherever RCC is used as a base in a composite pavement, the need to control reflective cracking is the major factor governing joint spacing.

14.3.2   Using concrete block paving

Concrete block paving has been widely used in urban projects since about the early 1970s. Its use became widespread throughout Europe at a time when the continent was undergoing major reconstruction after World War II, and equipment was developed that was capable of mass-producing blocks of high-compressive strength and close dimensional tolerance. In Germany, the Netherlands and Denmark, block paving was initially favoured for its ease of laying, relative low cost, durability and good skid resistance. Its usage spread to the UK and, today, concrete block paving is commonly found in areas where its aesthetically pleasing appearance compliments urban regeneration projects, in residential roads and car parking areas that service housing developments and shopping malls, and in local-area traffic management schemes where pedestrians and slow-moving vehicles share the road space.

The concrete blocks can be rectangular or shaped to provide greater lateral interlock: a typical rectangular block is 200 mm × 100 mm × 80 mm deep. The blocks, which have a relatively high flexural strength ($>6\ N/mm^2$), are typically made with a high cement content, a low water : cement ratio and a mixture of fine and coarse aggregates. The coarse aggregate is most usually 6 mm in maximum size: while 12 mm has been used in the past, it results in an open-textured surface, and this has been associated with durability problems over the long term.

The general approach used with concrete blocks in flexible pavements is to substitute the block construction for the conventional surface course, and position the blocks on a 30–50 mm thick (laying) course of compacted bedding sand, which, depending on the loads to be carried, is placed on a layer of asphalt (which may also be a regulating layer) or on a lean concrete base. The bedding sand, which should have a maximum size of 5 mm and not contain more than 3% by mass of silt and/or clay, is placed loosely to a uniform level (typically, about 65 mm for a 50 mm compacted layer), following which it is compacted using a vibrating rubber-tyred roller or a vibrating-plate compactor. A further sand layer is then spread loosely across the compacted surface, and the blocks are laid between edge restraints and vibrated down to the desired level using plate vibrators, while ensuring that the additional bedding sand is forced up between the joints. Filling sand is then brushed into the joints to enhance the structural performance of the block pavement.

When the blocks are being laid, it is most important that they be positioned correctly. If the joint gaps are too tight, spalling may occur due to the creation of high local stresses; if the joints are too wide, the jointing sand is easily removed and the entry of rainwater into the foundation is facilitated, thereby compromising the structural stability of the pavement. The optimum joint width is 2–5 mm, and, generally, this is achieved with the aid of spacer nibs on the blocks. While the retention of the jointing sand is crucial to the structural effectiveness of the pavement, there are many ways in which the sand can be unintentionally removed (e.g. using vacuum road sweepers or by turbulent water flow around drainage gullies). However, experience has shown that, as time goes by, detritus will seal the sand voids so that the joints become practically impervious: thus, block pavements can usually be built to normal longitudinal and crossfall requirements, to avoid surface-ponding of rainwater.

The sand in a well-laid joint has a number of functions: (a) it provides a 'separation cushion' that minimises spalling of the block edges under load; (b) it permits the transfer of shear forces across joints and prevents individual blocks from moving up and down relative to adjacent ones; and (c) in conjunction with the laying pattern it provides the horizontal interlock that gives the pavement its lateral rigidity.

Concrete block paving is used in appropriate locations for a number of reasons: (a) its attractiveness of appearance and ease of construction; (b) the lack of need for curing time on completion of construction allows the pavement to be opened to traffic immediately; (c) contrasting colours can be used to differentiate areas of traffic usage and/or underground services; and (d) easy access is provided to underground services, and maintenance trenches can be readily reinstated without visible patching (i.e. using the same paving blocks). A challenge associated with the use of block paving on some roads includes choosing a suitable pattern (e.g. a 45° herringbone pattern) to achieve the interlocking required for stability under the anticipated traffic, that is, non-interlocking patterns can result in pavement 'creep' at locations where the horizontal forces associated with the acceleration and deceleration of heavy vehicles cause the blocks to move in the direction of traffic flow. Also, the interface between block paving and other pavement constructions can be a source of difficulty. Thus, regular inspection and maintenance is crucial to the long-term effective performance of concrete block paving, and ensuring the integrity of the jointing and bedding sand is paramount to this.

REFERENCES

ASTM (American Society for Testing and Materials) (2012) ASTM C618-12a: Standard specification for coal fly ash and raw or calcined natural pozzolan for use in concrete. ASTM, West Conshohocken, PA, USA.

Bannon CA (1991) Concrete roads: an overview of design, specification and construction issues. *Joint Meeting of The Institution of Engineers of Ireland and the Irish Concrete Society*. Dublin, Ireland, Paper.

BSI (British Standards Institution) (2006) BS EN 15167-1:2006: Ground granulated blast furnace slag for use in concrete, mortar and grout. Definitions, specifications and conformity criteria. BSI, London, UK.

BSI (2009) BS EN 12350-4:2009: Testing fresh concrete. Degree of compactability. BSI, London, UK.

BSI (2010) BS EN 13036-1:2010: Road and airfield surface characteristics. Test methods. Measurement of pavement surface macrotexture depth using a volumetric patch technique. BSI, London, UK.

BSI (2011) BS EN 197-1:2011: Cement. Composition, specifications and conformity criteria for common cements. BSI, London, UK.

BSI (2012) BS EN 450-1:2012: Fly ash for concrete. Definition, specifications and conformity criteria. BSI, London, UK.

BSI (2013) BS EN 206:2013: Concrete. Specification, performance, production and conformity. BSI, London, UK.

Donegan J (2013) *Britpave Guide to Roller Compacted Concrete Pavements*. Britpave, Wokingham, England.

HA (Highways Agency) (2009) *Interim Advice Note 73/06, Revision 1 (2009): Design Guidance for Road Pavement Foundations (Draft HD25)*. Stationery Office, London, UK.

HA, Transport Scotland, Welsh Government and Department for Regional Development Northern Ireland (2014) *Manual of Contract Documents for Highway Works*, vol. 1. *Specification for Highway Works*. Stationery Office, London, UK.

Petrie RE and Matthews SC (1990) Roller compacted concrete pavements: recent Australian developments and prospects for the 90's. *Proceedings of the 14th Australian Road Research Board Conference*. Australian Road Research Board, Vermont, Australia, part 8, pp. 66–79.

Quality Ash Association (2004) *Best Practice Guide No. 1: The Placing and Compaction of Concrete Containing PFA/Fly Ash*. Quality Ash Association, Wolverhampton, UK. See: http://www.ukqaa.org.uk/wp-content/uploads/2014/02/Best_Practise_Guide_1-0_Nov_2004.pdf (accessed 13/06/2015).

Highways
ISBN 978-0-7277-5993-1

http://dx.doi.org/10.1680/h5e.59931.469

Chapter 15
Pavement asset management

Seósamh Costello   Senior Lecturer, University of Auckland, New Zealand

Martin Snaith   Engineering Consultant, UK

15.1.   Introduction

The strategic road network in England alone comprises almost 7000 km of motorways and all-purpose trunk roads valued at approximately £108 billion (HA, 2013). Given that the whole of the UK has almost 400 000 km of public roads, this is less than 2% of the total network length. These statistics indicate the value of the highway asset and that it needs to be managed in an efficient and effective manner. Also, the UK, in common with most developed countries, has substantially completed the construction of its motorway and trunk road network and the focus, for many years into the future, will be on the ongoing management of that network so that it continues to meet the needs of its users.

While there are many assets within the highway reserve that require management, including bridges, tunnels, culverts, road markings, road signs and traffic signals, the pavement is the most extensive of these assets, and it therefore attracts most of the highway maintenance budget. Consequently, this chapter will focus solely on the management of the pavement asset; however, many of the principles presented are applicable to those other assets (e.g. see Costello *et al.*, 2011).

The term pavement asset management is commonly used to describe the activities associated with managing the road pavement as an asset, and the title pavement asset manager is often assigned to the engineer responsible for these activities. Similarly, the term pavement management system is used to describe the systems adopted, usually computerised, to help manage the pavement asset.

15.2.   What is pavement asset management?

A pavement starts to deteriorate as soon as it is constructed and opened to traffic: for example, the available friction on the surface reduces over time, longitudinal depressions develop in wheel paths, cracks develop in the pavement structure, and the riding quality of the surface decreases. Eventually, the extent and severity of these defects reach a point where the performance of the pavement is compromised. At that point the pavement requires some form of maintenance treatment to rectify its deficiencies and restore its performance to an acceptable level. The basic challenges that face the pavement asset manager are

- where to intervene (i.e. determining which sections of a road network need intervention in the form of a maintenance treatment)

■ when to intervene (i.e. determining the most appropriate time to intervene on each road section)

■ what intervention to apply (i.e. determining the most appropriate maintenance treatment to apply on these road sections).

While a purely engineering approach to the above can yield technically robust solutions, more often than not the budget is insufficient to carry out all identified maintenance treatment interventions. One solution is to leave some sections untreated, resulting in a less than desirable performance: if so, which sections should go untreated and what are the consequences of not treating them? Alternatively, a number of sections could undergo a less expensive, but technically inferior, treatment intervention, thereby allowing more sections to be treated: if so, which sections should undergo the less expensive intervention, what should the treatment be, and what are the consequences of implementing the inferior treatment?

Pavement asset management and pavement management systems help to answer the above questions, thereby providing support for pavement engineers and asset managers when making decisions. While relatively simple prioritisation techniques can be applied, the application of whole-life costing or life cycle analysis to answer such questions is now state of the practice. Such an approach, however, requires an understanding of pavement management information, deterioration modelling, maintenance treatments, associated costs and benefits, prioritisation and optimisation – and a good understanding of the decisions that road agencies need to make at their various management levels.

15.3. Pavement management levels

In the technical literature, the levels of pavement management are variously termed network level, project level, strategic level, tactical level or operational level. However, for the purposes of this chapter the management functions introduced by Paterson and Robinson (1991) will be used, namely planning, programming, preparation and operations, as follows:

■ Planning: long-term strategic planning on a network level to determine network needs, budget requirements, and the impact of policy decisions.

■ Programming: the development of short and medium-term multi-year works programmes and associated expenditure programmes, under budget constraints.

■ Preparation: the development of individual road schemes identified in the works programme to the point where they are ready for implementation, including determining the appropriate maintenance treatment, detailed pavement design, specifications and bills of quantities.

■ Operations: the daily to weekly works activities of an organisation, including scheduling of work and the organisation of labour, plant and materials. (Note: this item falls outside the remit of this chapter.)

15.4. Pavement management information

Information is central to the management of a pavement asset. At the most basic level, the asset manager needs to understand what they are tasked with managing, including its size, composition, location and condition. Such information is typically referred to as inventory data and condition data. Inventory data do not markedly change over time whereas condition data vary with time. Pavement assets deteriorate over time due to the damaging effects of both traffic and the environment. In the case of heavily trafficked pavements, the dominant cause of deterioration is heavy vehicle loading, whereas the environment increasingly has the dominant detrimental effect on lesser travelled pavements. Hence, usage and environmental data that affect pavement condition should also be gathered.

Ideally, all the above information is stored in a road database of some description. While many proprietary road database systems are available, some road agencies have developed their own systems in house. The ease with which engineers and asset managers can access such information is becoming increasingly important to help with decision-making in the office and, through mobile communication devices, on site.

Data collection is expensive and is often the largest cost associated with the running of a pavement management system. For high-volume routes, it is usual to collect appropriate condition data annually; however, for lesser trafficked routes, a data collection strategy that looks at collecting particular data items every second or third year may be adequate. Alternatively, a strategy that collects condition data on a half, or a third, of the network each year may be appropriate. To assist road agencies in making decisions on what data to collect, when to collect it and the level of detail to which to collect it, the World Bank has produced guidelines that address the relevance, appropriateness, reliability and affordability of data collection (Paterson and Scullion, 1990).

15.4.1 Inventory data
Pavement inventory items typically include information on

- Pavement structure. This includes information on the pavement layer thicknesses and composition, from as-built drawings, local knowledge, cores and/or trial pits, as well as information on maintenance history. Such information is key, as different pavement structures behave differently in response to traffic loading and the environment, resulting in different modes of failure and consequent lives.
- Location referencing. Some method for referencing the location of the asset is required, thereby providing a reliable means of assigning data (e.g. pavement structure data) to the network. Geographic information systems (GIS) require the information in the form of eastings and northings, as obtained from the global positioning system (GPS). However, a linear referencing system for assets such as a highway network is also required to reference the pavement location to physical features on the network. In such cases, the location is most usually defined with reference to the road class, road number and section: for example, M62/105 refers to section 105 on the M62 motorway (HA, 2009). Further information is needed to identify a specific location on the section, typically the linear distance travelled from the start of the section. In the case of a two-way two-lane road, the direction of travel is also required, and in the case of a multi-lane road (e.g. a motorway), the lane number is required. The start and/or end of a section is typically located at an intersection or at another physical feature (e.g. a change in the number of lanes).

15.4.2   Condition data

Understanding the condition of each pavement asset is an essential step towards a performance-based approach to managing that asset. Condition data can relate to the functional and/or structural performance of the pavement. Functional performance is the ability of the pavement to perform its function to road users; that is, to provide a safe, economic and comfortable ride. Structural performance is the ability of the pavement to continue carrying the imposed traffic loading. Typical condition data that are collected include information on the structural and functional performance of the pavement, as follows:

■   Surface deflection. This is a measure of the deflection of the surface of the pavement under an applied load, thereby providing an indication of the strength of the pavement: it is currently measured using the falling-weight deflectometer from a stationary vehicle and trailer, or the deflectograph from a slow-moving vehicle. Recent advances in the development and testing of traffic-speed deflectometers (Ferne *et al.*, 2013) will probably result in their increased usage to collect surface deflection data. With suitable research and modelling, a residual life can be calculated from the surface deflection, to provide an estimate of the remaining structural life of the pavement.

■   Roughness. This is defined as the longitudinal deviation of the pavement surface from its intended profile per distance travelled, and is sometimes referred to as mega-texture. Typically measured using a laser profilometer attached to a traffic-speed condition survey vehicle, it is reported as profile variance in the UK or, more commonly elsewhere, as the international roughness index (IRI), in units of metres per kilometre. The IRI, which is closely related to user comfort, is determined by simulating the accumulated vertical movement per distance travelled of the body of a 'standard' car with fixed suspension characteristics that is driven over the measured longitudinal road profile at 80 km/h.

■   Surface texture. This is a measure of the texture of the road surface; commonly referred to as macro-texture, it is recorded as the mean texture depth in millimetres. Macro-texture, which allows excess water to be removed from the contact area between a vehicle tyre and the road surface, is typically measured in the wheel tracks using a high-frequency laser attached to a traffic-speed condition survey vehicle.

■   Skid resistance. This, a measure of the in-service friction of the pavement surface when wet, is a function of the texture of the aggregate (referred to as micro-texture) and the macro-texture. It is typically measured in the UK using the sideway-force coefficient routine investigation machine (SCRIM), and is reported as the sideways force coefficient.

■   Wheel track rutting. This, the longitudinal depression in the wheel paths of the road, is typically measured using a laser-based transverse rut bar mounted on a traffic-speed condition survey vehicle. It is reported as the average depth, in millimetres, below the road surface. In some cases, the standard deviation of the rut depths is also provided.

■   Cracking. This is manifested as longitudinal, transverse, block or crocodile cracks on the pavement surface. While longitudinal, transverse and block cracking can be caused by reflection cracking from joints, shrinkage or overlaid cracks in the lower layers of the pavement, crocodile cracking in the wheel tracks is brought on by fatigue of the asphalt layer due to loading. Although traditionally recorded using visual surveys, the introduction of high-speed digital imaging, faster processing times and advanced algorithms has seen the industry move towards the use of traffic-speed condition surveys to collect these data.

■ Surface defects. Various other surface defects may also be recorded (e.g. potholes, ravelling or fretting and flushing). While these defects have been traditionally recorded using visual surveys, the industry is now moving towards the use of traffic-speed condition surveys, to gather such data in an objective and systematic manner.

15.4.3 Usage and environmental data
Usage and environmental data recorded include

■ Annual average daily traffic (AADT). This is a measure of the daily traffic traversing the pavement that is determined from traffic surveys.
■ Heavy commercial vehicle (HCV) content. This is the percentage of HCVs in the AADT, and is determined from traffic surveys.
■ Equivalent standard axle load (ESAL). This is a pavement design concept that converts the spectra of axle loads of different axle configurations to the number of repetitions of a 'standard' axle, usually taken as an 80 kN dual-wheel single axle that would impart the equivalent amount of damage to the pavement. The cumulative ESAL, which is of interest over a period of time, is determined from weigh-in-motion sites on the highway network, from periodic surveys at weighbridge sites, or estimated from knowledge of the average ESAL per HCV in the region.
■ Environmental data. These include measures of rainfall and temperature from weather stations (e.g. the mean monthly rainfall and the weighted mean annual air temperature).

15.5. Deterioration modelling
Pavement data collected 1, 2 or more years previously need to be updated to an estimate of their current condition, in order to allow 'current' condition data to be compared across the network as a basis for decision-making. In all but the most basic of pavement management systems, predictive modelling is also employed to estimate the future condition of the pavement asset. In particular, the implementation of a treatment today will have knock-on costs and benefits well into the future that need to be quantified. In order to determine those costs and benefits, the future condition of the pavement requires estimation.

In addition to the current functional and structural condition data that require prediction into the future, information on the structural composition of the pavement, traffic-loading data and environmental data are used as predictors from which to estimate future condition. Such information will ideally be available in a road database; however, asset managers often need to make predictions with less than ideal data – albeit with reduced confidence in the outputs.

Figure 15.1 The probabilistic nature of pavement deterioration

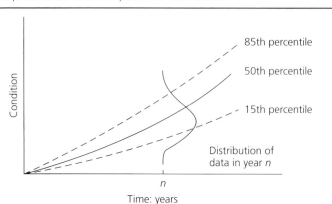

Approaches to predictive modelling can be grouped into two basic categories, namely deterministic models and probabilistic models, each of which have their advantages and disadvantages.

With deterministic models, a single condition value is predicted for a specific road section at a specific point in time. In such models, the stochastic nature and variability associated with the deterioration of pavement condition is not taken into account: instead, a representative value (e.g. the mean or median) is chosen. With probabilistic models, the probability of a condition parameter being at any particular value at a specific point in time is predicted. The stochastic nature and variability are thereby accounted for in the calculation, albeit the simplicity and convenience of using a single value is lost.

Both predictive methods are related, given that the mean, median or other representative value chosen for the deterministic model is taken from a probability distribution of condition, which represents the stochastic nature of the prediction (Ortiz-Garcia et al., 2006). Figure 15.1, which is based on hypothetical historical data for a generic condition measure, demonstrates this concept. Note that the 50th percentile is the chosen value for this particular example deterministic model; the 15th and 85th percentiles are also shown, to demonstrate the spread in the historical data from which the 50th percentile model was estimated. In any particular year, there is actually a distribution of values to represent the condition parameter. While a deterministic model will choose a 'representative' percentile, such as the 50th percentile, as its model, probabilistic modelling takes into consideration the whole distribution, thereby allowing a wide range of probabilities. Indeed, the appropriate choice of 'representative' value is a matter for debate. By definition, 50% of the network is in a worse condition than the 50th percentile 'representative' value. This has led many to consider, for example, the 85th percentile when taking risk into account for treatment selection.

15.5.1 Deterministic deterioration models

Deterministic models can be absolute or incremental in nature. Absolute models start from the 'as-new' condition value, and predict the value of the condition parameter at

Figure 15.2 Generic examples of deterministic deterioration models: (a) absolute; (b) incremental

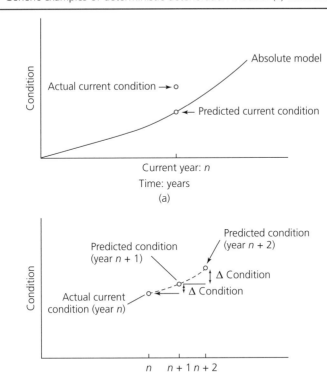

a particular point in time as a function of influencing parameters such as cumulative traffic loading to date and pavement strength. Taking a generic condition measure as an example, Figure 15.2(a) shows such an absolute model.

Issues arise with absolute models when the predicted current value is different from the actual current value; given the possible distribution of condition values this is a common occurrence (see Figure 15.2(a) for this scenario). In such cases, the absolute model needs to be adjusted to fit the actual condition value, which itself creates issues. Hence, most management systems use incremental models, as they take the current condition value as the starting point and predict the progression from that point forward, albeit following the same overall behaviour.

Incremental models predict the change in the condition parameter from one year to the next, based on similar influencing parameters. To determine the value of a condition parameter in the following year, the predicted incremental change is added to the current condition value, and the resulting condition forms the starting point for predicting the deterioration for the following year, and so on (Figure 15.2(b)).

Deterministic methods for predicting condition can also be categorised as empirical models and structured-empirical models.

Empirical models are based on the statistical analysis of historical deterioration trends. The dependent variable – in this case the condition parameter – is related to the independent variables (i.e. the influencing parameters), and a relationship is determined through regression analysis. Typical regression model forms can range from a simple linear model to complex non-linear models. Empirical models are, by definition, based on specifically 'local' data; thus, while capable of providing extremely useful predictions for the area in which they are developed, their application elsewhere is problematic. Also, issues can arise when extrapolation beyond the limits of the historical data set is attempted.

Structured-empirical models have a predetermined functional form (i.e. curve shape) that is based on engineering knowledge from previous studies of pavement performance. They can also range from a simple linear model to complex non-linear models such as 'S'-shaped curves (e.g. see the models reported in Odoki and Kerali (2000) for the International Study of Highway Development and Management (ISOHDM)). The independent variables are pre-determined, thereby restricting the data requirements to those of most importance. These models have default settings and require calibration to local conditions. The calibration coefficients included in the models can be altered to increase or decrease the rate of deterioration to match local conditions.

Figure 15.3(a) displays a structured-empirical model for cracking, based on the default ISOHDM model. Cracking is modelled as crack initiation, which predicts the length of time before cracks appear in the pavement, and crack progression, which predicts the rate at which cracking spreads once a crack has been initiated. If pavements in the network under scrutiny crack sooner or later than the default model, then the crack initiation calibration coefficient can be altered until the resulting model matches observed behaviour. Similarly, if pavements in the network concerned progress faster or slower than the default model, the crack progression calibration coefficient can be altered until the resulting model matches observed behaviour. In Figure 15.3(a), the calibrated model cracks sooner than the default, but then progresses at a slower rate once cracking has been initiated. (The reader is encouraged to read Bennett and Paterson (2000) for a description of how to calibrate these models.)

15.5.2 Probabilistic deterioration models

Probabilistic methods for predicting condition include survivor curves and Markov models. The stochastic nature and variability of pavement deterioration are built into these models.

The survivor curve model is a continuous function that predicts the probability of a condition parameter being at a specific value at a specific point in time or, alternatively, the probability of a particular event happening at a particular point in time. The example in Figure 15.3(b) (based on Henning et al. (2006)) illustrates the probability of a pavement being cracked for a given surface age. Two survivor curves are included in this

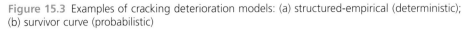

Figure 15.3 Examples of cracking deterioration models: (a) structured-empirical (deterministic); (b) survivor curve (probabilistic)

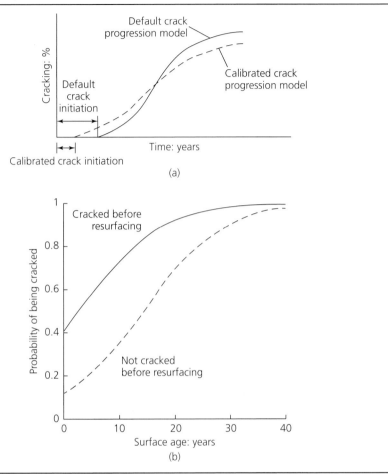

figure: one for pavements that were previously cracked, and one for pavements that were not. The plotted probabilities are based on a specific traffic loading, surface thickness and pavement strength.

Markov models, also referred to as stochastic models, are typically employed at the planning level to predict the condition of a road network, as distinct from individual road lengths, and to display the results as a condition histogram that details the proportions of the network in a range of condition bands. The method that uses these models employs transition probability matrices to represent the probabilities of moving from one band to another, and the models assess strategic funding policies by predicting the effects of various funding and maintenance scenarios on the medium to long-term performance of road networks. (See Costello *et al.* (2005) for a description of the methodology and a case study of its implementation.)

15.5.3 Pavement performance studies

In an effort to better understand pavement deterioration, a number of long-term pavement performance programmes have been established internationally to closely monitor pavement performance, based on detailed knowledge of the specific sites included in the studies. Prominent among these is the Strategic Highway Research Program study in North America, and readers are referred to Kerali *et al.* (1996) for information on data analysis procedures for the determination of deterioration models from studies such as this.

In addition, a number of accelerated pavement testing facilities (e.g. Brown, 2004) now exist that allow road pavements to be tested to destruction. Such facilities have yielded an improved understanding of pavement performance, especially when and how they fail, and what are the influencing factors: they can also be used to obtain indications of how new or marginal pavement materials, new construction techniques and/or new designs will perform, given that there are no historical data.

15.6. Maintenance treatment

Under the influence of both traffic loading and the environment, the surface and structure of a pavement start to deteriorate as soon as construction is completed, and, eventually, the extent and severity of the defects cause the performance of the road to be compromised. At that point the pavement requires intervention (i.e. maintenance treatment) to rectify the deficiencies and restore its performance to an acceptable level.

15.6.1 Selecting the maintenance treatment

Depending on the composition of the pavement, a number of treatment options can be selected to repair a deficiency in the surface or structure of the pavement. In broad terms, pavement treatments are either surface treatments, which repair or replace the surface but have little or no influence on the structure of the pavement, or structural treatments, which, in addition to replacing the surface, impart additional strength to the pavement and increase its residual life. The detailed design and selection of treatments are covered elsewhere (e.g. see Chapters 16–18), and, hence, only generic treatments are noted briefly here

- Resealing. This refers to the use of thin surfacings (e.g. a surface dressing or a thin asphalt overlay) that rejuvenate the pavement surface, increase skid resistance and macro-texture, and seal cracks.
- Rehabilitation. This includes the application of a structural overlay that strengthens the existing pavement and adds to its residual structural life, or an inlay in which damaged layers of the pavement are removed and replaced to improve the functional performance of the pavement and, potentially, its structural performance (if appropriate materials are used).
- Reconstruction. As the term implies, this involves the rebuilding of the pavement down to formation level at the end of its initial design life, that is, the pavement as a whole is redesigned and replaced.

15.6.2 When to intervene?

Figure 15.4 shows, generically, how the pavement condition parameter deteriorates over time until it reaches the intervention level at which, assuming an adequate budget, a

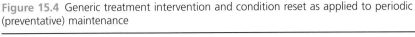

Figure 15.4 Generic treatment intervention and condition reset as applied to periodic (preventative) maintenance

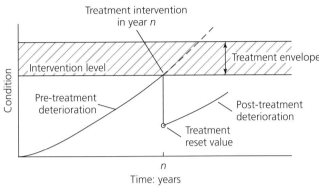

treatment should be applied that will result in the resetting of the condition parameter to a nearly new or another defined state after treatment (see Section 15.6.3). At this intervention level, there is a treatment envelope within which a particular treatment (e.g. rehabilitation) is suitable to repair the defect(s), and failure to seize the opportunity to carry out the treatment within that time-frame could result in the subsequent need for a more expensive treatment (e.g. reconstruction). The term preventive maintenance (also called planned or periodic maintenance) has been coined to explain the concept whereby a cheaper maintenance treatment that is applied at an opportune time can prevent the need for a more expensive treatment at a later date.

Intervention levels should be defined for every condition parameter under consideration, along with an associated treatment intervention. In practice, however, there is no definitive set of intervention levels, and the levels at which interventions do occur vary between road agencies. However, experience has shown that some condition parameters can experience an accelerated progression stage that represents rapid failure of the pavement towards the end of its design life (e.g. see Henning *et al.*, 2009). The current general rule, therefore, is to intervene just before this rapid failure occurs.

15.6.3 Maintenance effects

Following intervention, the effects of the applied treatment need to be modelled and, in sophisticated predictive models, taken into account in the post-treatment predictive modelling. For example, a pavement that receives a reseal can be expected to crack faster in the future if cracking existed before the reseal – hence the term reflective cracking – in comparison with a pavement that was not cracked but received the same reseal (see Figure 15.3(b)); however, the skid resistance and texture predictions would be essentially the same for both scenarios after treatment.

Figure 15.4 also shows how these maintenance effects are simulated in the pavement management system through the use of a reset value. The reset value can be simply reset to zero, as is the case for wheel-track rutting and cracking following a pavement reconstruction. Alternatively, it may be a function of the construction technique and materials (e.g. as in the case of skid resistance and texture) or it may be a function of the pretreatment condition value (e.g. as in the case of roughness, following the placement of an overlay). The roughness example is also a function of the overlay thickness as well as of construction quality and country-specific techniques. Surface deflection and residual life can be reset for all structural treatments, based on the available pavement design-life information.

15.7. Consideration of costs and benefits

The application of a maintenance treatment to a road section has consequences in terms of the costs and benefits to various stakeholders (e.g. the maintaining agency, road users, nearby residents and society in general). The purpose of determining the costs and benefits is to allow a comparison of the various treatment options, with the intention of choosing the option that results in the greatest return from the investment, however that is measured.

15.7.1 Costs

The direct costs incurred by a road agency in executing a maintenance treatment are well understood by engineers (i.e. the costs of materials, labour and plant as well as the costs of design and supervision). Such costs are commonly captured in the unit rates used to carry out a particular task.

Indirect (non-agency) costs are all those 'other' tangible and intangible costs arising from the use of the road. An example of tangible costs are those arising from delays to road users as a result of the maintenance activities on the road network. In the UK, delay costs are captured in the QUADRO (Queues And Delays at Roadworks) model (HA *et al.*, 2006). Intangible costs, typically, are the social costs incurred by the community living adjacent to the road (e.g. noise, dust or water pollution).

15.7.2 Benefits

Road agency benefits arise from future savings in direct costs that accrue when the cost of applying a treatment 'now' is less than that which would otherwise be required in the future (e.g. due to reduced routine maintenance costs or preventing the need for a more expensive treatment at a later date).

While user benefits accrue to various stakeholders as a result of the application of a maintenance treatment and the corresponding improvement in pavement condition, quantifying such benefits is not always easy. At the most basic level, a measure of the improvement in pavement condition can be considered a benefit. Including a measure of usage in the calculation that recognises that some roads carry more traffic than others and, hence, that greater benefits will accrue takes the process a step further. Again, such information is well understood and is relatively easily acquired from the road database.

The calculation of full economic benefits, however, is often less well understood, as they accrue to various stakeholders and are less tangible. These benefits are most easily grasped by discussing them in terms of those to whom they accrue, namely to the road user and to the wider community.

Economic benefits accruing to the road user include

■ Lower vehicle operating costs due to a reduction in road roughness. These can include reductions in fuel consumption, tyre wear, vehicle maintenance and so on.
■ Reductions in travel time due to, for example, a smoother pavement surface. (Note, however, that in developed countries it is argued that the most significant travel time benefits are generated from improvements to the road network, e.g. due to a road realignment or to increasing a road's capacity by adding extra lanes.)

Economic benefits accruing to the wider economy include

■ Lower accident rates due to improved skid resistance or the implementation of a road safety scheme. Included here are the social and economic savings associated with reduced numbers of injuries and fatalities on the road network.
■ Reduced environmental impacts of noise or vehicle emissions upon both nearby residents and road users due to, for example, changes in rolling resistance. (However, this can be a disbenefit if the new surface is noisier than the existing one.)

Which benefits are included in any given assessment depends on the road agency's policy and, indeed, on the availability of such data. For example, the UK Department for Transport does not, as a rule, include lower vehicle operating costs or travel time savings resulting from a smoother pavement surface as a benefit in its calculations; conversely, it is widely accepted that these should be used in developing countries where roads are, generally, in a poorer condition, and relatively minor maintenance improvements can lead to the accrual of major benefits to road users. Indeed, a calculation of full economic benefits on top of direct agency cost considerations is usually required to secure a road infrastructure loan from donors such as the World Bank or the Asian Development Bank. The World Bank's Highway Development and Management (HDM-4) model (Kerali, 2000) has been developed to assist with such analyses, and readers are also referred to Odoki et al. (2012) for an example of its application in the UK.

By contrast, in long-term performance-based contracts such as the design, build, finance and operate (DBFO) concessions employed in the UK (see Chapter 1) and the performance-specified maintenance contracts (PSMCs) used in Australia and New Zealand, the operating company or consortium is mainly concerned with minimising whole-life direct (i.e. agency) costs while operating within the predefined functional and structural-condition performance levels laid down by the controlling road agency.

15.7.3 Comparing costs and benefits
The purpose of determining the costs and benefits is to allow a comparison of the various treatment options, with the intention of implementing the one that results in the greatest

economic or financial return. This exercise involves a comparisons of options under-taken over an analysis period, typically 10–30 years, clearly defining what costs and benefits are incurred each year of the agreed period, with the costs and benefits (if they are defined in monetary terms) for each year discounted back to their present value (PV) using the standard economic formula

$$PV = \frac{c_i}{(1 + r)^i} \tag{15.1}$$

where c is the cost or benefit incurred in year i and r is the discount rate expressed as a fraction.

The discount rate represents the time value of money: it is sometimes simplistically explained as equating to the market interest rate (which is the term used when projecting forward). One approach used to establish an appropriate discount rate is to consider it as the opportunity cost of the investment if used for an alternative purpose, the assumption being that the investment under consideration displaces other investments that would have earned a return. In practice, the appropriate discount rate to use is normally provided by the government.

In order to determine the benefits of carrying out the proposed treatment, a baseline 'do-nothing' scenario also needs to be assessed. As well as providing a baseline comparator for options, the inclusion of the do-nothing option also allows an assessment of whether it is actually better to do nothing (i.e. it may show that the do-nothing option results in the greatest return). The do-nothing scenario usually involves the continuation of routine (i.e. reactive) maintenance, and, hence, is more correctly described as the 'do-minimum' scenario on the assumption that, at least, reactive maintenance has to be done in order to prevent the road from being closed to traffic. Each proposed periodic treatment option is then compared with the do-minimum option to determine the total return from carrying out each treatment versus not doing so.

To account for the fact that a pavement asset will still have value at the end of the analysis period – this can be large if the pavement had been rehabilitated or recon-structed – a residual value should ideally be included in the analysis, and this should also be discounted back to present-day values. Readers are referred to Snaith and Orr (2006) for an example of a condition-based valuation methodology that can be adapted for this purpose.

15.8. What treatment schemes should have priority?

In any given year, many maintenance treatment schemes with a positive return may be proposed across a road network. Essentially all road agencies, however, operate under budgetary constraints, so that not everything by way of maintenance that should be done in any one year can actually be included for action in the programme for that budget year. The consequence of this is that the road authority needs an objective process that allows the selection of a limited number of maintenance schemes from a large number, in a logical manner; that is, a process of prioritisation that determines which schemes will be funded and which will be delayed.

Prioritisation normally involves ranking all candidate maintenance schemes in their proposed order of importance and costing them; taken in ranked order, every scheme for the year in question is then scheduled into the works programme until the cumulative cost exceeds the annual budget. The remaining candidate schemes not scheduled by this method are then put aside to be subsequently reconsidered for inclusion in the next year's programme together with any others suggested by the decline in road condition in the intervening year. In practice, the prioritisation processes vary considerably between road agencies; however, for discussion purposes the methods used can be described in order of increasing sophistication, ranging from basic through intermediate to advanced.

15.8.1 Basic prioritisation methods

Basic prioritisation methods consider the current condition of the pavement only in their calculations. The simplest method is a 'worst first' approach that involves a simple ranking, in descending order, of the worst roads based on current condition only. Such a method is generally considered to be ineffective, albeit attempts to improve it have attempted to take into account the effect of such parameters as traffic volumes or road hierarchy, to ensure that important roads are given priority. While the latter methods have been found to be more effective, if relatively simplistic, they will not be discussed further here; instead, the reader is referred to Robinson *et al.* (1998) for further information.

15.8.2 Intermediate prioritisation methods

Intermediate prioritisation methods consider the costs and benefits that accrue over the analysis period due to each treatment scheme proposed for inclusion in the maintenance programme. This is carried out through a process, as discussed previously, whereby both the costs and benefits for each scheme are discounted back to their present-day values, and each do-something option is then compared with the do-minimum option to enable a rank to be developed for each proposed scheme. The ranking criteria most commonly used by road maintenance agencies include cost-effectiveness, net present value (NPV) and the NPV/cost ratio.

15.8.2.1 Cost-effectiveness method

The simplest cost-effectiveness method is the expected life of the proposed treatment divided by its unit cost: it is often also expressed as its reciprocal (i.e. the 'average annualised cost'). However, such a simple calculation neither considers the magnitude of the improvement in pavement condition nor the volume of traffic that will benefit from the improvement. Consequently, most cost-effectiveness methods use the area-under-the-curve concept, whereby the pavement condition is plotted against time for the proposed treatment scheme, as well as for the do-minimum treatment. The difference in areas between the two is then taken as a measure of the effectiveness of carrying out the treatment, and this is divided by the difference in the unit cost of carrying out both options. With this method, the cost-effectiveness is calculated for all treatment sections, and, to allow for the variation in traffic between sections, it is multiplied by the AADT.

The area-under-the-curve method rewards schemes that provide the greatest improvement in pavement condition for as long as possible and for as many road users as possible, while taking into account the unit cost of providing each treatment. There are many variants of this method, and the reader is referred to Haas *et al.* (1994) for further information.

15.8.2.2 NPV methods

The NPV is the difference between the discounted benefits and costs over the analysis period *n*, as expressed below

$$\text{NPV} = \sum_{i=0}^{n} \text{PV}_\text{B} - \sum_{i=0}^{n} \text{PV}_\text{C} \tag{15.2}$$

where *i* is the year (with $i = 0$ in the base year) and PV_B and PV_C are the present values of benefits and costs, respectively, as previously defined.

A positive NPV indicates that the investment in the treatment scheme (i.e. the do-something option) is worthwhile, based on the data considered and the given discount rate. Also, the greater the NPV returned, the greater is the return on the investment. While NPV can be used for prioritisation purposes, it is only moderately effective under budget constraints, as, all other things being equal, an expensive treatment such as a reconstruction will typically have a larger NPV than an inexpensive treatment such as a reseal. Hence, the ratio of the NPV divided by the cost of the maintenance treatment (i.e. the initial investment) provides a better solution.

The NPV/cost ratio determines the NPV per unit of investment, and is therefore used to rank projects under budgetary constraint. With this method, the many treatment schemes proposed across the road network can be ranked from the largest to the smallest; however, only schemes with positive values should be ranked, as negative NPV/cost values do not provide a return on the initial financial investment.

The ratio given by NPV/cost is akin to the cost-effectiveness method, the difference being that, in this case, the 'effectiveness' is measured by the NPV.

15.8.3 Advanced prioritisation methods

Advanced prioritisation methods are characterised by their use of a process of optimisation. While there is some debate as to whether optimisation is in fact a prioritisation technique, for the purpose of this discussion optimisation is assumed to be a sophisticated form of prioritisation.

Optimisation as applied to pavement management can be summarised as finding the set of treatment options across a road network, over a defined period, that provide the desired optimum solution for whatever parameter is deemed to be the most significant for the proper and efficient operation of the network under scrutiny – that is, the objective function parameter.

Typical objective functions, where the optimisation process aims to minimise or maximise the parameters over the analysis period, are (a) to maximise pavement condition, given a constrained budget; (b) to minimise road user costs, given a constrained budget; (c) to minimise total transport costs (i.e. the sum of construction, maintenance and road user costs), given a constrained budget; and (d) to minimise maintenance costs, while still achieving a certain minimum pavement condition.

The first three of the above are typical objective functions for government-owned road networks. The fourth, however, is a typical objective function for a performance-based contract such as a DBFO or PSMC (see Section 15.7.2). In these types of contract the financial costs to the managing consortia of carrying out the works (as opposed to the costs to the road users and the road authority) is the focus of the modelling, while still meeting the terms laid down in the contract relating to key performance indicators.

True optimisation, also known as total enumeration, is essentially the consideration of all possible solutions in order to select the particular treatment and timing thereof that best meet the objective function. In pavement management, ideally, this requires the consideration of every treatment combination for every section of the pavement for every year of the analysis period. While this would provide the 'optimum' solution, the number of possible solutions that need to be considered and the computing power required to complete the necessary calculations in a reasonable time-frame have resulted in other, suboptimal, techniques being applied (see Robinson *et al.*, 1998).

15.9. Pavement management as a tool to aid decision-making

Ultimately, pavement management is carried out to assist asset managers in making logical decisions. However, the output from any pavement management system is only one, albeit critical, element in the decision-making process. Many other factors can also influence decisions (e.g. environmental, social and political factors), so these are often considered alongside the outputs of pavement management systems.

The decisions that need to be taken depend on the management function being undertaken. In this respect, it is important to understand the relationship between the planning, programming, preparation and operations levels (see Section 15.3). This relationship is represented in Figure 15.5, where the outputs of the planning level (i.e. the budget constraints and performance levels) are inputs into the programming level that restrict the budget available each year and influence intervention decisions. In turn, the outputs from the programming level (the selected maintenance schemes) are inputs into the preparation level, as they dictate which schemes are developed further and evaluated for implementation on the road network. While the operational level is outside the scope of this chapter, it is included in Figure 15.5 for completeness, and to demonstrate the feedback loop to the road database when maintenance schemes are physically implemented on the network, thereby allowing the maintenance history to be updated.

15.9.1   Planning decisions

Planning and policy decisions are taken within an organisation by policy-makers and senior management, and, typically, the road network as a whole is considered when making strategic decisions. Pavement management systems can assist with this decision-making by producing estimates of the expenditure required to maintain the

Figure 15.5 The pavement management cycle

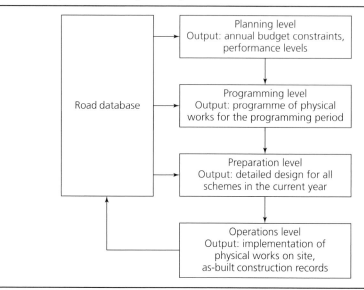

network to a certain condition level or, alternatively, by estimating the reduction or improvement in condition levels resulting from various budgetary scenarios. Typical applications include

- Estimating the effects of proposed decreases in maintenance funding on the condition of the road network over future years. Figure 15.6 shows how a reduction from £20 million to £15 million per annum changes the estimate of average roughness at the end of the analysis period from 4 m/km to almost 5.5 m/km, while a reduction to £10 million results in an average roughness of 6.5 m/km.
- Developing a long-term (typically 10 years or more) expenditure plan to assist local or central government to understand its budget requirements for future years.
- Estimating, for tender bidding purposes, the funding required to maintain a road network to a certain predefined level for the life (typically 10–30 years) of, say, a performance-based contract.

Setting appropriate treatment intervention levels can also be a policy issue. While intervention levels can be set that are based on engineering judgement or experience, these may not be affordable or provide the greatest benefit (e.g. as in determining the minimum condition levels to include in a performance-based contract, as the decision taken is binding for the life of the contract). Pavement management systems can assist with such decisions by comparing the costs and benefits that arise from the implementation of differing intervention levels. Such analyses can also include economic, social and environmental criteria in their determination, through multi-criteria analysis (Ortiz-Garcia et al., 2005).

Figure 15.6 Effect of different annual budget scenarios on the network condition

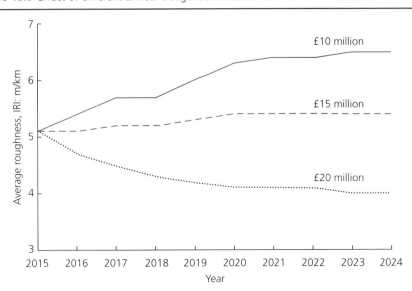

15.9.2   Programming decisions

Decisions at the programming level are normally taken by the pavement asset manager. At this level, a suggested programme of planned works is produced for, say, the next 1–3 years, based on known budgetary constraints for each of the years (as determined at the planning level). Given the budgetary constraints, each year's programme then has to undergo prioritisation to determine which of the maintenance schemes can be included for that year. The works programme is updated annually by re-running the pavement management system after including the latest condition surveys: this is why it is termed a rolling programme.

The proposed schemes included in the first year of the rolling programme are put forward for further investigation at the preparation stage, and, if approved, are implemented. While successive years' schemes are subject to decreasing confidence the further into the future they are programmed, they do provide a basis upon which to prepare budgets in terms of both finance and resources (e.g. labour, plant and materials). The more confidence the asset manager has in the predictive capabilities of the models incorporated in the pavement management system, then the more confidence they will have in proposing the programmed schemes. An example programme of maintenance works is given in Table 15.1.

15.9.3   Preparation decisions

Decisions at the preparation, or scheme evaluation, stage are typically taken by the pavement design engineer. Such decisions relate to which of the treatment alternatives on a specific project scheme will provide the greatest return, the specific scheme having been identified at the programming level. This is usually undertaken by consideration of the

Table 15.1 A prioritised 3 year rolling programme of maintenance schemes

Year	Rank	Road section	Length: km	Maintenance treatment	Project cost: £ million	Cumulative cost: £ million
1	1	A6-04	9.5	Resealing	2.4	2.4
	2	A2-09	5.5	Rehabilitation	2.8	5.2
	3	M3-01	7.5	Reconstruction	15.5	20.7
2	1	M2-03	8.1	Reconstruction	16.2	16.2
	2	A5-16	4.5	Rehabilitation	2.3	18.5
	3	A1-03	3.5	Resealing	0.9	19.4
3	1	A8-06	7.5	Resealing	1.9	1.9
	2	M2-02	8.5	Rehabilitation	4.3	6.2
	3	A4-11	2.5	Reconstruction	5.1	11.3
	4	A2-12	4.4	Reconstruction	8.9	20.2

whole-life cost of each of the treatment alternatives, and, in the UK, is currently under-taken for motorways and trunk roads under the stewardship of the Highways Agency using Software for the Whole-life Economic Evaluation of Pavements (SWEEP), a component within the Highways Agency pavement management system (HAPMS).

Various treatment alternatives are assessed over the analysis period against the do-minimum option (e.g. these may include reconstructing a pavement, carrying out a structural overlay or resealing the pavement surface (including pre-seal repairs as required)) – and the alternative that results in the greatest economic return is the preferred solution. The treatment alternatives have to be designed to a sufficient level to allow a detailed whole-life cost analysis to be undertaken. In addition to the immedi-ate costs for each treatment, all alternatives will have consequential treatment costs in later years of the analysis that will be relevant to the calculation of the whole-life cost: these may be minor in the case of the reconstruction but substantial in the case of the reseal.

Project evaluation should also consider the timing of project schemes. Although a comparison of various treatment alternatives for implementation in the budget year may result in a preferred alternative, the option of delaying this treatment for one or more years should also be considered. This can be accommodated in the whole-life cost calcu-lation as simply another 'do-something' option. However, a holding treatment would almost certainly need to be implemented immediately if, for example, a reconstruction or structural overlay is delayed until a future date. Such a scenario could potentially lead to a higher return, thereby indicating that the treatment intervention should be delayed rather than being carried out in the budget year.

15.10.   Final comments

The application of a complex or advanced pavement management system does not necessarily guarantee that the maintenance results are more accurate or that the pavement asset manager is likely to make better decisions. Outcomes depend largely on the degrees of confidence placed on the inputs, especially on the accuracy of the available condition and inventory data, as well as on the quality of the predictive models, knowing when to intervene, and knowledge of maintenance effects.

Pavement asset management is still in the process of evolution, and sound engineering knowledge is fundamental to decision-making relating to when to intervene with a particular maintenance treatment. Future challenges include the integration of the pavement asset decision-making process with other highway assets (e.g. bridges and drainage structures), in order to determine where and when limited maintenance funding is best spent in order to achieve the desired outcomes. However, while the pavement management systems currently available cannot be described as infallible, they undoubtedly provide valuable assistance to pavement asset managers, road agencies and governments in their respective decision-making roles.

REFERENCES

Bennett CR and Paterson WDO (2000) *The Highway Development and Management Series*, vol. 5. *A Guide to Calibration and Adaptation.* World Road Association, Paris, France, and World Bank, Washington, DC, USA.

Brown SF (2004) Accelerated pavement testing in highway engineering. *Proceedings of the Institution of Civil Engineers – Transport* **157(3)**: 173–180.

Costello SB, Snaith MS, Kerali HGR, Tachtsi VT, and Ortiz-Garcia JJ (2005) Stochastic model for strategic assessment of road maintenance. *Proceedings of the Institution of Civil Engineers – Transport* **158(4)**: 203–211.

Costello SB, Moss WF, Read CJ and Grayer S (2011) Life-cycle planning methodology for ancillary highway assets. *Proceedings of the Institution of Civil Engineers – Transport* **164(4)**: 251–257.

Ferne B, Langdale P, Wright MA, Fairclough R and Sinhal R (2013) Developing and implementing traffic-speed network level structural condition pavement surveys. *Paper presented at the 9th International Conference on the Bearing Capacity of Roads, Railways and Airfields*, Trondheim, Norway, 25–27 June 2013.

HA (Highways Agency) (2009) *Network Management Manual and Routine and Winter Service Code.* See http://www.dft.gov.uk/ha/standards/ghost/nmm_rwsc/index.htm (accessed 13/06/15).

HA (2013) *Highways Agency Business Plan 2013–2014.* See http://assets.highways.gov.uk/about-us/corporate-documents-business-plans/S120450_Highways_Agency_Business_Plan_2013-14.pdf (accessed 13/06/15).

HA, Transport Scotland, Welsh Government and Department for Regional Development Northern Ireland (2006) Section 1: the QUADRO manual. In *Design Manual for Roads and Bridges*, vol. 14. *Economic Assessment of Road Maintenance.* Stationery Office, London, UK.

Haas R, Hudson WR and Zaniewski J (1994) *Modern Pavement Management.* Krieger, Malabar, FL, USA, pp. 214–222.

Henning TFP, Costello SB and Watson T (2006) *A Review of the HDM/dTIMS Pavement Models Based on Calibration Site Data.* Land Transport New Zealand, Wellington, New Zealand, Research Report 303.

Henning TFP, Dunn RCM, Costello SB and Parkman CC (2009) A new approach for modelling rutting on the New Zealand state highways. *Road and Transport Research* **18(1)**: 3–18.

Kerali HGR (2000) *The Highway Development and Management Series*, vol. 1. *Overview of HDM-4*. World Road Association, Paris, France, and World Bank, Washington, DC, USA.

Kerali HR, Lawrance AJ and Awad KR (1996) Data analysis procedures for long-term pavement performance prediction. *Transportation Research Record* **1524**: 152–159.

Odoki JB and Kerali HGR (2000) *The Highway Development and Management Series*, vol. 4. *Analytical Framework and Model Descriptions*. World Road Association, Paris, France, and World Bank, Washington, DC, USA.

Odoki JB, Anyala M and Bunting E (2012) HDM-4 adaptation for strategic analysis of UK local roads. *Proceedings of the Institution of Civil Engineers – Transport* **166(2)**: 65–78.

Ortiz-Garcia JJ, Snaith MS and Costello SB (2005) Setting road maintenance standards by multicriteria analysis. *Proceedings of the Institution of Civil Engineers – Transport* **158(3)**: 157–165.

Ortiz-Garcia JJ, Costello SB and Snaith MS (2006) Derivation of transition probability matrices for pavement modeling. *Transportation Engineering* **132(2)**: 141–161.

Paterson WDO and Robinson R (1991) Criteria for evaluating pavement management systems. In *Pavement Management Implementation* (Holt FB and Granling WL (eds)). American Society for Testing and Materials, Conshohocken, PA, USA, pp. 148–163.

Paterson WDO and Scullion T (1990) *Information Systems for Road Management: Draft Guidelines on System Design and Data Issues*. World Bank, Washington, DC, USA, Report INU 77.

Robinson R Danielson U and Snaith M (1998) *Road Maintenance Management: Concepts and Systems*. Macmillan, London, UK.

Snaith MS and Orr DM (2006) Condition based capital valuation of a road network. *Proceedings of the Institution of Civil Engineers – Municipal Engineer* **159(2)**: 91–95.

Highways
ISBN 978-0-7277-5993-1

ICE Publishing: All rights reserved
http://dx.doi.org/10.1680/h5e.59931.491

Chapter 16
Structural strengthening of road pavements

Derek McMullen Pavement Engineer, Atkins Ltd, UK

16.1. Introduction

This chapter describes the techniques that are used to decide when a road pavement needs structural maintenance, and the strengthening measures that may be appropriate. The use of deflection measurements within the process is considered, and several overlay design systems are described. When selecting a design approach in a locale, it is not appropriate to 'mix and match' elements from different design procedures: the latest manuals or other sources relevant to that locale should be consulted for details of the selected procedure.

16.2. Concept of pavement strengthening

Road pavements generally deteriorate gradually under traffic loading. Evidence of structural deterioration in flexible pavements (i.e. that which affects the structural integrity of the pavement) is commonly seen as rutting and/or cracking in the wheel tracks. These defects may develop simultaneously but, in a properly designed pavement, neither their extent nor severity should become excessive before the design traffic is achieved.

The rate of pavement deterioration depends on several factors, including (TRL, 1993) the volume and type of traffic carried, the properties of the materials in each pavement layer, the surrounding environment and the maintenance strategy adopted.

Deterioration will progress more rapidly if the volumes of heavy vehicles and/or the vehicle wheel loads increase more rapidly than were taken into account in the original pavement design procedure. Wheel-track rutting of flexible pavements generally develops gradually due to the viscoelastic properties of the bitumen; however, rutting will develop faster on uphill road sections that carry slow-moving heavy wheel loads, especially in hot climates. Also, pavement cracking is often initiated at the surface, due to age hardening of the bitumen (Burt, 1986; Nunn *et al.*, 1997; Rolt *et al.*, 1986). Pavement deterioration may also be initiated or accelerated by poorly designed and/or maintained drainage. Drainage problems account for a significant number of premature pavement failures, generally caused by water trapped within the pavement layers or subgrade (Cedergren, 1987; Nicholls *et al.*, 2008).

Historically, two levels of pavement condition have been identified as triggering the need for a repair response: critical and failure. Ideally, failure should never be reached, while

Table 16.1 Road surface conditions that are broadly comparable with the condition of the whole pavement

Wheel-track cracking	Wheel-track rutting under a 2 m straight edge			
	<5 mm	5 to <10 mm	10 to <20 mm	≥20 mm
None	Sound	Sound	Investigatory	Failed
Less than half-width[a] or a single crack	Investigatory	Investigatory	Investigatory	Failed
More than half-width[a]	Failed	Failed	Failed	Failed

Based on Kennedy and Lister (1978) and HA *et al.* (1994a)
[a] The half-width is likely to be in the range 0.5–1.0 m; if there is no rutting to define the wheel tracks, use 0.5 m as half the wheel-track width

the critical level should only occur near the end of the initial design period of the road (Kennedy and Lister, 1978). The critical condition is now more appropriately referred to as an investigatory condition (HA *et al.*, 2008a). This implies that, as the end of the service life of a pavement is approached, it may be necessary to investigate signs of failure with a view to establishing a timely and economic remedial works programme, thereby actually avoiding failure with its consequent direct costs to the road authority and indirect costs to road users.

As an example of how this may be effected, the investigatory and failure levels used in the development of deflection-life relationships in the UK are given in Table 16.1.

A more recent variant of condition classification is given in Table 16.2. This is used in association with defined threshold values for each parameter measured with a TRACS (TRAffic speed Condition Survey) vehicle during network level surveys.

16.2.1 Timing of pavement strengthening

The treatment needed to restore a pavement to an acceptable condition depends on its current condition and the expected traffic loading. As the pavement condition worsens, the required treatment can be expected to be deeper and more extensive.

Typical treatments that might be applied should investigatory surveys show signs of incipient failure would be a programme of structural overlays or, more immediately, resealing works to preserve the structural integrity of the pavement. It is known from experience that these will only be effective, from an economic and engineering stand-point, if effected at the investigatory deterioration level. Clearly, as a pavement reaches the end of its designed service life, be it from new or after remedial works, the need for monitoring condition surveys will increase. When investigatory levels of deterioration are identified, additional detailed investigations will be required at the project level to determine the most appropriate rehabilitation strategy.

Table 16.2 Condition categories for texture depth, rut depth and ride quality

Category	Definition
1	Sound – no visible deterioration
2	Some deterioration – lower level of concern
	The deterioration is not serious, and more detailed (project level) investigations are not needed unless extending over long lengths, or several parameters are at this category at isolated positions
3	Moderate deterioration – warning level of concern
	The deterioration is becoming serious and needs to be investigated. Priorities for more detailed (scheme-level) investigations depend on the extent and values of the condition parameters
4	Severe deterioration – intervention level of concern
	This condition should not occur very frequently on the motorway and all-purpose trunk road network, as earlier maintenance must have prevented this state from being reached. At this level of deterioration, more detailed (scheme-level) investigations should be carried out on the deteriorated lengths at the earliest opportunity, and action taken if, and as, appropriate

HA *et al.* (2008a); reproduced by courtesy of the Highways Agency

16.3.   Structural assessment procedure

The aim of a structural assessment is to determine the ability of a pavement to carry the anticipated future traffic loading for a given design period. The results of the assessment will usually include an estimate of the remaining life of the pavement (expressed in terms of the cumulative number of standard axles) to a predefined pavement condition (e.g. as in Table 16.1). The assessment should also produce recommendations for any necessary remedial works (e.g. repairing the surface or drainage defects) and, additionally, for any strengthening works (e.g. overlaying or edge haunching). The causes of the observed defects must be identified, to ensure that the recommended remedial and strengthening works are appropriate and compatible.

Structural assessments usually include deflection surveys. The surface deflection is regarded as a quantitative indicator of the strength of a flexible pavement at a given road location: when an appropriate analysis (see Section 16.4) is carried out, it is possible to give a reasonable estimate of the remaining life of the pavement. However, strengthening works are not normally designed from the results of deflection analysis alone, as other evidence is needed to define the condition of the pavement and foundation layers and to understand the likely failure mechanism.

The structural assessment procedure generally follows a stepped approach (Figure 16.1). This comprises a coarse (functional) condition survey of the road or network under investigation, followed by a more detailed survey of those road sections targeted by the

Figure 16.1 The stepped approach to structural assessment

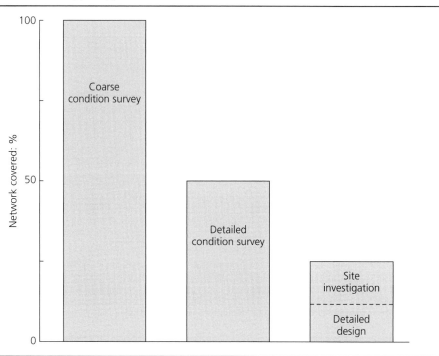

coarse survey as locations where remedial or strengthening works might be needed (i.e. where investigatory levels are breached); the detailed surveys are then undertaken at representative test lengths on the selected pavement sections.

16.3.1 Structural assessment procedure used on UK trunk roads

The structural assessment procedure currently recommended for trunk roads in the UK is summarised in Figure 16.2. The procedure is sequential, with progression to the detailed investigation dependent on the extent and severity of the pavement distress. Note, however, that the full process may not be necessary, depending on the needs of the project and the availability of relevant records.

Ideally, the network level surveys are undertaken annually using relatively coarse and economic techniques. This level of survey may identify a number of sections where the investigatory level is breached and a further detailed investigation is required. For example, network level surveys on English trunk roads are undertaken annually using vehicle-based TRACS and SCRIM (Sideways force Coefficient Routine Investigation Machine) equipment (see also Chapter 17). The output from TRACS includes the rut depth, texture depth, ride quality and intensity of cracking, while SCRIM gives an indication of the level of skid resistance. By comparing the results of these surveys with defined threshold values, it is possible to identify the road sections where further investigations are needed. Table 16.2 gives the condition categories for parameters measured

Figure 16.2 Overview of the structural assessment procedure for flexible pavements used in the UK, based on design standard HD 30 (HA *et al.*, 2008b)

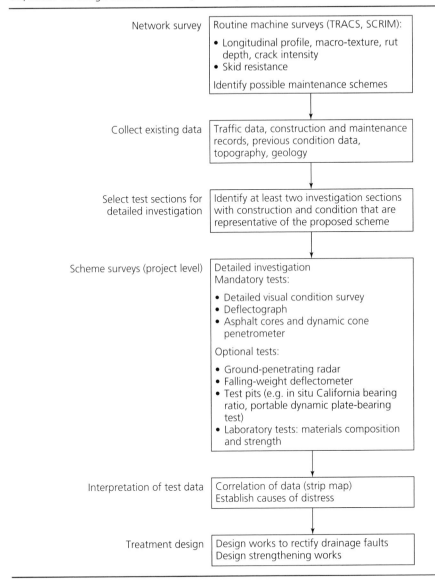

Network survey	Routine machine surveys (TRACS, SCRIM): • Longitudinal profile, macro-texture, rut depth, crack intensity • Skid resistance Identify possible maintenance schemes
Collect existing data	Traffic data, construction and maintenance records, previous condition data, topography, geology
Select test sections for detailed investigation	Identify at least two investigation sections with construction and condition that are representative of the proposed scheme
Scheme surveys (project level)	Detailed investigation Mandatory tests: • Detailed visual condition survey • Deflectograph • Asphalt cores and dynamic cone penetrometer Optional tests: • Ground-penetrating radar • Falling-weight deflectometer • Test pits (e.g. in situ California bearing ratio, portable dynamic plate-bearing test) • Laboratory tests: materials composition and strength
Interpretation of test data	Correlation of data (strip map) Establish causes of distress
Treatment design	Design works to rectify drainage faults Design strengthening works

by TRACS-type surveys; associated threshold values are given in design standard HD 29 (HA *et al.*, 2008a).

A detailed investigation is generally undertaken at project level, for those road sections where it has been shown that potential pavement problems exist. The main objectives of the detailed investigation are to establish the reasons for the condition of the pavement, and to collect sufficient data to facilitate the design of the improvement works so that appropriate remedial and strengthening measures may be applied.

At each test location, a programme of field tests is undertaken that should include deflection testing by an appropriate device (see Section 16.4). Associated measurements of rut depth and surface cracking should be made at each location.

A full inspection of the surface water drainage system is also recommended, including carriageway crossfall and gradient, gullies, manholes and catchpits. Examination of outfall pipes will indicate whether they are functioning correctly. Filter edge drains should also be inspected for excessive growth or detritus over the filter media, and, if necessary, a trial pit should be dug to expose the drainage pipe for inspection.

The thicknesses of the bound layers may be obtained by extracting cores (150 mm diameter): the cores can also be used to assess the condition of the asphalt layers and for laboratory testing. Ground-penetrating radar techniques offer a rapid means of assessing the layer thickness of bound layers, with only limited coring needed for calibration purposes.

The condition of the bound layers can be assessed by subjecting the cores to visual inspection and laboratory testing. The cores should be taken at locations exhibiting different surface defects (e.g. surface cracking or wheel-track rutting). Laboratory tests carried out on asphalt cores will generally include determination of the layer stiffnesses. The deformation resistance may be assessed by a wheel-tracking test, a cyclic compression test, or by comparing the asphalt layer thicknesses for a set of cores taken at close spacing across a wheel track. Tests on cement-bound materials should include the determination of their compressive strength.

Tests are also needed to confirm if tar is present in the existing asphalt layers, as some of the polycyclic aromatic hydrocarbons, phenols and cresols that may be present in tar are carcinogenic. In UK practice, when the proportions of these chemicals exceed specified threshold levels, any asphalt removed by planing is classified as a hazardous waste with consequent implications for its reuse in the works or disposal to landfill (ADEPT, 2013; EA, 2010). Recycling is possible (Merrill et al., 2004) by encapsulating the planings containing tar within cement-bound or cold-bitumen-bound materials (e.g. wet lean concrete, foamed bitumen), subject to Environment Agency agreement.

The condition of the granular layers and subgrade can be assessed using trial pits and associated laboratory testing such as particle size distribution, Atterberg limits, in situ moisture content and density, the dry density–moisture content relationship, and in situ California bearing ratio (CBR) determinations. In addition, the soaked (representing a saturated subgrade) CBR test, carried out on material compacted to 95–98% of the maximum dry density, will provide a lower estimate of subgrade strength.

An estimate of the relative strength of each granular layer and the subgrade can be obtained using the dynamic cone penetrometer (DCP). This device offers a relatively rapid means of identifying any weak layers beneath the bituminous surfacing. The DCP test is usually undertaken immediately after the extraction of an asphalt core (to assess the underlying unbound layers) or at the bottom of a trial pit (to assess the subgrade strength under thicker pavements). The test is suitable for most granular and weakly stabilised layers, but some difficulty may be experienced in penetrating materials with large particles, strongly stabilised layers or very dense crushed stone.

With favourable ground conditions the DCP test can provide information on the unbound pavement and foundation layer thicknesses and in situ CBR values. Details of the DCP and associated analysis are given in HD 29 (HA *et al.*, 2008a) and Overseas Road Note 18 (TRL, 1999).

Proper interpretation of the test data collected from the test sections should reveal the main contributions to the deterioration in the overall pavement strength. The survey data can be effectively presented in the form of a strip map showing the pattern of cracking and rutting in each 100 m length of carriageway, with other relevant data shown in graphical or tabular format below the strip map. Typical data would include deflection results, pavement layer thicknesses, the subgrade strength, the projected traffic over the design life and other relevant field and laboratory test results.

Current UK recommendations (HA *et al.*, 2008b) advise that overlay design should not be based on deflection results alone, and that surface defects and material condition should also be considered. It is often helpful to effect a quantitative structural analysis taking account of the relative condition of each layer to the overall pavement strength. Pavement model calibration may be based on measured deflection, adjusting the stiffness moduli to values appropriate to known environmental and loading conditions.

Differences in deflection measurements between test sections can indicate a change in the subgrade condition if there are no significant variations in the material properties of the pavement layers. The reasons for differences in deflection measurements between test sections (or variation in the deflection along a given section) should first be sought by comparing the information presented on the strip map. Higher deflections may be expected to coincide with areas of cracking or wheel-track rutting, but this is not always the case. The overriding aim is to explain the causes of the pavement distress and to design appropriate treatments. Deflections are used as an indicator of pavement performance with the knowledge that deflection–condition relationships (see Section 16.5) generally exhibit considerable scatter.

When the causes of the observed defects have been identified, consideration can be given to the design of any necessary treatment works. From a structural viewpoint, the timing of the works will be indicated by the condition of the road (confirmed by a site inspection) and any estimate of remaining life available. Additional factors such as existing and future traffic levels and the relative importance of the road will influence budget availability. However, for medium and heavily trafficked roads, whole-life costing invariably demonstrates that the remedial works needed to preserve the existing structure should not be delayed.

While structural overlay is generally the most efficient method of pavement strengthening, it may not be feasible at some locations due to road level constraints (e.g. where there is a central concrete barrier or, in urban areas, where kerbs cannot be raised). In these cases, inlay or reconstruction may be the only option. In some instances, reconstruction, or at least partial reconstruction, will be an option that is preferred to overlaying: for example, if only one lane of a multi-lane carriageway requires strengthening, and the additional cost of applying an overlay over extra lanes would be prohibitive. If the investigations indicate that there is a weakness in any of the pavement layers, it may be preferable to reconstruct to the bottom of the layer concerned rather than overlay the weakened structure.

In UK practice, the assessment of flexible pavements (termed 'flexible pavements with asphalt base' in current UK standards) and flexible composite pavements (now termed 'flexible pavements with hydraulically bound base') includes analysis of deflectograph data (see Section 16.6.4). For flexible pavements, a comprehensive treatment selection chart is given in HD 30 (HA *et al.*, 2008b), which was updated in IAN 158 (HA, 2012). Additional guidance for flexible composite pavements includes a condition classification based on the visual condition, deflection variability, crack severity, hydraulically bound mixture (HBM) strength, HBM condition (observed from cores) and surface modulus derived from falling weight deflectometer (FWD) measurements. Guidance on strengthening is given for four condition classes: the principal determining factor for each class is the type, extent and severity of cracking in the HBM base. A comprehensive treatment selection chart is given in TRL657 (Coley and Carswell, 2006), now also updated in IAN 158 (HA, 2012).

The assessment of rigid pavements includes the analysis of the results obtained from visual condition surveys, including the condition of joints and the load transfer efficiency between adjacent slabs, which may be assessed using the FWD. Maintenance and treatment options (Coley and Carswell, 2006; HA, 2012; HA and Britpave, 2001; HA *et al.*, 2008b) will depend on the condition of joints, the state of the concrete and the condition of the foundation.

16.4.   Use of deflection measurements

It is generally accepted that pavement deflections can be used as an indicator of the structural performance of a pavement; that is, the magnitude of the deflection under a known load can be related to the strength of the pavement layers and subgrade (HA *et al.*, 2008a; Kennedy and Lister, 1978; Lister, 1972). By comparison with other design parameters (e.g. strain in the pavement layers and subgrade), surface deflection can be measured relatively easily. With jointed concrete pavements, load transfer across joints can be assessed by loading the slab on one side of the joint and measuring the deflection at both sides of the joint.

In France and the UK, the deflectograph is widely used to measure deflections. This device is essentially an automated version of the deflection beam, both of which measure the surface deflection under a rolling wheel load. Alternatively, the FWD records the surface deflection profile under a stationary impact load. More recently, equipment for deflection measurement at traffic speeds has been under development: for example, the rolling-wheel deflectometer and the traffic-speed deflectometer (TSD).

The choice of the deflection measurement device and the operating procedure depends on the design method to be used to assess the remaining life and any strengthening measures. This is because design charts developed for deflections from a specific testing device generally cannot be used with deflections measured by other devices. In addition, design relationships developed for one country, particularly those developed empirically, are not necessarily applicable elsewhere. When a seemingly appropriate relationship is selected, it should therefore be calibrated to local conditions with considerable care.

16.4.1 Deflection beam, deflectograph and TSD

The deflection beam was originally developed by a Dr A. C. Benkelman in the USA. However, the variant used in the UK was designed by the Transport Research Laboratory (Kennedy, 1978; Kennedy et al., 1978): operating procedures appropriate for use in hot climates are described in the literature (Smith and Jones, 1980; TRL, 1999).

The deflection in the pavement that is measured by the deflection beam is induced by passing a loaded dual-wheeled axle past the beam tip. The load used on the axle varies from country to country (e.g. in the USA and UK it is commonly set at 6350 kg, whereas other users set it to 8160 or 10 000 kg). While the deflection–load relationship may be assumed to be linear, considerable care is required when comparing results under different axle loads.

There are two different ways of applying the axle loading. With one method the dual wheels are initially stationary at the beam tip, to give the point of maximum deflection, and, thereafter, the vehicle is moved away: this gives what is known as the rebound deflection (Asphalt Institute, 2000). With the other method, the dual wheels straddle the beam close to the pivot point at the start of the measurement cycle, and the dual-wheel assembly is then moved forward, passes the beam tip, to give the maximum deflection, and continues moving away from the beam: this gives what is termed the transient deflection (Kennedy, 1978). The two methods can yield very different results, particularly in hot climates where the bituminous layer changes from mainly elastic to viscous behaviour.

The Lacroix deflectograph, which was developed in France by the Laboratoire Central des Ponts et Chaussées, consists of two deflection beam-type mechanisms (one for each wheel track) that are mounted on a common frame located underneath a two-axle lorry. The deflectograph is driven along the road at a constant but slow speed of about 2.5 km/h, and measures the deflection in both wheel tracks at approximately 4 m intervals.

A correlation (Figure 16.3) has been established between deflections measured with the deflectograph and the deflection beam. The correlation is based on measurements on pavements containing crushed stone, cement-bound and bituminous-bound roadbases constructed on a range of subgrades with CBR values varying from 2.5% to 15.0%;

Figure 16.3 Correlation between deflections measured with the deflection beam and the deflectograph (Kennedy and Lister, 1978). (Reproduced by courtesy of the Transport Research Laboratory)

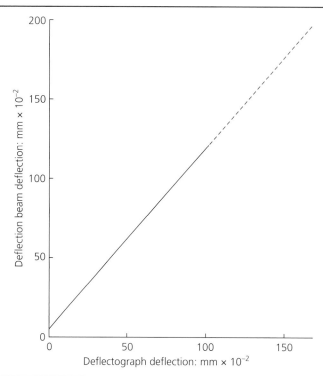

pavements with rolled asphalt, bitumen macadam and tarmacadam surfacings were also included, and the relationship was established within the temperature range 10–30°C. Figure 16.3 indicates that deflectograph deflections are generally less than those obtained with the deflection beam: this is because neither device gives an absolute measure of deflection, and the influence of the deflection bowl on the beam supports is greater with the deflectograph.

The Highways Agency TSD is a Danish-initiated device (formerly known as the high speed deflectograph) capable of measuring deflections in the nearside wheel path at speeds of up to 80 km/h. The TSD device, based on Doppler laser technology, was further developed for use as a routine survey tool capable of providing estimates of the structural condition of the UK's strategic road network (Ferne *et al.*, 2009a, 2009b). The TSD data will be used to define a network structural condition (NSC) category that will be reported in the Highways Agency pavement management system (HAPMS) for each 100 m length of road. The current approach to overlay design makes use of an equivalent deflectograph deflection that is derived from TSD measurements calibrated against a sample deflectograph survey; the development of a direct analysis approach is ongoing.

16.4.2   Falling-weight deflectometer

The FWD was originally designed in France, and introduced later in Denmark (Bohn *et al.*, 1972) and the Netherlands (Claessen and Ditmarsch, 1977; Claessen *et al.*, 1976; Shell IPC, 1978, 1985). With this equipment, an impulse load is applied to the road surface by dropping a weight onto a spring system; the weight and drop height are adjusted to give the required impact loading. The surface deflection of the pavement under the applied load is measured by geophones located directly under the loaded area and at several offset positions: this gives the surface profile of the deflection bowl in addition to the maximum deflection under the loaded area (Figure 16.4).

Considerable research effort has been spent seeking a reliable method of determining the pavement layer moduli from the FWD deflection bowl, using three- and four-layer pavement models. Empirical studies have shown that the outer FWD deflections (i.e. those furthest from the centre of loading) are influenced principally by the subgrade modulus. As the offset from the centre of loading decreases, the influence of the upper pavement layers increases, and the central deflection is a function of the moduli of all pavement layers and the subgrade (HA *et al.*, 2008a).

The analysis of FWD deflections (to determine likely weak pavement layers, estimate remaining life and, where required, the overlay design thickness) is normally based on analytical methods. The layer stiffnesses are derived by a back-analysis procedure based on the FWD deflection profile, ideally assisted by a comprehensive knowledge base and auxiliary testing of the constituent layers of the pavement (Evdorides and Snaith, 1996). Strain or deflection criteria are then used to estimate the residual life and any required overlay thickness.

Figure 16.4 FWD deflection bowl (based on HA *et al.*, 2008a). (Reproduced by courtesy of the Highways Agency)

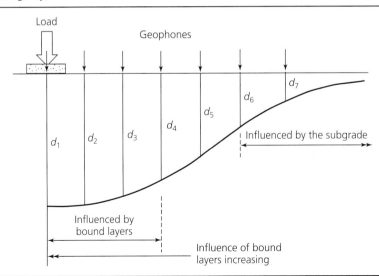

There is no direct correlation between FWD deflections and those measured with the deflectograph or deflection beam. Hence, FWD deflections cannot be used as input to design charts developed for these other measurement devices.

16.4.3   Effect of temperature on deflection

The influence of pavement temperature needs to be taken into account in all deflection-based design methods. The effect of higher temperature is to reduce the effective stiffness of the bituminous layer, resulting in higher deflections. (In the longer term, however, age hardening increases the effective stiffness and lessens the influence of temperature on measured deflections.) With some bituminous materials there is a tendency for the surfacing to flow plastically upward between the dual tyres of the deflectograph (or deflection beam vehicle) as the temperature increases. In practice, therefore, it is best to restrict deflection measurement to those months of the year and times of the day when temperatures are relatively low and the pavement response can be considered elastic. Also, it is necessary to establish a relationship between the measured deflection and the pavement temperature (normally measured at a depth of 40 mm) so that all deflections can be 'corrected' to that which would occur at a standard reference temperature.

In jointed concrete pavements, the effect of temperature increase is to close up the joints, which results in an overall reduction in the deflection at joint locations. A similar effect can be expected in flexible composite pavements, where the cement-bound layer generally has regular transverse cracks.

The need for a temperature correction can be assessed by taking repeated deflection measurements at several test positions as the pavement heats up during the morning of testing. For each position, a deflection–temperature relationship is plotted, from which a correction to a reference temperature can be deduced. A reference temperature of 20°C is normally adopted for moderate climates; 35°C has been suggested (Smith and Jones, 1980; TRL, 1999) as appropriate for tropical and sub-tropical conditions.

16.5.   Use of deflection–life relationships

For most pavement types it is possible to define a relationship between deflection (under a specific loading regime), traffic damage (expressed in cumulative standard axles carried) and structural condition (usually defined in terms of surface cracking and rutting). The main features of such a relationship (Figure 16.5) are (a) a deflection trend line that shows how the deflection varies as the cumulative traffic increases (i.e. as the damage increases) and (b) an envelope criterion curve plotted through points of similar condition at different levels of cumulative traffic. The relationship is based on the assumption that, at a given level of cumulative traffic damage, higher deflections will be recorded as the structural condition worsens.

By plotting the position – point P in Figure 16.5 – corresponding to the current deflection level and the traffic carried to date (N_P, in cumulative standard axles), the remaining life (N_R) to the condition defined by the criterion curve can be estimated by following the trend line. The criterion curve is used with certain deflection-based design procedures (Asphalt Institute, 2000; Kennedy and Lister, 1978) to define the limiting deflection

Figure 16.5 Elements of a deflection–life relationship (based on Lister, 1972)

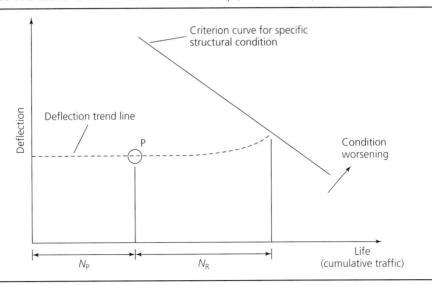

values for overlay design or new pavement design, based on the traffic anticipated during the desired design life.

Various relationships of this type have been prepared for deflection data stemming originally from the Benkelman beam (Asphalt Institute, 2000; Kennedy and Lister, 1978; Lister, 1972; Norman *et al.*, 1973). In the more recent variants, probability has been introduced, to increase the confidence with which the pavement engineer may use the system. These relationships are usually obtained from field-based observations of varying robustness that are specific to the device used for deflection measurement, its method of operation, and the materials and environment for which the relationships were originally derived. Hence, considerable care is required when assessing the suitability of a relationship in anything other than its original specification.

To illustrate how such methods have been derived and are used, the principles of the current UK process (Kennedy and Lister, 1978) are described here: these relationships were developed for deflections measured with the deflection beam. It should be noted, however, that the variant currently recommended for use on UK trunk roads (HA *et al.*, 2008a, 2008b) makes use of deflections measured with the deflectograph. In this context, note also that flexible pavements are broadly classified as having either a determinate life (i.e. a finite life) or a long life, based on current condition and asphalt thickness.

16.5.1 Deflection–life relationships for pavements with a determinate life

The development of deflection histories for UK pavements has involved regular monitoring of pavement deflection and the associated condition. Road condition has been classified as sound, critical or failed, based on the severity and extent of wheel-track rutting and wheel-track cracking.

Four performance charts have been published (Kennedy and Lister, 1978), based on the deflection trends observed from construction to the onset of 'critical' conditions. Different charts have been produced for pavements with granular, cement-bound and bituminous roadbases, and a further distinction has been made between granular bases that develop some form of cementing action and those that do not. The relationships are reported to be valid for flexible pavements with subgrade CBRs in the range 2.5–15% and surfaced with bituminous materials commonly used in the UK. The charts (e.g. see Figure 16.9) indicate deflection trend lines that lead to four envelope curves defining different levels of probability that the 'critical' life of the pavement will be achieved. As in the previously described generic system (see Figure 16.5), the remaining life is determined by following the deflection trend line from a point defined by the present deflection and past traffic, in cumulative standard axles, to the envelope of the selected probability level for the particular construction type.

It has been suggested (HA *et al.*, 1994a) that the life of a flexible pavement can be characterised as having four main phases (Table 16.3) that can be identified in a deflection–life relationship. During the early life of the pavement (phase 1), the deflection may decrease due to compaction of the pavement layers and also to moisture content changes in the granular layers and subgrade. Phase 2 is represented by a horizontal section of the deflection trend line, leading to a gradual increase in the deflection at the investigatory condition (phase 3), where it is likely that cracks are being initiated, with a more rapid deflection increase at phase 4. Investigation is needed at phase 3 so that the cause of the deterioration can be determined before the design of strengthening works.

Table 16.3 Phases in the design life of a determinate pavement

Phase	Structural behaviour
1	New or strengthened pavement is stabilising. Deflection is variable but generally decreasing
2	Structural behaviour may be predicted with confidence. Deflection remains stable
3	Structural behaviour becomes less predictable. Deflections may continue as phase 2, gradually increase or increase more rapidly. Formerly known as the 'critical' condition (Kennedy and Lister, 1978), the term 'investigatory' is now used to emphasise the need to monitor structural behaviour
4	Pavement deteriorates to a failed condition from which it can be restored only by total reconstruction

Based on HA *et al.* (1994a)

For pavements with asphalt bases, the oxidation of bitumen throughout the asphalt layer results in increasing stiffness and a reduction in deflection (Croney and Croney, 1991; Kerali *et al.*, 1996; Merrill, 2005). Where this occurs, it may be that phase 1 'spills over' into phase 2.

16.5.2 Deflection–life relationships for long-life pavements

Later evidence suggests that thick flexible pavements may not conform to the phased behaviour of Table 16.3. A study (Nunn, 1998) of ten heavily trafficked motorway sections in the UK showed a trend of decreasing deflection with age and traffic: at three sites, the trend of decreasing deflection continued well beyond the design life. The sections under study were of flexible construction and had carried up to 56 msa.

Thick pavements that show a trend of decreasing deflection with age and traffic have been termed 'long life'. In UK practice, the analysis of deflectograph data (HA *et al.*, 2008b) determines if a pavement has a long life, based on the thickness of asphalt (300 mm or greater) and a standard deflection. Further information on the concept of long-life flexible pavements in UK practice is available in the literature (Merrill, 2005; Nunn *et al.*, 1997).

There is still a need for the continuous monitoring of pavements, to observe structural behaviour under increasing traffic levels. The use of accelerated testing methods (Brown, 1998) could well assist in predicting such behaviour for different pavement constructions.

16.5.3 Site-specific relationships

As noted previously, published deflection–life relationships generally do not have application outside the region for which they were developed. However, a relationship can be established for any given road, based on a condition-monitoring programme at selected test sites.

The test sites should be selected to represent the full range of pavement condition (including no defects), traffic loading, construction thickness, subgrade strength, topography and elevation (i.e. cut/fill). Condition data collected periodically at each test site (Table 16.4) should include the deflection and the associated structural condition assessed in terms of surface cracking and wheel-track rutting. The test sites should be monitored at least two or three times each year to assist detection of seasonal variations in the measurements taken.

An accurate assessment of traffic loading (in terms of cumulative standard axles carried) is needed at each test site. Historic loading data may be available from records or previous road study reports; failing that, it may be necessary to project back in time from current survey data. The current traffic loading should be assessed by regular manual classified counts and axle load surveys (TRL, 2004): this will facilitate the assessment of the number of commercial vehicles (cv) and the commercial vehicle wear factor (i.e. the number of equivalent standard axles per cv, ESA/cv). In the longer term, this will also assist in verifying the annual vehicle growth rate.

Table 16.4 Data required at each test site to establish a site-specific deflection–life relationship

Traffic loading	Historic traffic data
	Manual classified counts
	Automatic traffic counts
	Annual vehicle growth rates for each vehicle class
	Axle load data (ESA/cv)
Surface condition	Deflection under specified load
	Verified deflection–temperature relationship
	Surface cracking (type and intensity)
	Rut depth

Table 16.5 A condition classification in terms of crack intensity

Crack index	Crack intensity: m/m^2	Condition
0	No crack	Sound
1	0–2	Critical
2	>2	Failed

Based on McMullen et al. (1990)

The structural condition should be classified according to the severity of the cracking and rutting in the vicinity of the deflection test position. While the classification of Table 16.1 (from UK practice) may be suitable in this respect, it may be more appropriate to adopt a classification that measures crack intensity alone, if this is the major mode of failure: this is shown in Table 16.5, where the crack intensity is defined as the length of crack per square metre. Alternatively, Table 16.6 gives a classification where the crack type is shown as a proxy for the measured intensity.

Table 16.6 A condition classification in terms of the crack description and the rut depth (Snaith and Hattrell, 1994)

Cracks		Ruts	
Index	Description	Index	Depth: mm
0	No crack	0	No rut
1	Single crack	1	1–5
2	Non-interconnected	2	6–10
3	Interconnected	3	11–15
4	Block	4	16–20
5	Disintegration	5	over 20

Pavement condition index = crack index + (rut index − 1*)
Note: * When the rut index is ≥1

Figure 16.6 Development of an idealised deflection–life relationship

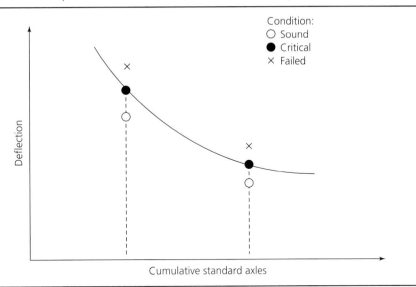

For each test site, the mean deflection corresponding to each condition classification should be plotted against the number of cumulative standard axles carried. Ideally, this will result in a sufficient range of deflection levels at which the road can be classified (e.g. as sound, critical or failed) (Figure 16.6).

However, in practice there may be a considerable scattering of results, in which case a set of criterion curves will need to be established to represent different levels of

Figure 16.7 Effect of overlay application on position within a deflection–life relationship

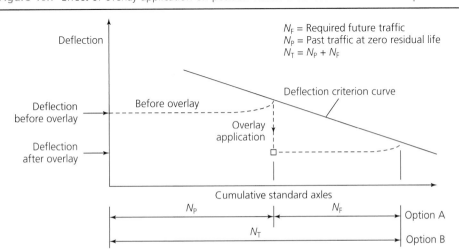

probability that a given life can be achieved before the onset of the defined 'critical' conditions.

In the longer term, the deflection data from the test sites should be used to establish deflection trend lines from the time of new construction to the onset of the predefined limiting condition. In addition, the influence of strengthening overlays should be assessed. The philosophy of the Transport Research Laboratory's LR833 report (Kennedy and Lister, 1978) suggests that a pavement may be treated 'as new' following a structural overlay (i.e. the deflection after overlaying becomes the early life deflection), and the overlay is designed for future traffic (N_F), as shown for option A in Figure 16.7. A more conservative approach (Snaith and Hattrell, 1994) would be to design overlays on the basis of the sum of past traffic at zero residual life (N_P) and required future traffic (N_F), as for option B in Figure 16.7.

16.6.   Overlay design methods for flexible pavements

Overlay design methods for flexible pavements can be broadly classified (Finn and Monismith, 1984) according to whether they are based on component analysis, on non-destructive empirical procedures or on analytically based procedures.

Early component analysis design methods involved comparing the existing pavement structure with a new pavement design, with the difference giving the estimate of required overlay thickness: examples of this approach are the Asphalt Institute's effective thickness procedure (Asphalt Institute, 2000) and the American Association of State Highway and Transportation Officials (AASHTO) structural number concept (AASHTO, 1993), both of which require an assessment of the condition of the individual layers in the existing pavement. Current non-destructive empirical design procedures include deflection-based methods such as the deflectograph analysis used in the UK (HA et al., 2008b) that was developed from the Transport Research Laboratory method (Kennedy and Lister, 1978). Analytically based procedures may also use deflection testing, and assume that the pavement can be analysed as a multi-layer elastic system; procedures based on the FWD are included in this category. Analytical procedures are also applicable to the design of concrete overlays on flexible pavements.

The different overlay design methods are demonstrated by the following examples.

16.6.1   Asphalt Institute effective thickness procedure

With the Asphalt Institute method (Asphalt Institute, 2000), each layer in the existing pavement is converted to an equivalent thickness of 'asphaltic concrete' by applying a conversion factor, based on the condition of the layer. The sum of the equivalent thicknesses for each layer is subtracted from the design thickness (Asphalt Institute, 2008) of a new full-depth asphalt pavement (designed for the future traffic loading), to give the required overlay thickness. The procedure is illustrated by the following example.

Problem. Assume that the construction details for an existing flexible pavement are as shown in Table 16.7. Determine the required asphaltic overlay thickness for a future life of 4 msa. The mean annual air temperature (MAAT) is 15.5°C.

Table 16.7 Details of an existing flexible pavement – overlay example using the effective thickness procedure

Layer	Thickness: mm	Material	Condition	Conversion factor[a]
1	100	Asphaltic concrete	Block cracking	0.5
2	150	Cement-stabilised base	Well graded, PI $= 4$	0.3
3	150	Granular sub-base	Well graded, PI $= 6$	0.2
4	–	Subgrade	Resilient modulus $= 60$ MPa	–

PI, plasticity index
[a] From MS-17 (Asphalt Institute, 2000)

Solution. From Figure 16.8, the design thickness of a new full-depth asphalt concrete pavement, T_n, is 290 mm, based on future traffic of 4 msa and a subgrade modulus of 60 MPa. The effective thickness of the existing pavement, T_e, is given by

$$T_e = (100 \times 0.5) + (150 \times 0.3) + (150 \times 0.2) = 125 \text{ mm}$$

Thus, the design overlay thickness, T_o, is

$$T_o = T_n - T_e = 290 - 125 = 165 \text{ mm}$$

16.6.2   Overlay design based on the AASHTO structural number

With this component analysis design method, the AASHTO structural number (SN) of a pavement is empirically related to the thickness and condition of the constituent layers

$$\text{SN} = a_1 D_1 + a_2 D_2 m_2 + a_3 D_3 m_3 \tag{16.1}$$

where a_1, a_2 and a_3 are layer strength coefficients and D_1, D_2 and D_3 are the thicknesses (in inches) of the surfacing, roadbase and sub-base, respectively, and m_2 and m_3 are drainage coefficients for the roadbase and sub-base, respectively. The basis for overlay design is to compare the design structural number that is required for future traffic loading (SN_f) with the effective structural number of the existing pavement (SN_{eff}), and to make up the difference with an asphalt concrete overlay: the design overlay thickness is then given by

$$\text{overlay thickness (in.)} = (\text{SN}_f - \text{SN}_{eff})/a_{ol} \tag{16.2}$$

where a_{ol} is the layer coefficient for the overlay material.

SN_f is obtained from the AASHTO design chart based on future traffic loading requirements. SN_{eff} may be estimated using Equation 16.1, with the coefficients a_1, a_2, a_3, m_2 and m_3 assigned on the basis of the visual condition and drainage characteristics of the

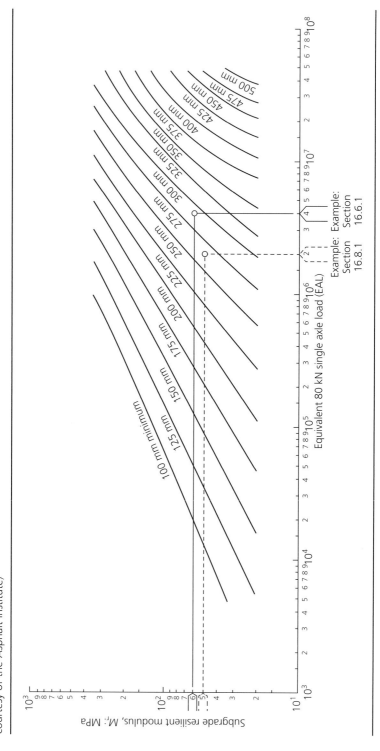

Figure 16.8 Asphalt Institute design chart for full-depth asphalt concrete pavements where MAAT = 15.5°C (Asphalt Institute, 2008). (Reproduced by courtesy of the Asphalt Institute)

existing layers. Suggested layer coefficients are included in the comprehensive AASHTO design guide (AASHTO, 1993).

Note also that the latest AASHTO design method is detailed in the *Mechanistic–Empirical Pavement Design Guide* (AASHTO, 2008): in the 2000 guide, this design procedure uses performance and distress prediction models to develop the required pavement thickness. A companion document (AASHTO, 2010) gives guidance on calibration to local conditions, policies and materials.

16.6.3 Transport Research Laboratory method

This non-destructive empirical design method is included to demonstrate the principles of the current deflectograph method recommended for use in the UK (see Section 16.6.4). The Transport Research Laboratory method is based on deflections measured with the deflection beam; with the variant currently adopted for UK trunk roads (see Section 16.6.4), deflections are input as deflectograph values.

The method developed by the Transport Research Laboratory (Kennedy and Lister, 1978) is based on deflections measured with the deflection beam, under a 3175 kg wheel load, and corrected to a standard temperature of 20°C. For overlay design purposes, a representative deflection is assigned to sections of road (i.e. design lengths that exhibit uniform deflection levels). The design lengths, which should have consistent construction and traffic levels, can be identified from a study of the deflection profile, which shows the individual deflections for the road pavement under investigation.

In UK practice, the representative deflection is generally taken as the 85th percentile value, which is used in association with a 0.5 probability criterion curve (e.g. see Figure 16.9) to give an estimate of the remaining life of the pavement, in terms of cumulative standard axles. (The 85th percentile is that value of deflection within a design length at or below which 85% of the deflections are found.) If the predicted remaining life is less than that required, a strengthening overlay will need to be considered: if this is practical (i.e. in terms of possible physical constraints such as kerb levels and bridge clearances), the required overlay thickness is determined by reference to an appropriate design chart (e.g. Figure 16.10). Note that Figure 16.10 specifies the thickness of rolled asphalt overlay needed to give the desired future life, after taking into account the remaining life of the existing pavement.

The Transport Research Laboratory procedure is illustrated by the following example.

Problem. Consider a flexible pavement with construction details as shown in Table 16.8. The highway has carried 3 msa since being newly constructed, and appears to require major maintenance. Determine the overlay requirements of a design length having an 85th percentile deflection beam deflection (corrected to 20°C) of 48 mm $\times$ 10^{-2}, in order to give a future life of 10 msa with a 0.50 probability of achieving that life.

Solution. The pavement is classified as having a bituminous roadbase, as the existing pavement has more than 150 mm of bituminous materials; this defines the appropriate design charts to be used.

Figure 16.9 Relation between standard deflection and life for pavements with bituminous roadbases (Kennedy and Lister, 1978). (Reproduced by courtesy of the Transport Research Laboratory)

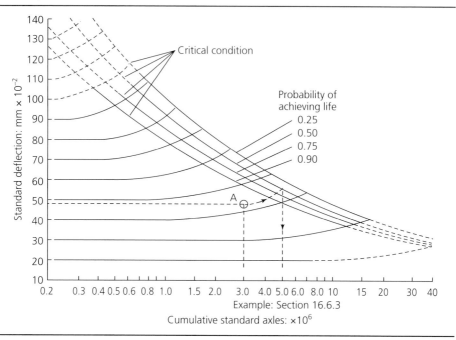

Figure 16.10 Overlay design chart (0.50 probability) for pavements with bituminous roadbases (Kennedy and Lister, 1978). (Reproduced by courtesy of the Transport Research Laboratory)

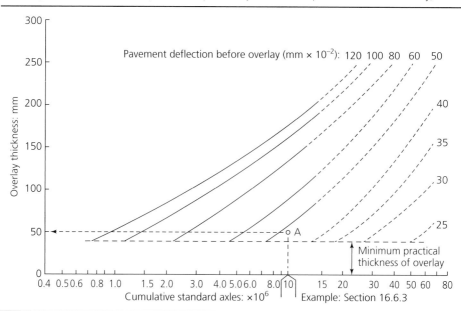

Table 16.8 Details of an existing flexible pavement – overlay example using the Transport Research Laboratory deflection method

Pavement course	Thickness: mm	Details
Surfacing	100	Rolled asphalt
Base	150	Asphaltic concrete
Sub-base	200	Crushed rock
Subgrade	–	CBR = 6%

First, determine the remaining life of the pavement section, by plotting the position corresponding to the 85th percentile deflection (48 mm $\times$ 10^{-2}) and the past traffic (3 msa) on the deflection–life relationship for pavements with bituminous roadbases (e.g. point A in Figure 16.9). Following the deflection trend line to the 0.50 critical curve, this gives a life of 5 msa: thus, the remaining life of the existing pavement section is $5 - 3 = 2$ msa. An overlay is therefore needed to achieve the desired life of 10 msa.

Next, determine the required overlay thickness of rolled asphalt from Figure 16.10, which is the overlay design chart for pavements with bituminous roadbases (0.5 probability). This figure shows that, for a future design life of 10 msa, an existing pavement with an 85th percentile deflection of 48 mm $\times$ 10^{-2} requires an overlay of 50 mm of rolled asphalt.

16.6.4 The deflectograph method used in the UK

The overlay design approach currently adopted for UK trunk roads (HA et al., 2008a, 2008b) makes use of peak deflections measured with the deflectograph under a 3175 kg wheel load, corrected to a standard temperature of 20°C. The deflection data are processed centrally within the HAPMS. Within HAPMS (HA, 2009), the road network is referenced by predefined sections; for overlay design purposes, the results of the deflectograph analysis are reported for every 100 m subsection length.

The analysis procedure for flexible and flexible-composite pavements is summarised in Figure 16.11. First, the standard deflection and the total thickness of bituminous material (TTBM) are used to determine if the pavement has the potential for a long life (Figure 16.12). The standard deflection is typically the 85th percentile of the deflectograph deflections recorded over a 100 m length, corrected to a standard temperature of 20°C. The TTBM is the combined thickness of all the contiguous intact asphalt layers in the existing pavement, and includes the asphalt surfacing layers (i.e. the top 100 mm of the existing pavement), regardless of their condition.

The pavement life categories derived from the deflection and thickness criteria are

- Potential long-life pavement (LLP): where the existing TTBM is at least 300 mm and future maintenance is expected to be resurfacing. This assumes timely maintenance will be undertaken to ensure that any surface deterioration is not allowed to compromise the pavement, and that drainage is maintained.

Figure 16.11 The deflectograph analysis method recommended for use in the UK, based on HD 29 (HA *et al.*, 2008a), HD 30 (HA *et al.*, 2008b) and TRL639 (Merrill, 2005)

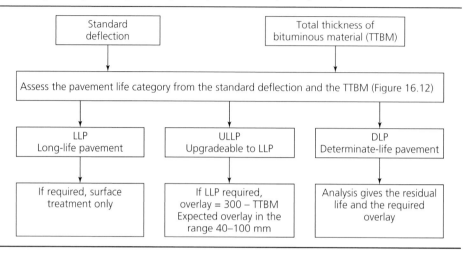

Figure 16.12 Pavement life categories from HD 30 (HA *et al.*, 2008b). (Reproduced by courtesy of the Highways Agency)

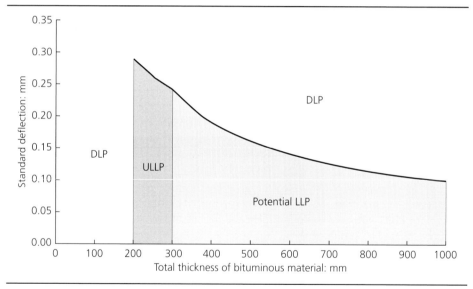

■ Upgradeable to long-life pavement (ULLP): where the TTBM is between 200 and 300 mm, and a moderate overlay will achieve long-life status, if that is required for the expected traffic loading.

■ Determinate-life pavement (DLP): where further investigation/analysis is needed to estimate the residual life to investigatory condition; if required for future traffic loading, the overlay thickness can also be determined, based on the measured deflections, the existing pavement type, the date of the last major treatment and the traffic loading. The existing pavement type is defined according to the roadbase material, namely unbound granular base with no cementing action (GNCA), flexible with hydraulically bound base (CEMT) and flexible with asphalt base (BITS).

For each 100 m subsection, the output includes the percentage of deflection results classified within each life category; for those deflections that are classified as DLP, the residual life (in years) and overlay required for a defined future life (normally a further 20 years) are determined. Caution is advised when interpreting the results, as a 100 m subsection may have overlay recommended based on a few DLP results while the remaining results within the subsection are LLP. The overlay thickness provided is for a traditional DBM125 material. If stiffer overlay materials are used, some reduction in the overlay thickness is possible for fully flexible pavements only.

Overlay design should not be based on deflection analysis alone; that is, the existing condition of the pavement layers and foundation materials should be considered. Hence, all pavement categories (including LLP and ULLP) should be investigated to ensure that defects are treated: this may involve local structural patching to the depth of defects such as rutting and cracking, or complete removal of the upper bituminous layers before replacement and application of the overlay.

16.6.5   Analytically based methods

Analytically based methods of overlay design generally make use of a layered structural analysis and the associated design criteria (e.g. deflection–life or strain–life relationships) appropriate to observed pavement behaviour. The basic components of this approach (McMullen et al., 1990) are as follows:

(a) select a suitable method of elastic analysis that is capable of applying a single or dual wheel load
(b) represent the existing pavement and subgrade by a layered elastic system in which each layer is characterised by an elastic modulus (E) and Poisson's ratio (v), as shown in Figure 16.13(a)
(c) calibrate the pavement model to existing site conditions (e.g. adjust the layer properties, within limits, until computed deflections/strains agree with measured values)
(d) calculate the critical parameters (deflection, strains) with increasing overlay thicknesses (Figure 16.13(b) shows a typical pavement model)
(e) use a deflection–life or strain–life relationship that is valid for the site conditions to estimate the future life given by each overlay increment
(f) present the results of the analyses in the form of overlay design charts, giving the overlay thickness required to achieve different values of future life.

The Shell design charts (Claessen and Ditmarsch, 1977; Shell IPC, 1985) are an early example of analytically based design charts produced for worldwide application. The more recent AASHTO mechanistic–empirical method (AASHTO, 2008) is based on an analytical pavement model that may be calibrated to different environments.

Figure 16.13 Typical flexible pavement models for elastic analysis: (a) an existing pavement; (b) an overlaid pavement

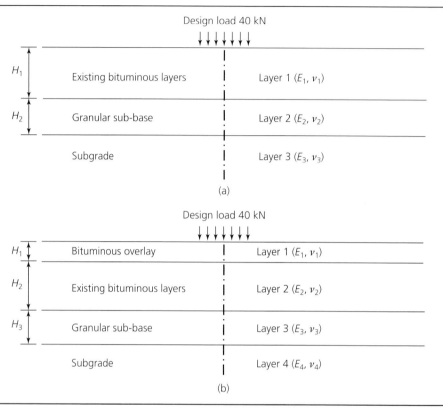

The analytical approach may also be used to extend (in terms of the number of cumulative standard axles) design charts produced by empirical means over a limited range of traffic loading.

An analytical overlay design method that is based on FWD deflections makes use of the full deflection profile and available layer thickness information, to assign layer properties, namely the elastic modulus (E) and Poisson's ratio (v), to an idealised elastic model of the pavement (Figure 16.13) in order to estimate the remaining life and to design the overlay requirement. The main elements of the FWD procedure (Figure 16.14) can be used for both flexible and rigid pavements, provided that appropriate design criteria, calibrated to local conditions, are available. In this case, information on existing layer thicknesses is usually obtained from asphalt cores, DCP tests and trial pit inspections.

The evaluation of the pavement layer moduli generally relies on a back-analysis procedure whereby the modulus value for each layer of the model is adjusted (within specified limits) until a close correlation is achieved between the deflection bowl measured under the FWD loading and that computed for the pavement model. In practice, FWD analysts may use

Figure 16.14 Overview of the FWD analytically based design procedure

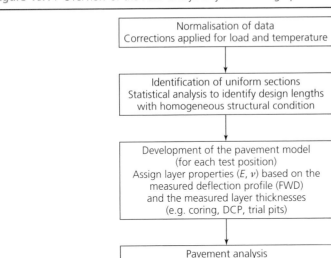

one or more software packages (proprietary or written to specification) to evaluate the moduli. However, sound engineering judgement is needed when using back-analysis software, as its main function is to introduce efficiency, logic and organisation to the evaluation procedure.

One early study (Rwebangira *et al.*, 1987) that compared the results produced by different back-analysis programs (all of which used curve-fitting procedures) found that the different programs gave significantly different results and that most methods were highly sensitive to the assumed layer thicknesses. Current UK advice (HA *et al.*, 2008a) states that a 15% underestimate of the thickness of a bound layer can result in a 50% overestimate of the modulus of that layer. If ground radar techniques are used to obtain thicknesses, regular physical measurements are still required for calibration purposes. It cannot be overemphasised that accurate measurement and the location of cores, DCP tests, trial pits and so on is essential. Also, whatever the analysis software, it is useful to undertake a sensitivity study (McMullen *et al.*, 1986) to assess the influence of variations in all input data on the critical output parameters.

As noted previously, the moduli assigned to each layer in the pavement model will have boundary limits imposed by the material type and its condition. In addition, the method of elastic analysis adopted may impose further conditions (e.g. by specifying a minimum layer thickness in terms of the radius of the applied circular wheel loading), or by specifying a minimum modular ratio between adjacent layers. The analysis process, therefore, comprises an analytical procedure, the results of which are modified in the light of engineering experience.

16.7. Concrete overlays for flexible pavements

Concrete overlays can be used as a strengthening measure for flexible pavements, provided that the existing pavement foundation is in good condition. In addition to the structural considerations, there has to be a minimum practical overlay thickness to allow for the use of steel reinforcement and appropriate cover.

The use of thick concrete overlays is suggested (HA *et al.*, 2008b) for flexible pavements and for most types of rigid pavements that would otherwise require reconstruction. In these cases, the existing pavement becomes the foundation for the new structure.

Continuously reinforced concrete pavement (CRCP) and continuously reinforced concrete base (CRCB) are the preferred types of rigid construction in the UK, and these are permitted as overlay materials. The use of unreinforced concrete or jointed reinforced concrete as overlay materials is not recommended in UK practice.

The overlay design procedure using CRCP or CRCB makes use of the design chart for new rigid pavements in HD 26 (HA *et al.*, 2006a). The chart indicates the thickness of CRCP or CRCB as a function of the required pavement life (expressed in millions of standard axles carried in the 'slow' lane), the flexural strength of the concrete, and the foundation class. The surface modulus of the pavement to be overlaid is used in place of the foundation class shown on the chart: thus, the representative values of surface modulus used for each foundation class are 100 MPa (foundation class 2), 200 MPa (foundation class 3) and 400 MPa (foundation class 4).

The surface modulus at the top of an existing pavement may be estimated from FWD measurements using the relationship (HA *et al.*, 2008a)

$$E_0 = 2(1 - v^2)\sigma_0 a/\delta_0 \tag{16.3}$$

where E_0 is the surface modulus at the top of the existing pavement (MPa), v is Poisson's ratio (=0.35), σ_0 is the contact pressure under the loading plate (kPa), a is the radius of the loading plate (mm) and δ_0 is the deflection at the centre of the loading plate (μm).

The CRCP overlay design procedure can be demonstrated by considering an existing flexible pavement that has a 15th percentile surface modulus value of 200 MPa: this represents the value exceeded by 85% of the values recorded for the treatment length. The design traffic is 140 msa, and reference to the design chart for new rigid pavements (Figure 16.15) shows that, for concrete with a flexural strength of $f_f = 4.5$ MPa, the required thickness of the CRCP overlay is 210 mm; assuming the pavement does not have a 1 m edge strip or a tied shoulder, the CRCP design thickness is increased by 30 mm to allow for edge loading. Finally, a bituminous surface course is added to satisfy UK requirements for surface course materials given in HD 36 (HA *et al.*, 2006b); in this case, a clause 942 thin surface course system is adopted (HA *et al.*, 2014).

From HD 26 (HA *et al.*, 2006a), the required longitudinal reinforcement is 0.6% of the concrete slab cross-sectional area, giving 1440 mm^2/m width for the 240 mm thick slab,

Figure 16.15 CRCP overlay design based on HD 26 design chart for new rigid pavements (HA *et al.*, 2006a). (Reproduced by courtesy of the Highways Agency)

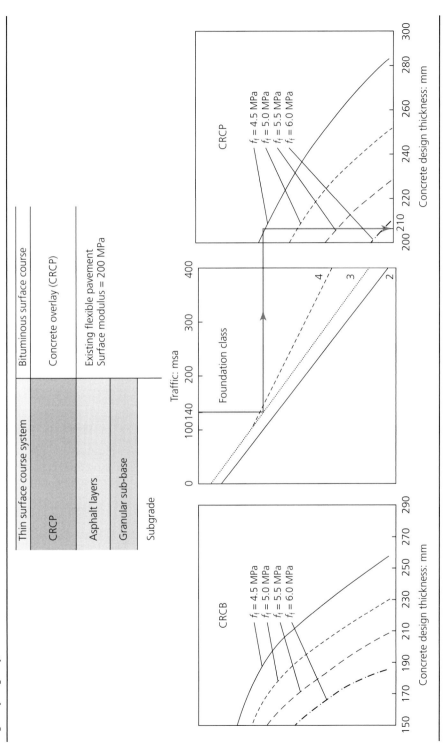

equivalent to T16 bars at a maximum spacing (centre-to-centre) of 139 mm. The required transverse reinforcement is T12 bars at 600 mm spacing.

16.8.   Overlay design methods for concrete pavements

Overlays on concrete pavements can be applied as a bituminous overlay or as a concrete overlay (overslabbing). The overlaying of concrete pavements generally presents greater problems than is the case with bituminous pavements, due to the presence of joints or wide cracks in the underlying concrete. With bituminous overlays, this can result in reflection cracking, where the cracking pattern in the existing concrete pavement propagates through the overlay material, and produces a similar pattern at the surface of the overlay. With concrete overlays, the new slab is required to accommodate differential movement across any existing joints or cracks; this is achieved (HA *et al.*, 1994b) by forming joints in the overlay at the same location as the existing joints or cracks in the underlying concrete pavement, or by separating (i.e. de-bonding) the overlay from the existing pavement.

With all overlay options, the existing concrete pavement must provide a stable platform of uniform strength to receive the overlay. As with bituminous pavements, drainage failures can lead to weakening of the lower unbound layers and subgrade, and, consequently, it is important that any drainage faults are rectified before assessing the overlay requirements. Thus, any loose or rocking slabs should be stabilised or replaced, any vertical movements at joints or cracks should be stabilised either by pressure grouting or vacuum grouting (HA and Britpave, 2001; Mildenhall and Northcott, 1986; TRRL, 1978) and, where necessary, surface levels should be regulated, and any spalling rectified at joints.

16.8.1   Bituminous overlays on rigid pavements

There is no uniformly accepted method for assessing the thickness of a bituminous overlay required to strengthen a rigid pavement. In addition to structural considerations, the overlay thickness will be governed by the need to delay the development of reflection cracking in the overlay.

Reflection cracking generally refers to a cracking pattern that appears at the surface of a bituminous overlay due to differential vertical movement under wheel loading, across joints or cracks in the underlying concrete layer. The surface cracks can take several years to develop, but they will eventually reflect the pattern of cracking in the underlying layer. If the differential vertical movement at joints/cracks is kept to a minimum, there is a lower probability of reflection cracking developing: thus, the Asphalt Institute advises that the mean deflection (as measured with the Benkelman beam) under a 4080 kg wheel load should not be more than 0.36 mm, and that the differential deflection should be limited to 0.05 mm.

Current UK guidance on strengthening concrete pavements (Coley and Carswell, 2006; HA *et al.*, 2008b) provides minimum overlay thicknesses associated with specific treatment options. The treatment options proposed (Coley and Carswell, 2006; HA *et al.*, 1994b; Jordan *et al.*, 2008; Sherman, 1982) include the following (a) to (f):

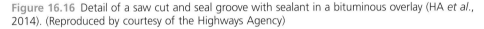

Figure 16.16 Detail of a saw cut and seal groove with sealant in a bituminous overlay (HA *et al.*, 2014). (Reproduced by courtesy of the Highways Agency)

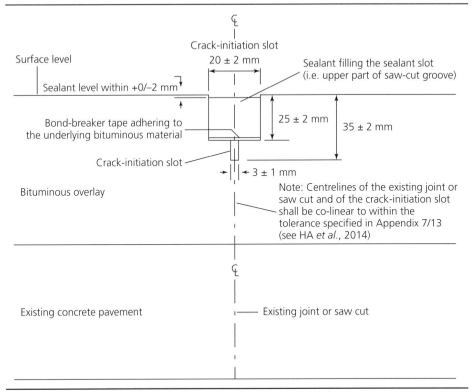

(a) Using a 'crack and seat' process whereby fine vertical transverse cracks are induced in the existing concrete slab before it is overlaid with asphalt. This reduces the load-spreading capability of the slab, but assists in controlling the development of reflection cracking in the overlay by two mechanisms, namely by reducing the horizontal strain level between adjacent concrete elements and by minimising the relative vertical movement between concrete elements under traffic loading. Following cracking and seating, the effective stiffness of the cracked concrete layer is assessed from FWD measurements; a minimum threshold stiffness is needed, previously determined by analytical design that takes into account the asphalt overlay thickness and future traffic requirements. Details of UK practice for 'crack and seat' are given in Series 700 of the *Specification for Highway Works* (HA *et al.*, 2014). A minimum asphalt overlay thickness of 150 mm is recommended (HA *et al.*, 2008b; Jordan *et al.*, 2008).

(b) Using an asphalt overlay with 'saw cut and seal', which involves creating a 'joint' in the asphalt overlay by cutting a slot in the asphalt above each joint in the concrete, inserting bond-breaker tape, and then applying a sealant. Figure 16.16 shows a detail from Series 700 of the *Specification for Highway Works* (HA *et al.*, 2014). It is reported (Coley and Carswell, 2006) that even with very accurate alignment of the saw cut in the asphalt with the joint in the concrete below, cracking can occur in the asphalt either side of the slot. For practical reasons, a minimum bituminous overlay thickness of 70 mm is recommended.

(c) Using an 'asphaltic crack-relief layer' to provide a medium through which differential movements in the underlying slabs are not easily transmitted. The Asphalt Institute recommends that this layer should be a 90 mm thick, coarse, open-graded hot-mix asphalt with 25–35% interconnecting voids placed on the prepared concrete pavement; the asphaltic overlay, which should be at least 90 mm thick, is then placed on top of the crack-relief layer in two lifts so that the overall minimum thickness of the overlay system is 180 mm. In UK trials, the use of thinner asphalt crack-relief layers on military airfields has shown (short-term) promising results (Coley and Carswell, 2006).

(d) Using a 'fabric interlayer' such as a polypropylene continuous filament between the pavement and the overlay. With this method, a heavy tack coat is applied to the prepared concrete surface, and the fabric rolled onto the tack coat; overlay paving operators then follow immediately behind. The bitumen-saturated interlayer prevents any horizontal movement of the cracked surface being transmitted to the overlying material. Various interlayers have been suggested, such as a stress-absorbing membrane interlayer (a thin rubber–asphalt layer that is applied to the surface of the pavement prior to overlaying).

(e) Using a modified binder to improve the elastic recovery and fatigue resistance of the bituminous overlay material.

(f) Using a pre-tensioned polymer reinforcement grid in the lower layer of the bituminous overlay to resist any horizontal movements in the cracked layer.

The Asphalt Institute effective thickness procedure (Asphalt Institute, 2000) – described previously for overlays to flexible pavements – can also be applied to rigid pavements. As with bituminous pavements, each layer of the existing rigid pavement is converted to an equivalent thickness of asphaltic concrete by applying an appropriate conversion factor. A minimum thickness of 100 mm is recommended for bituminous overlays placed directly on concrete pavements; if the design overlay thickness is 175–225 mm, the use of a crack-relief layer may also be considered. The procedure is illustrated by the following example.

Problem. Consider an existing rigid pavement with construction as defined in Table 16.9. Determine the required asphaltic overlay thickness for a future life of 2 msa. The mean annual air temperature is 15.5°C.

Solution. From Figure 16.8, the design thickness of a new full-depth asphaltic concrete pavement, T_n, is 265 mm, based on future traffic of 2 msa and a subgrade modulus of 50 MPa. The effective thickness of the existing pavement, T_e, derived using factors from the Asphalt Institute, is given by

$$T_e = (200 \times 0.5) + (100 \times 0.2) = 120 \text{ mm}$$

Then, the design overlay thickness, T_o, is

$$T_o = T_n - T_e = 265 - 120 = 145 \text{ mm.}$$

Table 16.9 Details of an existing rigid pavement – overlay example using the effective thickness procedure

Layer	Thickness: mm	Material	Condition	Conversion factor[a]
1	200	Jointed Portland cement concrete	Cracked, seated on sub-base	0.5
2	100	Sand and gravel sub-base	Well graded	0.2
3	–	Subgrade	Resilient modulus = 50 MPa	–

[a] From Asphalt Institute (2000)

The Asphalt Institute has also produced a design chart (Asphalt Institute, 2000) that provides an indication of the minimum thickness of the asphaltic concrete overlay needed on a Portland cement concrete pavement based on the slab length and the temperature differential. Suggested overlay thicknesses range from 100 to 230 mm for slab lengths that vary from 3 to 18 m, and for temperature differential values from 17 to 44°C. (Note that the temperature differential is defined as the difference between the highest normal daily maximum temperature and the lowest normal daily minimum temperature for the hottest and coldest months, respectively, based on a 30 year average.)

16.8.2   Concrete overlays on rigid pavements

Historically, the curing time required for concrete slabs tended to discourage their use as overlays in the UK, albeit concrete overlays had been successfully used in the USA. Factors that influence early strength development are the cement type and content, the water/cement ratio and the curing time (e.g. concrete with standard rapid-hardening Portland cements can achieve a strength of 25 N/mm^2 in less than 18 h, with a cement content of 400 kg/m^3 or more). Current UK advice on materials and methods for the rapid construction of concrete overlays on rigid pavements is given in HD 27 (HA et al., 2004).

Advice on strengthening rigid pavements is given in HD 32 (HA et al., 1994b) and the Concrete Pavement Maintenance Manual (HA and Britpave, 2001).

Fully bonded overlays are appropriate if the existing concrete pavement is in good condition and the required concrete overlay (generally unreinforced concrete) is relatively thin (e.g. 50–150 mm). With this form of construction, the two concrete layers should behave as a monolithic slab: careful preparation of the existing concrete surface is therefore needed to ensure that a good bond is achieved. Surface preparation usually consists of cleaning by grit blasting or shot blasting, and the existing joints are also cleaned and resealed prior to overlaying. With fully bonded overlays, the new joints should be located directly above the existing joints and any wide cracks in the underlying pavement, to prevent reflection cracking, and, to assist in creating the bond, cement grout may be applied immediately before the concrete overlay is placed.

The use of limestone aggregate in the overlay concrete mix and also the use of a fabric reinforcing mesh have also been suggested (HA *et al.*, 1994b) as means of controlling shrinkage-induced cracking in the overlay.

Unbonded or 'debonded' overlays are separated from the existing concrete pavement by a positive separation course (e.g. using a 40–50 mm thick bituminous regulating layer, or a polythene film in a double layer). Use of this type of concrete overlay is most appropriate when the existing concrete pavement is extensively cracked. The overlay may be either unreinforced, jointed reinforced or continuously reinforced concrete. The overlay joints need not be matched to the same locations as those in the underlying existing slab.

The concrete overlay design options currently permitted on trunk roads in the UK are CRCP, which may include a thin asphalt surface course, and CRCB, which includes a 100 mm asphalt overlay.

The required thicknesses for each type of concrete overlay may be obtained from the UK design charts for new rigid pavements in HD 26 (HA *et al.*, 2006a), using the methodology outlined in Section 16.7. In practice, the minimum practical concrete overlay thickness suggested is 200 mm for CRCP and 150 mm for CRCB (plus 100 mm asphalt overlay with CRCB).

REFERENCES

AASHTO (American Association of State Highway and Transportation Officials) (1993) *Guide for Design of Pavement Structures*. AASHTO, Washington, DC, USA.

AASHTO (2008) *Mechanistic–Empirical Pavement Design Guide: A Manual of Practice*. AASHTO, Washington, DC, USA.

AASHTO (2010) *Guide for the Local Calibration of the Mechanistic–Empirical Pavement Design Guide*. AASHTO, Washington, DC, USA.

ADEPT (Association of Directors of Environment, Economy Planning and Transport) (2013) *Managing Reclaimed Asphalt: Highways and Pavements*, version 2013.12. ADEPT, Buckinghamshire County Council, Aylesbury, UK.

Asphalt Institute (2000) *Asphalt Overlays for Highway and Street Rehabilitation, Manual Series No. 17 (MS-17)*, 3rd edn. Asphalt Institute, Lexington, KT, USA.

Asphalt Institute (2008) *Thickness Design: Asphalt Pavements for Highways and Streets, Manual Series No. 1 (MS-1)*, 9th edn. Asphalt Institute, Lexington, KT, USA.

Bohn A, Ullidtz P, Stubstad R and Sorensen A (1972) Danish experiments with the French falling weight deflectometer. *Proceedings of the 3rd International Conference on the Structural Design of Asphalt Pavements*. International Society for Asphalt Pavements, Lino Lakes, MN, USA, vol. 1, pp. 1119–1128.

Brown SF (1998) Developments in pavement structural design and maintenance. *Proceedings of the Institution of Civil Engineers – Transport* **129(3)**: 201–206.

Burt AR (1986) M4 Motorway, a composite pavement. The mechanism of failure. *Proceedings of the 2nd International Conference on Bearing Capacity of Roads and Airfields*, *Plymouth, England*. WDM, Bristol, UK, pp. 397–407.

Cedergren HR (1987) *Drainage of Highway and Airfield Pavements*. Krieger, FL, USA.

Claessen AIM and Ditmarsch R (1977) Pavement evaluation and overlay design – the Shell method. *Proceedings of the 4th International Conference on the Structural Design of Asphalt Pavements*. International Society for Asphalt Pavements, Lino Lakes, MN, USA, vol. 1, pp. 649–661.

Claessen AIM, Valkering CP and Ditmarsch R (1976) Pavement evaluation with the falling weight deflectometer. *Proceedings of the Association of Asphalt Paving Technologists* **45**: 122–157.

Coley C and Carswell I (2006) *Improved Design of Overlay Treatments to Concrete Pavements: Final Report on the Monitoring of Trials and Schemes*. Transport Research Laboratory, Crowthorne, UK, Report TRL657.

Croney D and Croney P (1991) *The Design and Performance of Road Pavements*, 2nd edn. McGraw Hill, London, UK.

EA (Environment Agency) (2010) *Guidance on Waste Acceptance Procedures and Criteria*, version 1. EA, Bristol, UK.

Evdorides HT and Snaith MS (1996) A knowledge-based analysis process for road pavement condition assessment. *Proceedings of the Institution of Civil Engineers – Transport* **117(3)**: 202–210.

Ferne B, Langdale P, Round N and Fairclough R (2009a) Development of the UK Highways Agency traffic speed deflectometer. *Proceedings of the 8th International Conference on the Bearing Capacity of Roads, Railways and Airfields, Champaign, Illinois, USA*. Taylor and Francis, London, UK, pp. 409–418.

Ferne B, Sinhal R and Fairclough R (2009b) Structural assessment of the English strategic road network – latest developments. *Proceedings of the 8th International Conference on the Bearing Capacity of Roads, Railways and Airfields, Champaign, Illinois, USA*. Taylor and Francis, London, UK, pp. 849–857.

Finn FN and Monismith CL (1984) *Asphalt Overlay Design Procedures: NCHRP Synthesis of Highway Practice 116*. Transportation Research Board, Washington, DC, USA.

HA (Highways Agency) (2009) *Network Management Manual*, part 2. *Asset Management Records*, issue 1, amnd 8. Stationery Office, London, UK.

HA (2012) *Interim Advice Note 158, Revisions to HD 30/08: Maintenance Assessment Procedure*. Stationery Office, London, UK.

HA and Britpave (2001) *Concrete Pavement Maintenance Manual*. The Concrete Society, Crowthorne, UK.

HA, Transport Scotland, Welsh Government and Department for Regional Development Northern Ireland (1994a) Section 3: pavement maintenance assessment. Part 2: Structural assessment methods. In *Design Manual for Roads and Bridges*, vol. 7. *Pavement Design and Maintenance*. Stationery Office, London, UK, HD 29/94.

HA, Transport Scotland, Welsh Government and Department for Regional Development Northern Ireland (1994b) Section 4: pavement maintenance methods. Part 2: maintenance of concrete roads. In *Design Manual for Roads and Bridges*, vol. 7. *Pavement Design and Maintenance*. Stationery Office, London, UK, HD 32/94.

HA, Transport Scotland, Welsh Government and Department for Regional Development Northern Ireland (2004) Section 2: pavement design and construction. Part 4: pavement construction methods. In *Design Manual for Roads and Bridges*, vol. 7. *Pavement Design and Maintenance*. Stationery Office, London, UK, HD 27/04.

HA, Transport Scotland, Welsh Government and Department for Regional Development

Northern Ireland (2006a) Section 2: pavement design and construction. Part 3: pavement design. In *Design Manual for Roads and Bridges*, vol. 7. *Pavement Design and Maintenance*. Stationery Office, London, UK, HD 26/06.

HA, Transport Scotland, Welsh Government and Department for Regional Development Northern Ireland (2006b) Section 5: surfacing and surfacing materials. Part 1: surfacing materials for new and maintenance construction. In *Design Manual for Roads and Bridges*, vol. 7. *Pavement Design and Maintenance*. Stationery Office, London, UK, HD 36/06.

HA, Transport Scotland, Welsh Government and Department for Regional Development Northern Ireland (2008a) Section 3: pavement maintenance assessment. Part 2: data for pavement assessment. In *Design Manual for Roads and Bridges*, vol. 7. *Pavement Design and Maintenance*. Stationery Office, London, UK, HD 29/08.

HA, Transport Scotland, Welsh Government and Department for Regional Development Northern Ireland (2008b) Section 3: pavement maintenance assessment. Part 3: maintenance assessment procedure. In *Design Manual for Roads and Bridges*, vol. 7. *Pavement Design and Maintenance*. Stationery Office, London, UK, HD 30/08.

HA, Transport Scotland, Welsh Government and Department for Regional Development Northern Ireland (2014) *Manual of Contract Documents for Highway Works*, vol. 1. *Specification for Highway Works*. Stationery Office, London, UK.

Jordan RW, Coley C, Harding HM, Carswell I and Hassan KE (2008) *Best Practice Guide for Overlaying Concrete*. Transport Research Laboratory, Wokingham, UK, Road Note 41.

Kennedy CK (1978) *Pavement Deflection: Operating Procedures for Use in the United Kingdom*. Transport Research Laboratory, Crowthorne, UK, Report LR835.

Kennedy CK and Lister NW (1978) *Prediction of Pavement Performance and the Design of Overlays*. Transport Research Laboratory, Crowthorne, UK, Report LR833.

Kennedy CK, Fevre P and Clarke C (1978) *Pavement Deflection: Equipment for Measurement in the United Kingdom*. Transport Research Laboratory, Crowthorne, UK, Report LR834.

Kerali HR, Lawrance AJ and Awad KR (1996) Data analysis procedures for the determination of long-term pavement performance relationships. *Transportation Research Record* **1524**: 152–159.

Lister NW (1972) *Deflection Criteria for Flexible Pavements*. Transport Research Laboratory, Crowthorne, UK, Report LR375.

McMullen D, Snaith MS and Burrow JC (1986) Back analysis techniques for pavement condition determination. *Proceedings of the 2nd International Conference on Bearing Capacity of Roads and Airfields, Plymouth, England*. WDM, Bristol, UK, pp. 335–344.

McMullen D, Snaith MS, May PH and Vrahimis S (1990) A practical example of the application of analytical methods to pavement design and rehabilitation. *Proceedings of the 3rd International Conference on Bearing Capacity of Roads and Airfields, Trondheim, Norway*. Tapir, Trondheim, Norway, pp. 1115–1124.

Merrill D (2005) *Guidance on the Development, Assessment and Maintenance of Long-life Flexible Pavements*. Transport Research Laboratory, Crowthorne, UK, Report TRL639.

Merrill D, Nunn M and Carswell I (2004) *A Guide to the Use and Specification of Cold Recycled Materials for the Maintenance of Road Pavements*. Transport Research Laboratory, Crowthorne, UK, Report TRL611.

Mildenhall HS and Northcott GDS (1986) *Manual for the Maintenance and Repair of Concrete Roads.* Stationery Office, London, UK.

Nicholls JC, McHale MJ and Griffiths RD (2008) *Best Practice Guide for Durability of Asphalt Pavements.* Transport Research Laboratory, Wokingham, UK, Road Note 42.

Norman PJ, Snowdon RA and Jacobs JC (1973) *Pavement Deflection Measurements and Their Application to Structural Maintenance and Overlay Design.* Transport Research Laboratory, Crowthorne, UK, Report LR571.

Nunn ME (1998) Structural design of long-life flexible roads for heavy traffic. *Proceedings of the Institution of Civil Engineers – Transport* **129(3)**: 126–133.

Nunn ME, Brown A, Weston D and Nicholls JC (1997) *Design of Long-life Flexible Pavements for Heavy Traffic.* Transport Research Laboratory, Crowthorne, UK, Report 250.

Rolt J, Smith HR and Jones CR (1986) The design and performance of bituminous overlays in tropical environments. *Proceedings of the 2nd International Conference on Bearing Capacity of Roads and Airfields, Plymouth, England.* WDM, Bristol, UK, pp. 419–431.

Rwebangira T, Hicks RG and Truebe M (1987) Sensitivity analysis of selected back calculation procedures. *Transportation Research Record* **1117**: 25–37.

Shell IPC (1978) *Shell Pavement Design Manual: Asphalt Pavements and Overlays for Road Traffic.* Shell IPC, London, UK.

Shell IPC (1985) *Addendum to the Shell Pavement Design Manual: Asphalt Pavements and Overlays for Road Traffic.* Shell IPC, London, UK.

Sherman G (1982) *Minimising Reflection Cracking of Pavement Overlays: NCHRP Synthesis of Highway Practice 92.* Transportation Research Board, Washington, DC, USA.

Smith HR and Jones CR (1980) *Measurement of Pavement Deflections in Tropical and Subtropical Climates.* Transport Research Laboratory, Crowthorne, UK, Report LR935.

Snaith MS and Hattrell DV (1994) A deflection based approach to flexible pavement design and rehabilitation in Malaysia. *Proceedings of the Institution of Civil Engineers – Transport* **105(3)**: 219–225.

TRL (Transport Research Laboratory) (1993) *A Guide to the Structural Design of Bitumen-surfaced Roads in Tropical and Sub-tropical Countries,* 4th edn. TRL, Crowthorne, UK, Overseas Road Note 31.

TRL (1999) *A Guide to the Pavement Evaluation and Maintenance of Bitumen-surfaced Roads in Tropical and Sub-tropical Countries.* TRL, Crowthorne, UK, Overseas Road Note 18.

TRL (2004) *A Guide to Axle Load Surveys and Traffic Counts for Determining Traffic Loading on Pavements.* TRL, Crowthorne, UK, Overseas Road Note 40.

TRRL (Transport and Road Research Laboratory) (1978) *A Guide to Concrete Road Construction,* 3rd edn. Stationery Office, London, UK.

Highways
ISBN 978-0-7277-5993-1

ICE Publishing: All rights reserved
http://dx.doi.org/10.1680/h5e.59931.529

Chapter 17
Wet skid resistance

David Woodward   Reader in Infrastructure Engineering, University of Ulster, Ireland

17.1.   Introduction

This chapter considers the wet skid resistance of asphalt surfacing materials. It considers the difference between the terms friction and skid resistance, explores tyre–road friction and the importance of the contact patch, and deals with road surface macro-texture and aggregate micro-texture. How road surface course materials are tested in the field and in the laboratory is then summarised. The chapter concludes by asking the question – skid resistance, what's next?

In this chapter, the word asphalt is used in the context of European standards, meaning a homogeneous mixture of coarse and fine aggregates, filler material and bituminous binder that is used in the construction of flexible pavement layers.

Friction, in the context of tyres and roads, represents the grip developed by a particular tyre on a particular road surface at a particular time. This is measured by means of a coefficient of friction, and is influenced by a large number of parameters relating to the road, such as the tyre, vehicle suspension, ambient conditions and contaminants (including water). In the context of a road accident, or a situation that may lead to one, it is the coefficient of friction available at the time and place of the incident that matters.

Skid resistance, which describes the contribution that the road makes to tyre–road friction, is a measurement of friction obtained under specified standardised conditions. It is generally chosen in order to fix the values of many variable factors so that the contribution that the road surface provides to tyre–road friction can be isolated. When building and maintaining roads, it is the initial-designed properties of the surfacing and its condition while in service that are important to road safety.

Water and other contaminants detrimentally affect the tyre–road surface interaction. For the same surface, the coefficient of friction will be lower in wet conditions. To reduce the criteria that increase the risk of a wet skidding accident occurring, the conditions must be such that the detrimental effect of water is reduced to acceptable levels.

17.2.   Tyre–road friction

Adequate grip is required between the tyres of a vehicle and the road surface to enable the vehicle to be driven safely. It is required when a vehicle is (a) under power, when a reaction force between the tyre and road is needed for the vehicle to accelerate or maintain speed; (b) braking, when forces are developed against the tyre and road that react against the action of the brakes so that the vehicle slows down; and (c) cornering, when reaction against side forces generated in response to steering action enables the vehicle to follow around a curve (Kane and Scharnigg, 2009). These forces are generated in the contact patch (see Section 17.3) as a result of friction between the tyre and the surface course. When these forces exceed the available friction, the contact patch will start to slide or slip over the road surface. Once this happens, control of the vehicle will be lost, and an accident may occur.

The mechanisms of tyre–road friction are still not fully understood. Friction is considered to be the sum of two main mechanisms: molecular adhesion and hysteresis losses. Adhesion occurs at the interface between the tyre-tread rubber and the road surface, and is considered to depend on the contact area and the roughness at the microscopic scale. Hysteresis relates to the deformation of the rubber as the tyre passes over the aggregate particles in the surface course, thereby creating an asymmetric pressure distribution.

When a road surface course is dry, the coefficient of friction is normally high and adequate for most driving, whereas, in wet conditions, tyre–road friction can decrease significantly. Therefore, for any given situation, the amount of friction depends on complex interactions involving the tyre (i.e. its tread depth and pattern, and contact patch area), the surface texture of the carriageway (i.e. the type of asphalt mixture and its ability to disperse water), the aggregate micro-texture (i.e. the rock type and degree of stone polish), the speed of the vehicle (which influences the contact time) and, importantly, the environmental conditions (i.e. the water film thickness). For example, a few drops of rain may be enough to influence friction measurements on a dry road. The presence of a thin water film or damp conditions may have only a limited negative effect at low vehicle speeds as a tyre's rubber has time to interact with the aggregate micro-texture: however, at higher speeds and greater rainfall intensities, macro-texture is needed to disperse greater amounts of water, as contact times decrease, for adequate interaction to occur between the tyre and the road surface.

17.3. Contact patch

The contact patch is that area of road surface that is in contact with a tyre as it rolls along the highway. The simplest method of measuring the contact patch is to apply paint to a tyre and load it onto cardboard placed on a steel plate to obtain a print (ASTM, 2010). Analysis of the print allows parameters such as the gross contact area, the groove or void area, the contact length and the contact width to be assessed. The static contact patch tends to have a circular shape at high tyre inflation pressures and lower load; it becomes elliptical at lower tyre inflation pressures and higher load. However, the amount and distribution of contact stress is not constant within the tyre–road surface contact patch. It is influenced by tyre factors such as the inflation pressure, the tread pattern, the load and speed, and by surface course factors such as the nominal aggregate size and the asphalt mix type and texture (e.g. whether it is positive or negative).

Tyre tread is used to remove water from the tyre–asphalt interface, and its quality is ensured by the imposition of a national requirement for a legal minimum tyre tread depth. Figure 17.1 compares the contact patch for two types of car tyre on a smooth surface, as measured using XSensor pressure mapping equipment. One tyre is a semi-slick Toyo and the other is a treaded Avon; however, the loading and inflation pressures of both tyres were the same, and both were fitted to the same vehicle for measurement purposes. Note that, although the contact areas are comparable, the tread pattern and distribution of stress is significantly different for the two tyres. (The semi-slick has been designed to maximise grip in dry conditions, while the treaded tyre has been designed for both dry and wet skid resistance.)

Figure 17.2 shows the variation in the contact pressure for a 10 mm stone mastic asphalt measured in the laboratory under simulated trafficking by a smooth GripTester tyre. A flexible XSensor pressure pad was placed between the friction tyre and the asphalt to measure the interface pressure conditions. This figure shows that the highest stresses are associated with individual aggregate particles, and that lower stresses correspond to mastic–aggregate conglomerations; it also shows significant areas where there is no tyre–asphalt contact.

The examples shown in Figures 17.1 and 17.2 show how each tread pattern–friction-measuring tyre is interacting with the pressure-measuring device. In practice, the stresses imposed by a vehicle tyre with tread rolling over a textured asphalt surface course will be a complex sum for each of the contact patches. Also, note that the amount of actual contact between the rubber of each tyre and the road surface will be significantly less than the gross contact patch area.

17.4. Road surface course parameters affecting skid resistance

There are two road surface course parameters that the road engineer can control in order to reduce the detrimental effect of water on skid resistance: (a) the provision of adequate surface course macro-texture so as to allow for the removal of water from within the contact patch, and (b) the use of aggregate with a micro-surface roughness (i.e. a micro-texture) that will break through the water film that develops between the road surface course and the tyre tread.

17.4.1 Surface course macro-texture

An asphalt surface course must have adequate texture to assist the tyre tread in the removal of water from the contact patch: this scale of texture is termed macro-texture. Typically, a minimum national requirement for the macro-texture on a new pavement surfacing is specified in the design documentation for the pavement. In the UK, the initial macro-texture on a new surface course is measured using the volumetric patch method (BSI, 2010). The initial texture-depth requirements, which vary according to the road type and the surface course type, have influenced the development and use of particular asphalt surface course materials. For example, on high-speed roads in the UK for which the posted speed limit is ≥ 50 mph, the average depth cannot be less than 1.3 mm/km for pavements with 14 mm thin surface course systems. For hot rolled asphalt surface courses, surface dressing and all other surfacings, the average depth per kilometre cannot be less than 1.5 mm. Irrespective of the surface course material, the macro-texture depth required at roundabouts on ≥ 50 mph roads cannot be less than

Figure 17.1 Contact patches showing stress distributions for (a) a semi-slick Toyo car tyre and (b) a treaded Avon car tyre

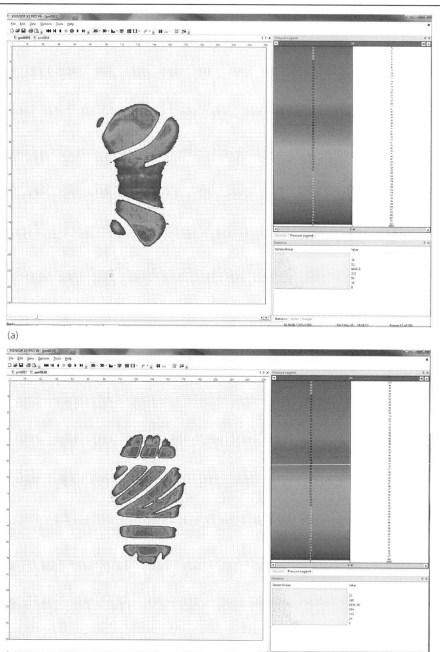

(a)

(b)

Figure 17.2 Contact stress distribution for a smooth GripTester tyre moving over a 10 mm stone mastic asphalt surface material

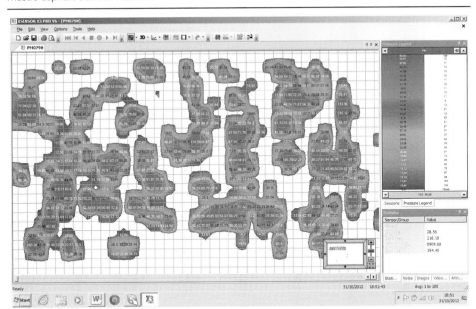

1.2 mm, and not less than 1.0 mm for roundabouts on roads with a posted speed limit ≤40 mph.

Most surface course asphalts belong to one of three generic macro-texture types: (a) those with positive texture (i.e. where the coarse aggregate protrudes from the plane of the surfacing, e.g. as with high friction surfacing, surface dressing, or 20 mm chippings applied to hot rolled asphalt), (b) those with negative texture (i.e. where the texture largely comprises surface voids between aggregate particles whose upper surfaces form a generally flat plane, e.g. as with stone mastic asphalt) and (c) those with a porous texture (i.e. where the coarse aggregate grading provides a high void content mixture, e.g. as with porous asphalt).

17.4.2 Aggregate micro-texture

Micro-texture is a measure of aggregate particle surface roughness that is measured in the laboratory using the polished stone value (PSV) test (BSI, 2009). It relates to the geological properties of the rock type and its condition or state when used. For various reasons not all types of rock, including those with good micro-texture, make a good surfacing aggregate and therefore micro-texture must always be considered in conjunction with other aggregate properties, for example, some types of rock may polish too quickly, or have such poor strength that they are susceptible to break-up and/or excess wear during pavement construction, in-service trafficking and adverse weather conditions.

Figure 17.3 shows the distribution of PSV data with respect to rock type from aggregate sources available in the UK (Woodward, 1995) and suggests a general PSV ranking

Figure 17.3 Distribution of PSV data with respect to rock type (Woodward, 1995)

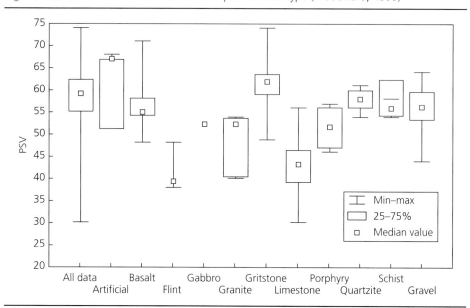

based on rock type. Today, it is recognised that of those rocks that are deemed suitable, the following are the best used in surface courses: hard limestones, fine-grained rocks such as quartzite and basalt, hard coarse-grained granites, and softer fine and medium–coarse sandstones and greywackes. Of these natural aggregates, the sandstone and greywacke types of rock provide the best values of wet skid resistance. These are composed of grains that produce a sandpaper-type of micro-texture that, under traffic, are plucked from the particle surface to expose new grains beneath. This plucking mechanism results in a renewable micro-texture that, in turn, maintains wet skid resistance.

However, as noted previously, the benefit derived from the renewable micro-texture must always be balanced against other essential properties such as resistance to wear and strength (i.e. the aggregate must neither wear away too quickly nor remain so intact as to become polished). This is the reason why hard aggregates with good resistance to abrasion and strength tend to have poorer levels of skid resistance. For most types of aggregate, increasing PSVs are usually achieved at the expense of almost every other property (e.g. strength, abrasion and soundness). Thus, for a given rock type, high PSVs require cautious and careful consideration to ensure that their usage does not lead to premature in-service failure. This is particularly important when dealing with aggregate sources for which historical in-service data do not exist.

How micro-texture affects skid resistance as measured by the pendulum test (see Section 17.6.1.1) is illustrated by the laboratory test data shown in Figure 17.4. The dry unpolished pendulum test results in this figure show that, when evaluated in a freshly crushed dry condition, almost all aggregate types have similar skid resistances whereas the wet unpolished pendulum values are reduced when the surface is wetted and a water film

Figure 17.4 Comparison of skid resistance for nine different aggregates, measured according to dry unpolished value (DUPV), wet unpolished value (WUPV) and PSV test conditions

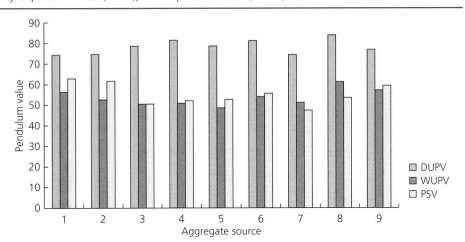

forms between the rubber and the aggregate surface. Then, after each aggregate is sub-jected to simulated trafficking in the PSV test, the results change as each aggregate reacts differently to the testing conditions (i.e. most decrease while some increase). It can be seen from these data that the presence of water, and the polishing conditions associated with the PSV test, illustrate the importance of wet skid resistance to road safety, particularly when a road surfacing is subjected to heavy traffic flows or to high stresses produced by braking or turning vehicles.

17.5. Measuring macro-texture

There are a number of different ways of measuring the macro-texture of a road surfacing, namely the volumetric, profilometry and drainability methods, as well as three-dimensional (3D) modelling.

The volumetric patch method (BSI, 2010) is most commonly used to measure the initial macro-texture of new surface courses and as a reference method in case of dispute. It uses a known volume of sand or glass beads that is spread out to form a circular patch on the road surface by filling the hollows in the surface up to the peaks. By dividing the distributed volume of sand by the area covered by the sand, a value is obtained that represents the average depth of the sand layer. This test, which is considered insensitive to micro-texture characteristics, is reported as the mean macro-texture depth (MTD).

The surfacing texture can also be characterised using surface profilers (BSI, 2004). This method, which is based on contactless two-dimensional (2D) surface profiling, deter-mines a mean profile depth (MPD): this can then be transformed into a quantity that estimates the MTD depth as measured using the volumetric patch method. For example, the WDM TM2 laser profiler device carries out a sequence of displacement measure-ments along the line of the surface profile and then the root mean square texture depth

Figure 17.5 MTD (volumetric patch method) versus MMTD (3D photogrammetry method)

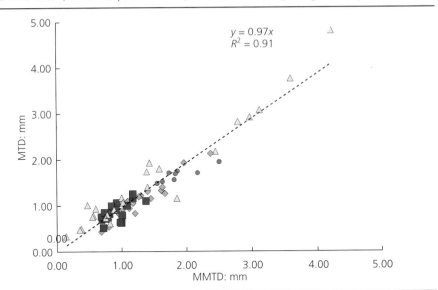

and MPD are calculated to provide averages at intervals of 10 and 50 m along the carriageway.

The circular texture (CT) meter is another road surface macro-texture profiler that uses a laser-displacement sensor to measure the vertical profile of a pavement surface (ASTM, 2009a): its software also calculates and reports the MPD and statistics. Another profiler device, SCRIMTEX, measures the surface macro-texture in front of the test wheel during network condition surveying for wet skid resistance.

The horizontal drainability of a road surface can also be assessed using an outflow meter (BSI, 2002) as a stationary device. This is used to provide a measure of the drainability in the road–tyre contact area for use on smooth non-porous road surfaces (i.e. those with MPD values of less than 0.4 mm) either in the field or in laboratory studies.

Macro-texture measuring techniques involving 3D modelling of road surface textures using stereo-photogrammetry (Millar, 2013) and laser-based techniques are currently being developed. These techniques, which involve the capture and processing of very large amounts of data, are mainly used at this time in laboratory-based studies to provide a wide range of 2D and 3D surface texture parameters (BSI, 2012). Figure 17.5 shows good correlation between the MTD (volumetric patch method) and mean measured texture depth (MMTD) derived from a contactless 3D photogrammetry method (Millar, 2013).

17.6. Measuring skid resistance

Skid resistance can either be measured in the field or in the laboratory. Selection of the most appropriate method depends on the reason for the assessment, for example, whether it is to evaluate the site of an accident, or as a participant in a national pavement maintenance programme, or to select an aggregate for use in a pavement's surface course. Each method has a set of standard conditions chosen to reflect the practicalities of carrying out the test in relation to the complex reality of friction at the tyre–road interface.

17.6.1 Measuring skid resistance on the roadway

While the skid resistance of a road surface can be measured either dry or in the wet, almost all specifications are based on testing the surface in a wet condition. At least 20 devices have been developed (Doe and Roe, 2008) to measure skid resistance: these can be divided into generic groups, that is, static and slow-moving, longitudinal friction, transverse friction, locked-wheel and decelerometer devices. Although all the devices measure skid resistance, the actual measurements recorded can differ widely for the same road surface. All the devices measure, in different ways, the frictional force developed between a moving tyre or a rubber slider and the road surface. Typically, they involve wetting the road surface and recording a quotient of the measured force and the applied vertical load, that is, a friction coefficient.

17.6.1.1 Static devices

Static (i.e. stationary) devices have typically been developed to be portable and suitable for localised road usage, as well as in the laboratory. There are a number of basic types, of which the pendulum test (BSI, 2011) is probably that which is most commonly used around the world to measure the slip and skid resistance of many different types of surface, including asphalt surfaces. As used here, the measured slip–skid resistance is the property of a trafficked surface that limits the relative movement between a tyre of a vehicle and the surface, or the contact patch of a pedestrian's footwear and the surface.

The pendulum test works on the same principle as a person sliding a foot along a surface to assess its slipperiness; that is, it involves the use of a stand-mounted tester that incorporates a spring-loaded standard rubber slider attached to the end of the pendulum arm. When in use the tester is adjusted and levelled so that when the pendulum arm is released from a horizontal position the rubber slider swings down and makes contact with the surface for a length of 126 mm. The loss of energy experienced by the pendulum assembly as it slides over the surface is measured by the reduction in length of the upswing, using a calibrated scale: this gives the pendulum test value (PTV) for the surface being tested. The PTV is typically specified, measured and reported for a wetted surface.

Two types of rubber slider are used with the pendulum tester: slider 57 (previously known as the Transport Research Laboratory slider) is normally used to measure highway surfaces for skid resistance, while slider 96 (originally known as 4S rubber) is a harder rubber slider that is typically used to measure the slip resistance of pedestrian indoor and outdoor surfaces. Slip–skid resistance is not a constant but varies with climate and trafficking.

When the pendulum test is used in conjunction with the PSV laboratory test (see Section 17.6.2.1), a narrower rubber slider is used.

As the pendulum test only measures a small area (about 0.01 m²) of the surface, this should be considered when deciding its applicability for usage on a surface that may have non-homogeneous surface characteristics such as ridges or grooves, or is rough textured so that the MTD exceeds 1.2 mm. Note also that the PTVs obtained with this test method are normally measured at 'spot' locations on the road surface and so they cannot be compared with results from mobile devices that measure slip–skid resistance over long lengths of road surface.

The dynamic friction tester (ASTM, 2009b) is a static tester assembly that provides a measure of surface friction as a function of sliding speed. This device, which is also capable of being used both in the field and in a laboratory, consists of a spinning horizontal disc that is fitted with three spring-loaded rubber sliders. As the disc is lowered onto the wet surface, its rotational speed decreases due to friction. The torque generated by the slider forces measured during the spin down is then used to calculate the friction as a function of speed.

17.6.1.2 Slow-moving devices

Skid-resistance-measuring devices are also available that may be pushed manually along the surface being evaluated: for example, the micro GripTester was developed to measure either dry or wet skid resistance at a walking pace in pedestrian areas, on road surfaces and paint markings, on reinstatements and manholes, and at accident sites. Figure 17.6

Figure 17.6 Transverse variations in the wet skid resistance across a four-lane road, as measured by the micro GripTester device

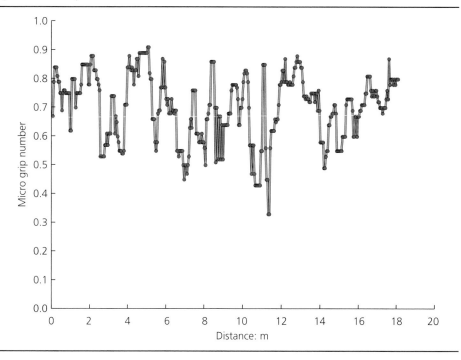

is an example of micro GripTester data, and shows the transverse variation in wet skid resistance measurements across a four-lane road.

The micro GripTester device is a scaled-down version of the GripTester (see the next section) and uses the same friction principle. An on-board computer measures the skid resistance every 48 mm and records test data relating to the friction coefficient (known as the grip number), chainage, speed, water application rate, latitude, longitude, altitude and temperature.

17.6.1.3 Longitudinal friction devices
Longitudinal friction devices attempt to measure skid resistance by simulating the interaction of a braked tyre with the road surface in a longitudinal direction.

Originally developed to assess the skid resistance of oil platform helidecks, the Grip-Tester involves the use of a three-wheel trailer that has now become widely used around the world. Weighing 85 kg, the trailer can be towed behind a van at speeds of up to 130 km/h on roads or airport runways, or pushed manually to take measurements in pedestrian areas. When it is in operation, a constant film of water is sprayed in front of the smooth test tyre, depending on the test speed. A simple fixed gear and chain system constantly brakes the test tyre, to give a fixed slip ratio of 15%. Continuous measurement of the slipping force and the vertical load allows the calculation of a friction coefficient known as the grip number.

The standard test speed is 50 km/h with water applied, to give a theoretical water film depth of 0.25 mm under the test tyre, and, in the UK, the skid resistance is typically measured in the inside wheel path of the left (most heavily trafficked) lane. An example of GripTester data measured at 1 m intervals is shown in Figure 17.7: this shows the variation in the dry and wet skid resistance for a straight section of surfacing with surface dressings composed of low-PSV limestone aggregate and high-PSV greywacke aggregate. Testing at a range of speeds shows the dry friction values for both rock types to be similar, with a grip number of about 1; however, when the surface dressing is wetted, the difference between the limestone and greywacke aggregates is evident.

17.6.1.4 Transverse friction devices
Transverse friction-measuring devices attempt to simulate the interaction between a braked tyre and the road surface as the angled tyre turns into a corner. SCRIM (sideway-force coefficient routine investigation machine) was introduced in the 1970s to provide a method to routinely measure the skid resistance of the UK road network. SCRIM evolved from motorcycle-based testing machines developed in the 1930s, which discovered that the force exerted on a wheel that is angled to the direction of travel, and held in a vertical plane with the tyre in contact with the road surface, could be correlated with the wet skid resistance of that surface. The sideways-force coefficient (SFC) determined in this way is defined as the force at right angles to the plane of the inclined wheel, expressed as a fraction of the vertical force acting on the wheel.

Figure 17.7 GripTester skid resistance data obtained at 1 m intervals on a road surface that is sealed with two different types of aggregate chipping (solid symbols, dry surface; open symbols, wet surface)

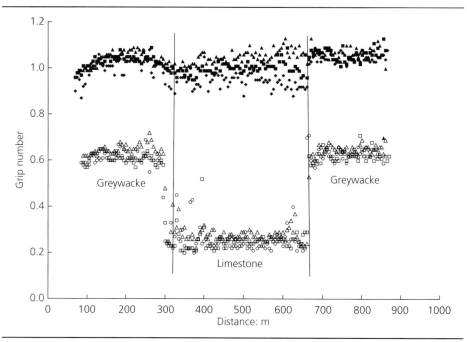

SCRIM is essentially a lorry with a water tank, and a smooth, inclined rubber tyre, under a known vertical load, that is mounted mid-machine in line with the nearside wheel track. It is possible to have two test wheels on the same vehicle – one for each of the wheel tracks. The test wheel is angled at 20° to the direction of travel of the vehicle. As the device moves forward, water is sprayed in front of the angled tyre, to give a film thickness of constant depth: the test wheel slides in the forward direction along the surface, and the force generated by the resistance to sliding is measured, and this, when divided by the vertical load, gives continuous SFC measurements of the road surface.

In the UK, SCRIM network testing is most commonly carried out between 1 May and 30 September, in order to reflect the seasonal influence on wet skid resistance. The test speed is typically 50 km/h, and the skid resistance is mostly measured in the inside wheel path of the outer left lane.

17.6.1.5 Locked-wheel devices

Both the GripTester and SCRIM devices measure 'slow' test speed skid resistance; that is, even though either apparatus can be used at higher speeds, the slip speed between the test tyre and road is actually relatively slow. More importantly, neither test replicates a high-speed braking event that causes the vehicle tyre to lock up.

The Pavement Friction Tester (Dynatest, 2014) was developed in the USA, and uses a locked-wheel principle to measure friction and so simulate these conditions. It comprises a towing vehicle and a purpose-built trailer with two wheels: one of the trailer wheels is fitted out as the test wheel, and has a test tyre that is typically smooth. The test wheel axle is fitted with a two-axis force transducer that measures the vertical force (load) and horizontal force (drag) on the wheel, while shaft encoders allow the speed of both wheels to be measured. During the test, water is pumped onto the surface of the road in front of the test wheel: the pump rate increases with speed so that a nominal water depth of 0.5 or 1.0 mm is maintained. A longitudinal friction coefficient is determined from the average frictional force over a 1 s period after the wheel has locked. The test speed can range from 20 to 130 km/h, and the tester device is used to characterise the friction–speed relationship for different surfaces.

17.6.1.6 Decelerometer devices

Decelerometer devices tend to be used by the police for crash investigations. Examples include the Skidman (Turnkey Instruments, 2014) and the Vericom (Vericom, 2014) devices, both of which involve the use of accelerometers. The Skidman is positioned in the front passenger foot-well of the police car, while the Vericom is typically mounted at windscreen level. Both devices measure the average deceleration of the vehicle through a skid, from a defined speed to a stop. The target speed is typically 50 km/h with the ABS system of the car disabled. This type of braking testing may also be done using GPS-based systems such as the Racelogic VBox3i (Racelogic, 2014). With a sample rate of 100 Hz, the VBox3i can give an accurate measurement of the braking time, distance and longitudinal deceleration.

17.6.2 Measuring skid resistance in the laboratory

The measurement of skid resistance in the laboratory is necessary to obviate the use of aggregates or asphalt mixes in surfaces that may polish and become dangerously slippery under heavy traffic. The laboratory test methods that are most widely used in the UK for this purpose are the PSV test for aggregate, the friction after polishing (FAP) test for aggregate and asphalt mixes, and the road test machine (RTM) wear test for high-friction surfacing systems and asphalt mixes.

17.6.2.1 Polished stone value (PSV) test

The PSV test (BSI, 2009), which is possibly the main laboratory method used throughout the world to measure the resistance of an aggregate to skidding, seeks to provide a measure of the resistance of coarse aggregates to the polishing action of vehicle tyres under conditions similar to those occurring on the surface of a road. The skid resistance of the aggregate relates to its surface roughness (i.e. its micro-texture). As discussed previously, some aggregates are able to maintain their surface roughness under heavy traffic, whereas others lose it as their surfaces become smooth and polished. In the dry, most aggregates can provide an adequate level of skid resistance; however, the ability to maintain adequate micro-texture is important for road safety, as this fine scale of texture is needed to cut through the water film beneath a moving tyre on a wet road.

Apart from some minor modifications, the PSV test – which is performed in two parts – has changed very little since it was devised in 1952. The first part of the test involves subjecting curved specimens of cubic-shaped 10 mm-sized aggregate chippings to a wet polishing action using the accelerated polishing machine. This part of the test lasts for 6 h, during which time each aggregate test specimen is subjected to 115 200 passes of a solid tyre under a force of 725 N: coarse corn emery abrasive and water is used as the polishing medium for the first 3 h, followed by air-floated fine emery flour and water for the remaining 3 h. In the second part of the test, the state of polish reached by each specimen is measured by means of a friction test using the pendulum tester. The result is expressed as a laboratory-determined PSV, and the higher the PSV measured, the greater is the resistance of the aggregate to polishing.

Most trunk road and motorway surfacing specifications in the UK require the use of aggregates with PSVs of 60–65, while critical sites may require aggregates with PSVs of 68 or above. (Practically, however, it might be noted that there are very few quarries in the UK that produce aggregates with PSVs above 65.) The need to maintain aggregate performance over time again highlights the importance of considering the PSV in relation to other properties because, as noted previously, PSV increases typically correspond to decreases in other properties (e.g. fragmentation, abrasion and soundness) (Woodward, 1995).

In the UK, the desire for very high values of PSV has favoured the use of one single type of aggregate (i.e. gritstone) as the main surfacing aggregate. Figure 17.8, which shows

Figure 17.8 PSV versus the aggregate abrasion value for basalt, gritstone and limestone rock aggregates (Woodward, 1995)

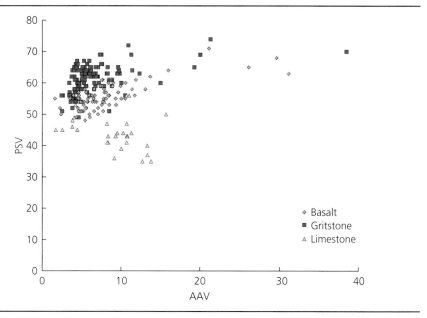

plots of the PSV versus the aggregate abrasion value for basalt, gritstone and limestone rock aggregates, clearly demonstrates that the gritstone aggregates have the best combination of wet skid resistance and abrasion resistance.

It is important to appreciate that the PSV test only assesses single-size 10 mm aggregate particles, whereas other British standard tests typically assess 14 mm-sized particles, and asphalt surface courses typically contain a range of aggregate sizes. Thus, while the PSV test attempts to simulate in-service factors such as rain, traffic loading and the presence of detritus, it cannot adequately simulate actual in-service conditions, given the considerable variations that there are in traffic conditions, stresses and types of asphalt surface courses.

Also, while the use of a solid tyre may have helped to improve the PSV test method, actual trafficking on the roadway is done by a pneumatic tyre – which means that different conditions are experienced within each contact patch. It must therefore be emphasised that the 6 h standard PSV test is not a true reflection of the performance of an aggregate under traffic but, rather, offers a ranking mechanism under simplified laboratory test conditions within which end use, climate and trafficking conditions are not considered. This must always be borne in mind when considering the use of PSV data specifying the values obtained using the standard test.

This limitation has prompted much research into how to improve the accuracy of predicting the wet skid resistance of an aggregate using the PSV test equipment. Thus, many testing protocols were developed to try to better simulate in-service conditions, such as using extended polishing cycles, different types of polishing agent, repetitive coarse and fine emery polishing cycles, low-temperature conditioning cycles, and sideways polishing to alter the stressing conditions during testing. These variations found that the standard 6 h PSV test does not give the lowest value of skid resistance for an aggregate (referred to as its ultimate state of polish) but, rather, that different aggregates respond differently to the differing test protocols assessed (i.e. some continue to lose wet skid resistance while others remain relatively unaffected).

Research has shown, however, that a modification of the standard PSV method that involves an additional period of 3 h offset polishing is capable of achieving the ultimate levels of polishing similar to that experienced in service at sites subjected to very heavy traffic flows and/or in-service stressing. Nonetheless, despite the important findings of this research, they have not yet been adopted.

17.6.2.2 Friction-after-polishing (FAP) test

The FAP test (BSI, 2014) is used to measure the skid resistance of both aggregate and asphalt surface course materials. Originally known as the Wehner–Schulze test, the FAP test has been used in Germany since the 1980s to evaluate surfacing materials (Huschek, 2004). The ability of this test to assess asphalt mixtures is regarded as an improvement over the PSV test.

In the FAP test, cylindrical specimens of 225 mm diameter are attached in the test device shown in Figure 17.9. Surface friction is determined by measuring the rate of

Figure 17.9 The friction-after polishing testing device

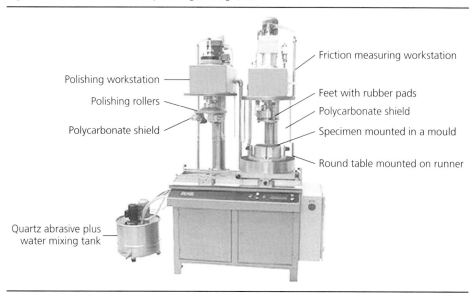

deceleration of a spinning disc with three rubber sliders that is lowered onto the surface of the specimen in the presence of water. A specimen is typically subjected to 90 000 revolutions of a polishing head containing three grooved rubber cones that are pressed down with a force of $0.4 \, \text{N/mm}^2$. The test specimen can also be blasted with grit, to

Figure 17.10 Examples of FAP data measured after varying periods of simulated time (Friel, 2013)

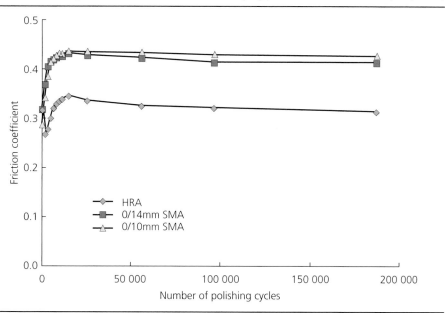

simulate weathering effects over varying periods of time, the friction remeasured, and the polishing regime repeated as desired.

Figure 17.10 shows graphs of FAP friction coefficient data plotted against the number of polishing passes. In this figure the same PSV 65 coarse aggregate was used to prepare a 14 mm stone mastic asphalt surfacing, a 10 mm stone mastic asphalt surfacing, and a 20 mm chipping layer applied to a hot rolled asphalt surfacing. The data in this figure show that the development of early life friction peaks to a maximum, and then is reduced as the test specimens become more polished. It also shows that the same PSV aggregate when used in different surface course materials can give different FAP friction coefficients.

FAP test specimens can be cored from laboratory-compacted slabs or from machine-laid and compacted road surfaces. An advantage of using test samples obtained from the pavements is that the FAP test results can then be compared with those obtained from SCRIM. If desired, test specimens can also be made using single-size aggregate particles.

17.6.2.3 RTM wear test

The RTM is accredited by the British Board of Agrément to assess the wear characteristics of high-friction surfacing systems for use in the highway authorities product approval scheme (Nicholls, 1997). The equipment consists of a device that is housed in an environmental chamber (to enable the test temperature to be held at 10°C) and a rotating circular table on which up to ten attached slabs – each 305 mm long × 305 mm wide × 50 mm deep – are subjected to accelerated trafficking from two full-size tyres attached to the RTM. The high-friction surfacing wear is assessed after 100 000 wheel passes by measuring changes in the macro-texture depth using the volumetric patch test (BSI, 2010), and changes in the wet skid resistance using the pendulum test (BSI, 2011).

While accredited to assess specialist high-friction surfacing systems, a wide range of asphalt surface course materials has been assessed using the RTM equipment (Friel, 2013). This has enabled asphalt slabs that have been sampled from on site, or mixed in the laboratory, to be assessed under controlled repeatable conditions without the need for full-scale road trials.

17.7. Effect of seasonal and environmental factors on skid resistance

The skid resistance of a road surface can vary throughout the year and from one year to the next. It is typically higher in prolonged wet or winter conditions due to the effects of detritus, frost, and wear by tyres on micro-texture and macro-texture. Also, the early-life skid resistance of a surface course in which the aggregate is coated with bitumen (e.g. a stone mastic asphalt) will change over time as the bitumen is worn away by tyres until, eventually, equilibrium conditions are achieved.

As noted above, the wet skid resistance in the UK is normally lower in summer than in other seasons of the year as a result of dry polishing by tyres in the presence of fine detritus. This seasonal variation, which has been known for many years and relates to equilibrium conditions between summer polishing and winter roughening, is affected by the volume and composition of traffic. The amount of polishing and/or wearing away differs for cars, buses and commercial vehicles of different sizes and different types of tyre. While the ambient temperature is not normally considered to directly influence the skid resistance, extreme temperatures can influence the deterioration mechanisms of road-surfacing materials. Extreme temperatures may also influence the rubber characteristics of tyre – and, consequently, friction or skid resistance measurements.

17.8. Wet skid resistance – what's next?

Currently, there is renewed interest in the many asphalt surface course materials that have been specifically designed to offer a wide range of enhanced performance characteristics, and in how these are measured and specified as part of asset management programmes. In this context, the future of skid resistance is about improving the use of materials and technologies to produce a longer-lasting, more sustainable, highway infrastructure asset without compromising safety.

The asphalt surface course standard TS2010 developed by Transport Scotland (TS, 2012) is a good example of this future. TS2010, which is based on a generic stone mastic asphalt, uses polymer-modified bitumen and is designed around void content so as to ensure long-term durability, and early-life wet skid resistance by the application of grit during the initial rolling at the time of construction. This specification has no macro-texture requirement, as this is deemed to be a function of proper design. The skid resistance is measured during early life using the GripTester test, and after 2 years using the SCRIM test.

The relationship between vehicle speed, wet skid resistance and type of surface course needs to be better understood. For example, the assumption that roads carrying traffic at higher speeds require surfaces with increasingly higher skid resistance is not necessarily valid: for example, research into high-speed skid resistance using the pavement friction tester to simulate locked-wheel conditions at motorway speeds has shown that low-textured, 6 mm thin surface course mixes can outperform larger nominal-size thin surface courses with greater macro-texture (Dunford, 2013; Dunford et al., 2014).

It is well recognised that the effect of water on tyre–road friction within the contact patch differs for different surface course mixtures: for example, well-compacted, low-void, low macro-texture, durable mixes give a greater tread–aggregate contact area and the means to disperse water at high speed. Why some aggregates and types of asphalt surfacings perform better than expected, while others give lower in-service skid resistance than predicted by the PSV test, requires greater understanding.

Skid resistance has long been regarded as the most important surfacing characteristic. Nowadays, however, traffic noise, rolling resistance, spray, the reduced use of virgin aggregates and the associated greater reuse of recycled materials in surface courses are receiving greater recognition as significant pavement design criteria. In the future, therefore, it can be expected that these will be given increased consideration, alongside the importance of skid resistance: for example, high-texture surface dressing may have good wet skid resistance but its greater rolling resistance will result in greater tyre wear, road noise and fuel efficiency, and, as a consequence, this type of texture, although safe, may not be considered suitable for use in a heavily trafficked urban environment.

Climatologists predict that global climate change will bring greater extremes of temperature and rainfall, and it can be expected that these will affect skid resistance in both positive and negative ways. Increasing energy and transport costs, increasing scarcity of certain natural resources, and the greater use of long-term maintenance contracts have all promoted greater interest in the concept of sustainable highway development. These changes will inevitability cause road engineers to reconsider their decision-making in relation to the relative importance of skid resistance in the design of pavement surfaces.

The issue of risk has highlighted that greater knowledge is required regarding the materials being used both in the structural layers of pavements and at the tyre–asphalt interaction level. While over-specifying aggregate with high levels of skid resistance may result in safer roads, it is not the most sustainable option if the aggregate or asphalt course fails prematurely after only lasting for a few years.

Perhaps, above all, it is essential to better understand what happens to skid resistance with time. Research needs to continue that is based on the use of existing recognised standard tests such as the PSV, FAP, SCRIM and GripTester tests. However, it also requires the development of new test methods, test equipment and better predictive modelling methods to better understand the fundamental reasons why different surface course mixtures may perform in different ways during their in-service lives.

REFERENCES

ASTM (American Society for Testing and Materials) (2009a) ASTM E2157-09: Standard test method for measuring pavement macrotexture properties using the circular track meter. ASTM, Philadelphia, PA, USA.

ASTM (2009b) ASTM Es11-09: Standard test method for measuring paved surface frictional properties using the dynamic friction tester. ASTM, Philadelphia, PA, USA.

ASTM (2010) ASTM F870-94: Standard practise for tread footprints of passenger car tires groove area fraction and dimensional measurements. ASTM, Philadelphia, PA, USA.

BSI (British Standards Institution) (2002) BS EN 13036-3:2002: Road and airfield surface characteristics. Test methods. Measurement of pavement surface drainability. BSI, London, UK.

BSI (2004) BS EN 13473-1:2004: Characterization of pavement texture by use of surface profiles. Determination of mean profile depth. BSI, London, UK.

BSI (2009) BS EN 1097-8:2009: Tests for mechanical and physical properties of aggregates. Determination of the polished stone value. BSI, London, UK.

BSI (2010) BS EN 13036-1:2010: Road and airfield surface characteristics. Test methods. Measurement of pavement macrotexture depth using a volumetric patch technique. BSI, London, UK.

BSI (2011) BS EN 13036-4:2011: Road and airfield surface characteristics. Test methods. Method for measurement of slip/skid resistance of a surface: the pendulum test. BSI, London, UK.

BSI (2012) BS EN ISO 25178-2: 2012: Geometric product specifications (GPS). Surface texture: areal terms, definitions and surface texture parameters. BSI, London, UK.

BSI (2014) BS EN 12697-49:2014: Bituminous mixtures. Test methods for hot mix asphalt. Determination of friction after polishing. BSI, London, UK.

Doe M and Roe P (2008) *Report on State of the Art of Test Methods: TYROSAFE Deliverable D04, 7th Framework Programme*. See http://tyrosafe.fehrl.org (accessed 16/06/2015).

Dunford A (2013) *Friction and the Texture of Aggregate Particles Used in the Road Surface Course*. PhD thesis, University of Nottingham, Nottingham, UK.

Dunford A, Viner H, Greene M and Brittain S (2014) High speed friction of thin surface course systems. *Proceedings of the 4th International Safer Roads Conference, Cheltenham, UK*.

Dynatest (2014) http://www.dynatest.com (accessed 16/06/2015).

Friel S (2013) *Variation of the Friction Characteristics of Road Surfacing Materials with Time*. PhD thesis, University of Ulster, Newtownabbey, UK.

Huschek S (2004) Experience with skid resistance prediction based on traffic simulation. *Proceedings of the 5th Symposium of Pavement Surface Characteristics, Toronto, Canada*.

Kane M and Scharnigg K (2009) *Report on Different Parameters Influencing Skid Resistance, Rolling Resistance and Noise Emissions: TYROSAFE Deliverable 10, 7th Framework Programme*. See http://tyrosafe.fehrl.org (accessed 16/06/2015).

Millar P (2013) *Non-Contact Evaluation of the Geometric Properties of Asphalt Surfacings Using Close Range Photogrammetry*. PhD thesis, University of Ulster, Newtownabbey, UK.

Nicholls JC (1997) *Laboratory Tests on High-friction Surfaces for Highways*. Transport Research Laboratory, Crowthorne, UK, Report TRL176.

Racelogic (2014) http://www.racelogic.co.uk (accessed 16/06/2015).

TS (Transport Scotland) (2012) TS2010: Surface course specification and guidance. TS, Glasgow, UK.

Turnkey Instruments (2014) http://www.turnkey-instruments.com (accessed 16/06/2015).

Woodward D (1995) *Laboratory Prediction of Surfacing Aggregate Performance*. DPhil thesis, University of Ulster, Newtownabbey, UK.

Vericom (2014) http://www.vericomcomputers.com (accessed 16/06/2015).

Highways
ISBN 978-0-7277-5993-1

ICE Publishing: All rights reserved
http://dx.doi.org/10.1680/h5e.59931.549

Chapter 18
Design and use of thin surface treatments

Hussain Khalid   Senior Lecturer, School of Engineering, Liverpool University, UK

18.1.   Surface treatment types and purposes

Surface treatments are used to improve the deteriorating surface condition of road pavements that are otherwise structurally sound. Numerous types of surface treatment are used worldwide to restore the skid resistance and texture depth of surfacings, seal surface cracks and correct the longitudinal profile of roads. This chapter considers various types of thin surface treatment, with special emphasis on surface dressings, which are widely used in the UK and are acknowledged to be one of the most common, and cost-effective, measures used to improve the quality and skidding resistance of road surfaces.

In its simplest sense, applying a surface dressing involves spraying a thin film of an emulsion binder onto an existing road surface, followed by the application of a layer of aggregate that is then compacted by a roller to promote contact between the chippings and the binder, and to initiate the formation of an interlocking mosaic. Prior to the turn of the 21st century, before stone mastic asphalt and thin surfacings came into greater usage in the UK, it was estimated that about 100 million square metres of road were surface dressed every year: now, around 6 million square metres of road are currently surface dressed each year in the UK, and the industry considers that this amount will increase significantly in future years (Robinson, 2014). Surface dressings are used on roads of all types, from those that carry very low volumes of traffic to motorways and trunk roads that carry tens of thousands of vehicles per day.

Road Note 39 (Roberts and Nicholls, 2008) and the *Code of Practice for Surface Dressing* (RSTA, 2014) provide basic advice regarding the selection of the appropriate type, design and application of the surface dressings used in the UK today. They are mainly used as a maintenance tool to fulfil the following functions:

- provide both texture and skid resistance to a road surface
- arrest the disintegration and loss of aggregate from a road surface
- seal the surface of the road against the ingress of water and thus protect the pavement structure from moisture damage
- provide a distinctive colour to the road surface
- improve the visual aesthetics of a road by providing a more uniform appearance to a heavily patched surface.

549

What a surface dressing will not do is restore the ride quality of an already deformed road.

Experience since the late-1960s has shown that high-friction surfacing (HFS) systems are highly effective in reducing traffic accidents at sites with high traffic flows and skidding risks: for example, at approaches to signal-controlled junctions, roundabouts and pedestrian crossings that are subject to heavy flows of vehicles (HA *et al.*, 1999). HFS treatments, which are essentially special types of surface dressing that use graded 1–3 mm aggregate and binders with special high-specification properties, are therefore discussed in this chapter.

Other types of thin surface treatment that are widely adopted in the UK, other European countries and elsewhere include slurry surfacings and micro-surfacings, so these are discussed here also. Cold-laid slurry surfacings offer a quick, efficient and cost-effective means of maintaining skid resistance, re-profiling the road surface and protecting roads against the damaging effects of water and air. Slurry surfacings range in thickness from about 2 to 8 mm, and micro-surfacings from about 10 to 20 mm. Slurry surfacings are suitable for footways, areas that are trafficked only occasionally and at low speeds, and for traffic delineation. Micro-surfacings, on the other hand, are targeted at all roads, including high-speed roads carrying significant traffic volumes and, hence, require appropriate levels of skid resistance and texture retention (HA *et al.*, 1999; RSTA, 2011a).

The chapter concludes with some brief discussions regarding proprietary hot-applied, very thin and ultra-thin surface treatments developed in Europe and used widely in the UK and USA (e.g. very thin asphalt concrete (BBTM, from its French name *béton bitumineux très mince*), and asphalt for ultra-thin layers (AUTL)), both of which were developed in France and subsequently franchised internationally. BBTMs are laid in 20–30 mm thick layers, whereas AUTLs are about 10–20 mm thick, and laid in single layers and compacted using specialist plant. Both BBTMs and AUTLs are considered as a cross between thin surface courses and surface dressings; that is, their functions include all those of surface dressings as well as being used to regulate the longitudinal road profile.

Thin surface treatments are currently in transition in terms of what material is used and where it should be used, and, hence, care has been taken to provide a comprehensive list of references that can be consulted for further detailed study.

18.2.   Surface dressings

For many years, surface dressings have been designed in the UK in accordance with Road Note 39 (Roberts and Nicholls, 2008). As noted elsewhere in relation to other bituminous materials, European Committee for Standardization (CEN) standards have now been introduced into the UK (BSI, 2006a), and, consequently, there have been significant changes in both surface dressing terminology and practice that are still in the course of transition. The sixth edition of Road Note 39 copes with these changes as far as possible at this time.

The CEN standards have introduced a new nomenclature for aggregates of d/D, where d and D are the minimum and maximum allowable stone sizes, respectively. These are applicable to all types of surface dressings, including those that use single-size aggregates.

A major change to the nomenclature of emulsions has also occurred since the introduction of CEN standards, and BS EN 13808 (BSI, 2013a) provides a framework specification to categorise bituminous emulsions, including both modified and unmodified varieties, for a wide range of applications. Emulsion terminology is based on chemical nature, the nominal binder content of the emulsion, binder type and chemical stability: hence, the former K1-70 category of bitumen emulsion (BSI, 1984) has now been replaced by different notations, such as C69B4, C69BF3 and C69BF4. (The reader is referred to the National Foreword of BS EN 13808 and to BS 434-2 (BSI, 2006b) for a detailed explanation.)

Cutback bitumens have been removed from Road Note 39 and are no longer available from suppliers, so that all surface dressing binders now used in the UK are bitumen emulsions.

Since the mid-2000s, the use of polymer-modified bitumen (PMB) emulsions in surface dressing works has increased significantly and, as a result of experience, Road Note 39 now recommends that the spread rates for these binders should be higher than for unmodified bitumen emulsions. PMB emulsions impart a number of benefits to surface dressings when compared with conventional emulsions: these include improved early grip on aggregate, low-temperature elasticity and long-term system cohesion.

The weight and, hence, the damaging power of passenger vehicles has increased since the turn of the 21st century, so that, when carrying out traffic assessments (HA *et al.*, 2006a), the weight limit indicating commercial vehicles is now taken as 3.5 t (it was 1.5 t) so as to accommodate the large increase in heavier private vehicles within the non-commercial range. As a consequence, the number of roads carrying low flows of medium to heavy weight vehicles has reduced, but the number of road categories carrying up to 500 cv/ lane/day – which cause most chipping embedment – has increased, thereby resulting in a closer correlation between the road traffic categories adopted in Road Note 39 and the road classification adopted in the New Road and Street Works Act 1991 (House of Commons, 1991).

18.2.1 Types of surface dressing

Surface dressing types vary according to the number of layers of chippings and binder used. Figure 18.1 is a schematic diagram of the main types of dressings currently described in Road Note 39.

A single-surface dressing consists of a single application of a binder followed by a single application of chippings. This system is used mainly at low shear stress sites.

A racked-in surface dressing system consists of a single application of binder laid as a thicker layer than in a single-surface dressing, followed by two applications of chippings

Figure 18.1 Types of surface dressings currently described in Road Note 39 (Roberts and Nicholls, 2008): (a) single-surface dressing; (b) racked-in surface dressing; (c) double-surface dressing; (d) inverted double-surface dressing; (e) sandwich surface dressing

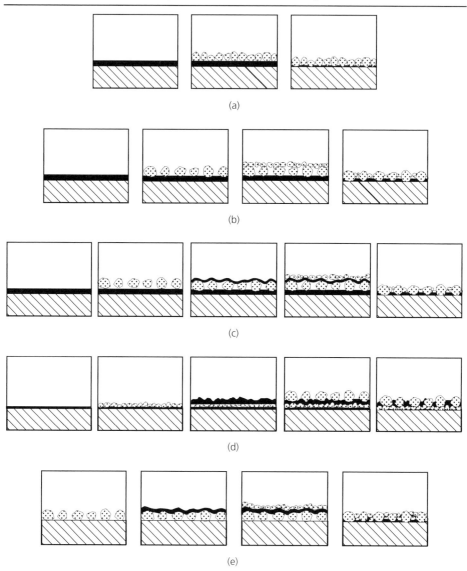

(a)

(b)

(c)

(d)

(e)

of different sizes. The first application consists of larger chippings, and constitutes about 90% cover, which leaves gaps in the mosaic that are filled by the second application of smaller-size chippings (i.e. the smaller chippings lock the larger chippings in place, and lead to a more stable mixture). This system is recommended for sites with heavy and/or fast traffic.

A double-surface dressing uses two layers of chippings and two applications of binder. The first covering of chippings is placed on top of the first binder application so that there are no gaps between the stones, and the second binder application is then placed between the two layers of chippings. While this system is primarily recommended for use on road surfaces that are binder lean, it is also used as an alternative to a racked-in system.

An inverted double-surface dressing, previously known as a pad coat and single-surface dressing, comprises a single-surface dressing with small-size chippings (i.e. the pad course) that is applied to a road with an uneven surface hardness, and this is then followed by a second surface dressing with larger-size chippings. The initial pad coat produces a more uniform surfacing, and this is subsequently surface dressed. Inverted double-surface dressings have been used on hard or very hard substrates (e.g. concrete) to reduce the effective hardness of the surface; however, a racked-in system is now the preferred option for use on very hard substrates.

A sandwich surface dressing involves the spreading of an unbound layer of chippings on the road surface prior to a single-surface dressing system being applied over it. The sandwich surface dressing is mainly used in situations where the substrate is binder rich: it is especially applicable in hot weather on single carriageways with heavy traffic flows. Both the sandwich and the double-surface dressing systems are characterised as having a higher texture depth than that produced with a single system with the same-sized chippings.

In some instances, special surface-dressing systems (e.g. HFS) are recommended (RSTA, 2011b) for use at sites that are defined as being high risk: for example, at (a) approaches to major junctions and pedestrian crossings, (b) sites with gradients steeper than 10% and (c) horizontal curves with radii <500 m on single carriageways.

There are three classes of HFS – types 1, 2 and 3 – and the use of each depends on the volume of commercial vehicles using the road (BBA, 2008): type 1 is suitable for a maximum traffic flow of 3500 cv/lane/day, type 2 for flows of up to 1000 cv/lane/day and type 3 for flows of up to 250 cv/lane/day. Essentially, each HFS system resembles a single-surface dressing, but utilises a very durable resin-based binder and very hard aggregates with a high polished stone value (PSV) and low aggregate abrasion value (AAV). The HFS systems can be applied either cold or hot, with the main difference being that, with the cold application, the resin-based binder is first sprayed onto the road surface, and the aggregates are then embedded into it, whereas with the hot-applied dressing the materials are pre-mixed and heated before being screeded onto the prepared surface. Notwithstanding its greater environmental impacts and temperature control requirements, the main benefit arising from using a hot-applied HFS lies in the reduced minimum time between completion of a new asphalt surfacing and applying an HFS (i.e. the time delay is significantly shorter with a hot-applied HFS than with the cold-applied variety).

18.2.2 Factors affecting the use and performance of surface dressings

In terms of performance, surface dressings are designed to provide skid resistance through adequate texture depth, while ensuring the acceptability of the noise levels generated at the tyre–road interface. The selection of an appropriate surface dressing system involves the selection of the type and spread rate of component materials, that is, binder and chippings, and the application of local adjustment criteria that are dependent on the site, the weather and traffic conditions. Typical factors influencing the choice of a system are as follows.

(a) Road hardness, which represents the resistance to chipping embedment by an existing road surface, is an important factor in determining the correct size of the chipping. Road Note 39 divides surface hardness into five categories, ranging from very hard (e.g. concrete roads) to very soft. The hardness of the road is normally measured (BSI, 2004) with a hardness probe when the surface temperature is between 15 and 35°C.

(b) The traffic category is a factor that relates to the anticipated traffic volume that the road is expected to carry. To assess the probable amount of chipping embedment, Road Note 39 divides traffic levels into eight categories, termed A to H, ranging in severity from >3250 cv/lane/day (class A) to <50 cv/lane/day (class H).

(c) The traffic speed is also accounted for, in that, on lower traffic category roads (e.g. classes F, G or H), on which vehicles are moving at relatively high speeds, it is recognised that the possibility of windscreen damage from loose chippings is increased.

(d) The surface condition is important in determining the appropriate type of surface dressing, as the existing binder content of the road surface affects the amount of binder required for the initial retention of chippings prior to their long-term embedment. Based on the binder content, Road Note 39 therefore divides the existing surface condition into six categories, ranging from very binder rich to very binder lean, for a given road hardness category.

(e) The highway layout addresses the geometric parameters along the length of road being considered that affect the stresses imposed by traffic on the road surface. Typically, these would include the longitudinal gradient, the radius of curvature and the superelevation. Other factors (e.g. turning movements and sharp decelerations at junctions) also affect the stresses imposed on road surfacings and influence the choice of binder used to retain the chippings at particular sites.

(f) The altitude of a site location above sea level, as well as its susceptibility to shade, influence the temperature range that a surface dressing can be expected to encounter when in service. These location parameters affect the embedment of chippings and the type and quantity of binder needed to retain them. While Road Note 39 adopts four national categories to cover geographical location and altitude above sea level, it recognises that the UK climate is not uniform, and therefore recommends that these categories be used as guides only, and that consideration should always be given to the local climate experienced at a proposed site.

(g) The key function of a surface dressing is the restoration of the (diminished) skid resistance of a surfacing on an otherwise structurally sound road pavement. Site requirements for skid resistance are decided by the skidding and road safety policies of an individual highway authority. Investigatory levels for skid resistance (formerly referred to as critical performance levels) for various types of site category (e.g. at approaches to roundabouts, steep gradients, etc.) are given in HD 28/04 (HA *et al.*, 2004) as measured by SCRIM (sideway-force coefficient routine investigation machine). (See Chapter 17 for details regarding factors affecting the measurement of skid resistance.) The relationship between skid resistance, traffic severity and the required aggregate PSV is provided in HD 36/06 (HA *et al.*, 2006b), which sets out the minimum PSV of chippings required for new construction and for maintenance works on major roads and motorways. Although the extent of the requirement to combat skidding may vary at different locales within a given site, the use of aggregates of varying PSV to resolve skidding problems is not considered practical. (Note also that PSV levels greater than 70 are only achieved with the special types of aggregate that are associated with HFS applications, and these are not considered in Road Note 39.)

(h) The seasonal weather variation is second only to traffic volume in its impact on the surface dressing performance, as the longevity of the treatment relies on chipping embedment in the pavement surfacing and/or its reorientation into more stable positions before the start of cold weather. If a stable mosaic is not established before the onset of low temperatures (e.g. $<10°C$), chippings are susceptible to being removed by traffic due to binder embrittlement. The general concept is that the larger the chipping size used, the earlier in the season the surface dressing treatment should be applied, to allow more time for chipping embedment and, hence, improved dressing stability before the onset of cold weather.

(i) Bituminous binders are viscoelastic materials, and are, therefore, temperature susceptible during both construction and service. Chipping embedment is also affected by temperature, as the hardness of the (bituminous) surfacing varies, becoming soft at high temperatures and hard at low ones. Persistent high temperatures often cause considerable chipping embedment, leading to bleeding or fatting-up.

(j) Bituminous emulsions break slowly in cold or wet conditions or when the humidity is high. When the humidity is greater than 80%, delays in the breaking of emulsions can be expected, and the addition of agents to accelerate the breaking may be necessary. Rain unfavourably affects the initial adhesion of a binder to chippings.

(k) The presence of moisture on the chippings and/or road surface affects the rate of the build-up of strength of newly laid surface dressings. Dampness on the chippings can reduce or delay the bond formation between the chippings and the binder.

18.2.3   Designing a surface dressing for a road surface

There are two principal objectives involved in the design of any surface dressing system: (a) the determination of the size and amount of chippings and (b) the determination of the type and amount of binder required. The two most critical parameters affecting the design are the traffic severity and the hardness and condition of the road surfacing.

18.2.3.1   Road Note 39 design process

Based on the above parameters, Road Note 39 gives recommendations for the type of surface dressing that is most likely to perform under different circumstances. These recommendations, which are entirely experience based, are derived from extensive road trials and practice. Figure 18.2, which is based on one of the diagrams embodied in the road note, deals with the selection of the type of surface dressing for the lower traffic category of road (i.e. roads on which surface dressing work is most commonly undertaken). (A similar flow chart is available in Road Note 39 for more heavily trafficked roads.)

The design methodology begins with the initial selection of the surface dressing type (see Section 18.2.1) based on traffic intensity, surface hardness and highway layout (see Section 18.2.2). The next step is the selection of the size of chippings from designated tables in Road Note 39: this is achieved by relating the number of commercial vehicles and the road hardness to the size of chipping required to achieve a desired texture depth as specified in HD 37 (HA *et al.*, 1999). For single-surface dressings used with traffic category E (251–500 cv/lane/day), and normal road surface hardness, the road note recommends a chipping size of 6.3/10 mm.

The selection of the size of chipping is followed by the selection of the chipping application rate to cover the binder film. Depending on the type of surface dressing selected, the chippings should be spread to achieve a designated percentage of road surface coverage by shoulder-to-shoulder aggregate particles. The adequacy of the chipping spread is checked by test procedures described in BS EN 12272-1 (BSI, 2002).

The next step involves the selection of the type of binder: based on experience, this takes into account the traffic category, time of year, weather conditions and the likely stresses to be encountered. Finally, the binder application rate is selected from designated tables in Road Note 39. The application rate is related to the size of the chippings, the nature (hardness) of the existing road surface and the envisaged degree of embedment of chippings by traffic (traffic category). For single-surface dressings and traffic category E (251–500 cv/lane/day), and normal road surface hardness, the road note recommends a spread rate of 1.6 litres/m^2 with a recommendation to use a PMB emulsion.

18.2.3.2 Appraisal of the Road Note 39 design method

The Road Note 39 approach, as with all empirical design methods, is most applicable to usage where it was developed (in the UK) at sites and under conditions similar to those experienced previously. Thus, the need to identify design parameters such as road hardness and surface condition tends to adversely affect the efficiency of the method because, in essence, it requires the situation at any selected site to be related to others with similar traits from which the design recommendations were derived.

Although the current edition of the road note includes recommendations for the use of modified binders such as PMB emulsions, insufficient detail is disclosed on the type and content of the modifiers used in the binders, so that the true nature of modified binders is still shrouded in the mystery of proprietary procurement. The only major shift in this regard is the focus in specifications on the cohesion of residual (emulsion) binder as measured in BS EN 13588 (BSI, 2008), and by the requirement in clause 956 of the Specifications for Highway Works (HA *et al.*, 2014) to report on the rheological

Figure 18.2 Selecting surface dressings for lower-trafficked roads, based on Road Note 39. (Adapted from Roberts and Nicholls (2008) and reproduced by courtesy of the Transport Research Laboratory)

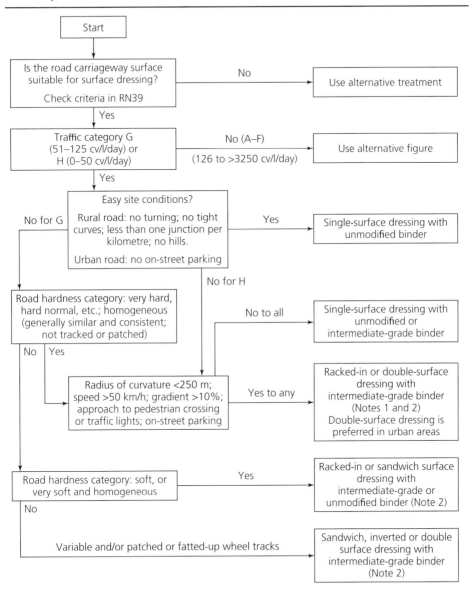

Note 1 High-friction surfacing may be considered, depending on the site difficulty and the quality of the substrate

Note 2 Where initial stability is required (junctions to major roads, pedestrian areas, fast commuter runs, on-street parking, etc.), intermediate-grade binder or above may be required. Double-surface dressings have greater stability than racked-in surface dressings and are more tolerant of varying surface conditions and road hardness. Racked-in and double-surface dressings may assist prevention of tearing at junctions, slip lanes, on hills, etc.

characteristics of the binder. This shift is likely to be intended to reflect the advantages of PMB emulsions, whose use in surface dressing works has undergone a significant increase, and the enhanced reliance of specifications on fundamental performance-related material properties.

With regard to the whole surface dressing system, Road Note 39 fails to include assessment tests, simulative or otherwise, to check on the adequacy of a design before its implementation: this is a major shortcoming. Also, important design factors such as the existing road surface texture, variations in the chipping compaction associated with aggregate type and shape, and the need to account for the resulting texture depth are still not considered adequately, notwithstanding the specification in HD 37 (HA *et al.*, 1999) of a minimum texture depth of 1.05 mm for major roads and 0.7–0.8 mm for lower traffic-category roads.

One major advantage of the Road Note 39 design approach, however, is its simplicity. Once the site and its prevailing conditions have been classified, the design process becomes fairly easy, as the binder and aggregate application rates are then easily obtained from pre-established tables associated with the different types of surface dressing.

18.2.4 Surface dressing distress modes

The performance levels of surface dressings are specified in terms of the macro-texture requirements and on the level of defects they exhibit in service, measured both qualitatively and quantitatively in accordance with BS EN 12272-2 (BSI, 2003). A surface dressing is considered to have failed when it is either (a) no longer able to meet the skid resistance and texture depth needs of the traffic using the surface or (b) no longer protecting the structure of the pavement from the ingress of water.

Premature failure is almost always the result of inadequacies in one or more of the four stages comprising the production of a surface dressing, namely specification, design, materials and construction (including aftercare). The following discussion is primarily based on a comprehensive overview of surface dressing failures, including definitions, evaluation, avoidance and remedies, that is available in the literature (HA *et al.*, 1999).

Chipping loss, known as scabbing, usually occurs either in the very early life of the dressing or in the first winter following placement. It results from traffic actions that break down the adhesive bond between the chippings and the binder, or the bond between the binder and the road surface. Scabbing is normally attributed to a binder failure due to the use of the wrong type of binder or the incorrect binder application rate. It is very often manifested as very little or no embedment of the chippings into the road surface because the surfacing is too hard and/or there is insufficient time between the laying of the dressing and the onset of cold weather, coupled with too brittle a binder to retain the chippings (Robinson, 1968).

Failure due to excessive chipping embedment is normally attributed to the road surface being too soft. It is caused by the actions of traffic, which force the chippings down into the surface and cause the binder to rise up so that there is a loss of texture depth and a reduction in the skid resistance. This failure mode has also been attributed to the selection of the wrong chipping size and an increase in traffic intensity. Research has shown (Abdulkarim, 1989) that the depth and rate of embedment depends on, among other things, the road surface temperature, the influence of heavy goods vehicles and the loading duration (i.e. the speed of traffic).

Failure due to bleeding is normally manifested as black patches of excess binder on the surface of the surface dressing. In other words, a bleeding surface dressing has a smooth and slick appearance where the aggregate particles are less visible. Bleeding is caused by either an excess of binder in proportion to the aggregate, or where the aggregate is forced by traffic to achieve levels of embedment beyond the design embedment depth (Gransberg and James, 2005). A highly temperature-susceptible binder has a tendency to become fluid at high ambient temperatures: this causes the binder to flow under the action of traffic, so that it rises to the road surface and flushes the chippings, with a consequent loss of texture depth – and of skid resistance. During warm summer months, the binder may rise to the top of the surface dressing as the chippings become embedded into the road surface.

Fatting-up is another failure mode that is considered as being due to either too heavy a rate of binder application or the use of smaller-size chippings than is necessary for the given traffic conditions and road hardness (Shuler *et al.*, 2011; Wright, 1978). It may also be caused by the subsequent embedment of chippings in the road surface, crushing of the chippings, or the absorption of dust by the binder leading to its increased effective volume (Southern, 1983).

Most chippings will wear out as a result of the abrading action of traffic over a long period of time. When this occurs, the surface dressing will have lost its ability to provide adequate skid resistance, even though the overall structure or stability of the surface dressing may be intact. The wearing out of chippings can be the result of using chippings of inadequate AAV. A good surface dressing design should ensure that, for the chipping selected, the design life of the dressing will have been exceeded before the traffic wears down the chippings to below acceptable AAV levels.

18.3. Slurry seals and micro-surfacings

Slurry seals and micro-surfacings are cold-mixed, thin surface treatments that are used as maintenance measures to restore skid resistance, improve the unevenness of roads and produce a good quality of ride for motor vehicles. To a limited extent they are also used to fill up surface depressions and provide a seal against the ingress of moisture into the pavement surfacing.

Slurry seals are made up of conventional or modified bitumen emulsion, fine aggregates, additional water, additive and mineral filler (typically either cement or hydrated lime): they are usually laid in thicknesses ranging from 2 to 8 mm, and are mostly used on footways and low-speed roads that only experience traffic occasionally. Micro-surfacings, which comprise graded aggregates up to 10 mm in size, PMB emulsion and fibre

Figure 18.3 Difference between (a) a slurry seal and (b) a micro-surfacing. (Adapted from Gransberg (2010) and reproduced by courtesy of the Transportation Research Board)

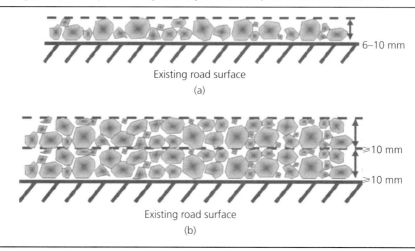

Existing road surface

(a)

Existing road surface

(b)

additives, water and cement filler, are commonly used for rut filling: they are usually laid in thicknesses ranging from 10 to 20 mm and are used on all classes of road (HA *et al.*, 1999).

Slurry seals and micro-surfacings are environmentally friendly solutions to some road surfacing problems. Being cold mixes, they require no more energy for mixing or application: also, they make use of bitumen emulsions that contain water, which evaporates into the atmosphere upon breaking and causes no harmful effects on the environment (Gransberg, 2010; RSTA, 2011a).

While slurry seals and micro-surfacings tend to be grouped for discussion purposes, micro-surfacings have three features that differentiate them from slurry seals: (a) their bitumen emulsions always contain polymer additives; (b) they cure more rapidly (through chemical reaction), as the quick-setting emulsions break and cure much more quickly due to the inclusion of the special additives, thereby permitting earlier traffic usage after maintenance; and (c) they can be placed in layers thicker than one-stone deep (Figure 18.3).

18.3.1 Design of slurry seals and micro-surfacings

Worldwide, the most common method used to design slurry seals and micro-surfacings is that developed by the International Slurry Surfacing Association (ISSA) and issued in Guideline Document A143 (ISSA, 2011). The empirically derived process, which is used for both materials, relies on using specified aggregate grading referred to as ISSA type II or type III: these gradations are of a nominal maximum aggregate size of 10 mm (9.5 mm in the USA), although it is permissible in the specification for type II to have a 5 mm maximum size. The type III envelope circumscribes coarser aggregate grading than that of type II.

Based on a selected grading, a matrix of mix recipes is created, and the manual mixing time for each mix is recorded. Using aggregate coating by the emulsion as the prime criterion, the best mix is selected from among those alternative designs whose mixing time exceeded the standard minimum of 120 s across the expected temperature range.

While the optimum water and emulsion contents are selected at the mixing stage, the optimum emulsion content is further examined and confirmed in the next stage, where the cohesion build-up of the mix is tested. Cohesion build-up is measured by a cohesion tester, which is a lightweight, portable device that can be adjusted to apply varying pressure to a slurry or micro-surfacing sample. Torque is applied with a manually operated torque wrench with the pressure supplied by compressed air, and the value measured is displayed on a torque meter. Cohesion values are measured at set times (after 30 and 60 min, and 24 h) and compared with minimum required values specified by the ISSA.

The optimum binder content is empirically determined by plotting the output from two tests, namely the abrasion loss against the emulsion content from the wet track abrasion test (WTAT) and the sand adhesion value against the emulsion content from the loaded wheel test (LWT), and determining where the two curves intersect. The WTAT identifies the minimum emulsion content against a specified maximum loss value, whereas the LWT identifies the maximum emulsion content against a specified maximum sand adhesion value. The optimum emulsion content represents the median of the evolved binder content range straddling the intersection point. In practice, the selected amount may be subsequently adjusted by an experienced designer to take account of climatic variables encountered during construction.

The WTAT (ISSA, 2012a) measures the resistance to wear of the micro-surfacing under wet abrasion conditions. With this test, a 6 mm thick and 279 mm diameter disc-shaped sample of the material is prepared by pouring a micro-surfacing sample into a circular opening of a polymethyl methacrylate template resting on a larger circlet of roofing felt, and allowed to set, after which it is oven dried and weighed. The sample is then placed in a water bath for 1 h, and is mechanically subjected to abrasion for 5 min with a standard rubber hose underwater. After this, the debris is washed off the sample, and it is then oven dried and weighed again. The resultant weight loss is then expressed in grams per square metre as the WTAT loss, and compared with ISSA permissible values.

With the LWT (ISSA, 2012b), the same size micro-surfacing sample as that in the WTAT is compacted by means of a loaded rubber tyre, simulating a wheel. From this test, the sand adhesion value, which is an indirect measure of the amount of excess binder in the mix, is obtained, and this is then compared with the acceptable limit for a mixture as given in the ISSA specifications.

18.4. Other thin surface treatments

While surface dressings have many advantages, they are also renowned for (a) requiring frequent adjustment (to suit local conditions) to the application rates of emulsion binder and aggregate, and constant attention in the early stages after laying; (b) their inability to restore evenness to the longitudinal profiles of deformed roads; (c) having loose chippings that cause damage to vehicles, especially to windscreens; and (d) the high noise levels generated by, especially, single-surface dressings (Morian, 2011; RSTA, 2008; Whiteoak and Read, 2003). The need for additional surface treatments that would fill the gap between surface dressings and slurry seals/micro-surfacings on the one hand and conventional thin surface courses (formerly known as wearing courses) laid at more than 40 mm in thickness has long been recognised. The performance requirements of these treatments were generally accepted to be easily and quickly laid, to reduce the chance of flying chips, to be quieter than surface dressings and, when possible, to enhance the major qualities of safety and impermeability.

18.4.1   Developments in the UK
The surface treatment gap in the UK has been bridged by the advent of thin asphalt surface course systems, now simply termed thin surfacings, which were originally developed in France in the 1980s.

Thin surfacing systems (Nicholls et al., 2010), some of which are hot mixed and hot laid, are currently the preferred surfacing option in the UK for both new construction and maintenance works on motorways and major trunk roads. Nominally laid at a thickness of 15–40 mm (Whiteoak, 2000), these materials are generally classified in the UK (HA, 2008) into three types, depending on the thickness at which they are laid: thin surfacings <18 mm thick are grouped as type A, those between 18 and 25 mm thick are type B, and type C thin surfacings are >25 mm but <50 mm thick.

18.4.2   Developments in the USA
The surface treatments most commonly used in the USA comprise hot-mixed, hot-laid thin overlays (used with or without milling of the underlying surfacing) and paver-laid, cold-mixed, cold-laid seal courses and micro-surfacings (Morian, 2011).

Thin overlays, which are laid at a thickness of up to 25 mm, are either ordinary asphalt mixtures with a dense or open-graded particle size distribution, or a stone mastic asphalt type of material, both of which provide the desired surfacing performance properties, such as restoring the texture depth, improving the skidding resistance, sealing against moisture ingress and arresting disintegration (Gransberg, 2010; Hall et al., 2009).

Chip seals, fog seals, sand seals and cape seals are cold-mixed and cold-laid surface treatments that are commonly used in the USA for maintenance purposes.

A chip seal can be visualised as akin to a surface dressing, in that it commonly involves the application of a bitumen emulsion – which may be polymer modified – onto an existing road surface, after which aggregates are evenly spread (usually one-stone thick), and a smooth-wheel roller is then used for compaction. Chip seals, which are also applied in multiple layers, are considered as an economically sustainable way of restoring skid resistance, of providing an impermeable seal against moisture ingress and of extending pavement life (Aktaş and Karaşahin, 2013; Gransberg and James, 2005; Moraes and Bahia, 2013; Shuler *et al.*, 2011).

A fog seal can be described as similar to a chip seal but without the dressing of the aggregate; that is, it makes use of an even spray of a diluted bitumen emulsion onto the distressed road surface, and, as such, is commonly used to prevent aggregate loss from chip seal treatments. It can be used to address ravelling and low-severity fatigue cracking, to improve aggregate retention in existing road surfacings and to extend the service lives of pavements, especially ones with chip seals, by increasing the impermeability of surfacings to both water and air (Im and Kim, 2013; Morian, 2011). Studies have also shown a fog seal to be a cost-effective, easy-to-construct technique that gives a pavement a desirable appearance.

Sand seals are similar to chip seals except that they make use of a covering of fine sand rather than aggregate chippings after the application of the bitumen emulsion. A pneumatic-tyred roller is used to roll the sand when it has been applied, after which any excess sand is often removed. Sand seals are mainly used to seal pavement surfaces, rejuvenate aged asphaltic courses, provide delineation and improve road surface friction (Morian, 2011).

A cape seal is an application of a chip seal with a reduced coverage of chippings that is then covered by a slurry seal. As the name suggests, the cape seal concept originated in Cape Town, South Africa, where it is still used as a maintenance tool.

A cape seal is applied when the pavement deterioration is greater than a slurry seal is expected to correct, but deterioration has not yet progressed to a point requiring an asphalt overlay. As well as providing a new wearing surface, a cape seal treatment eliminates the problem of loose aggregate, holds the stones of the chip seal firmly in place, reduces traffic noise compared with that emanating from a chip seal surface and prevents water penetration causing subsequent damage to the road foundation. The disadvantage of the cape seal treatment is that it requires different construction equipment to lay a chip seal and a slurry seal. In addition, the cape seal application process requires a much longer construction time than either a chip seal treatment or a slurry seal application, resulting in greater traffic delays (Morian, 2011).

18.4.3 Developments in France

Two main types of thin surfacings have emerged from France in recent years that are of particular interest: (a) a thin (20–30 mm) asphalt concrete BBTM, which is specified in BS EN 13108-2 (BSI, 2006c), and (b) a thinner (10–20 mm) material termed asphalt for ultra-thin layers (AUTL), which was developed initially as Euroduit and then as Novachip, and is specified in prEN 13108-9 (BSI, 2013b). Both BBTM and AUTL are hot-mix asphalt materials that are laid at the above-mentioned thicknesses, in which the aggregate particles are generally gap graded to form stone-to-stone contact and to provide an open-surface texture. Both require the prior application of a tack or bond coat to achieve adequate adhesion with, and to seal, the existing road surface. The recommended application rate of the emulsion binder is about 0.7 and 1.0 litre/m^2 for

BBTM and AUTL, respectively (Bellanger *et al.*, 1992; Serfass *et al.*, 1991). Neither of these hot-mix, hot-laid thin surfacings require the aftercare service necessary with surface dressings, and can be laid in weather conditions outside those suitable for surface dressing works.

Both Novachip and AUTL are now marketed as proprietary products. In the UK, Novachip was used under the name Safepave (Nicholls, 2001), whereas in the USA it retained its franchise name, and has been used extensively as a cost-effective surface rehabilitation treatment on heavily trafficked high-speed roads (Hall *et al.*, 2009; Kandal and Lockett, 1997; Morian, 2011).

While both BBTM and AUTL use a 10 mm nominal size aggregate mix with a gap between the 2 and 6 mm size fractions, the main differences between them lie in the percentages of coarse aggregate and the binder content. BBTM, on average, has between 65 and 70% 6/10 mm size fractions and 5.5–6.0% by mass of binder while AUTL has 75–80% 6/10 mm stones and 5.2–5.6% by mass of binder (Bellanger *et al.*, 1992). The balance of either mix comprises a combination of fine aggregates and filler, with AUTL being predominantly coarser. The binder is most often a 60/70 or 80/100 pen grade bitumen, which is often modified by polymers or fibres to improve the mechanical properties and temperature susceptibility of the material.

Derivatives of the cape seal concept have relatively recently been introduced into France with trials using higher-performing components (Deneuvillers *et al.*, 2012).

REFERENCES

Abdulkarim AJ (1989) *Investigation of the Embedment of Chippings for an Improved Road Surface Design Procedure*. PhD thesis, Heriot-Watt University, Edinburgh, UK.

Aktaş B and Karaşahin M (2013) Chip seal adhesion performance with modified binder in cold climates: experimental investigation. *Transportation Research Record* **2361**: 63–68.

BBA (British Board of Agrément) (2008) *Guidelines Document for the Assessment and Certification of High Friction Surfacing for Highways*. BBA, Watford, UK.

Bellanger J, Brosseaud Y and Gourdon JL (1992) Thinner and thinner asphalt layers for maintenance of French roads. *Transportation Research Record* **1334**: 9–11.

BSI (British Standards Institution) (1984) BS 434-2:1984. Bitumen road emulsions (anionic and cationic). Code of practice for use of bitumen road emulsions. BSI, London, UK.

BSI (2002) BS EN 12272-1:2002: Surface dressing. Test methods. Rate of spread and accuracy of spread of binder and chippings. BSI, London, UK.

BSI (2003) BS EN 12272-2:2003: Surface dressing. Test methods. Visual assessment of defects. BSI, London, UK.

BSI (2004) BS 598-112:2004: Sampling and examination of bituminous mixtures for roads and other paved areas. Method for the use of road surface hardness probe. BSI, London, UK.

BSI (2006a) PD 6689: Surface dressing. Guidance on the use of BS EN 12271. BSI, London, UK.

BSI (2006b) BS 434-2:2006: Bitumen road emulsions. Code of practice for the use of cationic bitumen emulsions on roads and other paved areas. BSI, London, UK.

BSI (2006c) BS EN 13108-2:2006. Bituminous mixtures. Material specifications. Asphalt concrete for very thin layers. BSI, London, UK.

BSI (2008) BS EN 13588:2008: Bitumen and bituminous binders. Determination of cohesion of bituminous binders with pendulum test. BSI, London, UK.

BSI (2013a) BS EN 13808:2013: Bitumen and bituminous binders. Framework for specifying cationic bituminous emulsions. BSI, London, UK.

BSI (2013b) BS EN 13108-9 (draft): Bituminous mixtures. Material specifications. Asphalt for ultra-thin layers (AUTL). BSI, London, UK.

Deneuvillers C, Harnois S and Priez C (2012) Technique de maintenance: la famille des Cape Seal. *Revue Générale des Routes et Aérodromes* **906**: 77–82.

Gransberg D (2010) *Micro-surfacing: NCHRP Synthesis 411.* Transportation Research Board, Washington, DC, USA.

Gransberg D and James DM (2005) *Chip Seal Best Practices: NCHRP Synthesis 342.* Transportation Research Board, Washington, DC, USA.

HA (Highways Agency) (2008) *Interim Advice Note 157: Thin Surface Course Systems – Installation and Maintenance.* Stationery Office, London, UK.

HA, Transport Scotland, Welsh Government and Department for Regional Development Northern Ireland (1999) Section 5: surfacing and surfacing materials. Part 2: bituminous surfacing materials and techniques. In *Design Manual for Roads and Bridges*, vol. 7. *Pavement Design and Maintenance.* Stationery Office, London, UK, HD 37/99.

HA, Transport Scotland, Welsh Government and Department for Regional Development Northern Ireland (2004) Section 3: pavement maintenance assessment. Part 1: skidding resistance. In *Design Manual for Roads and Bridges*, vol. 7. *Pavement Design and Maintenance.* Stationery Office, London, UK, HD 28/04.

HA, Transport Scotland, Welsh Government and Department for Regional Development Northern Ireland (2006a) Section 2: pavement design and construction. Part 1: traffic assessment. In *Design Manual for Roads and Bridges*, vol. 7. *Pavement Design and Maintenance.* Stationery Office, London, UK, HD 24/06.

HA, Transport Scotland, Welsh Government and Department for Regional Development Northern Ireland (2006b) Section 5: pavement materials. Part 1: surfacing materials for new and maintenance construction. In *Design Manual for Roads and Bridges*, vol. 7. *Pavement Design and Maintenance.* Stationery Office, London, UK, HD 36/06.

HA, Transport Scotland, Welsh Government and Department for Regional Development Northern Ireland (2014) *Manual of Contract Documents for Highway Works*, vol. 1. *Specification for Highway Works.* Stationery Office, London, UK.

Hall JW, Smith KL, Titus-Glover L *et al.* (2009) *Guide for Pavement Friction. NCHRP Web-only Document 108.* Transportation Research Board, Washington, DC, USA.

House of Commons (1991) *New Roads and Street Works Act 1991.* Stationery Office, London, UK.

Im JH and Kim YR (2013) Methods for fog seal field tests with polymer modified emulsions: Development and performance evaluation. *Transportation Research Record* **2361**: 88–97.

ISSA (International Slurry Surfacing Association) (2011) *Recommended Performance Guideline for Micro-surfacing: ISSA Document A143.* ISSA, Annapolis, MD, USA.

ISSA (2012a) *Wet Track Abrasion Test of Slurry Seals: ISSA Technical Bulletin 100.* ISSA, Annapolis, MD, USA.

ISSA (2012b) *Measurement of Excess Asphalt in Bituminous Mixtures by Use of a Loaded Wheel Tester and Sand Adhesion: ISSA Technical Bulletin 109*. ISSA, Annapolis, MD, USA.

Kandal PS and Lockett L (1997) *Construction and Performance of Ultrathin Asphalt Friction Course*. National Centre for Asphalt Technology, Auburn University, Auburn, AL, USA, Report 97-05.

Moraes R and Bahia HU (2013) Effect of curing and oxidative ageing on ravelling in emulsion chip seals. *Transportation Research Record* **2361**: 69–79.

Morian DA (2011) *Cost–benefit Analysis of Including Micro-surfacing in Pavement Treatment Strategies and Cycle Maintenance*. Pennsylvania Department of Transportation, Harrisburg, PA, USA, Report FHWA-PA-2011-001-080503.

Nicholls JC (2001) *A History of the Recent Thin Surfacing Revolution in the United Kingdom. TRL Report 522*. Transport Research Laboratory, Wokingham, UK.

Nicholls JC, Carswell I, Thomas C and Sexton B (2010) *Durability of Thin Asphalt Surfacing Systems. Part 4: Final Report After Nine Years' Monitoring*. Transport Research Laboratory, Wokingham, UK, Report 674.

Roberts C and Nicholls JC (2008) *Design Guide for Road Surface Dressing*, 6th edn. Transport Research Laboratory, Crowthorne, UK, Road Note 39.

Robinson DA (1968) Surface dressing – variations upon a theme. *Journal of the Institution of Highway Engineers* **April**: 17–21.

Robinson HR (2014) *Surface Dressing Statistics (private communication)*. Road Surface Treatment Association, Wolverhampton, UK.

RSTA (Road Surface Treatment Association) (2008) *Guidance Note on Quieter Road Dressings*. RSTA, Wolverhampton, UK.

RSTA (2011a) *Code of Practice for Slurry Surfacing Incorporating Micro-surfacing*. RSTA, Wolverhampton, UK.

RSTA (2011b) *Code of Practice for High Friction Surfacing*. RSTA, Wolverhampton, UK.

RSTA (2014) *Guidance Note on Types and Design of Surface Dressings: RSTA Code of Practice for Surface Dressing Pt 4*. RSTA, Wolverhampton, UK.

Serfass JP, Bense P, Bonnot J and Samanos J (1991) New type of ultrathin friction course. *Transportation Research Record* **1304**: 66–72.

Shuler S, Lord A, Epps-Martin A and Hoyt D (2011) *Manual for Emulsion-based Chip Seals for Pavement Preservation*. Transportation Research Board, Washington, DC, USA, NCHRP Report 680.

Southern D (1983) Premium surface dressing systems. *Shell Bitumen Review* **60**: 4–8.

Whiteoak D (2000) Specialist surface treatments. In *Asphalts in Road Construction* (Hunter RN (ed.)). Thomas Telford, London, UK, pp. 465–509.

Whiteoak D and Read JM (2003) Surface dressing and other specialist treatments. In *Shell Bitumen Handbook* (Hunter R (ed)), 5th edn. Thomas Telford, London, UK, pp. 371–417.

Wright N (1978) Recent developments in surface dressing in the UK. *Proceedings of Eurobitume, London, UK,* pp. 156–161.

Highways
ISBN 978-0-7277-5993-1

ICE Publishing: All rights reserved
http://dx.doi.org/10.1680/h5e.59931.567

Index

Page references in *italics* refer to figures separate from the corresponding text.